Results and Problems in Cell Differentiation

Series Editors:
W. Hennig, L. Nover, U. Scheer

26

Springer

Berlin
Heidelberg
New York
Barcelona
Hong Kong
London
Milan
Paris
Singapore
Tokyo

Dietmar Richter (Ed.)

Regulatory Peptides and Cognate Receptors

With 71 Figures

 Springer

Professor Dr. Dietmar Richter
Institut für Zellbiochemie
und Klinische Neurobiologie
Universität Hamburg
Martinistr. 52
D-20246 Hamburg
Germany

ISSN 0080-1844
ISBN 3-540-65325-2 Springer-Verlag Berlin Heidelberg New York

Regulatory peptides and cognate receptors / Dietmar Richter (ed.). p. cm. – (Results and problems in cell differentiation, ISSN 0080-144 ; 26) Includes bibliographical references and index. ISBN 3-540-65325-2 (hardcover : alk. paper) 1. Peptide hormones. 2. Peptide hormones – Receptors. 3. Neuropeptides. 4. Neuropeptides – Receptors. I. Richter, Dietmar, 1939- . II. Series. [DNLM: 1. Neuropeptides – physiology. 2. Receptors, Neuropeptide – physiology. W1 RE248X v.26 1999 / WL 104 R344 1999] OH607.R4 vol. 26 [QP572.P4] 571.8'35 s – dc21 [573.8'374] DNLM/DLC for Library of Congress

© Springer-Verlag Berlin Heidelberg 1999
Printed in Germany

The use of general descriptive names, registered names, trademarks, etc. in this publication does not imply, even in the absence of a specific statement, that such names are exempt from the relevant protective laws and regulations and therefore free for general use.

Production: PRO EDIT GmbH, D-69126 Heidelberg
Cover Concept: Meta Design, Berlin
Cover Production: design & production, D-69121 Heidelberg
Typesetting: Zechnersche Buchdruckerei, D-67346 Speyer
SPIN: 10674843   39/3137 – 5 4 3 2 1 0 – Printed on acid-free paper

Preface

In the last two decades our knowledge of bioactive peptides and their cognate receptors, most of which are members of the seven transmembrane receptor superfamily, has increased enormously. Bioactive peptides are small proteins which, besides their hormonal functions in regulating cellular metabolism in various tissues, may also act as neurotransmitters and therefore often carry the prefix 'neuro'. It is becoming increasingly clear that many of the cognate receptors involved in transducing the peptidergic signal across the cell membrane via a family of G-proteins exist in multiple forms and that the number of receptor subtypes frequently exceeds the number of corresponding peptide ligands. This poses a number of questions, particularly concerning the distinct physiological functions of the receptor subtypes of a given receptor family; even in the light of the results of gene knockout experiments, the functional identification of the various subtypes remains a major challenge. The recent success in identifying new peptidergic ligands and their receptors clearly demonstrates that our view of this field is still rather limited. The aim of this book is therefore to review a few selected examples out of the ever-growing number of peptides and receptors and to make it clear to the reader what is commonly accepted in each field and what unresolved problems remain. In the introductory chapter a brief account of aspects concerning the co-evolution of peptides and their receptors is given. The following 14 chapters deal with various peptide-receptor systems and describe examples of how new strategies such as 'reverse physiology' may help to uncover new peptides and their receptors and unravel their functions. It is the diversity of the peptide-receptor systems that makes this field exciting to those who work in it, and it is hoped that some of this excitement is conveyed by this book to the reader.

Hamburg, April 1999   *Dietmar Richter*

Contents

Corticotropin-Releasing Factor (CRF)
and its Role in the Central Nervous System
Masao Ito and Mariko Miyata

CRF and CRF Receptors
Jelena Radulovic, Thomas Blank, Klaus Eckart, Marko Radulovic,
Oliver Stiedl, and Joachim Spiess

Neural Oxytocinergic Systems as Genomic Targets for Hormones and as Modulators of Hormone-Dependent Behaviors
Donald W. Pfaff, Sonoko Ogawa, and Lee-Ming Kow

Vasopressin Receptors: Structural Functional Relationships and Role in Neural and Endocrine Regulation
Oscar Schoots, Fernando Hernando, Nine V. Knoers, and J. Peter H. Burbach

The Oxytocin Receptor
Tadashi Kimura and Richard Ivell

Targeted Mutagenesis of the Murine Opioid System
Michael D. Hayward and Malcolm J. Low

Galanin and Galanin Receptors
Tiina P. Iismaa and John Shine

The Cholecystokinin – Gastrin Family of Peptides and Their Receptors
Jens F. Rehfeld

Function of the Neuropeptide Head Activator for Early Neural
and Neuroendocrine Development
Wolfgang Hampe, Irm Hermans-Borgmeyer, and H. Chica Schaller

Invertebrate Neurohormones and Their Receptors
Cornelis J. P. Grimmelikhuijzen, Frank Hauser,
Kathrine Krageskov Eriksen, and Michael Williamson

The 'Chicken and Egg' Problem of Co-evolution of Peptides and Their Cognate Receptors: Which Came First?

Mark G. Darlison and Dietmar Richter

1
Introduction

As will be evident from the other chapters in this Volume, small peptide molecules regulate a wide variety of biological processes in both vertebrate and invertebrate species. For each bioactive peptide there exists one or more specific membrane-bound receptors, which transduce(s) the signal of peptide binding into a cellular response. The majority of these receptors share a common topology with seven membrane-spanning domains, an extracellular amino terminus and a cytoplasmically located carboxy terminus. Since this class of receptors translates the process of peptide binding into an intracellular response through an interaction with one or more of a family of GTP-binding proteins (G-proteins), they have been named G-protein-coupled receptors (see Probst et al. 1992; Meyerhof et al. 1993). Other types of peptide receptor are known, including those for growth factors, such as epidermal growth factor, which have a single membrane-spanning domain and an intracellular ligand-activated tyrosine kinase domain (see McInnes and Sykes 1997), that for the peptide Phe-Met-Arg-Phe-amide which contains an integral ligand-gated sodium channel (Lingueglia et al. 1995), and the 200-kDa head-activator receptor of hydra which exhibits sequence similarity to members of the low-density lipoprotein receptor family (Hampe et al., this Vol.). The role of the latter may be that of a carrier protein, binding and presenting head-activator, which is a small hydrophobic peptide, to the 'true' head-activator receptor.

Biologically active peptides derive from larger precursors which often contain more than one type of peptide molecule. So, for example, the human pre-proenkephalin gene encodes a precursor of 267 amino acids which contains six copies of Met-enkephalin (Tyr-Gly-Gly-Phe-Met) and one copy of Leu-enkephalin (Tyr-Gly-Gly-Phe-Leu) (Noda et al. 1982). In contrast, each type of peptide receptor is encoded by a distinct gene, although alternative splicing of primary gene transcripts can give rise to receptor variants. When one considers the large number of bioactive peptides, and the multiplicity of receptors to

Institut für Zellbiochemie und Klinische Neurobiologie, Universitätsklinikum Hamburg-Eppendorf, Universität Hamburg, Martinistr. 52, 20246 Hamburg, Germany

which they bind, one wonders at how the specificity of peptide–receptor inter-actions unfolded. Surprisingly, very little is known about how the genes that encode the various peptides and their cognate receptors evolved to enable the two types of product to interact so precisely. Even less is understood about how this specificity is maintained.

In this chapter, which develops ideas that have been introduced elsewhere (Darlison and Richter 1999), we will consider: (1) the order in which peptide genes and receptor genes might have arisen; (2) the likely role played by genome doublings (tetraploidization events) in the evolution of peptides and their corresponding receptors; and (3) the possible advantage(s) to an organ-ism of possessing families of receptors and peptides. Note that although this discussion will be largely based on studies on neuropeptides and their cognate G-protein-coupled receptors, many of the concepts advanced may also apply to other peptide–receptor systems.

2
Appearance of Peptide Genes and Peptide Receptor Genes

As noted above, peptide receptors exhibit great selectivity in the binding of peptide ligands. Yet the genes that encode the two types of component which interact must have arisen independently of one another, even if the period of time between the emergence of a given peptide gene and the corresponding receptor gene was short. When considering this 'chicken and egg' problem, three scenarios can be envisaged.

2.1
Co-evolution of Peptide and Receptor Genes

Perhaps the most attractive idea is that peptide genes and their corresponding receptor genes arose at about the same time and then evolved together. However, this scenario implies that once the specificity of a given peptide–receptor interaction was established, then constraints were imposed not only upon the active portion of the peptide but also on the structure of the peptide-binding domain of the cognate receptor. Since it is hard to imagine what kind of genetic mechanism could have influenced such co-evolution, this proposal seems unlikely. Nevertheless, it is evident that features of certain peptides, such as the disulphide bridge that is present in all members of the vasopressin (VP)–oxytocin (OT) family (see Darlison and Richter 1999; Kimura and Ivell, this Vol.; Pfaff et al., this Vol.; Schoots et al., this Vol.), have been highly con-served during evolution. Similarly, the presence of seven membrane-spanning domains in G-protein-coupled receptors has been absolutely preserved.

2.2
Did Peptide Genes Arise Before Receptor Genes or Vice Versa?

The two remaining scenarios predict: (1) that peptide genes came into existence early in evolution before the emergence of receptor genes; and (2) that peptide genes appeared later, as a source of ligands for pre-existing peptide receptors. We do not know for certain which proposal is correct. Indeed, to properly address this issue, we would need to know how the sequence of a peptide and its receptor changed over a period of a million years or so, which is clearly impossible. So, which sequence of events is the more likely?

It is evident that many peptide sequences are much smaller than those of the receptors to which they bind; the most extreme example of this is thyrotropin-releasing hormone which comprises only three amino acids (Glu-His-Pro) (Bauer et al., this Vol.). This might suggest that receptors required more time to emerge than peptides and, hence, that peptide genes appeared earlier in evolution than receptor genes. However, as mentioned above, peptides are embedded within much larger precursor molecules, and these must be processed (for example, by removal of the amino-terminal signal peptide sequence and other 'spacer' sequences) to release the active components. Furthermore, each polypeptide precursor has a defined tertiary structure, and this is necessary for correct recognition by the appropriate processing endopeptidase(s). On the other hand, each receptor also has a defined tertiary structure, which in the case of G-protein-coupled receptors includes seven membrane-spanning domains. This three-dimensional structure enables the receptor to bind the correct peptide ligand and translate this binding step into a change in receptor conformation so that the appropriate G-protein, which binds to the receptor on the cytoplasmic side of the membrane, is activated.

It is difficult to give a clear answer to the 'chicken and egg' question of which came first (the peptide gene or the receptor gene) and, indeed, the situation may not be identical for all peptide–receptor pairs. However, because the sequences of peptide receptors are, generally, considerably larger in size, and arguably more complex, than the precursor molecules that harbour the corresponding ligands, it seems likely that peptide genes arose earlier in evolution than receptor genes. Support for the idea that peptides emerged very early in the animal kingdom comes from studies on cnidarians, a phylum which includes jellyfish and sea anemones. There, a wide variety of neuropeptides (and processing endopeptidases) have been characterized, suggesting that this class of molecules arose, as transmitters, at an early stage in evolution, perhaps before the appearance of classical neurotransmitters such as acetylcholine (Grimmelikhuijzen et al. 1996 and this Vol.).

3
Consequences of Genome Doubling
During Vertebrate Evolution

3.1
Multiple Receptors for a Given Peptide Ligand

As alluded to above, a given peptide is often able to bind to a family of sequence-related receptors (called receptor subtypes), each of which is encoded by a separate gene, rather than to a single receptor type. This is most evident, perhaps, in the case of mammalian neuropeptide receptors. Thus, to date, five somatostatin (SST) receptors (sst_1 to sst_5; see Epelbaum et al. 1994; Schindler et al. 1996; Kreienkamp, this Vol.), five neuropeptide Y receptors (Y_1, Y_2, Y_4, Y_5 and y6; see Blomqvist and Herzog 1997), four opioid or opioid-like receptors (the µ-, δ- and κ-types and ORL1; Kieffer 1995; Hayward and Low, this Vol.; Reinscheid et al., this Vol.), three VP receptors (V_{1A}, V_{1B} and V_2; see Burbach et al. 1995; Schoots et al., this Vol.), and two main types of corticotropin-releasing factor receptor (CRFR1 and CRFR2; see Ito and Miyata, this Vol.; Radulovic et al., this Vol.) have been identified through the cloning of complementary DNAs (cDNAs). Note that the different receptor subtypes, for a given peptide ligand, usually exhibit a high degree of sequence similarity with one another (for example, the human V_{1A} receptor displays 49 and 37% identity to the human V_{1B} and V_2 receptors, respectively) and it is generally assumed, therefore, that the corresponding genes arose by one or more duplications of a common ancestral gene.

When did all of these duplications take place? Clearly, during the course of evolution, individual duplication events may occur at random. However, in vertebrates at least, it is thought that two, possibly three, genome doublings (tetraploidization events) occurred at an early stage in their evolution (Lundin 1993). It seems likely that these events made a major contribution to the total number of receptor genes that we can recognize today.

3.2
Existence of Peptide Families

If the evolution of the vertebrate genome did involve two or more tetraploidization events, then the number of peptide genes should also have increased, resulting in the genesis of families of peptides. Such families do, indeed, exist. For example, the two nonapeptides VP and OT are closely related in sequence, and both contain a highly conserved disulphide bond (see Darlison and Richter 1999). Furthermore, cloning and analysis of the corresponding rodent genes have revealed that they are similar in sequence, have comparable intron–exon organizations, and are situated next to one another in the genome in a 'tail-to-tail' configuration (Ivell and Richter 1984; Schmitz et al. 1991; Mohr et al. 1995). These data strongly suggest that the VP and OT genes arose by duplication of a common ancestral gene.

The recent sequencing of a rat cDNA, which derives from an mRNA that is expressed in brain, has identified a novel peptide that has been named cortistatin (CST; de Lecea et al. 1996; Kreienkamp, this Vol.; Sutcliffe and de Lecea, this Vol.). In fact, the CST precursor predicts two peptides, CST-14 and CST-29, which exhibit significant sequence identity at their common carboxy-terminal end to SST, which also exists in two forms (SST-14 and SST-28; Schindler et al. 1996). Furthermore, both CST and SST contain a pair of cysteine residues, which in SST are known to form a disulphide bond, in similar relative positions. Taken together, these data suggest that the CST and SST precursor genes share a common ancestral origin.

4
A Model for the Evolution
of Specific Peptide–Receptor Interactions

Although we do not know how the specificity of peptide–receptor interactions emerged, a plausible scheme has been put forward by Moyle et al. (1994). This is based on their studies on the binding of glycoprotein hormones to their cognate receptors. Luteinizing hormone (LH), human chorionic gonadotropin (hCG) and follicle-stimulating hormone (FSH) all consist of two polypeptides, a unique β subunit and a common α subunit. These glycoproteins bind in a very specific fashion to their corresponding receptors which, although they belong to the G-protein-coupled receptor family, have very long (~350-residue) amino-terminal domains; hCG and LH interact selectively with the LH receptor, while FSH binds selectively to the FSH receptor. Structure–function studies on the LH and FSH receptors, and their cognate ligands, have pinpointed unique regions within both the receptor sequences and the glycoprotein sequences that are responsible for their specificity of interaction. Thus, a unique region of the hCG β subunit has been identified which prevents the binding of hCG to the FSH receptor, and two different domains of the FSH β subunit have been delineated that restrict the binding of FSH to the LH receptor. In addition, unique sequences within the long amino-terminal extracellular domains of the LH and FSH receptors have been identified which impede binding of the inappropriate hormone. To account for these data, Moyle et al. (1994) suggested that, originally, there existed a single primordial β-subunit gene (Fig. 1). At some point in evolution, this gene was duplicated and the binding selectivities of the resultant hormones arose by the introduction of residues that blocked binding to one of the two receptors. The proposed model also assumes the existence of an ancestral receptor gene, the product of which could bind multiple glycoproteins. Following duplication of this gene, it is suggested that sequence determinants were introduced into different regions of the

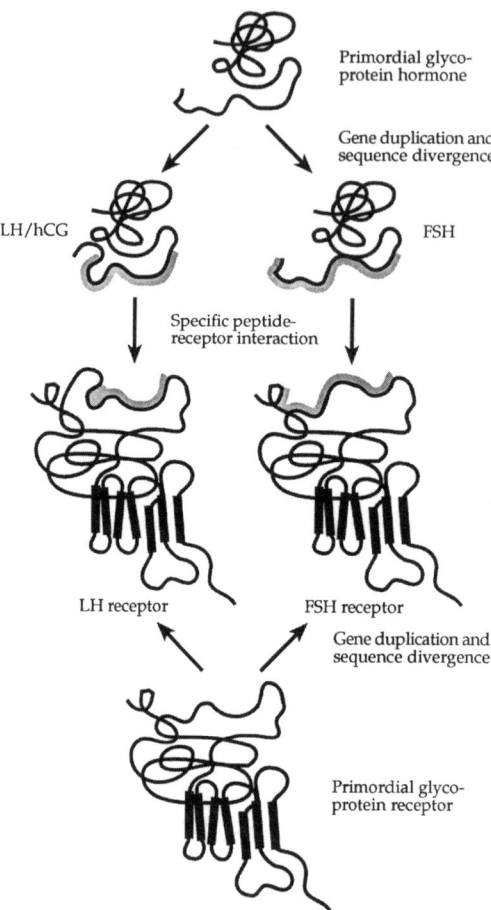

Fig. 1. Schematic representation of how the specificity of interaction between glycoprotein hormones and their cognate G-protein-coupled receptors may have evolved. This figure is based on the model originally proposed by Moyle et al. (1994). Areas of interaction between a glycoprotein and its receptor are shaded grey. FSH follicle-stimulating hormone; hCG human chorionic gonadotropin; LH luteinizing hormone. For further details, see text

two resultant receptors, thus preventing the binding of inappropriate ligands.

Interestingly, a model similar to that for glycoproteins and their receptors has been put forward to account for the agonist selectivity of the μ-, δ- and κ-opioid receptors (Metzger and Ferguson 1995). In this scheme, it is proposed that the extracellular loop regions of opioid receptors confer agonist selectivity, primarily by a mechanism of exclusion.

5
Possible Reasons for the Existence of Families of Receptors and Peptides

What advantage might be conferred on an organism by the existence of multiple receptor subtypes, each of which can bind the same peptide molecule?

Also, since it is evident that different peptides can bind to the same receptor molecule (for example, both SST and CST can bind to all five human SST receptors; Fukusumi et al. 1997; Kreienkamp, this Vol.), what purpose is fulfilled by this surplus of activating ligands?

5.1
Advantages of Having Multiple Receptors for a Given Peptide

In the case of receptors, several ideas can be put forward: (1) different subtypes could have dissimilar biochemical properties, such as differences in agonist selectivity, rates of desensitization and/or internalization, coupling to effector systems, and subcellular localization; (2) gene duplication and subsequent sequence divergence (of the promoter region), through natural mutation, could bring about a change in the expression pattern and/or regulation of one or both of the resultant receptor genes; (3) the duplication of a receptor gene, that was essential to an organism, would substantially lower the risk of lethality as a result of spontaneous mutation; and (4) duplication of a receptor gene, and ensuing evolution, might culminate in the gain of a new biological function in an organism.

It is difficult to provide experimental evidence in support of all of the above suggestions. Nevertheless, studies on mammalian SST receptors have provided evidence that the different subtypes have distinct biochemical properties. As with many G-protein-coupled receptors, SST receptors are subject to desensitization, and subsequent internalization, upon prolonged agonist exposure. However, while the rat sst_1, sst_2 and sst_3 subtypes are each rapidly internalized in the presence of either SST-14 or SST-28, sst_5 is internalized only in the presence of SST-28, and sst_4 is not internalized by either agonist (Roth et al. 1997). Furthermore, it is known that activation of SST receptors can result in the opening of G-protein-gated inward-rectifying potassium (GIRK) channels via a direct interaction with free $\beta\gamma$ subunits of heterotrimeric G-proteins which dissociate from the α subunit upon agonist binding. Studies on *Xenopus laevis* oocytes have indicated that the different SST receptors can be distinguished by their ability to couple to GIRK channels. Thus, SST can induce GIRK1 (Kir3.1)-mediated potassium currents by activation of either sst_2, sst_3, sst_4 or sst_5, but not sst_1 (Kreienkamp et al. 1997). In addition, in this cell system, sst_2 was shown to couple to GIRK1 more efficiently than any of the other SST receptors.

Indirect evidence also exists in support of the idea that gene duplication can result in a change in the expression pattern and/or regulation of one or both of the resultant receptor genes. Thus, the freshwater snail *Lymnaea stagnalis* possesses a peptide (called [Lys[8]]conopressin) that is a member of the VP–OT family of peptides (Van Kesteren et al. 1995a). This molecule, which is synthesized in the central nervous system, binds to two different receptors, LSCPR1 (van Kesteren et al. 1995b) and LSCPR2 (van Kesteren et al. 1996). Although the exact physiological functions of the two conopressin receptors are currently unclear, the corresponding genes are known to be expressed, in the *Lymnaea* brain, in mutually exclusive populations of neurons (van Kesteren et al. 1996). One plausible interpretation of these data is that the two conopressin receptors arose by duplication of a common ancestral gene, and that subsequent mutation of the promoter region of one (or both) of the progenitor genes resulted in their different spatial patterns of expression.

5.2
Advantages of Having Multiple Peptides for a Given Receptor

As suggested above for receptors, if families of peptide genes also arose as a consequence of two or three tetraploidization events during the course of vertebrate evolution, then this may have: (1) culminated in a change in the pattern of expression and/or regulation of one or both of the resultant peptide genes; (2) provided protection against the occurrence of spontaneous lethal mutations; and/or (3) resulted in the generation of greater functional diversity. Indirect evidence in support of the first idea can be found in studies on the genes that encode VP and OT. As mentioned earlier, these two genes have similar nucleotide sequences and comparable intron–exon organizations, indicating that they arose by a process of gene duplication. However, although both genes are expressed in hypothalamic magnocellular neurons, they are transcribed in mutually exclusive populations of cells (Mohr et al. 1988). This is presumably a consequence of sequence differences in the promoter regions of the VP and OT genes.

There is also, perhaps, indirect support for the notion that gene duplication could provide a means by which to protect an animal against the occurrence of fatal mutations. This comes from the surprising finding that mice lacking the preprosomatostatin gene grow to adulthood, and have no obvious physiological or behavioural abnormalities (Juarez et al. 1997). Since CST can bind, as well as SST, to the five different SST receptors (Fukusumi et al. 1997; Kreienkamp, this Vol.), it may be that CST can mediate some of the more important functions of SST in SST-deficient animals.

6
Concluding Remarks

In this short chapter, we have discussed how peptide genes and their cognate receptor genes may have arisen, and have suggested reasons why families of peptides and receptors may have evolved. Based on the available data, we propose that bioactive peptides came into existence before the receptors with which they interact. However, this is only one of three possible evolutionary schemes. The problem with studying the evolution of peptides and/or receptors is that when examining a given sequence, what is actually seen is a sequence that is still evolving. That is to say, one observes only a 'snap-shot' of evolution.

Lastly, in general, scientists tend to expect all observable data to fit some predefined model. A good example of this is the study of orphan G-protein-coupled receptors, that is to say receptors whose activating ligand is unknown. The sequences of a number of such receptors have been determined and, based on sequence comparisons with known receptors, some of them are predicted to be peptide receptors (see, for example, O'Dowd et al. 1995; Kolakowski et al. 1996). In due course, it is likely that the activating ligand for a number of current orphan peptide receptors will be identified, as was the case for the opioid receptor-like ORL1 receptor, which binds nociceptin/orphanin FQ (Meunier et al. 1995; Reinscheid et al. 1995; Darland et al. 1998; Reinscheid et al., this Vol.), and the receptors for orexin-A and orexin-B, two peptides that stimulate food consumption (Sakurai et al. 1998; Sutcliffe and de Lecea, this Vol.). However, if a given peptide gene has not evolved sufficiently to enable the encoded ligand to recognize the product of an orphan receptor gene, or if the gene(s) for a given orphan receptor and/or its corresponding peptide has/have evolved such that a specific interaction no longer exists, then it is evident that it will not be possible to identify the orphan receptor in question. What we must remember is that evolution is a dynamic process and that not every aspect of what we study will necessarily make sense.

References

Blomqvist AG, Herzog H (1997) Y-receptor subtypes – how many more? Trends Neurosci 20: 294–298

Burbach JPH, Adan RAH, Lolait SJ, van Leeuwen FW, Mezey E, Palkovits M, Barberis C (1995) Molecular neurobiology and pharmacology of the vasopressin/oxytocin receptor family. Cell Mol Neurobiol 15:573–595

Darland T, Heinricher MM, Grandy DK (1998) Orphanin FQ/nociceptin: a role in pain and analgesia, but so much more. Trends Neurosci 21:215–221

Darlison MG, Richter D (1999) Multiple genes for neuropeptides and their receptors: co-evolution and physiology. Trends Neurosci 22:81–88

de Lecea L, Criado JR, Prospero-Garcia Ó, Gautvik KM, Schweitzer P, Danielson PE, Dunlop CLM, Siggins GR, Henriksen SJ, Sutcliffe JG (1996) A cortical neuropeptide with neuronal depressant and sleep-modulating properties. Nature 381:242–245

Epelbaum J, Dournaud P, Fodor M, Viollet C (1994) The neurobiology of somatostatin. Crit Rev Neurobiol 8:25–44

Fukusumi S, Kitada C, Takekawa S, Kizawa H, Sakamoto J, Miyamoto M, Hinuma S, Kitano K, Fujino M (1997) Identification and characterization of a novel human cortistatin-like peptide. Biochem Biophys Res Commun 232:157–163

Grimmelikhuijzen CJP, Leviev I, Carstensen K (1996) Peptides in the nervous systems of cnidarians: structure, function, and biosynthesis. Int Rev Cytol 167:37–89

Ivell R, Richter D (1984) Structure and comparison of the oxytocin and vasopressin genes from rat. Proc Natl Acad Sci USA 81:2006–2010

Juarez RA, Rubinstein M, Chan EC, Low MJ (1997) Increased growth following normal develop-
ment in middle-aged somatostatin-deficient mice. Soc Neurosci Abstr 659.10
Kieffer BL (1995) Recent advances in molecular recognition and signal transduction of active
peptides: receptors for opioid peptides. Cell Mol Neurobiol 15:615–635
Kolakowski LF Jr, Jung BP, Nguyen T, Johnson MP, Lynch KR, Cheng R, Heng HHQ, George SR,
O'Dowd BF (1996) Characterization of a human gene related to genes encoding somatostatin
receptors. FEBS Letts 398:253–258
Kreienkamp H-J, Hönck H-H, Richter D (1997) Coupling of rat somatostatin receptor subtypes
to a G-protein gated inwardly rectifying potassium channel (GIRK1). FEBS Letts 419:92–94
Lingueglia E, Champigny G, Lazdunski M, Barbry P (1995) Cloning of the amiloride-sensitive
FMRFamide peptide-gated sodium channel. Nature 378:730–733
Lundin LG (1993) Evolution of the vertebrate genome as reflected in paralogous chromosomal
regions in man and the house mouse. Genomics 16:1–19
McInnes C, Sykes BD (1997) Growth factor receptors: structure, mechanism, and drug discovery.
Biopolymers 43:339–366
Metzger TG, Ferguson DM (1995) On the role of extracellular loops of opioid receptors in con-
ferring ligand selectivity. FEBS Letts 375:1–4
Meunier J-C, Mollereau C, Toll L, Suaudeau C, Moisand C, Alvinerie P, Butour J-L, Guillemot
J-C, Ferrara P, Monsarrat B, Mazarguil H, Vassart G, Parmentier M, Costentin J (1995)
Isolation and structure of the endogenous agonist of opioid receptor-like ORL$_1$ receptor.
Nature 377:532–535
Meyerhof W, Darlison MG, Richter D (1993) The elucidation of neuropeptide receptors and their
subtypes through the application of molecular biology. In: Hucho F (ed) Neurotransmitter
receptors. New comprehensive biochemistry, vol 24. Elsevier, Amsterdam, pp 339–357
Mohr E, Bahnsen U, Kiessling C, Richter D (1988) Expression of the vasopressin and oxytocin
genes in rats occurs in mutually exclusive sets of hypothalamic neurons. FEBS Letts 242:
144–148
Mohr E, Meyerhof W, Richter D (1995) Vasopressin and oxytocin: molecular biology and evolution of the peptide hormones and their receptors. Vitamins and hormones, vol 51. Academic
Press, San Diego, pp 235–266
Moyle WR, Campbell RK, Myers RV, Bernard MP, Han, Y, Wang X (1994) Co-evolution of
ligand–receptor pairs. Nature 368:251–255
Noda M, Teranishi Y, Takahashi H, Toyosato M, Notake M, Nakanishi S, Numa S (1982)
Isolation and structural organization of the human preproenkephalin gene. Nature 297:
431–434
O'Dowd BF, Scheideler MA, Nguyen T, Cheng R, Rasmussen JS, Marchese A, Zastawny R, Heng
HHQ, Tsui L-C, Shi X, Asa S, Puy L, George SR (1995) The cloning and chromosomal map-
ping of two novel human opioid-somatostatin-like receptor genes, GPR7 and GPR8,
expressed in discrete areas of the brain. Genomics 28:84–91
Probst WC, Snyder LA, Schuster DI, Brosius J, Sealfon SC (1992) Sequence alignment of the
G-protein coupled receptor superfamily. DNA Cell Biol 11:1-20
Reinscheid RK, Nothacker H-P, Bourson A, Ardati A, Henningsen RA, Bunzow JR, Grandy DK,
Langen H, Monsma FJ Jr, Civelli O (1995) Orphanin FQ: a neuropeptide that activates an opi-
oidlike G protein–coupled receptor. Science 270:792–794
Roth A, Kreienkamp H-J, Nehring RB, Roosterman D, Meyerhof W, Richter D (1997)
Endocytosis of the rat somatostatin receptors: subtype discrimination, ligand specificity, and
delineation of carboxy-terminal positive and negative sequence motifs. DNA Cell Biol 16:
111–119
Sakurai T, Amemiya A, Ishii M, Matsuzaki I, Chemelli RM, Tanaka H, Williams SC, Richardson
JA, Kozlowski GP, Wilson S, Arch JRS, Buckingham RE, Haynes AC, Carr SA, Annan RS,
McNulty DE, Liu W-S, Terrett JA, Elshourbagy NA, Bergsma DJ, Yanagisawa M (1998)
Orexins and orexin receptors: a family of hypothalamic neuropeptides and G protein-coupled
receptors that regulate feeding behavior. Cell 92:573–585

Schindler M, Humphrey PPA, Emson PC (1996) Somatostatin receptors in the central nervous system. Prog Neurobiol 50:9–47

Schmitz E, Mohr E, Richter D (1991) Rat vasopressin and oxytocin genes are linked by a long interspersed repeated DNA element (LINE): sequence and transcriptional analysis of LINE. DNA Cell Biol 10:81–91

Van Kesteren RE, Smit AB, De Lange RPJ, Kits KS, Van Golen FA, Van Der Schors RC, De With ND, Burke JF, Geraerts WPM (1995a) Structural and functional evolution of the vasopressin/oxytocin superfamily: vasopressin-related conopressin is the only member present in *Lymnaea*, and is involved in the control of sexual behavior. J Neurosci 15 : 5989–5998

van Kesteren RE, Tensen CP, Smit AB, van Minnen J, van Soest PF, Kits KS, Meyerhof W, Richter D, van Heerikhuizen H, Vreugdenhil E, Geraerts WPM (1995b) A novel G protein–coupled receptor mediating both vasopressin- and oxytocin-like functions of Lys-conopressin in Lymnaea stagnalis. Neuron 15 : 897–908

van Kesteren RE, Tensen CP, Smit AB, van Minnen J, Kolakowski LF Jr, Meyerhof W, Richter D, van Heerikhuizen H, Vreugdenhil E, Geraerts WPM (1996) Co-evolution of ligand-receptor pairs in the vasopressin/oxytocin superfamily of bioactive peptides. J Biol Chem 271 : 3619–3626

Thyrotropin Releasing Hormone (TRH), the TRH-Receptor and the TRH-Degrading Ectoenzyme; Three Elements of a Peptidergic Signalling System

Karl Bauer[1], Lutz Schomburg[1], Heike Heuer[1], and Martin K.-H. Schäfer[2]

1
Historical Background

The isolation and structural identification of thyrotropin releasing hormone (TRH) in 1969 (Bøler et al. 1969; Burgus et al. 1969; see also Guillemin 1978; Schally 1978) ushered in the new area of modern neuroendocrinology because for the first time a hypothalamic hypophysiotrophic hormone had been deciphered at the molecular level. This discovery opened new avenues for studying the ever interesting question "How does the brain talk to the body?". Since science can be defined as "knowledge put in order" the scientific milestones leading to this highlight of discovery should be acknowledged.

Based on careful anatomical studies, Ernst and Berta Scharrer formulated as early as the 1920s the revolutionary concept of neurosecretion, suggesting that specialized nerve cells might be able to synthesize and secrete true hormones into the blood stream to act at target sites remote from their point of origin. The field of neuroendocrinology was firmly established in the early 1950s by the pioneering work of Vincent Du Vigneaud and coworkers when they succeeded in isolating oxytocin and vasopressin and elucidating their structure as nonapeptides, thereby giving birth to a new class of substances: neuropeptide hormones. In parenthesis, due to their vulnerability to digestive enzymes, peptides had not been considered before as a class of substances that could act as hormones.

Although experimental data suggested that functions of the anterior pituitary were also governed by the hypothalamus, the connections between these structures were unclear, since careful anatomical and embryological studies had clearly established that the anterior pituitary is not innervated by hypothalamic neurons. Again, anatomical studies provided a likely explanation by demonstrating the existence of capillary vessels extending from the median eminence to the anterior pituitary. Based on a wealth of physiological experi-

[1] Max-Planck-Institut für experimentelle Endokrinologie, Feodor-Lynen-Str. 7, 30625 Hannover, Germany
[2] Institut für Anatomie und Zellbiologie, Philipps-Universität Marburg, Robert-Koch-Str. 6, 35037 Marburg/Lahn, Germany

ments G.W. Harris put forward the concept of a hypophyseal portal blood-chemotransmitter system. Although intriguing to some neuroendocrinologists, this hypothesis was met with great scepticism by traditional endocrinologists. Only the discovery of the chemical nature of TRH finally validated the concept of the hypothalamic control of anterior pituitary hormone secretion and opened new vistas to the development of the field of neuroendocrinology.

For young scientists it is difficult to imagine the problems of this enterprise. To illustrate the courage of the research groups some of the difficulties and technical problems should be mentioned (for a more detailed review see Reichlin 1989). First of all, reliable bioassays had to be established since radioimmunoassays were not yet available. Once a relatively simple and robust biological assay had been devised for TRH, the initial attempts to isolate corticotropin-releasing factor were abandoned and all efforts were concentrated on TRH. Since it was clear right from the beginning that only incredibly small amounts of these substances are present in the hypothalamus, the research groups headed by R. Guillemin and A. V. Schally, respectively, collected and processed enormous numbers (about 5 million each group) of hypothalamic fragments. (The extraordinary efforts and the rivalry between both groups are documented in detail by Wade 1981.) Following extraction, the analytical methods available were very limited (gel filtration, electrophoresis, partition chromatography, TLC, etc.; HPLC systems with the advanced support materials had not yet been developed). Nevertheless, both groups succeeded in obtaining purified material in milligram quantities, but at that time analytical methods (NMR, mass spectrometry, etc.) were not yet sophisticated enough to decipher the structure. Since the biological activity of the material was not destroyed by any of the many peptidases tested (inconceivable then for a peptide hormone) it was even speculated that TRH might not be a peptide hormone at all. However, hydrolysis of highly purified material then clearly demonstrated that the three amino acids identified (glutamic acid, histidine and proline) almost accounted for the total weight of the preparation. Subsequently, peptides of all possible sequence combinations (including glutamine) were synthesized, but all were biologically inactive. Chemical modification studies finally revealed that the peptide pyroGlu-His-Pro-NH$_2$ exhibited full biological activity. The existence of the pyroGlu residue was a surprise since this residue was not considered as a naturally occurring modification but as an extraction artefact (as shown for several proteins, such as immunoglobulins), all the more since at that time pyroGlu had just been disproved to be the initiating amino acid in protein synthesis, previously a favoured hypothesis.

2
TRH, a Widespread and Multifunctional Signal Molecule

2.1
The Neuropeptide Hormone TRH

The neuropeptide hormone TRH is synthesized by specialized nerve cells located within a distinct region of the hypothalamus, the paraventricular nucleus. After axonal transport to the nerve endings which terminate at the median eminence, TRH is released into a system of fenestrated capillary vessels that extend from the floor of the hypothalamus through the pituitary stalk to the anterior pituitary. Like classical hormones, TRH is transported by this blood stream to reach its target sites in the anterior pituitary. Thyrotrophic cells, equipped with TRH-receptors, are thus stimulated to secrete the pituitary hormone thyrotropin (thyroid-stimulating hormone; TSH) which then stimulates the thyroid gland to release thyroid hormones. Unexpectedly, subsequent studies in man (Jacobs et al. 1971) and with cultured rat tumor cells (GH_3 cells) (Tashjian et al. 1971) demonstrated TRH to be very potent also in releasing prolactin from the pituitary, whereby prolactin release precedes the release of TSH. Surprisingly, release of growth hormone following TRH is also observed in many acromegalic patients and subjects with disturbed metabolic functions (for review see Harvey 1990), but not in healthy human individuals.

2.2
TRH, a Neuromodulator and/or Neurotransmitter

With the development of sensitive radioimmunoassays and immunhistochemical methods, it could be demonstrated that TRH is not only confined to the thyrotrophic area of the hypothalamus, but widely distributed in other hypothalamic areas and throughout the central nervous system (CNS). Although the hypothalamus contains the highest concentration, over 70% of the total brain TRH is found in extrahypothalamic brain areas, in the pineal gland, spinal cord, posterior pituitary and retina (Winokur and Utiger 1974; for review see Prasad 1984). Next to the hypothalamus the posterior pituitary contains the highest concentrations of TRH. Since hypothalamic lesions lead to almost total depletion of posterior-lobe TRH (Jackson and Reichlin 1977) it is reasonable to assume that TRH is stored in the nerve endings of magnocellular neurons together with oxytocin and/or vasopressin. Correspondingly, there is some evidence that TRH can affect vasopressin secretion in man (Sowers et al. 1976). In the retina TRH content varies with illumination, being low during the night and high during the day (Schaeffer et al. 1977). The wide distribution of high affinity binding sites throughout the CNS (see below) and a wealth of biochemical, pharmacological, electrophysiological and behavioural studies strongly suggest that TRH most likely acts as a neuromodulator that amplifies

Table 1. Extrapituitary Effects of TRH

Behavioural and physiological effects	Pharmacological effects
Stimulates locomotor activity	Antagonizes endogenous opiate effects e.g. hypothermia, growth hormone and prolactin release
Arouses from sleep and hibernation	
Induces (wet dog) shaking behaviour and head-to-tail rotation	Opposes actions of sedative drugs e.g. barbiturates, ethanol, chlorpromazine
Decreases food and water consumption	Potentiates action of acetylcholine and dopamine
Increases blood pressure, heart rate, cerebral blood flow, body temperature	Increases noradrenaline turnover
Increases glucagon release and gastric acid secretion	Potentiates lethality to strychnine
Induces ulcer formation	

From reviews by Vale et al. (1977), Yarbrough (1979), Morley (1981), Prasad (1984), Griffiths (1985) and Reichlin (1986)

or dampens other neurotransmitter systems. For example, TRH is known to potentiate the excitatory action of the primary neurotransmitter acetylcholine on cortical neurons (Yarbrough 1976) and affects also the release of catecholamines from presynaptic endings. Some of the extrapituitary effects of TRH are listed in Table 1.

Whether TRH itself is released like a neurotransmitter directly into the synaptic cleft to be recognized at the postsynaptic membrane still remains an open question which warrants detailed investigation.

2.3
TRH in the Periphery

The existence of authentic TRH has now also been demonstrated in many other tissues. In rat pancreas TRH content is very high at birth and declines thereafter very rapidly (Morley et al. 1977; Martino et al. 1978). Since TRH is colocalized with insulin in the same granules (Madsen et al. 1991), TRH is thought to modulate insulin secretion. In the gastrointestinal tract TRH and TRH receptors are found in all segments and intestinal TSH has recently been reported to be an important factor within the intestinal immune system (Wang et al. 1997). In the heart TRH and TRH transcripts are mainly localized in the atrium and seem to support muscle contraction (Carnell et al. 1992; Shi et al. 1996). In the reproductive system TRH and TRH mRNA are confined to the Leydig cells of the testis, whereas TRH is not detectable in the ovaries

(Feng et al. 1993; Satoh et al. 1994). Besides TRH, the TRH-like peptide pyroGlu-Glu-Pro-NH$_2$ is also present in testis (Cockle et al. 1989) and it has been suggested that this peptide may be important for the process of capacitation and thus has been termed fertilization-promoting peptide (FPP) (Cockle 1995). Substantial amounts of pyroGlu-Glu-Pro-NH$_2$ are also found in the anterior pituitary, but not in the posterior pituitary. In contrast, authentic TRH is barely detectable in normal anterior pituitaries, but considerable amounts are found in some tumors, especially from acromegalic patients (Le Dafniet et al. 1990) where TRH mRNA can also be detected by Northern blot analysis (J. Ehrchen, A. Peters, K. Bauer, D.K. Lüdecke, unpubl.). It remains to be investigated whether under these conditions TRH might function as an autocrine or paracrine regulator.

Phylogenetic studies have shown that in some frogs such as *Rana pipiens* (Jackson and Reichlin 1972) TRH is not only present in brain but also abundantly expressed in skin glands. Furthermore, TRH also occurs in species such as larval lamprey (which lacks TSH) as well as in amphioxus and snails (species without pituitaries). These findings led to the concept that the peptidergic signal substance has been "co-opted" by the pituitary as a releasing hormone (Jackson 1982).

3
Biosynthesis of TRH

In view of the small size of TRH it was initially proposed that TRH may be formed by an enzymatic, nonribosomal mechanism like glutathione, carnosine and some peptide antibiotics. The original reports supporting this hypothesis could not be verified (Dixon and Acres 1975; Bauer and Lipmann 1976) and subsequent experiments by Rupnow et al. (1979) provided strong evidence favouring TRH synthesis following posttranslational processing of a precursor protein. Based on the findings that TRH is found in abundant amounts in frog skin (Jackson and Reichlin 1972), Richter et al. (1984) first succeeded in cloning a partial cDNA coding for the amino terminal region of preproTRH that contained three copies of the sequence Lys-Arg-Gln-His-Pro-Gly-Lys/Arg-Arg. As we now know, TRH is formed from this precursor by the successive action of prohormone convertases (cleaving the precursor at dibasic amino acid residues), carboxypeptidase E and D, peptidylglycine α-amidating monooxygenase (PAM) and glutaminyl cyclase.

3.1
Molecular Aspects

The strategy used by Richter et al. (hybridization screening of cDNA libraries by use of oligonucleotides) to identify the partial and subsequently the full length cDNA of *Xenopus* TRH (Fig. 1) was not successful in attempts to clone the cDNA of mammalian TRH. Only after an antiserum was generated that recognized the central core of an assumed prohormone sequence (Jackson et al. 1985) was the mammalian TRH precursor finally identified by screening a rat hypothalamic λgt 11 expression library with this antibody (Lechan et al. 1986). The nucleotide sequence of the largest clone contained an open reading frame of 765 nucleotides flanked by 102 base pairs of 5' untranslated sequence and 450 base pairs of 3' untranslated sequence. It encoded a protein of 255 amino acids (molecular weight of 29.247) that contained five TRH progenitor sequences (Gln-His-Pro-Gly) flanked by pairs of basic amino acids with an additional pair in the amino-terminal leader sequence (Fig. 1).

Lee et al. (1988) elucidated the genomic structure of rat preproTRH which is 2.6 kb in size with 3 exons and 2 introns. Exon 1 encodes the 5' untranslated region of the mRNA, exon 2 the signal peptide and the amino terminal peptide, while exon 3 contains all five copies of the TRH progenitor sequence and the 3' untranslated region. This structural organization of three discrete and functionally distinct exons is shared by other multivalent peptide hormone genes such as preproopiomelanocortin and preproenkephalin A and B, respectively. As expected, the same intron-exon architecture was found for the gene encoding the human TRH precursor (Yamada et al. 1990). This gene is 3.3 kb in size and encodes a protein of 242 amino acids with six copies of the TRH progenitor sequence. Recently, the cDNA of sockeye salmon TRH precursor was reported (Ohide et al. 1996). The cDNA encodes a protein of 259 amino acids with eight copies of the TRH progenitor sequence (Fig. 1).

Since the complete processing of the TRH precursor protein not only generates the TRH progenitor sequences but also liberates six cryptic peptides in the rat and seven connecting peptides in the human it has been suggested that these peptides may be biologically important. For rat preproTRH 160–169 it has been reported that this peptide is released from the hypothalamus by depolarizing agents (Valentijn et al. 1991), potentiates *in vitro* TRH-dependent pituitary TSH secretion (Bulant et al. 1990) and increases TSHβ as well as prolactin mRNA *in vitro* and *in vivo* (Carr et al. 1992). Rat preproTRH 178–199 has been considered to be a potent corticotropin-release inhibiting factor (Redei et al. 1995), but this effect could not be reproduced by others (Nicholson and Orth 1996). Whether these fragments are indeed of biological importance still warrants further investigation. This is especially true in light of the sequence information available. Between species the overall homology is strikingly low. At the amino acid level there is only a 50% homology between the rat and human precursor. Since for preproTRH 160–169 only 4 out of 10 amino acids and 14 out of 26 amino acids of preproTRH 178-199 are conserved between rat and human, the sequence data do not support the view that these fragments are of general endocrine importance. In retrospect, the sequence data also explain why the mammalian cDNA of preproTRH could not be identified by library screening with the frog cDNA

Fig. 1. Comparison of the TRH precursor proteins from human, rat, salmon and *Xenopus*. Ordinate and abscissa represent the hydrophobicity and peptide sequence, respectively. Hydrophobicity profiles were calculated according to Kyte and Doolittle algorithm using a window size of 11 amino acids. TRH prosequences are indicated by *black solid boxes*. Sequences of human, rat, salmon and *Xenopus* preproTRH are from Yamada et al. (1990), Lechan et al. (1986), Kuchler et al. (1990) and Ohide et al. (1996), respectively. A similar analysis has been presented before by Ohide et al. (1996)

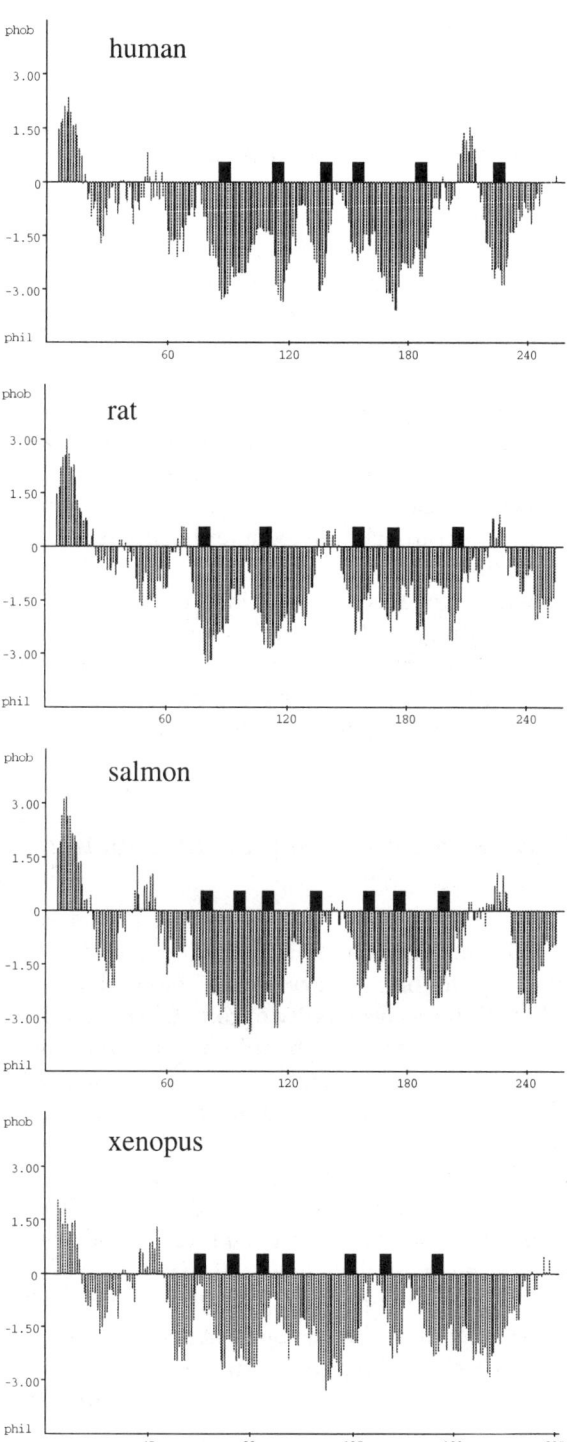

probes; only the TRH progenitor sequences are conserved but not the intervening sequences.

3.2
Regulation of TRH Biosynthesis

TRH gene expression in neurons of the hypothalamic paraventricular nucleus
is subject to multifactorial control mechanisms. The transcriptional downregulation by thyroid hormones is a specific feature of this brain area and is not
observed in other brain regions expressing TRH (Koller et al. 1987; Segerson
et al. 1987). Analysis of the transcriptional starting point of the rat TRH gene
and the promotor sequences revealed several sequences with homologies to
half sites of thyroid hormone response elements (TREs) as well as consensus
sequences for cAMP response element (CRE). So far, however, the exact mechanisms of the cell specific negative regulation of the TRH gene by thyroid hormones are not completely understood. The same is true for the regulation of
TRH gene transcription by other hormones (e.g. negative regulation by glucocorticoids), various factors (e.g. stimulation of the TRH gene by epidermal
growth factor recently described by Ren et al. 1998) and various conditions
(e.g. suckling, food deprivation and cold exposure mediated by several neurotransmitter systems). For more details we refer the reader to the reviews by
Jackson et al. (1990), Rondeel and Jackson (1993), Toni and Lechan (1993),
Stevenin and Lee (1995) and Wilber et al. (1996).

4
TRH Receptor and Signal Transduction

Early binding studies with [^{3}H]TRH (Grant et al. 1972; Gourdji et al. 1973;
Hinkle and Tashjian 1973) clearly demonstrated the existence of high-affinity
binding sites on pituitary membranes and a rat clonal pituitary cell line (GH$_3$
cells). With the availability of radiolabeled [N-3-methyl-His]TRH, a TRH agonist with eight times higher binding affinity compared to TRH, TRH-receptors
could also be demonstrated in brain, retina and spinal cord. By autoradiography the distinct localization of TRH receptors throughout the brains of several species could then be visualized (for review see Sharif 1989). A widespread
distribution was noticed (see below).

TRH binding has been studied extensively using tumor models of thyrotropes (mouse thyrotropic Tt/T cells) and somatomammotropes (GH$_3$ and
related cell lines) since these cells have a relatively high receptor density, about
100 000 receptors per cell. [^{3}H]TRH appears to bind to a single class of non-
interacting sites with an apparent dissociation constant of 10 nM. This concentration is close to the concentration of TRH causing half-maximal stimulation
of hormone secretion, 2–3 nM when measured under the same conditions. The

TRH-receptor exhibits strict structural specificity in all three amino acid positions. With the exception of the N-3-methyl-His derivative mentioned above, all structural modifications (even supposedly minor alterations such as addition of a methyl group to the amide nitrogen or deamidation) cause a dramatic decrease in binding affinity. Surprisingly, the N-1-methyl-His derivative is absolutely inactive (for review see Hinkle 1989). Unfortunately, none of the several hundreds of TRH analogs synthesized acts as a classical antagonist of the pituitary TRH-receptor. The only known competitors at the TRH-receptor belong to a completely different class of substances, namely the benzodiazepines. Especially chlordiazepoxide and midazolam exhibit high affinity for the TRH-receptor (Ki values 0.94 and 0.07 µM, respectively), but, unfortunately, only *in vitro* and not *in vivo* (for review see Drummond et al. 1989). Pharmacologically the TRH-receptor still remains ill-defined and warrants further investigation.

Various TRH analogs that exhibit no or very low endocrine activities were found to elicit a variety of pharmacological and behavioural effects centrally with actions longer-lasting and of greater potency than those of TRH itself (Monden et al. 1995). In vitro, some of these analogs also affect cellular activities such as stimulation of acetylcholine release (Oka et al. 1996) and increase in cAMP formation (Mori et al. 1991). Together with reports that in brain intracellular signalling via cAMP or IP_3 varies with brain regions (Iriuchijima and Mori 1989), these data seem to indicate that besides the pituitary receptor for TRH additional receptors may be discovered. However, for the pituitary TRH-receptor originally it was also thought that the interaction of TRH with this receptor leads to an increase in cAMP and protein kinase A activation. In 1981 the first reports then described that TRH rapidly stimulates the turnover of inositol phospholipids in GH_3 cells (Rebecchi et al. 1981; Schlegel et al. 1981; Sutton and Martin 1982). Meanwhile, extensive studies by several laboratories firmly established that at the pituitary level TRH acts predominantly via the inositol-phospholipid-calcium-protein kinase C transductory pathway (for reviews see Drummond 1986; Gershengorn 1986; Hinkle 1989; Gershengorn and Osman 1996; Hinkle et al. 1996). Nevertheless, TRH signalling via other pathways cannot be excluded. Ohmichi et al. (1994) and recently Palomero et al. (1998) reported that TRH stimulates MAP kinase activity in GH_3 cells and transfected COS-7 cells by pathways only partially dependent on PKC activity. The relevance of MAPK activation in pituitary cell physiology is still unknown but may be related to the potential action of TRH as a proliferative factor.

4.1
Molecular Aspects of the Pituitary TRH-Receptor

TRH-receptor (TRH-R) studies were greatly stimulated when the laboratory of M. C. Gershengorn succeeded in identifying the cDNA encoding the mouse pituitary TRH-receptor by expression cloning (Straub et al. 1990). Their results clearly demonstrated that the pituitary TRH-receptor belongs to the family of G protein coupled seven transmembrane receptors and provided the basis for studying in detail the structure, function and regulation of this receptor at the molecular level. The isolated cDNA predicts a protein of 393 amino acids with a calculated molecular weight of 44.5 kDa. Several groups reported thereafter the isolation and analysis of rat (De la Pena et al. 1992a; Zhao et al. 1992; Sellar et al. 1993) and human TRH-R cDNA clones (Duthie et al. 1993; Matre et al. 1993; Yamada et al. 1993). In contrast to TRH these receptors are highly conserved among different species. Only in the C-terminal tail are major variations noticed. The gene of the human TRH-R spans more than 30 kb and contains three exons (Iwasaki et al. 1996). Exon I is entirely untranslated. Intron I is 541 nucleotides in length and ends 88 base pairs upstream from the translational start ATG. Exons 2 and 3 are separated by a second intron which extends

a

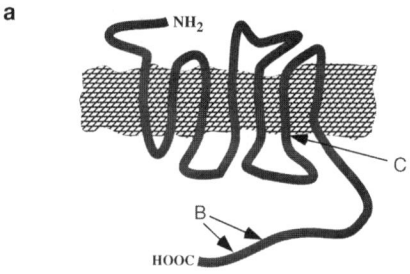

b

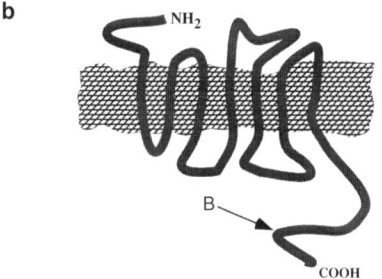

c

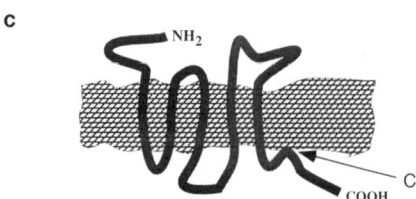

Fig. 2. The described isoforms of the TRH-receptor are not drawn to scale and arrangement of hydrophilic sequences is arbitrary. A Seven-transmembrane model of human, mouse or rat TRH-R. B Alternatively spliced TRH-R as found in rat and mouse anterior pituitary. The protein is the result of an alternative splice event of a retained intron in exon 3. C Five-transmembrane spanning isoform of the TRH-R as found in human pituitary and pituitary tumors. Alternative splicing at an internal acceptor site in exon 3 yields a 95-residue-shorter variant that lacks sequences of the sixth transmembrane spanning domain and presents a C terminus different from both TRH-R illustrated above

for more than 25 kb and separates the regions encoding the transmembrane domains 5 and 6. Tumor specific alternative splicing has been described at this intron/exon boundary. The splice variant lacks the 5'-end of exon 3. Due to a shift in frame, the C terminus is completely different and consequently this protein does not bind TRH (Yamada et al. 1997).

Alternative splicing of a retained intron in the primary TRH-R transcripts has also been described in rat and mouse pituitary (De la Pena et al. 1992b; Jones et al. 1996). In rat the shorter variant lacks 25 amino acids in the C-terminal cytoplasmic domain. The last 12 amino acids differ completely, but functional expression experiments demonstrated that both forms show indistinguishable electrophysiological responses to TRH (De la Pena et al. 1992b). A structural survey of these TRH-R variants is presented in Fig. 2. For more detailed information, we refer to the review by Gershengorn and Osman (1996).

4.2
Regulation of the Pituitary TRH-Receptor

Initial binding studies with pituitary cell lines have demonstrated that the concentration of TRH-receptors is regulated both homologously and heterologously. These effects were also observed *in vivo*. Homologous downregulation of TRH-receptors leads first to a rapid and extensive internalization of TRH-receptors via clathrin coated vesicles (Ashworth et al. 1995). As much as 80% of the TRH-receptor complex is sequestered in pituitary cells with a half-life of about 2–3 min (Hinkle 1989; Gershengorn and Osman 1996; Yu and Hinkle 1997, 1998). In the second phase, exposure to TRH leads to a reduction in the rate of transcription and a decrease in the mRNA stability of the TRH-receptor. With GH_3 cells stably expressing the cloned TRH-receptor the apparent half-life of the TRH-receptor transcripts decreased from 3 to 0.75 h after stimulation with 1 µM TRH for 1.5 h (Fujimoto et al. 1992a). Interestingly, a rather reversed effect was observed with transfected COS-1 cells (Fujimoto et al. 1992b). A destabilizing potential stem-loop structure with an AT-rich sequence in the 3' untranslated region and the activation of a specific RNAse are considered to be responsible for this cell specific posttranscriptional effect (Narayanan et al. 1992).

Heterologously the density of TRH-receptors is regulated by various hormones and other factors. Thyroid hormones exert a powerful negative control over the pituitary response to TRH. T_3 effectively downregulates TRH-R mRNA and decreases the receptor density. Estrogens, by contrast, increase TRH-R levels. The upregulation of TRH-R mRNA seems to involve both transcriptional activation and transcript-stabilization (Kimura et al. 1994). Glucocorticoids and calcitriol also increase TRH-R density whereas EGF and all drugs that lead to elevated cAMP concentrations cause a decline in the concentration of this TRH-R (for reviews see Hinkle 1989; Gershengorn 1993).

The analysis of the TRH-R gene and the promotor region revealed that neither a TATA box nor a CAAT box or a GC-rich sequence were present in close proximity 5' to the transcriptional start site (Iwasaki et al. 1996). Putative regulatory elements, however, were found within 1 kb of the 5' flanking region: a perfect PEA-3 site, a GATA motif, two almost perfect Pit-1 sites, a sequence very similar to the consensus AP1-site, two almost perfect TRE half sites and a 12/15 match of a glucocorticoid response element. The basal activity of this promotor has been demonstrated in GH4C1 cells (Iwasaki et al. 1996), but a detailed analysis of the regulatory important sequences has not yet been reported.

5
Inactivation of TRH

5.1
TRH-Metabolism

Although TRH was found to be resistant to a variety of proteolytic enzymes (Burgus et al. 1966), rapid inactivation by blood and tissue enzymes could be demonstrated even before the structure of TRH had been elucidated. Meanwhile it could be shown that the degradation is catalyzed by several enzymes (Fig. 3) (for reviews on the catabolism of TRH see O'Cuinn et al. 1990; O'Leary and O'Connor 1995).

Since the post proline cleaving enzyme and the pyroglutamate aminopeptidase are actually cytosolic enzymes, these peptidases certainly cannot be important for the inactivation of TRH after its release. In contrast, the membrane bound TRH-degrading enzyme that is preferentially associated with synaptosomal membranes (O'Connor and O'Cuinn 1984) has been identified as a true ectoenzyme. This enzyme is localized on the surface of neuronal and adenohypophyseal cells (preferentially on lactotrophic cells) but not on glial cells (Bauer et al. 1990; Cruz et al. 1991), indicating that this peptidase most likely serves specialized functions rather than general scavenger functions.

Membrane bound TRH-degrading enzyme

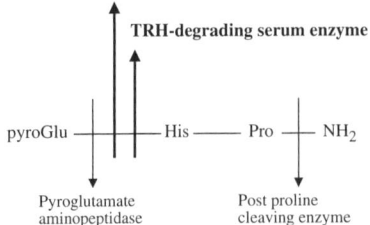

Fig. 3. Cleavage of TRH by TRH-degrading enzymes

5.2
Characteristics of the TRH-Degrading Ectoenzyme (TRH-DE) and the TRH-Degrading Serum Enzyme

Initial studies with the TRH-degrading serum enzyme had already revealed that this enzyme exhibits an extraordinarily high degree of substrate specificity, a feature uncommon to most peptidases (Bauer et al. 1981). Further studies with membrane preparations and partially purified enzyme preparations then demonstrated that the membrane-bound enzyme possesses identical properties with respect to substrate specificity and other characteristics (e.g. inhibitors, pH and temperature profiles). Both enzymes degrade only TRH but none of the other naturally occurring neuropeptides tested so far (O'Connor and O'Cuinn 1984; Garat et al. 1985; Wilk and Wilk 1989; Elmore et al. 1990). Even the hypothalamic hypophysiotrophic hormone GnRH (gonadotropin-releasing hormone), a decapeptideamide with the aminoterminal sequence pyroGlu-His, is not degraded by this enzyme, although it inhibits the degradation of TRH quite effectively.

After solubilization by limited proteolysis under very mild conditions we succeeded in purifying the TRH-degrading ectoenzyme (TRH-DE) from rat and pig brain about 200 000-fold by conventional chromatographic methods (Bauer 1994). A highly active enzyme preparation could thus be obtained in electrophoretically homogenous form. For the truncated enzyme SDS-PAGE revealed a molecular mass of 116 kDa while by gel filtration a molecular mass of 230 kDa was estimated, suggesting that the enzyme exists in homodimeric form.

With the purified enzyme we could extend previous studies (O'Connor and O'Cuinn 1987) and convincingly demonstrate that the TRH-DE is indeed a metallopeptidase (Czekay and Bauer 1993). As expected for a cell surface protein, this enzyme could be identified as a glycoprotein and lectin blot analysis indicates that its glycostructure belongs to the complex type (Bauer 1994 and unpubl. data).

5.3
Molecular Aspects

Peptide fragments were generated from the purified enzyme and sequenced. PCR experiments with degenerated oligonucleotides that were constructed according to the sequence information amplified a fragment of rat TRH-DE cDNA. This fragment allowed the isolation of a 6-kb cDNA clone by screening a rat pituitary cDNA library (Schauder et al. 1994). Sequence analysis predicts an open reading frame of 3075 nt giving rise to a protein of 1025 amino acids followed by a 3′ untranslated region of 3 kb. The cDNA-deduced sequence information is presented in Fig. 4. Sequence comparison with the databases suggests that the enzyme belongs to the M1 family of Zn-dependent metallopeptidases (Rawlings and Barrett 1995) like aminopeptidases A and N, the puromycin-sensitive aminopeptidase and others. The consensus sequence for coordination of the catalytically essential Zn^{2+}-ion (HEXXH) and several putative N-glycosylation sites are found in the large extracellular domain. After transfection, COS-7 cells expressed a recombinant ectoenzyme cleaving TRH with indistinguishable properties to the native TRH-DE.

Southern analysis suggests that the TRH-DE gene is present as a single copy and distinct hybridization signals are obtained with human, monkey, rat, mouse, dog, bovine, rabbit and chicken DNA. Using the rat cDNA as a probe we recently succeeded in isolating the cDNA encoding the human TRH-DE. The deduced primary sequence reveals an unexpected high degree of similarity to the rat enzyme (see Fig. 4). Notably human and rat TRH-DE share a central stretch of 225 identical amino acids encompassing the aforementioned Zn-binding motif (Schomburg et al. 1998). The notion that the TRH-DE is very highly conserved between species is further strengthened by our recent analysis of a partial cDNA clone from zebrafish. For the deduced sequence of 741 amino acids in length the homology to human or rat TRH-DE was 75% (unpubl. data). The high degree of substrate specificity of the TRH-DE and the conserved TRH structure might have provided the evolutionary pressure to maintain central parts of the enzyme's primary structure. The analysis of the genomic organization of the human TRH-DE gene revealed that the coding sequence is interrupted by 19 introns (unpubl. data). All intron/exon boundaries conform to the consensus sequence and can be perfectly aligned to the respective sites in the genes encoding human aminopeptidase N (Lerche et al.

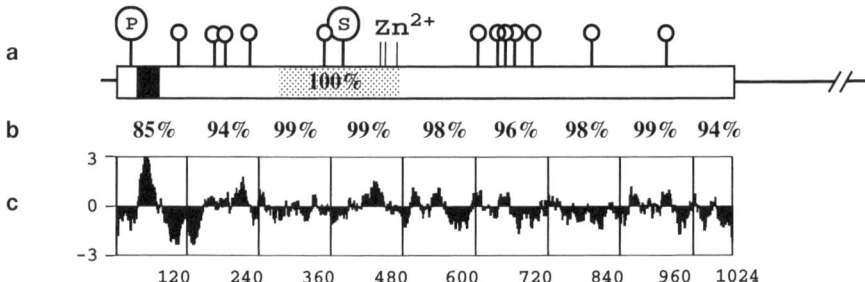

Fig. 4. TRH-DE: schematic structure of the cDNA-deduced protein, interspecies comparison and hydrophobicity profile. A *Boxed area* represents the open reading frame of rat and human TRH-DE, a predicted type II membrane protein. Indicated are the short and presumably cytosolic N-terminal sequence with a potential protein kinase C phosphorylation site (*P*) followed by a transmembrane spanning domain (*black box*) and the large extracellular domain with 12 potential N-glycosylation sites, a putative tyrosine sulfation site (S) and the Zn-binding motif HEXXH. **B** *Numbers* denote degree of sequence conservation (identical amino acids) between rat and human TRH-DE. *C* Hydrophobicity profile of the amino acid sequence encoded by human TRH-DE cDNA as calculated with an average window size of 11 residues. Ordinate and the abscissa represent the hydrophobicity and protein sequence, respectively

1996) and mouse aminopeptidase A (Wang et al. 1996). The comparison of the genomic sequences indicates further that the genes of human aminopeptidase N and mouse aminopeptidase A contain an additional intron that is not present in the human TRH-DE gene. This finding suggests that the M1 family of Zn-dependent metallopeptidases arose by divergent evolution from a common ancestor.

5.4
The TRH Degrading Serum Enzyme

Based on the unusual characteristics and the almost identical properties of the TRH-degrading ectoenzyme and the TRH-degrading serum enzyme it has been suggested that both enzymes are derived from the same gene (O'Cuinn et al. 1990). This notion is strongly supported by our recent findings (unpubl.). We also succeeded in purifying the TRH-degrading serum enzyme to homogeneity. After trypsin digestion the sequence data of the peptide fragments generated were found to be identical to the sequences identified previously with the purified ectoenzyme from rat and porcine brain. Moreover, both enzymes are immunologically identical since they are recognized by monoclonal and polyclonal enzymes generated against the TRH-degrading ectoenzyme isolated from brain.

Ontogenic studies demonstrated that in rats the activities of the serum enzyme (Neary et al. 1976) and the TRH-degrading liver enzyme are very low at birth (Scharfmann and Aratan-Spire 1991; Bauer and Schulz 1992) but rapidly increase in parallel at the age of 2–3 weeks. This development is quite different from that of other tissues such as the brain where expression of the TRH-degrading enzyme is already detectable at the embryological age of 14 days. In brain the specific enzymatic activities increase rapidly, reaching almost adult levels at birth. These findings strongly indicate that the serum enzyme is synthesized by the liver. This hypothesis is further supported by data from lectin blot analysis, since the serum enzyme and the enzyme from liver are recognized by the same lectins whereas the TRH-degrading ectoenzyme from brain exhibits different patterns. Finally, from data on the genomic organization of the gene it is clear that the serum enzyme is not the product of an alternative splice variant (unpubl. data). Taken together, the information presently available truly supports the concept that the TRH-degrading serum enzyme is synthesized by the liver and released into the blood by a yet unknown shedding mechanism.

5.5
Hormonal Regulation of TRH-Inactivating Enzymes

Since the TRH-inactivating enzymes (TRH-DE and the TRH-degrading serum enzyme) exhibit such unusual characteristics, we asked whether these enzymes might serve not only very specialized but possibly even regulatory functions. The mandatory prerequisite for a regulatory function necessarily implies that the activity of such enzymes be controlled in turn by some other mechanisms. Within the hypothalamus-pituitary-thyroid axis one would obviously ask whether the TRH-inactivating enzymes are controlled by thyroid hormones.

Previous studies on the degradation of TRH by serum have demonstrated that the activity of the TRH-degrading serum enzyme is influenced by the thyroid status of the animals (Bauer 1976; Dupont et al. 1976; White et al. 1976). The activity decreased when the animals were treated with the goitrogenic agent PTU (propylthiouracil) and increased in a dose-related manner after injection of thyroid hormones. Moreover, a significant sex difference was noticed. Euthyroid female rats showed only 85% of the enzymatic activity of euthyroid males (Bauer 1976).

For the TRH-degrading ectoenzymes from rat brain, spinal cord, retina and lung, the enzymatic activities were found not to be influenced by the thyroid status of the animals (not even after multiple injections of pharmacological doses of triiodothyronine) and also no sex difference could be observed. In contrast, the activity of the adenohypophyseal TRH-degrading enzyme was found to be stringently regulated by thyroid hormones (Bauer 1987; Ponce et al. 1988; Suen and Wilk 1989). The rapid increase in the enzymatic activity 4–6 h after injection of triiodothyronine (T_3) is preceded by a dramatic increase in the mRNA levels (Schomburg and Bauer 1995). Conversely, after PTU treatment a rapid decrease in the activity as well as in the transcript levels could be demonstrated.

In agreement with the observation that the activity of the adenohypophyseal enzyme of normal male rats was three times higher than that of normal female rats (Bauer 1988), subsequent studies clearly demonstrated that the activity and the mRNA levels of the adenohypophyseal enzyme rapidly decrease after injection of estradiol (Schomburg and Bauer 1997), but not after injection of other steroid hormones. Conversely, after ovariectomy an increase in the enzymatic activity could be observed (Bauer 1988).

Interestingly, after the hormonal manipulation of the animals the levels of TRH-receptor transcripts also changed within a few (2–4) hours in mirror image to the mRNA levels of TRH-DE. In contrast, the TSHβ-mRNA levels decreased relatively slowly after T_3 injection. After an initial reduction in the size of these transcripts the nadir of TSHβ-mRNA levels with very low transcript concentrations was reached only 48 h after T_3 injection (Krane et al. 1991; Schomburg and Bauer 1995).

The kinetic data thus indicate that in the adenohypophysis the early events apparently take place at the signal receiving site by mechanisms regulating (1) the number of TRH receptors and (2) the activity of the TRH-degrading ectoenzyme, which seem to operate in mirror image to each other at specific TRH target sites of anterior pituitary cells. We therefore assume that the adenohypophyseal TRH-DE may represent a regulatory element that may serve an integrative function in modulating the response of adenohypophyseal target cells to TRH and thus pituitary hormone secretion.

6
Distribution of TRH, TRH-R and TRH-DE and Cellular Localization

The T_3 induced changes in the transcript levels of the adenohypophyseal TRH-degrading ectoenzyme and the TRH-receptor was also demonstrated recently by *in situ* hybridization histochemistry (Heuer et al. 1998). In agreement with the previous studies which demonstrated that the TRH-degrading ectoenzyme is preferentially localized on lactotrophic cells (Bauer et al. 1990) a widespread distribution was noticed in the anterior pituitary. Whereas signals were not detected in the intermediate lobe, expression of the enzyme was observed in the posterior pituitary, confirming previous studies which reported that in untreated rats the enzymatic activities are considerably higher in the posterior pituitary compared to the anterior pituitary (Bauer 1989). In contrast to the adenohypophyseal enzyme, however, T_3 affected neither the activity nor the transcript levels of the neurohypophyseal enzyme. The physiological function of the tripeptideamide and the TRH-degrading ectoenzyme in the posterior pituitary still remains obscure since in this tissue neither could TRH-binding sites be detected by autoradiography nor could expression of TRH-R mRNA be visualized by *in situ* histochemistry.

In brain, *in situ* hybridization studies in general confirmed previous immunocytochemical findings of TRH-immunoreactive perikarya in many extrahypothalamic regions including the olfactory system, the reticular thalamic nucleus, the amygdaloid complex, throughout the caudate-putamen complex, the diagonal band of Broca and the bed nucleus of stria terminalis, the midbrain periaqueductal gray, the raphe nuclei and the dorsal motor nucleus of the vagus (Segerson et al. 1987). The relatively high TRH mRNA levels in many of the TRH synthesizing neurons most likely reflect their high biosynthetic activity which is quite typical for neurons having long and widespread projection systems. This has also been observed for other hypothalamic neuropeptide systems such as POMC (pro-opiomelanocortin) neurons in the arcuate nucleus and MCH (melanin-concentrating hormone) in the lateral hypothalamus. Both neuropeptide systems have only one (MCH) or two (POMC) sites of synthesis in the entire brain, but they have very widespread projections in many brain areas with densely innervated target regions.

While TRH is well characterized at the level of the perikarya, less well investigated are the terminal fields of TRH neurons. As to the site of storage and release of a neuropeptide, the *in situ* hybridization techniques can provide only very limited information. The immunocytochemical studies have, unfortunately, been hampered by several drawbacks. Not only did this small neuro-peptide cause problems in raising specific and sensitive antibodies, but also even more severe technical problems were encountered with regard to the optimal fixation procedure. In order to avoid diffusion and extensive loss of TRH during the fixation procedure, stronger aldehyde fixatives such as acrol-ein and fixation at lower pH values using picric acid were employed with greater success (Tsuruo et al. 1987; Merchenthaler et al. 1988). High densities of TRH innervation were observed in the eminentia mediana, the preoptic area and other hypothalamic areas, the nucleus of the solitary tract, the olfactory bulb, lateral septum, amygdalohippocampal area and spinal cord. Nevertheless, the information on the extent of the terminal fields of TRH is still incomplete in many brain regions and the target areas of TRH neurons are most likely greatly underestimated.

To further elucidate TRH target sites, studies on the localization and distri-bution of TRH-receptors may provide some information to correlate receptor densities in different brain regions and the physiological effects mediated by TRH. Unfortunately, since suitable antibodies are not available yet, the distri-bution of TRH-receptors has been studied mainly by autoradiography (Rostène et al. 1984; Manaker et al. 1985; Mantyh and Hunt 1985; Pazos et al. 1985; Sharif 1989) and more recently by *in situ* hybridization histochemistry. Both techniques document a very uneven expression pattern of the TRH-receptor and support the concept of a complex TRHergic innervation system in the CNS.

As illustrated by *in situ* hybridization histochemistry the TRH-receptor is preferentially expressed in limbic structures. Highest transcript levels were found in the accessory olfactory bulb, in the amygdalohippocampal area as well as in most of the amygdaloid nuclei, in numerous hypothalamic subre-gions, septal areas, rhinal and perirhinal cerebral cortex and in the hippocam-pal formation, in several brainstem nuclei and in motoneurons of the spinal cord (Cálza et al. 1992; Wu et al. 1992; Zabavnik et al. 1993; Heuer et al. 1998). In most areas, the distribution of TRH-receptor transcripts essentially match-es the TRH binding sites, suggesting that TRH-receptor proteins are located in the vicinity of cell bodies synthesizing the receptor. However, mismatches could also be observed, particularly in the cerebral cortex and the hippocam-pus, where low but detectable levels of TRH binding contrasts with few TRH-receptor expressing cells. In this situation it was of special interest to study the localization of the TRH-degrading ectoenzyme (the third element of the sig-nal system), with the anticipation that the distribution of TRH-DE could pos-sibly reflect pathways of TRHergic innervation.

Since TRH is the only known substrate of TRH-DE and TRH is selectively inactivated only by TRH-DE one would expect that the enzyme is localized specifically at TRH target sites in the vicinity of the sites of TRH release and/or target interaction (pre- or postsynaptic membranes) with high enzyme con-centrations in densely TRH innervated brain regions. Therefore, TRH-DE may be considered as a marker to map the TRHergic pathways. This is in analogy to the cholinergic system where acetylcholinesterase has served for many decades as a marker enzyme to unravel cholinergic pathways and cholinergic terminal fields in the CNS and peripheral nervous system. Unlike acetylcholinesterase, however, a histochemical method that is suitable for monitoring TRH-DE activity at the cellular level has not been developed so far. Furthermore, antibodies that are sufficiently specific for TRH-DE are not yet available for immunocytochemical studies. Therefore, *in situ* hybridization histochemistry was the method of choice to obtain detailed information on the expression of TRH-DE at the cellular level (Heuer et al. 1998). The region-specific distribution of TRH-DE transcripts correlates well with enzyme activity data. Highest transcript levels were found in neo- and allocortical regions with strongest *in situ* hybridization signals in the olfactory bulb, the piriform cortex, the cerebral cortex, the granular layer of the cerebellar cortex and the pyramidal cells of the Ammon's horn. In the diencephalon, the highest TRH-DE mRNA levels were observed in the medial habenulae followed by several hypothalamic nuclei. In the mesencephalon and brainstem, only low levels of TRH-DE expression could be observed with moderate signals in the superior colliculi, substantia nigra, dorsal raphe and in the periolivar region. In the spinal cord, TRH-DE mRNA positive neurons were present in all layers with slightly higher transcript levels in the superficial dorsal horn and the motoneuron area of the ventral horn. Enzyme-specific hybridization signals were completely absent over white matter regions, supporting an exclusive neuronal localization of this ectopeptidase in the CNS. These findings correlate well with cell culture experiments, where the enzyme could only be detected on the surface of neuronal cells, but not of glial cells.

Although there is some overlap between TRH-DE mRNA distribution and that of preproTRH and TRH-R mRNA, the general distribution patterns of all three components appear strikingly different. Figure 5 exemplarily illustrates the divergent mRNA distribution pattern of the receptor and the degrading enzyme in septal regions which not only are densely innervated by TRH neurons (Ishikawa et al. 1984) but also contain high levels of preproTRH transcripts as well (Segerson et al. 1987). Highest signal intensities of the enzyme could be observed in the dorsal part of the lateral septum, whereas the TRH-receptor is predominantly expressed in the medial septal nuclei and in the diagonal band of Broca excluding a coexpression of receptor and enzyme in most neurons.

Mismatches are also observed in other brain regions, particularly in the cerebral cortex where the low but detectable levels of TRH binding contrasts with the few TRH-R expressing cells and the strong expression of TRH-DE. Similarly, in the hippocampal formation the high TRH-DE mRNA levels of all pyramidal neurons contrasts with the restricted TRH-R mRNA. There are several possible explanations. As pointed out by Cálza et al. (1992) technical problems might lead to an underestimation of TRH binding sites and transcript

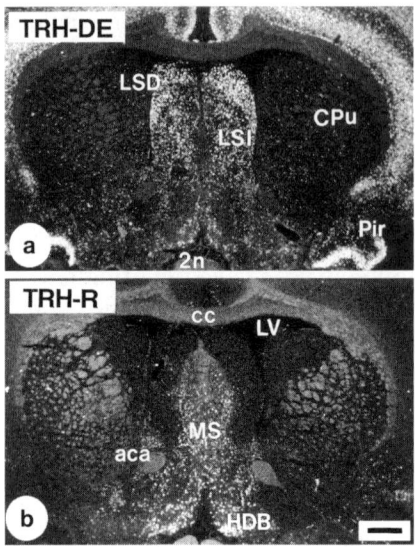

Fig. 5. Comparison of TRH-degrading enzyme (TRH-DE) and TRH-receptor (TRH-R) mRNA distribution in septal regions of the CNS. TRH-DE as well as TRH-R were expressed in septal nuclei, although the subnuclear distribution pattern of these two species differs substantially. High levels of TRH-DE mRNA (**a**) were observed in the dorsal lateral septum (*LSD*) and with slightly lower signal intensity in the intermediate lateral septum (*LSI*) and in medial septal nuclei (*MS*). In contrast, the strongest TRH-R hybridization signals (**b**) were observed in medial septal nuclei, whereas in the LSI the TRH-R mRNA levels were low but detectable. The diagonal band of Broca (*HDB*) represents another region with high TRH-R mRNA expression, whereas enzyme-specific hybridization signals were limited to scattered neurons (scale bar 500 μm). *aca* Commissura anterior; *cc* corpus callosum; *CPu* caudatus-putamen; *LV* lateral ventricle; *Pir* piriform cortex; *2n* optic nerve

levels. As another likely possibility we assume the existence of other TRH-receptors not yet identified. Alternatively, these discrepancies could also be explained by a differential localization of these proteins at specific sites of the synaptic cleft. Together with the TRH binding data one might postulate a presynaptic localization of the receptor protein and a postsynaptic orientation of the enzyme. For example, the TRH-R protein might possibly be located on presynaptic terminals of ventral forebrain neurons projecting to cortical regions. This interpretation is supported by the demonstration of high TRH-R and TRH mRNA levels in presumably cholinergic neurons of the diagonal band of Broca (Segerson et al. 1987; Cálza et al. 1992; Heuer et al. 1998). In the hippocampal area TRH-R may also very well be present on cholinergic terminals originating from the medial septal area. These anatomical considerations seem to be supported by numerous physiological and neuropharmacological studies, providing evidence that TRH stimulates cholinergic mechanisms (for review see Yarbrough 1983), being extremely potent in activating forebrain cortical pathways (Horita et al. 1986). For example, TRH has been shown to potentiate the excitatory actions of cholinergic agents on neurons and to increase the release of brain acetylcholine as well as the high affinity uptake of choline (Giovannini et al. 1991; Okada 1991). Based on these reports and the TRH binding data it is tempting to speculate that the neuromodulator TRH might be recognized by TRH-receptors localized on presynaptic cholinergic terminals, to be inactivated subsequently by TRH-DE on postsynaptic dendritic or somal membranes of cholinoceptive neurons. However, this intriguing hypothesis still remains speculative and warrants further investigation. In any case and notwithstanding any interpretation, the predominant expression of TRH-DE throughout the neocortex and in allocortical regions including the hippocampal formation implicates an important role of the TRH-degrading ectoenzyme in the TRH-mediated modulation of higher sensory, locomotor and cognitive functions of the CNS.

7
Future Aspects

Despite the important function of TRH within the hypothalamus-pituitary-thyroid axis and its effects on TSH synthesis and posttranslational modification, early studies in rats with specific hypothalamic lesions already demonstrate that the major control of TSH secretion appears to be through direct feedback of thyroid hormones at the pituitary level (Reichlin and Utiger 1967). It is therefore not surprising that mice lacking TRH (generated by targeted disruption of the TRH gene) exhibit only mild to moderate hypothyroidism (Yamada et al. 1997). Correspondingly, in man a recently published case report described mutations in the gene of the known TRH-receptor leading to inactivation of this gene and consequently to symptoms comparable to central hypothyroidism. Otherwise, the only presenting symptoms were short stature with markedly delayed bone maturation (Collu et al. 1997). TRH is obviously important as a central modulator for the dynamic regulation of TSH secretion but hypothalamic control of TSH by TRH cannot be considered as the "thyroidstat" per se.

While the endocrine activities of TRH are fairly well understood, the central actions of this tripeptideamide still remain rather mysterious despite all efforts and a vast body of literature. So far, we do not even understand the regional distribution of the three elements of TRH signalling.

Due to the technical problems mentioned, the projections of TRH neurons are most likely greatly underestimated. Possibly other methods of fixation could be developed to overcome some of these problems. Alternatively, transgenic mice could be generated in which the disrupted TRH gene is replaced by a lacZ gene in order to trace TRH projections by an extremely sensitive method. With regard to the pituitary, TRH-receptor and TRH-DE specific antibodies (not yet available) are urgently needed to define the specific localization of these proteins at their specific sites. In addition, transgenic animals should be generated that lack the known TRH-receptor (containing possibly the lacZ gene again for tracing the distribution of this receptor in the CNS). With these animals, binding studies (preferably by autoradiography) should provide definite answers as to the existence of other TRH-receptors, especially in the CNS. Additional TRH-receptors might be anticipated since for almost all neuropeptides and neurotransmitters multiple and distinct receptors have been defined and, as mentioned, indirect evidence suggests that such receptors may exist (cAMP signalling in brain; TRH analogs devoid of endocrine activities mediating effects in the CNS). Studies along these lines might provide clues as to the biological function of TRH in brain, which, possibly, might be of clinical importance.

Stimulated by early reports on the therapeutical potential of TRH for certain clinical conditions, especially in depression and schizophrenia (also in Alzheimer's disease and motor neuron disease), many clinical trials have been performed in man (Yarbrough 1979; Griffiths 1985; Horita et al. 1986; Kelly 1995; Callahan et al. 1997; Horita 1998). So far, however, a therapeutic indication for TRH or any of the TRH analogs has not been established. This is not surprising since such studies are hampered by severe problems. First of all, TRH given intravenously or orally is rapidly inactivated by the TRH-degrading serum enzyme and will cross the blood-brain barrier very poorly, if at all. Even after intrathecal administration TRH itself elicits only short lasting effects, probably due to the high activites of TRH-DE in brain. Therefore, only peptidase resistant TRH analogs desirably devoid of endocrine activities (to prevent endocrine disorders) might have therapeutic potential provided they cross the blood-brain barrier sufficiently and are recognized selectively by receptors in the CNS. The fundamental question to be answered thus remains: are there yet unknown TRH receptors in brain which can be activated specifically by these analogs?

Provided the expectations of the potential therapeutic application of TRH hold true, we imagine that an alternative might be more promising, namely the synthesis of specific inhibitors against the TRH-inactivating enzymes that are sufficiently lipophilic to penetrate the blood-brain barrier. We would expect that these inhibitors would considerably increase the duration of action of endogeneously released TRH at its specific target sites. With such inhibitors we could envisage a scenario comparable to the effects of some acetylcholinesterase inhibitors such as physiostigmine which is used clinically to facilitate recovery from anesthesia and reportedly has analeptic actions in animals. Based on the analeptic action of TRH and some TRH analogs and the reported interactions of TRH with cholinergic mechanisms, the synthesis of enzyme specific inhibitors as potential psychostimulant agents appears, to us, a very intriguing concept. Whether such inhibitors may disclose the role of central TRH (in particular in varying neurobehavioural functions such as sleep, anxiety, depression, learning and memory) remains to be studied in great detail. So far, however, highly selective and potent inhibitors against the TRH-inactivating enzymes are not available, but with the identification of these enzymes as metallopeptidases new strategies can be explored which have been successful for the design of active site directed inhibitors of other metallopeptidases such as angiotensin converting enzyme.

Although 30 years have passed since TRH was first discovered, the biological role of this tripeptideamide in the CNS still remains rather enigmatic, but perhaps we are just at the beginning.

Acknowledgements. We thank Prof. Dr. H. Jäckle, Prof. Dr. P.W. Jungblut and Prof. Dr. E. Weihe for continuous support and encouragement, V. Ashe for typing and especially for linguistic help, together with D. Murnane. This study was supported by grants from the Deutsche Forschungsgemeinschaft.

Note added in proof:
Recently a novel, pharmacologically distinct receptor for TRH (TRH-R2) has been identified and characterized (Cao et al. 1998; Itadani et al. 1998).

References

Ashworth R, Yu R, Nelson EJ, Dermer S, Gershengorn MC, Hinkle PM (1995) Visualization of the thyrotropin-releasing-hormone receptor and its ligand during endocytosis and recycling. Proc Natl Acad Sci USA 92:512–516

Bauer K (1976) Regulation of degradation of thyrotropin releasing hormone by thyroid hormones. Nature 259:591–593

Bauer K (1987) Adenohypophyseal degradation of thyrotropin-releasing hormone regulated by thyroid hormones. Nature 330:375–377

Bauer K (1988) Degradation and biological inactivation of thyrotropin releasing hormone (TRH): regulation of the membrane-bound TRH-degrading enzyme from rat anterior pituitary by estrogens and thyroid hormones. Biochimie 70:69–74

Bauer K (1989) Multihormonal regulation of thyrotropin releasing hormone-degrading ectoenzyme from rat anterior pituitary. In: Casanueva FF, Dieguez C (eds) Recent advances in basic and clinical endocrinology. Elsevier, Amsterdam, pp 135–140

Bauer K (1994) Purification and characterization of the thyrotropin-releasing hormone-degrading ectoenzyme. Eur J Biochem 224:387–396

Bauer K, Lipmann F (1976) Attempts toward biosynthesis of the thyrotropin-releasing hormone and studies on its breakdown in hypothalamic tissue preparations. Endocrinology 99:230–242

Bauer K, Schulz M (1992) Inactivation of the thyrotropin releasing hormone. Acta Med Austriaca 1:79–82

Bauer K, Nowak P, Kleinkauf H (1981) Specificity of a serum peptidase hydrolysing thyroliberin at the pyroglutamyl histidine bond. Eur J Biochem 118:173–176

Bauer K, Carmeliet P, Schulz M, Baes M, Denef C (1990) Regulation and cellular localization of the membrane-bound thyrotropin-releasing hormone-degrading enzyme in primary cultures of neuronal, glial and adenohypophyseal cells. Endocrinology 127:1224–1233

Bøler J, Enzmann K, Bowers CY, Schally AV (1969) The identity of chemical and hormonal properties of the thyrotropin-releasing hormone and pyroglutamyl-histidyl-proline-amide. Biochem Biophys Res Commun 37:705–710

Bulant M, Roussel J-P, Astier H, Nicolas P, Vaudry H (1990) Processing of thyrotropin-releasing hormone prohormone (pro-TRH) generates a biologically active peptide, prepro-TRH-(160–169), which regulates TRH-induced thyrotropin secretion. Proc Natl Acad Sci USA 87: 4439–4443

Burgus R, Ward DN, Sakiz E, Guillemin R (1966) Action des enzymes proteolytiques sur les preparations purifiees de l'hormone hypothalamique TSH-hypophysiotrope, TRF. CR Acad Sci Paris 262:2643–2645

Burgus R, Dunn T, Desiderio D, Guillemin R (1969) Structure moleculaire du facteur hypothalamique hypophysiotrope TRF d'origine ovine: mise en évidence par spectrometrie de masse de la sequence PCA-His-Pro-NH$_2$. CR Acad Sci (Paris) 269:1870–1873

Callahan AM, Frye MA, Marangell LB, George MS, Ketter TA, L'Herrou T, Post RM (1997) Comparative antidepressant effects of intravenous and intrathecal thyrotropin–releasing hormone: confounding effects of tolerance and implications for therapeutics. Biol Psychiatry 41:264–272

Cao J, O'Donnell D, Vu H, Payza K, Pou C, Godbout C, Jakob A, Pelletier M, Lembo P, Ahmad S, Walker P (1998) Cloning and characterization of a cDNA encoding a novel subtype of rat thyrotropin-releasing hormone receptor. J Biol Chem 273 (48):32281–32287

Cálza L, Giardino L, Ceccatelli S, Zanni M, Elde R, Hökfelt T (1992) Distribution of thyrotropin-releasing hormone receptor messenger RNA in the rat brain: an *in situ* hybridization study. Neuroscience 51:891–909

Carnell NE, Feng P, Kim UJ, Wilber JF (1992) Preprothyrotropin-releasing hormone mRNA and TRH are present in the rat heart. Neuropeptides 22:209–212

Carr FE, Fein HG, Fisher CU, Wessendorf MW, Smallridge RC (1992) A cryptic peptide (160–169) of thyrotropin-releasing hormone prohormone demonstrates biological activity in vivo and in vitro. Endocrinology 131:2653–2658

Cockle SM (1995) Fertilization-promoting peptide: a novel peptide, structurally similar to TRH, with potent physiological activity. J Endocrinol 146:3–8

Cockle SM, Aitken A, Beg F, Smyth DG (1989) A novel peptide, pyroglutamylglutamylproline amide, in the rabbit prostate complex, structurally related to thyrotropin-releasing hormone. J Biol Chem 264:7788–7791

Collu R, Tang J, Castagné J, Lagacé G, Masson N, Huot C, Deal C, Delvin E, Faccenda E, Eidne KA, Van Vliet G (1997) A novel mechanism for isolated central hypothyroidism: inactivating mutations in the thyrotropin-releasing hormone receptor gene. J Clin Endocrinol Metab 82: 1361–1365

Cruz C, Charli J-L, Vargas MA, Joseph-Bravo P (1991) Neuronal localization of pyroglutamate aminopeptidase II in primary cultures of fetal mouse brain. J Neurochem 56:1594–1601

Czekay G, Bauer K (1993) Identification of the thyrotropin-releasing-hormone-degrading ectoenzyme as a metallopeptidase. Biochem J 290:921–926

De la Pena P, Delgado LM, Del Camino D, Barros F (1992a) Cloning and expression of the thyrotropin-releasing hormone receptor from GH$_3$ rat anterior pituitary cells. Biochem J 284:891–899

De la Pena P, Delgado LM, del Camino D, Barros F (1992b) Two isoforms of the thyrotropin-releasing hormone receptor generated by alternative splicing have indistinguishable functional properties. J Biol Chem 267:25703–25708

Dixon JE, Acres SC (1975) The inability to demonstrate the non-ribosomal biosynthesis of thyrotropin-releasing hormone in hypothalamic tissue. Fed Proc 34:658

Drummond A (1986) Inositol lipid metabolism and signal transduction in clonal pituitary cells. J Exp Biol 124:337–358

Drummond AH, Hughes PJ, Ruiz-Larrea F, Joels LA (1989) Use of receptor antagonists in elucidating the mechanism of action of TRH in GH$_3$ cells. Ann NY Acad Sci 1989:197–204

Dupont A, Labrie F, Levasseur L, Dussault JH, Schally AV (1976) Effect of thyroxine on the inactivation of thyrotropin-releasing hormone by rat and human plasma. Clin Endocrinol 5:323–328

Duthie SM, Taylor PL, Anderson L, Cook J, Eidne KA (1993) Cloning and functional character-isation of the human TRH receptor. Mol Cell Endocrinol 95:R11–R15

Elmore MA, Griffiths EC, O'Connor B, O'Cuinn G (1990) Further characterization of the sub-strate specificity of a TRH hydrolysing pyroglutamate aminopeptidase from guinea-pig brain. Neuropeptides 15:31–36

Feng P, Gu J, Kim UJ, Carnell NJ, Wilber JF (1993) Identification, localization and developmen-tal studies of rat prepro thyrotropin-releasing hormone mRNA in the testis. Neuropeptides 24:63–69

Fujimoto J, Narayanan CS, Benjamin JE, Heinflink M, Gershengorn MC (1992a) Mechanism of regulation of thyrotropin-releasing hormone receptor messenger ribonucleic acid in stably transfected rat pituitary cells. Endocrinology 130:1879–1884

Fujimoto J, Narayanan CS, Benjamin JE, Gershengorn MC (1992b) Posttranscriptional up-regu-lation of thyrotropin-releasing hormone (TRH) receptor messenger ribonucleic acid by TRH in cos-1 cells transfected with mouse pituitary TRH receptor complementary deoxyribonu-cleic acid. Endocrinology 131:1716–1720

Garat B, Miranda J, Charli J-L, Joseph-Bravo P (1985) Presence of a membrane bound pyrogluta-myl amino peptidase degrading thyrotropin releasing hormone in rat brain. Neuropeptides 6:27–40

Gershengorn MC (1986) Mechanism of thyrotropin releasing hormone stimulation of pituitary hormone secretion. Annu Rev Physiol 48:515–526

Gershengorn MC (1993) Thyrotropin-releasing hormone receptor: cloning and regulation of its expression. Recent Prog Horm Res 48:341–363

Gershengorn MC, Osman R (1996) Molecular and cellular biology of thyrotropin-releasing-hor-mone receptors. Physiol Rev 76:175–191

Giovannini MG, Casamenti F, Nistri A, Paoli F, Pepeu G (1991) Effect of thyrotropin releasing hormone (TRH) on acetylcholine release from different brain areas investigated by micro-dialysis. Br J Pharmacol 102:363–368

Gourdji D, Tixier-Vidal A, Morin A, Pradelles P, Morgat JL, Formageot P, Kerdelhué B (1973) Binding of a tritiated thyrotropin-releasing factor to a prolactin secreting clonal cell line (GH₃). Exp Cell Res 82:39–46

Grant G, Vale W, Guillemin R (1972) Interaction of thyrotropin releasing factor with membrane receptors of pituitary cells. Biochem Biophys Res Commun 46:28–34

Griffiths EC (1985) Thyrotropin releasing hormone: endocrine and central effects. Psychoneuro-endocrinology 10:225–235

Guillemin R (1978) Peptides in the brain: the new endocrinology of the neuron. Science 202: 390–402

Harvey S (1990) Thyrotropin-releasing hormone: a growth hormone-releasing factor. J Endo-crinol 125:345–358

Heuer H, Ehrchen J, Bauer K, Schafer MKH (1998) Region-specific expression of thyrotropin-releasing hormone-degrading ectoenzyme in the rat central nervous system and pituitary gland. Eur J Neurosci 10:1465–1478

Hinkle PM (1989) Pituitary TRH receptors. Ann NY Acad Sci 553:176–187

Hinkle PM, Tashjian AH (1973) Receptors for thyrotropin-releasing hormone in prolactin pro-ducing rat pituitary cells in culture. J Biol Chem 248:6180–6186

Hinkle PM, Nelson EJ, Ashworth J (1996) Characterization of the calcium response to thyro-tropin-releasing hormone in lactotrophs and GH cells. Trends Endocrinol Metab 7 (10): 370–374

Horita A (1998) An update on the CNS actions of TRH and its analogs. Life Sci 62:1443–1448

Horita A, Carino MA, Lai H (1986) Pharmacology of thyrotropin-releasing hormone. Annu Rev Pharmacol Toxicol 26:311–332

Iriuchijima T, Mori M (1989) Regional dissociation of cyclic AMP and inositol phosphate forma-tion in response to thyrotropin-releasing hormone in the rat brain. J Neurochem 52: 1944–1946

Ishikawa K, Inoue K, Tosaka H, Shimada O, Suzuki M (1984) Immunohistochemical characterization of TRH-containing neurons in rat septum. Neuroendocrinology 39:448–452

Itadani I, Nakamura T, Itoh J, Iwaasa H, Kanatani A, Borkowski J, Ihara M, Ohta M (1998)
Cloning and characterization of a new subtype of thyrotropin-releasing hormone receptors.
Biochem Biophys Res Commun 250:68–71

Iwasaki T, Yamada M, Satoh T, Konaka S, Ren Y, Hashimoto K, Kohga H, Kato Y, Mori M (1996)
Genomic organization and promoter function of the human thyrotropin-releasing-hormone
receptor gene. J Biol Chem 271:22183–22188

Jackson IMD (1982) Thyrotropin-releasing hormone. N Engl J Med 306:145–155

Jackson IMD, Reichlin S (1972) Thyrotropin-releasing hormone: abundance in the skin of the
frog, Rana pipiens. Science 198:414–415

Jackson IMD, Reichlin S (1977) Brain thyrotropin-releasing hormone is independent of the
hypothalamus. Nature 267:853–854

Jackson IMD, Wu P, Lechan RM (1985) Immunohistochemical localization in the rat brain of the
precursor for thyrotropin-releasing hormone. Science 229:1097–1099

Jackson IMD, Lechan RM, Lee SL (1990) TRH prohormone: biosynthesis, anatomic distribution
and processing. Front Neuroendocrinol 11:267–312

Jacobs LS, Snyder PJ, Wilber JF, Utiger RD, Daughaday WH (1971) Increased serum prolactin
after administration of synthetic thyrotropin releasing hormone (TRH) in man. J Clin Endocrinol 33:996–998

Jones KE, Brubaker JH, Chin WW (1996) An alternative splice variant of the mouse TRH receptor mRNA is the major form expressed in the mouse pituitary gland. J Mol Endocrinol 16:
197–204

Kelly JA (1995) Thyrotropin-releasing-hormone – basis and potential for its therapeutic use.
Essays Biochem 30:133–149

Kimura N, Arai K, Sahara Y, Suzuki H, Kimura N (1994) Estradiol transcriptionally and post-
transcriptionally up-regulates thyrotropin-releasing hormone receptor messenger ribonucleic acid in rat pituitary cells. Endocrinology 134:432–440

Koller KJ, Wolff RS, Warden MK, Zoeller RT (1987) Thyroid hormones regulate levels of thyrotropin-releasing hormone mRNA in the paraventricular nucleus. Proc Natl Acad Sci USA 84:
7329–7333

Krane IM, Spindel ER, Chin WW (1991) Thyroid hormone decreases the stability and the
poly(A) tract length of rat thyrotropin β-subunit messenger RNA. Mol Endocrinol 5.469–475

Kuchler K, Richter K, Trnovsky J, Egger R, Kreil G (1990) Two precursors of thyrotropin-releasing hormone from skin of Xenopus laevis. J Biol Chem 265:11731–11733

Lechan RM, Wu P, Jackson IMD, Wolf H, Cooperman S, Mandel G, Goodman RH (1986)
Thyrotropin-releasing hormone precursor: characterization in rat brain. Science 231:159–161

Le Dafniet M, Lefebvre P, Barret A, Mechain C, Feinstein MC, Brandi AM, Peillon F (1990) Normal
and adenomatous human pituitaries secrete thyrotropin-releasing hormone in vitro: modulation by dopamine, haloperidol and somatostatin. J Clin Endocrinol Metab 71:480–486

Lee SL, Stewart K, Goodman RH (1988) Structure of the gene encoding rat thyrotropin-releasing
hormone. J Biol Chem 263:16604–16609

Lerche C, Vogel LK, Shapiro LH, Noren O, Sjostrom H (1996) Human aminopeptidase-n is
encoded by 20 exons. Mammal Genome 7:712–713

Madsen OD, Nielsen JH, Michelsen B, Westermark P, Betsholtz C, Nishi M, Steiner DF (1991)
Islet amyloid polypeptide and insulin expression are controlled differently in primary and
transformed islet cells. Mol Endocrinol 5:143–148

Manaker S, Winokur A, Rostene WH, Rainbow TC (1985) Autoradiographic localization of
thyrotropin-releasing hormone receptors in the rat central nervous system. J Neurosci
5:167–174

Mantyh PW, Hunt SP (1985) Thyrotropin-releasing hormone (TRH) receptors: localization by
light microscopic autoradiography in rat brain using [³H] [3-Me His²] TRH as the radioligand. J Neurosci 5:551–561

Martino E, Lernmark A, Seo H, Steiner DF, Refetoff S (1978) High concentration of thyrotropin-releasing hormone in pancreatic islets. Proc Natl Acad Sci USA 75:4265–4267

Matre V, Karlsen HE, Wright MS, Lundell I, Fjeldheim AK, Gabrielsen OS, Larhammar D, Gautvick KM (1993) Molecular cloning of a functional human thyrotropin-releasing hormone receptor. Biochem Biophys Res Commun 195:179–185

Merchenthaler I, Csernus V, Csontos C, Petrusz P, Mess B (1988) New data on the immunocytochemical localization of thyrotropin-releasing hormone in the rat central nervous system. Am J Anat 181:359–376

Monden T, Mizuma H, Yamada M, Murakami M, Mori M (1995) A novel analog of TRH, YM14673, causes a decrease in brain TRH receptors in-vitro. Endocr Res 21:803–814

Mori M, Iriuchijima T, Yamada M, Murakami M, Kobayashi S (1991) A novel TRH analog, YM14673, stimulates intracellular signaling systems in the brain more potently than predicted by its pituitary actions. Res Commun Chem Pathol Pharmacol 71:17–26

Morley JE (1981) Neuroendocrine control of thyrotropin secretion. Endocr Rev 2:396–436

Morley JE, Garvin TJ, Pekary E, Hershman JM (1977) Thyrotropin-releasing hormone in the gastrointestinal tract. Biochem Biophys Res Commun 79:314–318

Narayanan CS, Fujimoto J, Geras-Raaka E, Gershengorn MC (1992) Regulation by thyrotropin-releasing hormone (TRH) of TRH receptor mRNA degradation in rat pituitary GH_3 cells. J Biol Chem 267:17296–17303

Neary JT, Kieffer JD, Federico P, Mover H, Maloof F (1976) Thyrotropin releasing hormone: development of inactivation system during maturation of the rat. Science 193:403–405

Nicholson WE, Orth DN (1996) Preprothyrotropin-releasing hormone-(178–199) does not inhibit corticotropin release. Endocrinology 137:2171–2174

O'Connor B, O'Cuinn G (1984) Localization of a narrow-specificity thyroliberin hydrolyzing pyroglutamate aminopeptidase in synaptosomal membranes of guinea-pig brain. Eur J Biochem 144:271–278

O'Connor B, O'Cuinn G (1987) Active site studies on a narrow-specificity thyroliberin-hydrolysing pyroglutamate aminopeptidase purified from synaptosomal membrane of guinea-pig brain. J Neurochem 48:676–680

O'Cuinn G, O'Connor B, Elmore M (1990) Degradation of thyrotropin-releasing hormone and luteinising hormone-releasing hormone by enzymes of brain tissue. J Neurochem 54:1–13

Ohide A, Ando H, Yanagisawa T, Urano A (1996) Hydropathy profiles of predicted thyrotropin-releasing hormone precursors are highly conserved despite low similarity of primary structures. J Neuroendocrinol 8:695–701

Ohmichi M, Sawada T, Kanda Y, Koike K, Hirota K, Miyake A, Saltiel AR (1994) Thyrotropin-releasing hormone stimulates MAP kinase activity in GH_3 cells by divergent pathways: evidence of a role for early tyrosine phosphorylation. J Biol Chem 269:3783–3788

Oka M, Itoh Y, Ukai Y, Yoshikuni Y, Kimura K (1996) Protein-kinases are involved in prolonged acetylcholine-release from rat hippocampus induced by thyrotropin-releasing-hormone analog ns-3. J Neurochem 66:1889–1893

Okada M (1991) Effects of a new thyrotropin releasing hormone analogue, YM-14673, on the in vivo release of acetylcholine as measured by intracerebral dialysis in rats. J Neurochem 56:1544–1547

O'Leary R, O'Connor B (1995) Thyrotropin-releasing-hormone. J Neurochem 65:953–963

Palomero T, Barros F, Del Camino D, Viloria CG, de la Pena P (1998) A G protein βγ dimer-mediated pathway contributes to mitogen-activated protein kinase activation by thyrotropin-releasing hormone receptors in transfected COS-7 cells. Mol Pharmacol 53:613–622

Pazos A, Cortex R, Palaxios J (1985) Thyrotropin releasing hormone receptor binding sites: autoradiographic distribution in the rat and guinea pig brain. J Neurochem 45:1448–1463

Ponce G, Charli J-L, Pasten JA, Aceves C, Joseph-Bravo P (1988) Tissue-specific regulation of pyroglutamate aminopeptidase II activity by thyroid hormones. Neuroendocrinology 48:211–213

Prasad C (1984) Thyrotropin-releasing hormone. In: Lajtha A (ed) Handbook of neurochemistry, vol 8. Plenum, New York, pp 175–200

Rawlings ND, Barrett AJ (1995) Evolutionary families of metallopeptidases. Methods Enzymol 248:183–228

Rebecchi MJ, Monaco ME, Gershengorn MC (1981) Thyrotropin-releasing hormone rapidly enhances [^{32}P]orthophosphate incorporation into phosphatidic acid in cloned GH$_3$ cells. Biochem Biophys Res Commun 101:124–130

Redei E, Hilderbrand H, Aird F (1995) Corticotropin release inhibiting factor is encoded within prepro-TRH. Endocrinology 136:1813–1816

Reichlin S (1986) Neural functions of TRH. Acta Endocrinol 276 (Suppl):21–33

Reichlin S (1989) TRH: historical aspects. Ann NY Acad Sci 553:1–6

Reichlin S, Utiger RD (1967) Regulation of the pituitary-thyroid axis in man: relationship of TSH concentration to concentration of free and total thyroxine in plasma. J Clin Endocrinol Metab 27:251–255

Ren Y, Satoh T, Yamada M, Hashimoto K, Konaka S, Iwasaki T, Mori M (1998) Stimulation of the preprothyrotropin-releasing hormone gene by epidermal growth factor. Endocrinology 139:195–203

Richter K, Kawashima E, Egger R, Kreil G (1984) Biosynthesis of thyrotropin releasing hormone in the skin of *Xenopus laevis*: partial sequence of the precursor deduced from cloned cDNA. Embo J 3:617–621

Rondeel JMM, Jackson IMD (1993) Molecular biology of the regulation of hypothalamic hormones. J Endocrinol Invest 16:219–246

Rostène WH, Morgat JL, Duissaillant M, Rainbow TC, Sarrieau A, Vial M, Rosselin G (1984) In vitro biochemical characterization and autoradiographic distribution of ^{3}H-thyrotropin-releasing hormone binding sites in rat brain sections. Neuroendocrinology 39:81–86

Rupnow JH, Hinkle PM, Dixon JE (1979) A macromolecule which gives rise to thyrotropin releasing hormone. Biochem Biophys Res Commun 89:721–728

Satoh T, Feng P, Kim UJ, Wilber JF (1994) Identification of thyrotropin-releasing-hormone receptor in the rat testis. Neuropeptides 27:195–202

Schaeffer JM, Brownstein MJ, Axelrod J (1977) Thyrotropin-releasing hormone-like material in the rat retina: changes due to environmental lightning. Proc Natl Acad Sci USA 74:3579–3581

Schally AV (1978) Aspects of hypothalamic regulation of the pituitary gland. Science 202:18–28

Scharfmann R, Aratan-Spire S (1991) Ontogeny of two topologically distinct TRH-degrading pyroglutamate aminopeptidase activities in the rat liver. Regul Pept 32:75–83

Schauder B, Schomburg L, Köhrle J, Bauer K (1994) Cloning of a cDNA encoding an ectoenzyme that degrades thyrotropin-releasing hormone. Proc Natl Acad Sci USA 91:9534–9538

Schlegel W, Roduit C, Zahnd G (1981) Thyrotropin releasing hormone stimulates metabolism of phosphatidyl inositol in GH$_3$ cells. A possible mechanism in stimulus-response coupling. FEBS Lett 134:47–49

Schomburg L, Bauer K (1995) Thyroid hormones rapidly and stringently regulate the messenger RNA levels of the thyrotropin-releasing hormone (TRH) receptor and the TRH-degrading ectoenzyme. Endocrinology 136:3480–3485

Schomburg L, Bauer K (1997) Regulation of the adenohypophyseal thyrotropin-releasing hormone-degrading ectoenzyme by estradiol. Endocrinology 138:3587–3593

Schomburg L, Turwitt S, Prescher G, Horsthemke B, Bauer K (1998) cDNA cloning of the human TRH-degrading ectoenzyme. Exp Clin Endocrinol Diabet 106 (Suppl):S5

Segerson TP, Kauer J, Wolfe HC, Mobtaker H, Wu P, Jackson IMD, Lechan RM (1987) Thyroid hormone regulates TRH biosynthesis in the paraventricular nucleus of the rat hypothalamus. Science 238:78–80

Sellar RE, Taylor RF, Zabvnik J, Anderson L, Eidne KA (1993) Functional expression and molecular characterization of the thyrotropin-releasing hormone receptor from the rat anterior pituitary gland. J Mol Endocrinol 10:199–206

Sharif NA (1989) Quantitative autoradiography of TRH receptors in discrete brain regions of different mammalian species. Ann NY Acad Sci 553:147–175

Shi ZX, Xu W, Mergner WJ, Li QL, Cole KH, Wilber JF (1996) Localization of thyrotropin-releasing-hormone m-RNA expression in the rat-heart by in-situ hybridization histochemistry. Pathobiology 64:314–319

Sowers JR, Hershman JM, Skowsky WR, Carlson HE (1976) Effects of TRH on serum arginine vasopressin in euthyroid and hypothyroid subjects. Horm Res 7:232–237

Stevenin B, Lee SL (1995) Hormonal regulation of the thyrotropin releasing hormone (TRH) gene. Endocrinologist 5:286–296

Straub RE, Frech GC, Joho RH, Gershengorn MC (1990) Expression cloning of a cDNA encoding the mouse pituitary thyrotropin-releasing hormone receptor. Proc Natl Acad Sci USA 87:9514–9518

Suen C-S, Wilk S (1989) Regulation of thyrotropin releasing hormone degrading enzymes in rat brain and pituitary by L-3,5,3'-triiodothyronine. J Neurochem 52:884–888

Sutton CA, Martin TFJ (1982) Thyrotropin-releasing hormone (TRH) selectively and rapidly stimulates phosphatidylinositol turnover in GH pituitary cells: a possible second step of TRH action. Endocrinology 110:1273–1280

Tashjian AH, Barowsky NJ, Jensen DK (1971) Thyrotropin releasing hormone: direct evidence for stimulation of prolactin production by pituitary cells in culture. Biochem Biophys Res Commun 43:516–523

Toni R, Lechan RM (1993) Neuroendocrine regulation of thyrotropin-releasing hormone (TRH) in the tuberoinfundibular system. J Endocrinol Invest 16:715–753

Tsuruo Y, Hökfelt T, Visser T (1987) Thyrotropin releasing hormone (TRH)-immunoreactive cell groups in the rat central nervous system. Exp Brain Res 68:213–217

Vale W, Rivier C, Brown M (1977) Regulatory peptides of the hypothalamus. Annu Rev Physiol 39:473–527

Valentijn K, Trachand BD, Liao N, Pelletier G, Vaudry H (1991) Release of pro-thyrotropin-releasing hormone connecting peptides PS4 and PS5 from perifused rat hypothalamic slices. Neuroscience 44:223–233

Wade N (1981) The Nobel Duel. Anchor Press, Doubleday, Garden City, New York

Wang J, Walker H, Lin Q, Jenkins N, Copeland NG, Watanabe T, Burrows PD, Cooper M (1996) The mouse BP-1 gene: structure, chromosomal localization and regulation of expression by type I interferons and interleukin-7. Genomics 33:167–176

Wang J, Whetsell M, Klein JR (1997) Local hormone networks and intestinal T cell homeostasis. Science 275:1937–1939

White N, Jeffcoate SL, Griffiths EC, Hooper KC (1976) Effect of thyroid status on the thyrotropin-releasing hormone-degrading activity of rat serum. J Endocrinol 71:13–19

Wilber JF, Feng P, Li QL, Shi Z (1996) The thyrotropin-releasing hormone gene. Trends Endocrinol Metab 7:93–100

Wilk S, Wilk EK (1989) Pyroglutamyl peptidase II, a thyrotropin releasing hormone degrading enzyme: purification and specificity studies of the rabbit brain enzyme. Neurochem Int 15:81–89

Winokur A, Utiger RD (1974) Thyrotropin-releasing hormone: regional distribution in rat brain. Science 185:265–267

Wu W, Elde R, Wessendorf MW, Hökfelt T (1992) Identification of neurons expressing thyrotropin releasing-hormone receptor mRNA in spinal cord and lower brainstem of rat. Neurosci Lett 142:143–146

Yamada M, Radovick S, Wondisford FE, Nakayama Y, Weintraub BD, Wilber JF (1990) Cloning and structure of human genomic DNA and hypothalamic cDNA encoding human prepro thyrotropin-releasing hormone. Mol Endocrinol 4:551–556

Yamada M, Monden R, Satoh T, Satoh N, Murakami M, Iriuchijima T, Kakegawa T, Mori M (1993) Pituitary adenomas of patients with acromegaly express thyrotropin-releasing hormone receptor messenger RNA: cloning and functional expression of the human thyrotropin-releasing hormone receptor gene. Biochem Biophys Res Commun 195:737–745

Yamada M, Saga Y, Shibusawa N, Hirato J, Murakami M, Iwasaki T, Hashimoto K, Satoh T, Wakabayashi K, Taketo MM, Mori M (1997) Tertiary hypothyroidism and hyperglycemia in mice with targeted disruption of the thyrotropin-releasing hormone gene. Proc Natl Acad Sci USA 94:10862–10867

Yarbrough GG (1976) TRH potentiates excitatory actions of acetylcholine on cerebral cortical neurons. Nature 263:523–524

Yarbrough GG (1979) On the neuropharmacology of thyrotropin releasing hormone (TRH). Prog Neurobiol 12:291–312

Yarbrough GG (1983) Thyrotropin releasing hormone and CNS cholinergic neurons. Life Sci 33:111–118

Yu R, Hinkle PM (1997) Desensitization of thyrotropin-releasing-hormone receptor-mediated responses involves multiple steps. J Biol Chem 272:28301–28307

Yu R, Hinkle PM (1998) Signal-transduction, desensitization, and recovery of responses to thyrotropin-releasing-hormone after inhibition of receptor internalization. Mol Endocrinol 12:737–749

Zabavnik J, Arbuthnott G, Eidne KA (1993) Distribution of thyrotropin-releasing hormone receptor messenger RNA in rat pituitary and brain. Neuroscience 53:877–887

Zhao D, Yang J, Jones KE, Gerald C, Suzuki Y, Hogan PG, Chin WW, Tashjian JAH (1992) Molecular cloning of a complementary deoxyribonucleic acid encoding the thyrotropin-releasing hormone receptor and regulation of its messenger ribonucleic acid in rat GH cells. Endocrinology 130:3529–3536

Corticotropin-Releasing Factor (CRF) and its Role in the Central Nervous System

Masao Ito[1] and Mariko Miyata[2]

1
Introduction

Corticotropin-releasing factor, commonly termed CRF, was originally described as a neurohumoral factor in the hypothalamus-pituitary system that regulates the synthesis and secretion of adrenocorticotropic hormone (ACTH). Since the complete peptide sequence of ovine CRF was reported by Vale et al. in 1981, it has also been termed corticotropin-releasing hormone or CRH. In all the animal species investigated so far (sheep, rats, humans, cattle, goats) CRF has been shown to contain 41 amino acids, and to share structural homology with some other peptides including urotensin-I and sauvagine (Chang et al. 1993; Vaughan et al. 1995). A large number of reports suggest that CRF in the hypothalamus plays a major role in the responses of the body to stress and is involved in affective disorders (as reviewed by Koob et al. 1993; De Souza 1995; Heinrichs et al. 1995; Gray and Bingaman 1996; Heit et al. 1997).

Besides the hypothalamus, however, CRF-like immunoreactivity and CRF mRNA expression are also distributed widely elsewhere in the central nervous system. Two major types of CRF receptors, CRFR1 and CRFR2, and their mRNAs are likewise expressed in many areas of the central nervous system (De Souza et al. 1985; Sawchenko et al. 1993; De Souza 1995; Chalmers et al. 1955). CRF peptide and its receptors may thus be involved in diverse central nervous system functions. Involvement in fear and anxiety has been suggested in the context of the presence of CRF in the amygdala (Gray 1993; Heilig et al. 1994; Gray and Bingaman 1996). A distinct role in cerebellar functions has been proposed in the context of the localization of CRF in the climbing fiber afferents to the cerebellum (Miyata et al. 1996,1998). It is imperative to identify the respective roles of CRF in individual regions of the central nervous system, and, based on integration of such knowledge, to elucidate the general functional significance of the CRF peptide in the central nervous system.

---
[1] Laboratory for Memory and Learning, Brain Science Institute, Institute of Physical and Chemical Research (RIKEN), Wako, Saitama 351-0198, Japan;
[2] Department of Physiology, Tokyo Women's Medical University, 8-1 Kawada-cho, Shinjuku, Tokyo 162-8666. Japan

This chapter reviews recent data on the distribution, cellular mechanisms of action, and functional roles, of the CRF peptide and its receptors in the central nervous system. While focusing on the CRF systems in the hypothalamus, amygdala, locus coeruleus and cerebellum, efforts will be made to identify specific features of each of the CRF systems and then to develop an expanded concept of stress responses that includes all of these CRF systems.

2
Distribution of CRF and its mRNA

Neurons and fibers exhibiting CRF-like immunoreactivity are widely distributed throughout the central nervous system (Merchenthaler et al. 1982; Chappell et al. 1986; Millan et al. 1986; Sakanaka et al. 1987), including the cerebral cortex, limbic system, anterior commissure, thalamus, hypothalamus, nucleus accumbens, superior and inferior colliculi, central gray, reticular formation (mesencephalic, pontine and medullary), cerebellum, and various brainstem nuclei.

In the hypothalamus, CRF-like immunoreactivity is detected in the paraventricular nucleus (PVN), medial, lateral, supraoptic, periventricular preoptic, medial preoptic, and premamillary nuclei, and the perifornical area. In the limbic system, it is found in the olfactory bulb, septal nucleus, hippocampus, amygdala and bed nucleus of the stria terminalis. In the cerebellum, CRF-like immunoreactivity is expressed specifically in climbing fiber afferents arising from the inferior olive, but some mossy fiber afferents, partly arising from the vestibular nuclei (Ikeda et al. 1992) are also immunoreactive (Cummings et al. 1988; Bishop and King 1992; King et al. 1997). CRF-like immunoreactivity is present in a number of brainstem nuclei including the raphe nucleus, laterodorsal tegmental nucleus, locus coeruleus, parabrachial nucleus, trigeminal nucleus, superior and inferior olives, vestibular nuclei, prepositus nucleus, nucleus of the solitary tract, cuneiform nucleus and external cuneate nucleus.

CRF-like immunoreactivity is a primary index for the presence of CRF, but caution must be exercised against possible cross-reactivity of the antibodies used, to other peptides. Such cross-reactivity has been reported in the dorsolateral hypothalamus (Nahon et al. 1989). It must also be noted that the CRF-like immunoreactivity detected in tissue extracts consists of multiple, heterogenous components (Yasuda and Nakamura 1997).

In situ hybridization to determine CRF mRNA expression provides another tool for mapping CRF-synthesizing neurons. It must be noted, however, that CRF mRNA expression does not always parallel CRF-like immunoreactivity. For example, CRF mRNA is expressed widely within most major cell types, whereas only a few widely scattered cells and fibers exhibit CRF-like immmunoreactivity in the olfactory bulb (Imaki et al. 1989). This discrepancy may be attributable to several possible factors; for example, the CRF peptide could be bound or sequestered in a relatively inaccessible form or post-translationally modified.

CRF also exists outside the central nervous system; placenta has a high concentration of CRF, especially during the 5 weeks before completion of gestation, although the physiological role of placental CRF is yet to be determined (Frim et al. 1988). CRF is also an antireproductive hormone of the testis (Dufau et al. 1993).

3
Distribution of CRF Receptors and Their mRNAs

CRFR1 receptors are composed of 415 amino acids. CRFR2 receptors in rats occur in two forms, containing either 411 (CRFR2α) or 431 (CRFR2β) amino acids. All of these receptor subtypes have five N-linked glycosylation sites. CRFR2α receptors exhibit 71% homology to CRFR1 receptors. The patterns of expression of CRFR1 and CRFR2 receptor mRNAs in the brain, although overlapping, are different (Sutton et al. 1994; Chalmers et al. 1995). CRFR1 receptor mRNA is expressed widely, predominantly in the cerebellum and pituitary, but also to a significant extent in the neocortex and olfactory bulb. By contrast, CRFR2 receptor mRNA expression in the central nervous system is more localized, being prominent in the septum, hypothalamus, raphe nucleus and olfactory bulb, although it is also expressed in peripheral tissues including the heart, skeletal muscle, gastrointestinal tract and the epididymis.

Besides these CRF receptors being expressed widely, a CRF-binding protein also exists in the blood, liver, placenta and central nervous system (Potter et al. 1991; Chalmers et al. 1995), which may regulate the availability of CRF. Mapping by both immunohistochemical and hybridization techniques shows that CRF-binding proteins are distributed predominantly in the rat telencephalon and that their expression overlaps with CRF-like immunoreactivity in the olfactory bulb.

4
CRF-Mediated Mechanisms
in the Hypothalamus-Pituitary System

Neurons exhibiting CRF-like immunoreactivity in the parvocellular portion of the PVN are best identified as neurons that contain and secrete CRF (Fig. 1). These CRF neurons, as they may be called, project their axon terminals to the medial eminence (Swanson et al. 1983). CRF, after being released from these terminals, is humorally transported via the hypothalamohypophysial portal system to the anterior pituitary, where CRF acts on corticotrophs to stimulate both the synthesis and secretion of ACTH (Owens and Nemeroff 1991). The

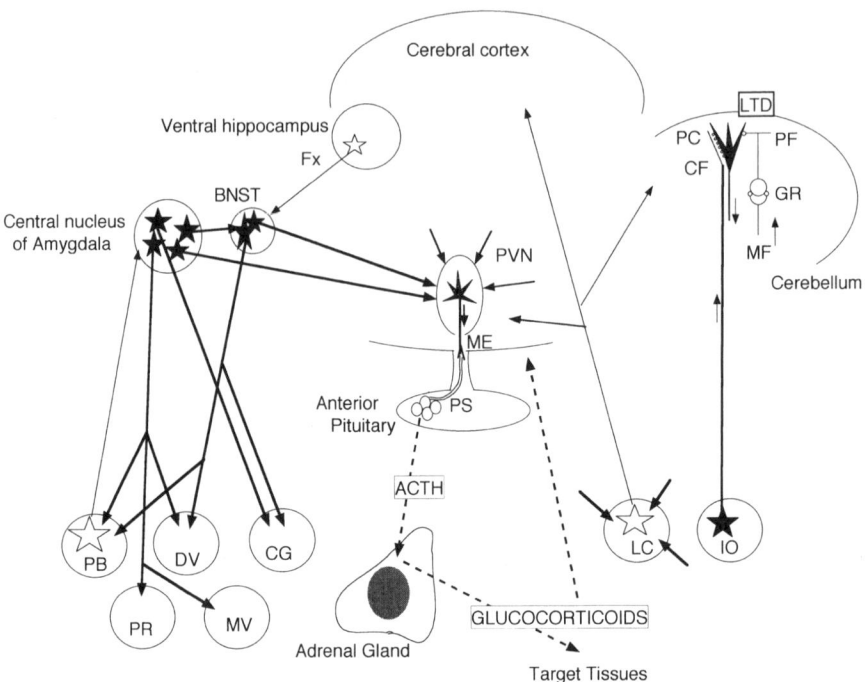

Fig. 1. CRF neuron circuit. *PVN* Paraventricular nucleus of the hypothalamus; *ME* medial eminence; *PS* hypothalamo-hypophisial portal system; *Fx*, fornix; *BNST*, bed nucleus of the stria terminals; *PB* parabrachial nucleus; *ACTH* adrenocorticotropic hormone; *DV* dorsal vagal complex; *MV* mesencephalic nucleus of the trigeminal nerve; *CG* central gray; *PR* pontine reticular nucleus; *LC* locus coeruleus; *IO* inferior olive; *LTD* long-term depression; *PC* Purkinje cell; *GR* granule cell; *PF* parallel fiber; *CF* climbing fiber; *MF* mossy fiber

secreted ACTH acts on cells located remotely in the adrenal cortex to release glucocorticoids (principally, cortisol in primates and corticosterone in rodents). The circulating glucocorticoids not only act on peripheral tissues but also exert feedback effects upon the central CRF neurons; for example, CRF transcription in PVN is markedly reduced by glucocorticoids (Lightman and Harbuz 1993). CRFR1-gene-deficient mutant mice exhibit agenesis of the corticosterone-producing zona fasciculata of the adrenal gland, presumably due to insufficient ACTH production during the neonatal period (Smith et al. 1998). Transgenic mice expressing antisense mRNA against glucocorticoids exhibit a significant reduction in the number of CRF-like immunoreactive neurons in the PVN and CRF stores in the medial eminence (Dijkstra et al. 1998).

CRF stimulates both the synthesis of and release from corticotrophs of ACTH, enhances the expression of ACTH and of its β-endorphin precursor and increases corticotroph cell volume (Owens and Nemeroff 1991). These effects may be mediated by CRF-induced elevation of intracellular Ca^{2+} concentration as demonstrated in cultured human ACTH-secreting pituitary adenoma cells, corticotrophs cultured from normal rat pituitaries (Guerineau et al. 1991), and human epidermoidal A-431 cells (Kiang 1994). Intracerebroventricular administration of CRF caused upregulation in the transcription of CRF receptor mRNA in the PVN, which was undetectable in control rats (Imaki et al. 1996). Ethanol challenge (intraperitoneal injection of 3 g/kg in rats) rapidly induces release of ACTH, and this induction is reproducible *in vitro* (Redei et al. 1986, 1988). Since this release is inhibited by CRF antibodies and antagonists, it is likely that ethanol induces CRF release from the PVN neurons, which in turn causes ACTH release from the anterior pituitary; ethanol in fact upregulated expression of CRF in the PVN (River and Lee 1996).

The release of CRF is modulated by several factors. In hypothalamic organ cultures, it is stimulated by serotonin and inhibited by both $GABA_A$ and $GABA_B$ receptor agonists (Calogero et al. 1988a). Acetylcholine stimulates CRF secretion from hypothalamic organ cultures in a dose-dependent manner (Calogero et al. 1988b). Microdialysis in the rat median eminence revealed that KCl-induced depolarization causes CRF secretion, which is inhibited by verapamil, an L-type Ca^{2+}-channel blocker (Gabr et al. 1994). It is likely that CRF is released from presynaptic terminals either as a neurotransmitter or as a neuromodulator.

5
Role of Hypothalamic CRF Neurons in Stress Responses and Depression

CRF neurons in the PVN have been postulated to play a role in the stress responses of animals based on the following three major reasons. Firstly, intracerebroventricular injection of CRF produces physiological effects such as increases in heart rate, blood pressure and plasma catecholamine concentration, and behavioral alterations such as enhanced locomotor activity and fighting, effects which are virtually identical to those in laboratory animals in response to stress. These effects were blocked by the CRF antagonist, α-h CRF(9-41) (abbreviated to α-h CRF hereafter), indicating that the behavioral effects of CRF are receptor-mediated (Britton et al. 1986b). Rats, following long-term infusion of CRF, showed a number of defective central nervous system and immune system responses to an acute peripheral endotoxic (inflammatory) challenge as occurs during chronic stress (Reul et al. 1998). Intracerebraventricular administration of CRF induced *c-fos* mRNA expression in the same brain regions as those involved in the responses to stress, including the PVN, limbic structures and locus coeruleus (Imaki et al. 1993). Secondly, stress stimuli such as swimming, restraint, injection of hypertonic saline and foot shock enhance CRF mRNA expression in the PVN (Lightman and Harbuz 1993). This effect is significant after 1 h restraint (Hsu et al. 1998). It is attenuated by benzodiazepines (Imaki and Vale 1993). Stress-induced upregulation of CRF mRNA develops postnatally; it is not induced during the first postnatal week, but it occurs by the ninth postnatal day (Baram et al. 1997a). Thirdly, PVN receives afferent inputs from diverse sources involved in stress responses, including the nucleus of the solitary tract that relays visceral sensory information, the amygdala and bed nucleus of the stria terminalis that mediate limbic signals (see Sect. 6), the lamina terminalis at the rostral margin of the third ventricle located outside the blood-brain barrier, that transduces information about the chemical environment of blood, and other nuclei of the hypothalamus involved broadly in the control of endocrine, autonomic and behavioral functions (Sawchenko et al. 1993).

Intracerebroventricular administration of CRF to adult rats and other laboratory animals induces, in addition to the aforementioned stress-response-like effects, anxiogenic and depressogenic effects including altered locomotor activity, decreased food intake, decreased sexual activity, and enhanced acoustic startle response (Koob et al. 1993). CRF-infused monkeys exhibited symptoms of behavioral despair, similar to their response to long-term separation (Kalin 1990). In contrast, CRFR1-deficient mice display markedly reduced anxiety, as represented by their visiting and spending more time in the open arms of the maze apparatus, compared to control mice (Smith et al. 1998). These observations on animals suggest that in humans, hypersecretion of CRF is involved in major depression. In fact, patients with major depression show increased CRF levels in the cerebrospinal fluid (Banki et al. 1992), increased number of neurons exhibiting CRF-like immunoreactivity and increased CRF mRNA expression in the PVN (Raadsheer et al. 1994, 1995). CRF hypersecretion is believed to cause a number of other anxiety-related disorders with similar features such as panic disorder and generalized anxiety disorder, and anorexia nervosa, an eating disorder characterized by severe weight loss in the pursuit of a slim appearance (De Souza 1995).

Since patients with major depression exhibit signs of dysfunction of the hypothalamic-pituitary-adrenal system such as elevated basal cortisol levels, resistance to glucocorticoid-mediated feedback and a blunted ACTH response to challenge with exogenous CRF, they may have defects in the central drive to the hypothalamic-hypophysial-adrenal system (see Plotsky et al. 1998). This view is consistent with the observation that antidepressant medications exert effects on CRF neurons and the activity of the hypothalamic-hypophysial-adrenal system (Heit et al. 1997). It is, however, unclear whether the hypothalamic CRF system is directly involved in the pathogenesis of major depression or merely mediates the consequences of other primary defects, for example, those arising from a limbic system dysfunction (see Sect. 6).

Links between responses to stress and depression have recently been investigated in the context of the importance of early environmental factors in regulating the development of the hypothalamic CRF system. When rat pups

2–14 days of age were exposed daily to 180 min separation from the mother and home cage, as adults they showed increased hypothalamic CRF mRNA levels compared with rats left entirely undisturbed (Plotsky and Meaney 1993). Adult male rats isolated for 6 hours daily during the first 2–20 postnatal days exhibited a 125% increase in CRF immunoreactivity in the median eminence and reduction in the density of CRF-receptor binding in the anterior pituitary compared to non-isolated rats (Ladd et al. 1996). In comparison to monkeys reared by mothers foraging under predictable conditions, infant monkeys raised by mothers foraging under unpredictable conditions exhibited persistently elevated cerebrospinal fluid concentrations of CRF (Coplan et al. 1996). A challenging hypothesis is that that stress in early life produces persistent alterations in CRF neuronal systems, which result in adults with hyperactive CRF neurons, and consequently increased vulnerability to depression (Heit et al. 1997).

6
CRF Neurons in the Amygdala and Bed Nucleus of the Stria Terminalis, and Their Role in Fear and Anxiety

The amygdala contains numerous neurons exhibiting CRF-like immunoreactivity, and their axons extend to the PVN as well as to the brainstem (Fig. 1). The central nucleus of the amygdala projects mainly to the caudal lateral and medial parvocellular portion of the PVN, while the medial nucleus of the amygdala projects to the rostral parvocellular portion of the PVN (Gray et al. 1989). The neurons exhibiting CRF-like immunoreactivity and projecting to the PVN are located mainly in the medial subdivision of the central nucleus of the amygdala and may comprise less than one-fifth of the descending pathway (Marcilhac and Siaud 1997).

The bed nucleus of the stria terminals is an amygdaloid nucleus that has similar efferent and afferent connections to the central nucleus of the amygdala, and provides a major input to PVN. This nucleus has been traditionally divided to the lateral and medial nuclei, and there are two distinct populations of neurons exhibiting CRF-like immunoreactivity, one in the dorsal, and the other in the ventral, regions of the lateral bed nucleus of the stria terminalis of rat brains. The neurons of the dorsal group have a smaller diameter and more primary dendrites than those of the ventral group, have both somatic and dendritic spines, smooth contoured nuclei and many dense and alveolate vesicles in their cytoplasm; neurons of the ventral group, however, had only dendritic spines, irregularly shaped indented nuclei and contained only alveolate vesicles in their cytoplasm (Phelix and Paull 1990). Anterolateral lesions of the bed nucleus of the stria terminalis resulted in a 30% decrease, while posterior intermediate or posterior medial lesions resulted in a small but significant increase, in CRF mRNA expression across the rostrocaudal extent of the medial parvocellular PVN, suggesting differential regional regulation of CRF in the PVN by limbic inputs via the bed nucleus of the stria terminalis (Herman et al. 1994). In the lateral subdivision of the central nucleus of the amygdala and in the dorsal part of the bed nucleus of the stria terminalis, a population of GABAergic neurons cosynthesize CRF or metenkephalin, but not both (Veinante et al. 1997).

In primary cultures of amygdala neurons, depolarization by KCl was shown to induce Ca^{2+}-dependent CRF secretion (Cratty and Birkle 1994). In dissociated amygdalal cultures, activators of protein kinases A (PKA) and C (PKC) increased CRF mRNA in a time- and dose-dependent manner; secretion and intracellular content of CRF also increased (Kasckow et al. 1997). Transfection of the primary cultures with a rat CRF promoter-luciferase reporter construct, followed by treatment with PKA and PKC activators, produced alterations in luciferase expression that were consistent with the changes observed at the level of CRF mRNA and peptide.

The amygdala and bed nucleus of the stria terminalis are probably involved conjointly in stress responses with the hypothalamic-pituitary-adrenal system because of their close anatomical connections. The projection from the bed nucleus of the stria terminalis shows a marked reduction in its fiber content following adrenalectomy (Mulders et al. 1997). Restraint for 20 min as a stress stimulus increased CRF release from the region of the central nucleus of the amygdala (Merali et al. 1998). The KCl-induced Ca^{2+}-dependent CRF release from the amygdala neurons was reported to be significantly increased by 42%, as compared with controls, in adult male offspring (8–16 weeks of age) of dams given daily saline injection (0.1 ml, s.c.) as a stress stimulus from gestational day 14 to 21 (Cratty et al. 1995). The prenatally stressed offspring also showed a 49% increase in the CRF content of the amygdala. It is possible that a long-lasting upregulation of the CRF-mediated neurotransmission occurs in the amygdala after exposure to prenatal stress, which may play a role in the generation of a hyperemotional state and increased anxiety.

Both the amygdala and bed nucleus of the stria terminalis send axons not only to PVN but also to certain brainstem structures (Fig. 1). These axons are a major forebrain source of CRF-immunoreactivity in the midbrain central gray (Gray and Magnuson 1992). Another target is the dorsal vagal complex that includes both the dorsal motor nucleus and the nucleus of the solitary tract. Approximately 22% of the axons projecting from the central nucleus of the amygdala and 15% from the lateral bed nucleus to the dorsal vagal complex are immunoreactive to CRF or other peptides (Gray and Magnuson 1987). The lateral part of the central nucleus of the amygdala projects to restricted parts of the bed nucleus and also to the parabrachial nucleus in the brainstem. The parabrachial nucleus projects back calcitonin-gene-related-peptide (CGRP)-like immunoreactive fibers to the central nucleus of the amygdala. Over 35% of the CRF neurons in the central nucleus of the amygdala receive CGRP terminals on their cell bodies and dendrites. CGRP administration into the amygdala produces effects similar to those induced by CRF administration into the central nervous system, which are probably mediated by an amygdala-CRF pathway (Harrigan et al. 1994). The central nucleus of the amygdala also projects CRF-like immunoreactive axons to the caudal pontine reticular nucleus (Fendt et al. 1997). CRF-like immunoreactive neurons in the central nucleus of the amygdala were confirmed as being labeled retrogradely by horse radish peroxidase injected to the dorsal and ventral parabrachial nuclei, mesencephalic nucleus of the trigeminal nerve and the lateral bed nucleus of the stria terminalis (Sakanaka et al. 1986). CRF-like immunoreactive neurons in the bed nucleus of the stria terminalis were also labeled by fast blue injected into the parabrachial nucleus (Moga et al. 1989).

The amygdala would directly influence brainstem functions through these connections. Stimulation of the amygdala produces signs of fear in the absence of an external source, while lesions in this region abolish normal fear responses (Heilig et al. 1994). Animals with amygdalal lesions showed complete blockade of fear-potentiated startle (Lee and Davis 1997a). Since a CRF antagonist injected into the caudal pontine reticular nucleus blocked fear-potentiated startle but had no effect on the baseline startle amplitude (Fendt et al. 1997), the effect of amygdalal stimulation could also be mediated via the pontine reticular nucleus. Rats preconditioned by a single intraoral perfusion with sodium-saccharin followed by intraperitoneal injection of LiCl rejected saccharin, and this conditioned taste aversion was abolished by bilateral electrolytic lesions of the amygdala (Schafe and Bernstein 1996). Microinjection of CRF into the ventral parabrachial region of the pons (pneumotaxic center) caused a receptor-mediated reduction of neuronal tidal volume and respiration rate, suggesting that the parabrachial nucleus is involved in the respiratory response to fear and anxiety (Bohmer et al. 1990). Microinfusion of CRF into the central nucleus of the amygdala caused a long-lasting increase in heart rate in stress-free rats, probably as a result of reduction in cardiac parasympathetic outflow, because no sign of sympathetic activation such as increase in plasma adrenaline or noradrenaline levels was noted (Wiersma et al. 1993).

Owing to the presence of direct pathways from the amygdala to the brainstem, the effects of intracerebroventricular administration of CRF, described in Section 5, cannot be ascribed solely to activation of the pituitary-adrenal system. Actually, the behavioral effects of CRF appear to be independent of the pituitary-adrenal system (Heinrichs et al. 1995). Enhancement of the startle reflex by CRF is likely to be mediated by fornix fibers arising from the ventral hippocampus and passing through the medial septum to the bed nucleus of the stria terminalis since transection of the fimbria/fornix blocked this effect (Lee and Davis 1997b). Since both chemical lesions of and infusion of a CRF antagonist into, the bed nucleus of the stria terminalis abolished CRF-enhanced startle, it appears that CRF in the cerebrospinal fluid activates the bed nucleus of the stria terminalis and thence the hypothalamic and brainstem target areas involved in anxiety and stress responses. Neither the suppression of responses in the conflict test nor the marked locomotor activation, both induced in rats by intracerebroventricular administration of CRF, was altered by prior administration of dexamethasone at a dose that blocked pituitary-adrenal activation in response to CRF, indicating that the behavioral effects of CRF are not mediated via the pituitary-adrenal system (Britton et al. 1986a). Intracerebroventricular administration of CRF induced locomotor activity both in hypophysectomized rats and in sham-operated rats, again indicating that this effect is not mediated by the pituitary-adrenal system (Eaves et al. 1985). Fawn-hooded rats which are innately hyperaroused, exhibit more freezing behavior in response to stress stimuli, develop adult-onset hypertension, and show significantly increased CRF mRNA expression in the central nucleus of the amygdala, in contrast to the downregulated CRF mRNA expression in PVN accompanying a loss of adrenal weight (Altemus et al. 1994).

A prominent effect of intracerebroventricular administration of CRF in immature rats is the induction of tonic-clonic seizures after an interval of 3–7 hours (Ehlers et al. 1983). Epileptiform activity began in the amygdala and then spread to the dorsal hippocampus and cerebral cortex (Baram et al. 1992). The CRF-induced seizures reduced in duration and increased in latency following pretreatment with the CRFR1 antagonist, NBI 27914 (Baram et al. 1997b). CRF has been shown to inhibit slow afterhyperpolarization and thereby excite hippocampal neurons, and this could be the cause of these seizures (Aldenhoff et al. 1983; Siggins et al. 1985). Kindling induced by once-daily stimulation of the amygdala increased expression of both CRF and CRF-binding proteins in GABAergic neurons of the dentate hilus, suggesting that CRF contributes to the development of amygdalal kindling (Smith et al. 1997).

7
Locus Coeruleus and its Role
in Stress Responses and Anxiety

The locus coeruleus is the source of the dorsal noradrenergic system which widely extends in the brain, including the entire cerebral cortex, hypothalamus, and cerebellum. Within the rat locus coeruleus, CRF-like immunoreactivity is mainly restricted to axons and axon terminals, which vary in size, synaptic vesicle content and formation of asymmetric or symmetric synaptic specialization (Van Bockstaele et al. 1996). Seventy percent of the axon terminals exhibiting CRF-like immunoreactivity synapse directly with dendrites, and a part of the receiving dendrites exhibit tyrosine hydroxylase activity, a catecholamine-synthesizing enzyme. These CRF-like immunoreactive axons extending to the locus coeruleus appear to originate from various sources: areas adjacent to the locus coeruleus, including the pedunculopontine tegmental nucleus, the nucleus paragigantocellularis of the caudal medulla, and the dorsal cap of the PVN (Valentino et al. 1992; Austin et al. 1995; Fig. 1).

Intracerebroventricular administration of CRF increased the spontaneous discharge rate of the locus coeruleus neurons (Valentino and Foote 1988) but inhibited their responses to sensory stimuli (Valentino et al. 1991). The effect of CRF increasing the spontaneous discharge rate was observed only in adult male rats, and not in female or immature male rats, suggesting complex sex- and age-dependent factors influencing CRF actions on locus coeruleus neurons (Borsody and Weiss 1996). The activity of the locus coeruleus neurons enhanced by intracerebroventricular administration of CRF was reduced by injection of α-h CRF into the locus coeruleus, but the reduction was greater when α-h CRF was injected into the ipsilateral parabrachial nucleus (Borsody and Weiss 1996). This may suggest that the CRF administered into the ventricles acts on locus coeruleus neurons indirectly via the parabrachial nucleus rather than directly. Nevertheless, microfusion of CRF directly into the locus coeruleus increased the discharge rates of locus coeruleus neurons at a dose 200 times smaller than that required to achieve the equivalent effect with intracerebroventricular administration, and concomitantly increased the norepinehrine levels in the prefrontal cortex (Curtis et al. 1997). Microfusion of CRF into the locus coeruleus also increased the levels of noradrenaline and its metabolites in the parietal cortex (Schulz and Lehnert 1996). Similar results obtained in the medial frontal cortex could not be reproduced by microfusion of CRF into the parabrachial nucleus (Smagin et al. 1995). These observations suggest that CRF directly activates the noradrenergic neurons in the locus coeruleus.

The CRF-responsiveness of locus coeruleus neurons has been proposed to play a role in stress responses, based on a variety of data (Valentino et al. 1998). Stress stimuli, either acute or chronic, caused a twofold increase in CRF content in the locus coeruleus (Chappell et al. 1986), while administration of benzodiazepines reduced it (Owens et al. 1991). Microfusion of CRF into the locus coeruleus attenuated the defensive withdrawal evoked by 30 min immobilization as a stress stimulus in rats (Smagin et al. 1996). Hypotensive challenge or colon distention increased the discharge rate of locus coeruleus neurons, and this effect was blocked or greatly attenuated by intracerebroventricular or direct administration of CRF to the locus coeruleus (Valentino et al. 1993; Page and Valentino 1994). Stress stimuli induced a prolonged increase in the release of aspartate and to a lesser degree of glutamate, as did superfusion of the locus coeruleus with CRF. α-h CRF abolished noise stress-induced amino acid release, but had no effect on tail-pinch-induced amino acid release, suggesting the involvement of different neuronal mechanisms in these two types of stress responses (Singewald et al. 1996). Marked desensitization occurred in the facilitatory effect of both stress stimuli and intracerebroventricular CRF administration on locus coeruleus neuron discharge rates, when these stimuli were repeated, and cross-desensitization developed between the stress-and CRF-induced effects (Conti and Foote 1996); the effect of acute hemodynamic stress caused by intravenous infusion of sodium nitroprusside was attenuated when CRF administration was repeated, and the effect of CRF administration was attenuated when white-noise stress stimuli were repeated. Attenuation of the effect of CRF was also observed in rats stressed, either acutely or repeatedly, by exposure to one or five daily 30-min sessions of foot shock (Curtis et al. 1995).

Involvement of the locus coeruleus in stress responses, however, seems to be linked with anxiety responses differently from the hypothalamic and amygdala systems. Infusion of rat locus coeruleus with drugs which increase locus coeruleus neuron discharges (idazoxan, substance P) decreased anxious behavior, and a drug which decreases locus coeruleus neuron discharges (desipramine) increased anxious behavior (Weiss et al. 1994). Integrating these observations also with other data, Weiss et al. (1994) proposed that while being activated by stress stimuli, the dorsal noradrenergic system originating from the locus coeruleus exerts a counterbalancing anti-anxiety influence on animal behavior.

8
CRF-Containing Climbing Fibers and Their Role in Cerebellar Functions

Inferior olivary neurons send afferent fibers to the cerebellar cortex, which eventually terminate at the dendrites of Purkinje cells to form a characteristic structure of the cerebellum, the climbing fibers (Fig. 1). CRF-like immunoreactivity is expressed in all divisions of the inferior olivary nucleus and at all levels from the cells of origin of the olivocerebellar afferents in the inferior olive to their climbing fiber terminals on Purkinje cell dendrites (Fig. 1; Palkovits et al. 1987; Sakanaka et al. 1987; Cha and Foote 1988; Kitahama et al. 1988; Sawchenko and Swanson 1990). The early appearance of CRF-like immunoreactive fibers in the cerebellum prior to synaptogenesis and migration of granule and Purkinje cells suggests a developmental role for CRF in target recognition and synaptic organization (Cummings et al. 1994a). Significant reduction in CRF content has been reported in the cerebella of patients with olivopontocerebellar atrophy (Mizuno et al. 1995), presumably reflecting the presence of CRF in climbing fibers. CRF mRNA expression in inferior olivary neurons increased following activation of these neurons by harmaline (Cummings et al. 1994b) or by sensory signals (Barmack and Young 1990; Barmack and Ericco 1993). Purkinje cells highly express CRFR1 mRNA (Potter et al. 1994), suggesting that CRF released from climbing fibers are received by CRFR1 receptors in Purkinje cells.

KCl challenge induces CRF release from rat cerebellar slices (Bishop et al. 1997). The major neurotransmitter of climbing fibers is glutamate or a related amino acid, and therefore, climbing fibers provide a typical example of the coexistence and simultaneous release of an amino acid neurotransmitter and a peptide neurotransmitter from the same synaptic terminals, which play complementary roles in synaptic mechanisms (Hökfelt 1991). A distinct effect of CRF is the reduction of afterhyperpolarization following a current-evoked spike train in Purkinje cells, presumably due to closure of Ca^{2+}-dependent K^+ channels responsible for generation of the afterhyperpolarization (Fox and Gruol 1993). Iontophoretic application of CRF to Purkinje cells also enhances their spontaneous activity and sensitivity to glutamate and aspartate, and antagonizes the inhibitory actions of γ-aminobutyric acid and enkephalins (Bishop 1990; Bishop and Kerr 1992; Bishop and King 1992). Differing from what has been demonstrated in cultured human ACTH-secreting pituitary adenoma cells (Guerineau et al. 1991) and human epidermoidal A-431 cells (Kiang 1994), CRF does not seem to act upon Ca^{2+} channels or Ca^{2+} stores in Purkinje cells, since α-h CRF has no effect on the membrane Ca^{2+} conductance or intracellular Ca^{2+} concentration in Purkinje cells (Miyata et al. 1998).

A unique role of CRF has recently been demonstrated in the induction of long-term depression (LTD) in Purkinje cells. When a Purkinje cell receives input signals both from parallel fibers and climbing fibers, conjunctively and repeatedly, the transmission efficacy at the parallel fiber synapses is persistently reduced (Ito 1989; Daniel et al. 1998). In *in vitro* cerebellar slices, LTD can be determined by measuring the initial upslope of parallel fiber-evoked excitatory postsynaptic potentials (EPSPs) before and after conjunctive stimulation of both parallel fibers and climbing fibers at 1 Hz for 5 min (Karachot et al. 1994). Two lines of evidence indicate that CRF plays a role in the induction of LTD. Firstly, perfusion of a cerebellar slice with CRF (0.5 µM) induces LTD when combined with repetitive parallel fiber stimulation; CRF replaces climbing fiber stimulation (Miyata et al. 1996). Secondly, in the presence of a CRF antagonist, namely α-helical CRF or astressin, another peptide CRF receptor antagonist, which is 100 times more potent than α-h CRF (Gulyas et al. 1995), LTD is no longer induced (Miyata et al. 1998).

The responses evoked in Purkinje cells by impulses of climbing fibers are characterized by Ca^{2+} spikes superposed on large EPSPs. LTD can be induced by stimulating parallel fibers in conjunction with membrane-depolarization-evoked Ca^{2+} spikes, even without stimulating climbing fibers. Since α-h CRF blocks this form of LTD also, CRF which is spontaneously released from climbing fibers appears to be enough to induce LTD. CRF may be required at certain background levels for maintaining the complex receptor and messenger processes. In cerebellar tissues in which climbing fibers had degenerated due to 3-acetylpyridine intoxication, LTD did not occur unless CRF was replenished (Miyata et al. 1998).

CRFR1 receptors are known to be positively coupled through Gs proteins with adenylate cyclase (Battaglia et al. 1987), but there is no evidence indicating that adenylate cyclase or its product, cyclic AMP, or cyclic AMP-dependent kinase (PKA), has any role in LTD. Since activation of human and mouse CRFR1 facilitates phosphoinositide hydrolysis in cDNA-transfected COS-7 cells (Xiong-y et al. 1995), and since CRF activates protein kinase C (PKC) in pituitary cells and A-431 cells (Kiang et al. 1994; Ishizuka et al. 1996), it is possible that CRF acts to maintain phosphoinositide hydrolysis and/or PKC activity in Purkinje cells, at levels required for the induction of LTD (Linden and Connor 1991; Daniel et al. 1998; De Zeeuw et al. 1998; Inoue et al. 1998).

CRF-like immunoreactivity and CRF mRNA expression in the inferior olivary nucleus were enhanced by optokinetic stimulation which activates climbing fiber afferents to the vestibulocerebellum (Barmack and Young 1990; Barmack and Errico 1993). The climbing fiber-mediated optokinetic signals are expected to play a role in the induction of LTD in Purkinje cells, and thereby contribute to adaptive control of the vestibuloocular reflex and optokinetic eye movement responses (Ito 1998). Since CRF-like immunoreactive climbing fibers exist throughout the cerebellum (King et al. 1997), and since LTD is a general phenomenon in the cerebellar cortex, the role of CRF in the cerebellum is probably all-pervasive. The cerebellum is involved in the adaptive control of various reflexes and compound movements such as locomotion and saccade, and in exercising voluntary movements (Ito, 1984, 1998). In addition, recent evidence suggests that the cerebellum helps learning in thought which is primarily a function of the cerebral association cortex (Leiner et al. 1993; Ito 1997; Schmahmann and Pandya 1997). CRF is therefore essential for the functions of climbing fibers which play a key role in cerebellar learning mechanisms for bodily as well as mental functions.

While the above considerations are limited to the possible role of CRF released from climbing fibers to Purkinje cells, the expression of CRFR1 mRNA at a high level as well CRFR2 mRNA at limited levels in granule cells (Chalmers et al. 1995; Primus et al. 1977) suggest a role of CRF at some mossy fiber-granule cell synapses, which, however, has not yet been identified (see discussion in Setc. 9).

9
Comments

Over the last three decades, our knowledge about the neuronal mechanisms and functional roles of CRF in the central nervous system has advanced greatly; yet, some fundamental questions still remain unanswered. At cellular levels, signal transduction pathways following activation of CRF receptors, and related gene regulatory mechanisms need to be clarified in greater detail. A difference has been found in the effects of CRF between cultured human ACTH-secreting pituitary adenoma cells (Guerineau et al. 1991) or human epidermoidal A-431 cells (Kiang 1994), and cerebellar Purkinje cells; CRF enhances Ca^{2+} entry into the former cells but not into the latter; stimulation of CRFR1 activates PKA in the former cells, but activates PKC in the latter (Miyata et al. 1998).

The roles of the CRF peptide distributed in the hypothalamus, limbic system, and brainstem, as reviewed in this chapter, have been related to the current concept of stress, which broadly encompasses physical stress due to pain, heat, cold and toxins, psychological stress due to defeat/frustration, conflict and fear, and arousal due to novelty, uncertainty and hunger/thirst (Hennessy and Levin 1979; Koob et al. 1993). First, CRF neurons in PVN drive the pituitary-adrenal system in response to physical stress, and via release of ACTH from pituitary and then glucocorticoids from the adrenocortex, influence adaptation of animals to stress stimuli. Second, CRF neurons in the central nucleus of the amygdala and the bed nucleus of the stria terminalis may mediate responses to psychological stress by activating the hypothalamic-pituitary-adrenal system and also brainstem nuclei which produce behavioral expressions of fear and anxiety. However, since food intake induces release of CRF from the central nucleus of the amygdala to the same extent as restraint does, amygdalal CRF is probably associated widely with events or cues of biological significance such as food availability or a threat to survival (Merali et al. 1998). Third, CRF fibers originating from various sources and passing into the locus coeruleus activate noradrenergic neurons to supply more noradrenaline widely to the cerebral cortex, cerebellum and hypothalamus, which would enhance arousal and may enable an animal to cope with anxiety and fear. Fourth, CRF in climbing fibers is required for induction of LTD which is a key process for cerebellar learning. Data on Barrington's nucleus are not reviewed in this chapter, but this pontine micturition center is rich in CRF neurons and receives afferent inputs from diverse supraspinal structures including the hypothalamus and limbic system (Valentino et al. 1994). Barrington's nucleus could be involved in responses to stress which might arise from bladder distention, or from tension which might arise when an animal marks its territory with urine.

An interesting recent finding concerns with the role of CRF in functions of climbing fibers (Sect. 8). It is a current thought that mossy and climbing fibers in the cerebellum play distict roles: mossy fibers convey input signals for execution of neural control, while climbing fibers representing errors between the intended and actually performed controls act to modify the transmission of mossy fiber signals via Purkinje cells by inducing LTD, in the manner of learning. Climbing fiber error signals arise from either physical stimuli such as nociceptive stimuli caused by unsuccessful movements and retinal slips caused by inadequate eye movements, or discrepancy between the intended and actual reaching movements (see Ito, 1998). Hence, the climbing fiber error signals may represent physical stress such as nociceptive stimuli or even psychological stress such as demand for precision and discontent about inaccurate performance. However, it is to be noted that some mossy fibers may release CRF to granule cells (Chalmers et al. 1955; Primus et al. 1997). Their functions have not been analysed, but they may well be related to errors, and consequently to stress, because neural control is often executed by error signals as input signals (Ito 1984). The presence of CRF in both climbing and mossy fibers therefore does not contradict the notion that CRF is related to stress, but it excludes the possibility that CRF is specifically related to learning. Thus, CRF in the central nervous system may generally be related to an extended range of stress stimuli, which in addition to conventional physical and psychological stressors would include any potential stimuli demanding a certain degree of tension, such as food intake, errors in motor performance or even urination.

The roles of CRF in other structures, however, have not yet been well analyzed. Abundance of CRF mRNA in the olfactory bulb (Imaki et al. 1989) suggests a role for CRF in olfactory mechanisms, which is unclear at present. The systematic distribution of CRF mRNA in the auditory system including the nucleus of the lateral lemniscus, the shell of the inferior olive, the medial division of the medial geniculate body, and the primary auditory cortex, in particular, in layers II, III and V (Imaki et al. 1991), suggest an as yet unknown role for CRF in auditory function. The vast majority of neurons tested in layers II–VI of the rat somatosensory cortex were unaffected by prolonged current- or pressure-ejection of CRF (Cahusac et al. 1998). It is also noted that a marked decrease in CRF content occurs in the cerebral cortex in connection with dementia, in Alzheimer's disease, Parkinson's dementia and progressive supranuclear palsy, a rare neurodegenerative disease with Alzheimer-type pathology, and also in the basal ganglia in Huntington's disease, suggesting an as yet unknown role of CRF in the pathogenesis of neurological disorders (De Souza 1995). The above-discussed concept of extended stress may provide an insight into the as yet unidentified functional implications of these CRF systems.

An interesting feature of CRF in the central nervous system is that while it is released from synaptic structures as either a neurotransmitter or a neuromodulator, it is also diffused through the extracellular fluid of brain tissues by so-called volume transmission (Zoli et al. 1998) and further remotely transported by the cerebrospinal fluid. CRF released from axons of PVN neurons in the medial eminence could be transported by this route to PVN, the bed nucleus of the stria terminalis, and amygdala and further to brainstem nuclei including the locus coeruleus, parabrachial nuclei, inferior olive and Barrington's nucleus. This could be the mode by which CRF harmoniously coordinates diverse stress-related activities in the endocrinological, immunological, autonomic and behavioral domains.

How can CRF be a common key factor for these diverse activities in different sites of the central nervous system? It is interesting to speculate that CRF cells emerging from the same evolutional origin have been incorporated into separate structures of the vertebrate brain and so are involved in different central nervous functions, yet maintaining the same unique functional role in adaptation to stressful stimuli which exceed normal physiological and psychological capacities of the body unless the body appropriately adapts to them.

References

Aldenhoff JB, Gruol DL, River J, Vale W, Siggins G (1983) Corticotropin-releasing factor decreases postburst hyperpolarizations and excites hippocampal neurons. Science 221:875–877

Altemus M, Smith MA, Diep V, Aulakh CS, Murphy DL (1994) Increased mRNA for corticotrophin releasing hormone in the amygdala of fawn-hooded rats: a potential animal model of anxiety. Anxiety 1:251–257

Austin MC, Rice PM, Mann JJ, Arango V (1995) Localization of corticotropin-releasing hormone in the human locus coeruleus and pedunculopontine tegmental nucleus: an immunocytochemical and in situ hybridization study. Neuroscience 64:713–727

Banki CM, Karmcsi L, Bissette G, Nemeroff CB (1992) Cerebrospinal fluid neuropeptides in mood disorder and dementia. J Affective Disord 25:39–46

Baram TZ, Hirsch E, Smead III OC, Schultz L (1992) Corticotropin-releasing hormone-induced seizures in infant rats originate in the amygdala. Annu Neurol 31:488–494

Baram TZ, Yi S, Avishai-Eliner S, Schultz L (1997a) Development neurobiology of the stress response: multilevel regulation of corticotropin-releasing hormone function. Ann NY Acad Sci 814:252–265

Baram TZ, Chalmers DT, Chen C, Loutsoukos Y, De Souza EB (1997b) The CRF1 receptor mediates the excitatory actions of corticotropin releasing factor (CRF) in the developing rat brain: in vivo evidence using a novel, selective, non-peptide CRF receptor antagonist. Brain Res 770: 89–95

Barmack NH, Errico P (1993) Optokinetically evoked expression of corticotropin-releasing factor in inferior olivary neurons of rabbits. J Neurosci 13:4647–4659

Barmack NH, Young WS III (1990) Optokinetic stimulation increases corticotropin-releasing factor mRNA in inferior olivary neuron of rabbit. J Neurosci 10: 631–640

Battaglia G, Webster E, De Souza EB (1987) Characterization of corticotropin-releasing factor receptor-mediated adenylate cyclase activity in the rat central nervous system. Synapse 1:572–581

Bishop GA (1990) Neuromodulatory effects of corticotropin releasing factor on cerebellar Purkinje cells: an in vivo study in the cat. Neuroscience 39:251–257

Bishop GA, Kerr CW (1992) The physiological effects of peptides and serotonin on Purkinje cell activity. Prog Neurobiol 39:475–492

Bishop GA, King JS (1992) Differential modulation of Purkinje cell activity by enkephalin and corticotropin releasing factor. Neuropeptides 22:167–174

Bishop GA, Burry RW, Benz B, Do KQ (1997) Release of corticotropin releasing factor (CRF) in the rat cerebellum is dependent upon depolarization. Abstr Soc Neurosci 56.8

Bohmer G, Schmid K, Ramsbott M (1990) Effects of corticotropin-releasing factor on central respiratory activity. Eur J Pharmacol 182:405–411

Borsody MK, Weiss JM (1996) Influence of corticotropin-releasing hormone on electrophysiological activity of locus coeruleus neurons. Brain Res 724:149–168

Britton KT, Lee G, Dana R, Risch SC, Koob GF (1986a) Activating and 'anxiogenic' effects of corticotropin releasing factor are not inhibited by blockade of the pituitary-adrenal system with dexamethasone. Life Sci 39:1281–1286

Britton KT, Lee G, Vale W, Rivier J, Koob GF (1986b) Corticotropin releasing factor (CRF) receptor antagonist blocks activating and 'anxiogenic' actions of CRF in the rat. Brain Res 369(1–2): 303–306

Cahusac PMB, Castro MG, Robertson L, Lowenstein PR (1998) Electrophysiological evidence against a neurotransmitter role of corticotropin-releasing hormone (CRH) in primary somatosensory cortex. Brain Res 793:73–78

Calogero AE, Gallucci WT, Chrousos GP, Gold PW (1988a) Interaction between GABAergic neurotransmission and rat hypothalamic corticotropin-releasing hormone secretion in vitro. Brain Res 463:28–36

Calogero AE, Gallucci WT, Bernardini R, Saoutis C, Gold PW, Chrousos GP (1988b) Effect of cholinergic agonist and antagonist on rat hypothalamic corticotropin-releasing hormone secretion in vitro. Neuroendocrinology 47:303–308

Cassell MD, Gray TS, Kiss JZ (1986) Neuronal architecture in the rat central nucleus of the amygdala: a cytological, hodological, and immunocytochemical study. Comp Neurol 246:478–499

Cha C, Foote SL (1988) Corticotropin-releasing factor in olivocerebellar climbing-fiber system of monkey (Saimiri sciureus and Macaca fascicularis): Parasagittal and regional organization visualized by immunohistochemistry. J Neurosci 8:4121–4137

Chalmers DT, Lovenberg TW, De Souza EB (1995) Localization of novel corticotropin-releasing factor receptor (CRF2) mRNA expression to specific subcortical nuclei in rat brain: comparison with CRF receptor mRNA expression. J Neurosci 15:6340–6350

Chang C-P, Pearse II RV, O'Connell S, Rosenfeld MG (1993) Identification of a seven transmembrane helix receptor for corticotropin-releasing factor and suvagine in mammalian brain. Neuron 11:1187–1195

Chappell PB, Smith MA, Kilts CD, Bissette G, Ritchie J, Anderson C, Nemeroff CB (1986) Alterations in corticotropin-releasing factor-like immunoreactivity in discrete rat brain regions after acute and chronic stress. J Neurosci 6:2908–2914

Conti LH, Foote SL (1996) Reciprocal cross-desensitization of locus coeruleus electrophysiological responsivity to corticotropin-releasing factor and stress. Brain Res 722:19–29

Coplan JD, Andrews MW, Rosenblum LA, Owens MJ, Friedman S, Gorman JM, Nemeroff CB (1996) Persistent elevations of cerebrospinal fluid concentrations of corticotropin-releasing factor in adult nonhuman primates exposed to early-life stressors: implications for the pathophysiology of mood and anxiety disorders. Proc Natl Acad Sci USA 20:1619–1623

Cratty MS, Birkle DL (1994) Depolarization-induced release of corticotropin-releasing factor (CRF) in primary neuronal cultures of the amygdala. Neuropeptides 26:113–121

Cratty MS, Ward HE, Johnson EA, Azzaro AJ, Birkle DL (1995) Prenatal stress increases corticotropin-releasing factor (CRF) content and release in rat amygdala minces. Brain Res 27: 297–302

Cummings S, Sharp B, Elde R (1988) Corticotropin-releasing factor in cerebellar afferent systems: a combined immunohistochemistry and retrograde transport study. J Neurosci 8:543–554

Cummings SL, Young WS III, King JS (1994a) Early development of cerebellar afferent systems that contain corticotropin-releasing factor. J Comp Neurol 350:534–549

Cummings SL, Hinds D, Young WS III (1994b) Corticotropin-releasing factor mRNA increases in the inferior olivary complex during harmaline-induced tremor. Brain Res 660:199–208

Curtis AL, Pavcovich LA, Grigoriadis DE, Valentino RJ (1995) Previous stress alters corticotropin-releasing factor neurotransmission in the locus coeruleus. Neuroscience 65:541–550

Curtis AL, Lechner SM, Pavcovich LA, Valentino RJ (1997) Activation of the locus coeruleus noradrenergic system by intracoerulear microinfusion of corticotropin-releasing factor: effects on discharge rate, cortical norepinephrine levels and cortical electroencephalographic activity. J Pharmacol Exp Ther 281:163–172

Daniel H, Levenes C, Crepel F (1998) Cellular mechanisms of LTD. Trends Neurosci 21:401–407

De Souza EB (1995) Corticotropin-releasing factor receptors: physiology, pharmacology, biochemistry and role in central nervous system and immune disorders. Psychoneuroendocrinology 20:789–819

De Souza EB, Insel TR, Perrin MH, River J, Vale WW, Kuhar MJ (1985) Corticotropin-releasing factor receptors are widely distributed within the rat central nervous system: an autoradiographic study. J Neurosci 5:3189–3203

De Zeeuw CI, Hansel C, Bian F, Koekkoek SKE, van Alpen AM, Linden DJ, Oberdick J (1998) Expression of a protein kinase C inhibitor in Purkinje cells blocks cerebellar LTD and adaptation of the vestibulo-ocular reflex. Neuron 20:1–20

Dijkstra I, Tilders FJH, Aguilera G, Kiss A, Rabadan-Diehl C, Barden N, Karanth S, Holsboer F, Reul MHM (1998) Reduced activity of hypothalamic corticotropin-releasing hormone neurons in transgenic mice with impaired glucocorticoid receptor function. J Neurosci 18:3909–3918

Dufau ML, Tinajero JC, Fabbri A (1993) Corticotropin-releasing factor: an antireproductive hormone of the testis. FASEB J 7:299–307

Eaves M, Thatcher-Britton K, Rivier J, Vale W, Koob GF (1985) Effects of corticotropin releasing factor on locomotor activity in hypophysectomized rats. Peptides 6:923–926

Ehlers CL, Henriksen SJ, Wang M, River J, Vale WW, Kuhar MJ (1983) Corticotropin releasing factor produces increases in brain excitability and convulsive seizures in rats. Brain Res 278:332–336

Fendt M, Koch M, Schnitzler HU (1997) Corticotropin-releasing factor in the caudal pontine reticular nucleus mediates the expression of fear-potentiated startle in the rat. Eur J Neurosci 9:299–305

Fox EA, Gruol DL (1993) Corticotropin-releasing factor suppresses the afterhyperpolarization in cerebellar Purkinje neurons. Neurosci Lett 149:103–107

Frim DM, Emanuel RL, Robinson BG, Smas CM, Adler GK, Majzoub JA (1988) Characterization and gestational regulation of corticotropin-releasing hormone messenger RNA in human placenta. J Clin Invest 82:287–292

Gabr RW, Gladfelter WE, Birkle DL, Azzaro AJ (1994) In vivo microdialysis of corticotropin releasing factor (CRF): calcium dependence of depolarization-induced neurosecretion of CRF. Neurosci Lett 169:63–67

Gray TS (1993) Amygdaloid CRF pathways. Role in autonomic, neuroendocrine, and behavioral responses to stress. Ann NY Acad Sci 697:53–60

Gray TS, Bingaman EW (1996) The amygdala: corticotropin-releasing factor, steroids, and stress. Crit Rev Neurobiol 10:155–168

Gray TS, Magnuson DJ (1992) Peptide immunoreactive neurons in the amygdala and the bed nucleus of the stria terminalis project to the midbrain central gray in the rat. Peptides 13:451–460

Gray TS, Carney ME, Magnuson DJ (1989) Direct projections from the central amygdaloid nucleus to the hypothalamic paraventricular nucleus: possible role in stress-induced adrenocorticotropin release. Neuroendocrinology 50:433–446

Gray TS, Magnuson DJ (1987) Neuropeptide neuronal efferents from the bed nucleus of the stria terminalis and central amygdaloid nucleus to the dorsal vagal complex in the rat. J Comp Neurol 262:365–374

Guerineau N, Corcuff J-B, Tabarin A (1991) Spontaneous and corticotropin-releasing factor-induced cytosolic calcium transients in corticotrophs. Endocrinology 129:409–420

Gulyas J, Rivier C, Perrin M, Koerber SC, Sutton S, Corrigan A, Lahrichi SL, Craig AG, Vale W, Rivier J (1995) Potent, structurally constrained agonists and competitive antagonists of corticotropin-releasing factor. Proc Natl Acad Sci USA 92:10575–10579

Harrigan EA, Magnuson DJ, Thunstedt GM, Gray TS (1994) Corticotropin releasing factor neurons are innervated by calcitonin gene-related terminals in the rat central amygdaloid nucleus. Brain Res Bull 33:529–534

Heilig M, Koob GF, Ekman R, Britton KT (1994) Corticotropin-releasing factor and neuropeptide Y: role in emotional integration. Trends Neurosci 17:80–85

Heinrichs SC, Menzaghi F, Merlo Pich E, Britton KT, Koob GF (1995) The role of CRF in behavioral aspects of stress. Ann NY Acad Sci 771:92–104

Heit S, Owens MJ, Plotsky P, Nemeroff CB (1997) Corticotropin-releasing factor, stress, and depression. Neuroscientist 3:186–194

Hennessy JW, Levine S (1979) Stress, arousal, and pituitary-adrenal system: a psychoendocrine hypothesis. In: Sprague JM, Epsitein AN (eds) Progress in psychobiology and physiological psychology, 8th ed n. Academic Press, New York, pp 255–271

Herman JP, Cullinan WE, Watson SJ (1994) Involvement of the bed nucleus of the stria terminalis in tonic regulation of paraventricular hypothalamic CRH and AVP mRNA expression. Neuroendocrinology 6:433–442

Hökfelt T (1991) Neuropeptides in perspectives: the last ten years. Neuron 7:867–879

Hsu DT, Chen FL, Takahashi LK, Kalin NH (1998) Rapid stress-induced elevations in corticotro-pin-releasing hormone mRNA in rat central amygdala nucleus and hypothalamic paraven-tricular nucleus: an in situ hybridization analysis. Brain Res 788:305–310

Ikeda M, Houtani T, Ueyama T, Sugimoto T (1992) Distribution and cerebellar projections of cholinergic and corticotropin-releasing factor-containing neurons in the caudal vestibular nuclear complex and adjacent brainstem structures. Neuroscience 49:635–651

Imaki J, Imaki T, Vale W, Sawchenko PE (1991) Distribution of corticotropin-releasing factor mRNA and immunoreactivity in the central auditory system of the rat. Brain Res 547:28–36

Imaki T, Vale W (1993) Chlordiazepoxide attenuates stress-induced accumulation of corticotro-pin releasing factor mRNA in the paraventricular nucleus. Brain Res 623:223–228

Imaki T, Nahon JL, Sawchenko PE, Vale W (1989) Widespread expression of corticotropin-releasing factor messenger RNA and immunoreactivity in the rat olfactory bulb. Brain Res 496:35–44

Imaki T, Shibasaki T, Hotta M, Demura H (1993) Intracerebroventricular administration of corticotropin-releasing factor induces c-fos mRNA expression in brain regions related to stress responses: comparison with pattern of c-fos mRNA induction after stress. Brain Res 616:114–135

Imaki T, Naruse M, Harada S, Chikada N, Imaki J, Onodera H, Demura H, Vale W (1996) Corticotropin-releasing factor up-regulates its own receptor mRNA in the paraventricular nucleus of the hypothalamus. Mol Brain Res 38:166–170

Inoue T, Kato K, Kohda K, Mikoshiba K (1998) Type 1 inositol 1,4,5-triphosphate receptor is required for induction of long-term depression in cerebellar Purkinje neurons. J Neurosci 18: 5366–5373

Ishizuka T, Morita H, Mune T, Daidoh H, Hanafusa J, Yamamoto M, Shibata T, Yasuda K (1996) Growth hormone secretion in human acromegalic pituitary adenomas: cyclic adenosine monophosphate and protein kinase C responses. Metabolis 45:206–210

Ito M (1984) The cerebellum annd neural control. Raven Press, New York, pp 335–338

Ito M (1989) Long-term depression. Annu Rev Neurosci 12:85–102

Ito M (1997) Cerebellar microcomplexes. Int Rev Neurobiol 41:475–485

Ito M (1998) Cerebellar learning in the vestibuloocular reflex. Trends Cog Sci 2:313–321

Kalin NH (1990) Behavioral and endocrine studies of corticotropin-releasing hormone in pri-mates. In: De Souza EB, Nemeroff CB (eds) Corticotropin-releasing factor: basic and clinical studies of a neuropeptide. CRC Press, Boca Raton, pp 275–298

Karachot L, Kado RT, Ito M (1994) Stimulus parameters for induction of long-term depression in in vitro rat Purkinje cells. Neurosci Res 21:161–168

Kasckow JW, Regmi A, Gill PS, Parkes DG, Geracioti TD (1997) Regulation of corticotropin-releasing factor (CRF) messenger ribonucleic acid and CRF peptide in the amygdala: studies in primary amygdalar cultures. Endocrinology 138:4774-4782

Kiang JG (1994) Corticotropin-releasing factor increases $[Ca^{2+}]_i$ via receptor-mediated Ca^{2+} channels in human epidermoid A-431 cells. Eur J Pharmacol 267:135–142

Kiang JG, Wang X, McClain D (1994) Corticotropin-releasing factor increases protein kinase C activity by elevation membrane-bound a and b isoforms. Chin J Physiol 37:105–110

King JS, Madtes Jr P, Bishop GA, Overbeck TL (1997) The distribution of corticotropin-releasing factor (CRF), CRF binding sites and CRF1 receptor mRNA in the mouse cerebellum. In De Zeeuw CI, Strata P, Voogd J (eds) The Cerebellum: from structure to control. Prog Brain Res 114:55–66

Kitahama K, Luppi P-H, Tramu G, Sastre J-P, Buda C, Jouvet M (1988) localization of CRF-immunoreactive neurons in the cat medulla oblongata: their presence in the inferior olive. Cell Tissue Res 251:137–143

Koob GF, Heinrichs SC, Pich EM, Menzaghi F, Baldwin H, Miczek K, Britton KT (1993) The role of corticotropin-releasing factor in behavioral responses to stress. In: Chadwick DJ, Marsh J, Ackrill K (eds) Coritcotropin-releasing factor. John Wiley, New York, pp 277–295

Ladd CO, Owens MJ, Nemeroff CB (1996) Persistent changes in corticotropin-releasing factor neuronal systems induced by maternal deprivation. Endocrinology 137:1212–1218

Lee Y, Davis M (1997a) Role of the septum in the excitatory effect of corticotropin-releasing hormone on the acoustic startle reflex. J Neurosci 17:6424–6433

Lee Y, Davis M (1997b) Role of the hippocampus, the bed nucleus of the stria terminalis, and the amygdala in the excitatory effect of corticotropin-releasing hormone on the acoustic startle reflex. J Neurosci 17:6434–6446

Leiner HC, Leiner AL, Dow RS (1993) Cognitive and language functions of the human cerebellum. Trends Neurosci 16:444–447

Lightman SL, Habuz MS (1993) Expression of corticotropin-releasing factor mRNA in response to stress. In: Chadwick DJ, Marsh J, Ackrill K (eds) Coritcotropin-releasing factor. John Wiley, New York, pp 173–196

Linden DJ, Connor JA (1991) Participation of postsynaptic PKC in cerebellar long-term depression in culture. Science 254:1656–1659

Marcilhac A, Siaud P (1997) Identification of projections from the central nucleus of the amygdala to the paraventricular nucleus of the hypothalamus which are immunoreactive for corticotrophin releasing hormone in the rat. Exp Physiol 82:273–281

Merali Z, McIntosh J, Kent P, Michaud D, Anisman H (1998) Aversive and appetitive events evoke the release of corticotropin-releasing hormone and bombesin-like peptides at the central nucleus of the amygdala. J Neurosci 18:4758–4766

Merchenthaler I, Vigh S, Petrusz P, Schally AV (1982) Immunocytochemical localization of corticotropin-releasing factor (CRF) in the rat brain. Am J Anat 165:385–396

Millan MA, Jacobowitz DM, Hauger RL, Catt KJ (1986) Distribution of corticotropin-releasing factor receptors in primate brain. Proc Natl Acad Sci USA 83:1921–1925

Miyata M, Hashimoto K, Kano M, Ito M (1996) Corticotropin releasing factor (CRF) induces long lasting depression of parallel fiber to Purkinje cell synaptic transmission in rat cerebellum. J Physiol (Paris) 90:419–420

Miyata M, Okada D, Hashimoto K, Kano M, Ito M (1998) Corticotropin releasing factor plays a permissive role in cerebellar long-term depression. Neuron 22:763–775, 1999

Mizuno Y, Takahashi K, Totsune K, Ohneda M, Konno H, Murakami O, Satoh F, Sone M, Takase, S, Itoyama Y, Mouri T (1995) Decrease in cerebellin and corticotropin-releasing hormone in the cerebellum of olivo-pontocerebellar atrophy and Shy-Drager syndrome. Brain Res 686:115–118

Moga MM, Saper CB, Gray TS (1989) Bed nucleus of the stria terminalis: cytoarchitecture, immunohistochemistry, and projection to the parabrachial nucleus in the rat. J Comp Neurol 283:315–332

Mulders WHAM, Meek J, Hafmans TGM, Coots AR (1997) Plasticity in the stress-regulating circuit: decreased input from the bed nucleus of the stria terminals to the hypothalamic paraventricular nucleus in Wistar rats following adrenalectomy. Eur J Neurosci 9:2462–2471

Nahon JL, Presse F, Bittencourt JC, Sawchenko PE, Vale W (1989) The rat melanin-concentrating hormone messenger ribonuclei acid encodes multiple neuropeptides coexpressed in the dorsolateral hypothalamus. Endocrinology 125:2056–2065

Owens MJ, Nemeroff CB (1991) Physiology and pharmacology of corticotropin-releasing factor. Pharmacol Rev 43:425–473

Owens MJ, Vargas MA, Knight DL, Nemeroff CB (1991) The effects of alpazolam on corticotropin-releasing factor neurons in the rat brain: acute time course, chronic treatment and abrupt withdrawal. J Pharmacol Exp Ther 258:349–356

Page ME, Valentino RJ (1994) Locus coeruleus activation by physiological challenges. Brain Res Bull 35:557–560

Palkovits M, Leranth C, Gorcs T, Young III WS (1987) Corticotropin-releasing factor in the olivocerebellar tract of rats: demonstration by light-and electron-microscopic immuno-histochemistry and in situ hybridization histochemistry. Proc Natl Acad Sci USA 84:3911–3915

Pavcovich LA, Valentino RJ (1997) Regulation of a putative neurotransmitter effect of corticotropin-releasing factor: effects of adrenalectomy. J Neurosci 17:401–408

Phelix CF, Paull WK (1990) Demonstration of distinct corticotropin releasing factor-containing neuron populations in the bed nucleus of the stria terminalis. A light and electron microscopic immunocytochemical study in the rat. Histochemistry 94:345-364

Plotsky PM, Meaney MJ (1993) Early, postnatal experience alters hypothalamic corticotropin-releasing factor (CRF) mRNA, median eminence CRF content and stress-induced release in adult rats. Mol Brain Res 18:195-200

Plotsky PM, Owens MJ, Nemeroff CB (1998) Psychoneuroendocrinology of depression. Hypothalamic-pituitary-adrenal axis. Psychiatr Clin North Am 21:293-307

Potter E, Beham DP, Fischer WH, Linton EA, Lowry PJ, Vale WW (1991) Cloning and characterization of the cDNAs for human and rat corticotropin-releasing factor-binding proteins. Nature 349:423-426

Potter E, Sutton S, Donaldson C, Chen R, Perrin M, Sawchenko PE, Vale W (1994) Distribution of corticotropin-releasing factor receptor mRNA expression in the rat brain and pituitary. Proc Natl Acad Sci USA 91:8777-8781

Primus RJ, Yevich E, Baltazar C, Gallager DW (1997) Autoradiographic localization of CRF1 and CRF2 binding sites in adult rat brain. Neuropsychopharmacology 17:308-316

Raadsheer FC, Hoogendijk WJG, Stam FC, Tilders FJH, Swaab DF (1994) Increased numbers of corticotropin-releasing hormone-expressing neurons in the hypothalamic paraventricular nucleus of depressed patients. Neuroendocrinology 60:436-444

Raadsheer FC, Van Heerikhuize JJ, Lucassen PJ, Hoogendijk WJ, Tilders FJ, Swaab D (1995) Corticotropin-releasing hormone (CRH) mRNA levels in paraventricular nucleus of patients with Alzheimer's disease and depression. Am J Psychiatr 152:1372-1376

Redei E, Branch BJ, Taylor AN (1986) Direct effects of ethanol on adrenocorticotropin (ACTH) release in vitro. J Pharmacol Exp Ther 237:59-64

Redei E, Branch BJ, Gholami S, Lin EYR, Taylor AN (1988) Effects of ethanol on CRF release in vitro. Endocrinology 123:2736-2743

Reul JM, Labeur MS, Wiegers GJ, Linthorst AC (1998) Altered neuroimmunoendocrine communication during a condition of chronically increased brain corticotropin-releasing hormone drive. Ann NY Acad Sci 840:444-455

River C, Lee S (1996) Acute alcohol administration stimulates the activity of hypothalamic neurons that express corticotropin-releasing factor and vasopressin. Brain Res 726:1-10

Sakanaka M, Shibasaki T, Lederis K (1986) Distribution and efferent projections of corticotropin-releasing factor-like immunoreactivity in the rat amygdaloid complex. Brain Res 382:213-238

Sakanaka M, Shibasaki T, Lederis K (1987) Corticotropin releasing factor-like immuno-reactivity in the rat brain as revealed by a modified cobalt-glucose oxidase-diaminobenzidine method. J Comp Neurol 260:256-29

Sawchenko PE, Swanson L (1990) Organization of CRF immuno-reactive cells and fibers in the rat: immunohistochemical studies. In: De Souza EB, Nemeroff CB (eds) Corticotropin-releasing factor: basic and clinical studies of a neuropeptide. CRC Press, Boca Raton, pp 29-51

Sawchenko PE, Imaki T, Potter E, Kovacs K, Imaki J, Vale W (1993) The functional anatomy of corticotropin-releasing factor. In: Chadwick DJ, Marsh J, Ackrill K (eds) Coritcotropin-releasing factor. John Wiley & Sons, New York, pp 5-29

Schafe GE, Bernstein IL (1996) Forebrain contribution to the induction of a brainstem correlate of conditioned taste aversion: I. The amygdala. Brain Res 741:109-116

Schmahmann JD, Pandya D (1997) The cerebrocerebellar system. Int Rev Neurobiol 41:31-60

Schulz C, Lehnert H (1996) Activation of noradrenergic neurons in the locus coeruleus by corticotropin-releasing factor. A microdialysis study. Neuroendocrinology 63:454-458

Siggins GR, Gruol D, Aldenhoff J, Pittman Q (1985) Electrophysiological actions of corticotropin-releasing factor in the central nervous system. Fed Proc 44:237-242

Singewald N, Zhou GY, Chen F, Philippu A (1996) Corticotropin-releasing factor modulates basal and stress-induced excitatory amino acid release in the locus coeruleus of conscious rats. Neurosci Lett 204:45-48

Smagin GN, Swiergiel AH, Dunn AJ (1995) Corticotropin-releasing factor administered into the locus coeruleus, but not the parabrachial nucleus, stimulates norepinephrine release in the prefrontal cortex. Brain Res Bull 36:71–76

Smagin GN, Harris RB, Ryan DH (1996) Corticotropin-releasing factor receptor antagonist infused into the locus coeruleus attenuates immobilization stress-induced defensive withdrawal in rats. Neurosci Lett 220:167–170

Smith GW, Aubry J-M, Dellu F, Contarino A, Bilezikjian LM, Gold LH, Chen R, Marchuk Y, Hauser C, Bentley CA, Sawchenko PE, Koob GF, Vale W, Lee K-F (1998) Corticotropin releasing factor receptor 1-deficient mice display decreased anxiety, impaired stress response, and aberrant neuroendocrine development. Neuron 20:1093–1102

Smith MA, Weiss SRB, Berry RL, Zhang L-X, Clark, M, Massenburg G, Post RM (1997) Amygdala-kindled seizures increase the expression of corticotropin-releasing factor (CRF) and CRF-binding protein in GABAergic interneurons of the dentate hilus. Brain Res 745: 248–256

Sutton P, Donaldson C, Chen R, Perrin M, Lewis K, Sawchenko PE, Vale W (1994) Distribution of corticotropin-releasing factor receptor mRNA expression in the rat brain and pituitary. Proc Natl Acad Sci USA 91:8777–8781

Swanson LW, Swachenko PE, River J, Vale W (1983) Organization of ovine corticotropin-releasing factor immunoreactive cells and fibers in the rat brain: an immunohistochemical study. Neuroendocrinology 36:165–186

Vale W, Spiess J, Rivier C, Rivier J (1981) Characterization of a 41-residue ovine hypothalamic peptide that stimulates secretion of corticotropin and beta-endorphin. Science 213: 1394–1397

Valentino RJ, Foote SL (1988) Corticotropin-releasing factor increases tonic but sensory-evoked activity of noradrenergic locus coeruleus neurons in unanesthetized rats. J Neurosci 8: 1016–1025

Valentino RJ, Page E, Curtis AL (1991) Activation of noradrenergic locus coeruleus neurons by hemodynamic stress is due to local release of corticotropin-releasing factor. Brain Res 555:25–34

Valentino RJ, Page M, van Bockstaele E, Astone-Jone G (1992) Corticotropin-releasing factor innervation of the locus coeruleus region: distribution of fibers and sources of input. Neuroscience 48:689–705

Valentino RJ, Foote SL, Page ME (1993) The locus coeruleus as a site for integrating corticotropin-releasing factor and noradrenergic mediation of stress responses. Ann NY Acad Sci 697:173–188

Valentino RJ, Page ME, Luppi PH, Zhu Y, Van Bockstaele E, Aston-Jones G (1994) Evidence for widespread afferents to Barrington's nucleus, a brainstem region rich in corticotropin-releasing hormone neurons. Neuroscience 62:125–143

Valentino RJ, Curtis AL, Page ME, Pavcovich LA, Florin-Lechner SM (1998) Activation of the locus coeruleus brain noradrenergic system during stress: circuitry, consequences, and regulation. Adv Pharmacol 42:781–784

Van Bockstaele EJ, Colago EE, Valentino RJ (1996) Corticotropin-releasing factor-containing axon terminals synapse onto catecholamine dendrites and may presynaptically modulate other afferents in the rostral pole of the nucleus locus coeruleus in the rat brain. J Comp Neurol 364:523–534

Vaughan J, Donaldson C, Bittecourt J, Perrin MH, Lewis K, Sulton S, Chan R, Turnbull AV, Lovejoy D, River C, River J, Swachenko PE, Vale W (1995) Urocortin, a mammalian neuropeptide related to fish urotensin I and to corticotropin-releasing factor. Nature 378:287–292

Veinante P, Stoeckel ME, Freund-Mercier MJ (1997) GABA- and peptide-immunoreac-tivities co-localize in the rat central extended amygdala. Neuroreport 8:2985–2989

Weiss JM, Stout JC, Aaron MF, Quan N, Owens MJ, Butler PD, Nemeroff CB (1994) Depression and anxiety: role of the locus coeruleus and corticotropin-releasing factor. Brain Res Bull 35: 561–572

Wiersma A, Bohus B, Koolhaas JM (1993) Corticotropin-releasing hormone microinfusion in the central amygdala diminishes a cardiac parasympathetic outflow under stress-free conditions. Brain Res 625:219–227

Xiong Y, Xie Ly, Abou-Samra AB (1995) Signaling properties of mouse and human corticotropin-releasing factor (CRF) receptors: decreased coupling efficiency of human type II CRF receptor. Endocrinology 136:1828–1834

Yasuda N, Nakamura K (1997) Heterogeneity of corticotropin-releasing factor (CRF). Jpn J Physiol 47:147–159

Zoli M, Torri C, Ferrari R, Jansson A, Zini I, Fuxe K, Agnati LF (1998) The emergence of the volume transmission concept. Brain Res Rev 26:136–147

CRF and CRF Receptors

Jelena Radulovic, Thomas Blank, Klaus Eckart, Marko Radulovic,
Oliver Stiedl, and Joachim Spiess[1]

1
Introduction

Corticotropin-releasing factor (CRF), a 41-residue polypeptide, is one of the hypothalamic hypophysiotropic peptides. In mammals, CRF is synthesized in the hypothalamic paraventricular nucleus PVN, transported to the median eminence and secreted in the portal circulation reaching the corticotrophs of the anterior pituitary (Cummings et al. 1983; Swanson et al. 1983), where it stimulates the release of corticotropin which then enhances the release of cortisol from the adrenal cortex. Hypothalamus, pituitary and adrenal are thus functionally connected and constitute the hypothalamus pituitary adrenal (HPA) axis. Activation of this axis represents an important indicator of the response to stress. On this basis, the investigation of the CRF biology has been important for both basic scientists and clinicians. It has been speculated that CRF may play a role in depressive illness (Mitchell 1998) and cognitive disease (reviewed by De Souza 1995). The characterization of the primary structure of CRF (Spiess et al. 1981) purified on the basis of its hypophysiotropic activity (Vale et al. 1981) enabled rapid progress in the understanding of the physiology, biochemistry and anatomy of CRF in the brain of numerous animal species.

2
The CRF Family, a Chemical Consideration

In the meantime, CRF has also been purified and characterized from rat (Rivier et al. 1983; Spiess et al. 1983), human (Jingami et al. 1985), porcine (Patty et al. 1985), and bovine hypothalamus (Esch et al. 1984), and from frog brain (Stenzel-Poore et al. 1992). Recently, a novel CRF-like peptide, urocortin (Ucn), has been characterized from rat brain (Vaughan et al. 1995) and later from human tissue (Donaldson et al. 1996). Besides these CRFs, closely related polypeptides have been identified. From frog skin and fish, sauvagine (Mon-

[1] Max-Planck Institute for Experimental Medicine, Department of Molecular Neuroendocrinology, Hermann-Rein-Str. 3, 37075 Goettingen, Germany

tecucchi and Henschen 1981) and urotensin I (Lederis et al. 1982), respectively, were purified and characterized. The sequences of CRF from different origin (Fig. 1) show high homology of 81 to 95%. The primary structures of human and rat CRF turned out to be identical. No variation was found for residues 3–21, 24, 26–32, and 34–37. Despite this high homology, remarkable differences in the biological properties of h/rCRF and oCRF were found, as will be explained later. In contrast to the high similarity between different CRF forms, Ucn, urotensin I and sauvagine differ significantly from h/r CRF.

CD analysis indicated that oCRF and h/rCRF consist of 70% and 85%, respectively, of α-helical structures, whereas β-structures are not largely present (Dathe et al. 1996). The secondary structure of h/rCRF analyzed with NMR revealed a well-defined α-helix between residues 6 and 36 with an amphipatic α-helix between residues 6 and 20 (Gulyas et al. 1995). The five N-terminal and C-terminal residues were found to be predominantly disordered.

The influence of secondary structure on the biological potency was investigated by different types of amino acid replacement sets such as single (Rivier et al. 1993) and double D-amino acid replacements (Rothemund et al. 1997), Ala series replacements (Kornreich et al. 1992), and homologous amino acid replacements (Beyermann et al. 1996). Comparison of the retention times in reversed-phase (RP)-HPLC of double D-amino acid replaced peptides revealed the presence of an amphipatic helix between residues 7 and 22 of oCRF (Krause et al. 1995).

The importance of α-helicity in several regions of CRF for the biological potency was studied with i-(i+3) Glu-Lys lactam bridges. The CRF antagonist astressin cyclo(30–33)[D-Phe^{12},Nle^{21,38},Glu^{30},Lys^{33}]h/rCRF-(12-41) was 30 times more potent than the antagonist [D-Phe^{12},Nle^{21,38}]h/rCRF-(12-41) (Gulyas et al. 1995). It was speculated that the lactam bridge reinstates the

human/rat CRF	SEEPPISLDLTFHLLREVLEMARAEQLAQQAHSNRKLMEII■	100%
ovine CRF	SQEPPISLDLTFHLLREVLEMTKADQLAQQAHSNRKLLDIA■	83%
bovine CRF	SQEPPISLDLTFHLLREVLEMTKADQLAQQAHNNRKLLDIA■	81%
porcine CRF	SEEPPISLDLTFHLLREVLEMARAEQLAQQAHSNRKLMENF■	95%
frog CRF	AEEPPISLDLTFHLLREVLEMARAEQIAQQAHSNRKLMDII■	95%
human urocortin	DNPSLSIDLTFHLLRTLLELARTQSQRERAEQNRIIFDSV■	39%
rat urocortin	DDPPLSIDLTFHLLRTLLELARTQSQRERAEQNRIIFDSV■	44%
urotensin I	NDDPPISIDLTFHLLRNMIEMARIENEREQAGLNRKYLDEV■	61%
sauvagine	ZGPPISIDLSLELLRKMIEIEKQEKEKQQAANNRLLLDTI■	44%

Fig. 1. Homology of the CRF peptide family. *Shaded* residues are conserved in the entire peptide family; residues in *italics* are not conserved in the CRF analogs of different species

structural constraints in the antagonist that is presumably induced by the N-terminal part in the agonist. Scanning the entire antagonist with this type of a lactam bridge showed the highest increase in potency for the cyclizations in positions 20–23, 26–29, 28–31, 29–32, and 30–33 (Miranda et al. 1997). Similar experiments performed with the agonist [Ac-Pro^4, D-Phe^{1}2,Nle^{21,38}]h/rCRF indicated an increased protency for analogs after cyclization between residues 30 and 33 (Rivier et al. 1998).

The amphipatic helix may also play an important role in the oligomerization of CRF analogs. Therefore, the most potent antagonists (Fig. 2) contain the entire α-helical domain of CRF. The amphipatic helix may also play an important role in the oligomerization of the CRF analogs. The β-structure of oCRF is associated with a deeper insertion of the region 6–12 in lipid bilayers, whereas h/rCRF is assumed to be located on the lipid bilayer surface (Rothemund et al. 1997). In addition, the selectivity for different receptor subtypes can be controlled by amino acid exchanges in the region 12–40.

The CRF receptors (CRFR) bind oCRF and h/rCRF with high affinity in the low nanomolar range, whereas the CRF binding protein (CRF-BP) binds h/rCRF with high affinity but not oCRF. Alteration of the h/rCRF residues 22, 23, and 25 did not decrease the affinity for CRF-BP (Sutton et al. 1995). The antagonist [D-Phe^{12},Nle^{21,38}]h/rCRF-(12–41) binds with reduced affinity to CRF-BP. The structural requirements for CRF-BP binding can be clearly distinguished from those for CRFR recognition of both antagonists and agonists.

Urotensin and sauvagine are CRF related peptides from fish and frog, respectively. Urocortin, the mammalian urotensin analog, was recently discovered (Vaughan et al. 1995; Donaldson et al. 1996). These peptides exhibit a homology of 39 and 61% to h/rCRF. They bind with high affinity to the CRF receptor, but their physiological roles may differ. For example, in animal experiments urocortin showed strong appetite-suppressing effects, which were not observed to that extent with CRF (Spina et al. 1996). These differences were explained by the higher affinity of urocortin to CRFR2 and CRF-BP.

```
h/rCRF           SEEPP ISLDL TFHLL REVLE MARAE QLAQQ AHSNR KLMEI I
                 hh    ββhββ ββ hh hhhhh hhhhh hhhhh h ttt hhhhh
                       ββhββ ββ hh hhhhh hhhhh hhhhh h ttt hhβhh
oCRF             SQEPP ISLDL TFHLL REVLE MTKAD QLAQQ AHSNR KLLDI A

α-helical CRF(9-41)  DL TFHLL REMLE MAKAE QEAEQ AASNR LLLEE A
astressin              fHLL REVLE MARAE QLAQE AHKNR KLMEI I
antisauvagine          fHLL RKMIE IEKQE KEKQQ AANNR LLLDT I
```

Fig. 2. Prediction of α-helical (*h*), β-sheet (*β*), and turn (*t*) regions on the basis of the rules of Chou and Fasman (1978), and comparison with three antagonist structures. *Shaded* residues are conserved in the antagonist group

3
Molecular Analysis of CRF Binding Sites

CRF and CRF-like peptides exert their biological activity through binding to G-protein-coupled CRF receptors (CRFRs). By cloning of cDNAs from different species and tissues, two receptor subtypes, subtype 1 (CRFR1) (Chang et al. 1993; Chen et al. 1993; Vita et al. 1993; Dautzenberg et al. 1997) and subtype 2 (CRFR2) (Kishimoto et al. 1995; Lovenberg et al. 1995a; Perrin et al. 1995; Stenzel et al. 1995; Liaw et al. 1996; Yu et al. 1996; Dautzenberg et al. 1997), have been identified. The primary structures have been deduced from the cDNA coding for the receptor subtypes. Accordingly, CRF-R1 is a 415-amino acid protein mainly expressed in brain and pituitary (Chen et al. 1993). Three functional splice variants of CRF-R2, α (411–413 amino acids), β (431–438 amino acids) and γ (397 amino acids) (Kishimoto et al. 1995; Lovenberg et al. 1995a; Stenzel et al. 1995; Sperle et al. 1997), have been observed. CRF-R2β appears to be in rodents a peripheral receptor produced in the heart and in vessels, whereas CRF-R2α has been found only in the CNS. In humans, CRF-R2α and CRF-R2β are co-expressed in peripheral organs and the CNS (Valdenaire et al. 1997), whereas CRF-R2γ isolated only from humans was only found in the brain (Sperle et al. 1997). On the basis of the cDNA data (Chen et al. 1993), the molecular weight of unmodified CRF-R1 is approximately 44,000. The CRF-R1 protein contains several putative sites for N-glycosylation in the N-terminal domain (Chalmers et al. 1996), and therefore larger molecular weights are commonly found *in vivo* due to tissue-specific posttranslational modifications (see Spiess et al. 1998 *for review*). Sizes of 76–80 kDa were determined for CRF-R1 from different regions of rat brain (Sydow et al. 1997; Radulovic et al. 1998). Interestingly, the corresponding mouse proteins were larger by 4–3 kDa (Radulovic et al. 1998).

CRF and CRF-like peptides bind not only with different affinities to CRF-R1 and CRF-R2, but also to a water-soluble 37-kDa CRF binding protein (CRF-BP) (Vaughan et al. 1995) characterized from rat and human sources (Potter et al. 1991). The interaction with CRF-BP is assumed to reduce the availability of free CRF or CRF-like peptides at their receptor sites (Behan et al. 1996).

The binding to CRFR1 is mainly dependent on residues 12–41 of CRF, whereas the N-terminal residues 4–8 [the greatest effect was observed with residue 8 (Rivier et al. 1998)] are mostly important for G-protein activation (Rivier et al. 1993). Thus, [D-Phe^{12},Nle^{21,38}]h/rCRF-(12-41) binds with high affinity to CRFR1, but no stimulation of cAMP occurs. Similarly, residues 9–28 of h/rCRF were found to be crucial for binding to the CRF-BP (Sutton et al. 1995). A specific high affinity ligand, h/rCRF(96-33)-OH, that does not bind to CRF-R1 or CRF-R2 has been developed (Sutton et al. 1995).

Mammalian CRF-R1 indiscriminately binds CRF peptides such as Ucn, h/r CRF, ovine CRF (oCRF), urotensin I and sauvagine with high affinity (Vaughan et al. 1995; Donaldson et al. 1996), whereas mammalian CRF-R2 exhibits a different substrate specificity (Vaughan et al. 1995; Donaldson et al. 1996). CRF peptides such as h/rCRF or oCRF and urotensin I bind with lower affinity than Ucn or sauvagine. In contrast, frog CRF-R1 differs significantly from its mammalian counterpart by binding sauvagine and oCRF with lesser affinity (Dautzenberg et al. 1997).

Experiments using chimeric molecules between CRF-R and other receptor proteins have provided evidence for intramolecular interactions facilitating ligand binding. In binding experiments employing Ucn or the CRF antagonist astressin as ligands and chimeric molecules derived from rCRF-R1 and either domains of rat growth hormone-releasing factor receptor or activin receptor, it was observed that mainly the N-terminal domain of CRF-R1 was required for high affinity binding (Perrin et al. 1998). Following a similar strategy (Perrin et al. 1998), it was recognized that the fourth extracellular loop of rCRF-R1 enhanced binding of astressin by a factor of 2–3. A similar approach was performed with oCRF as ligand and chimeric receptor molecules consisting of rCRF-R1 except for its extracellular fourth loop and the fourth extracellular loop from the glucagon receptor. Such a chimeric receptor binds CRF with low affinity only (Sydow et al. 1998). High affinity ligand binding requires under these experimental conditions the presence of the fourth extracellular loop of rCRF-R1. In agreement with these data, it was observed that rCRF-R1 constructs without the nucleotide sequence coding for the fourth extracellular loop were expressed as membrane-bound receptors binding oCRF only with low affinity (Sydow et al. 1997).

When chimeric receptor molecules derived partially from human CRFR1, frog CRFR1 and frog CRFR2α were modified by point mutations, the ligand binding site of frog CRFR1 was mapped to the extracellular sequence of frog CRFR1 between residues 70–89 (Dautzenberg et al. 1997).

On the basis of the data reported to date, it is probable that the intramolecular interactions of the CRF receptor vary depending on the ligand involved.

The role of cysteine residues of CRF-R1 for ligand binding and cAMP production has been investigated by reduction of disulfide bridges with dithiothreitol and mutation of cysteine into serine residues (Qi et al. 1997). On the basis of these data, only extracellular cysteines in the N-terminal domain are important for ligand binding and cAMP generation.

4
Distribution of CRF-like Peptides
and Their Binding Sites in the CNS

The presence and distribution of CRF, Ucn, CRF receptors and CRF-BP in the brain has been investigated by radioimmunoassay, autoradiography, immunohistochemistry and *in situ* hybridization histochemistry. Conclusive evidence about the distribution of the known CRF-like peptides and their binding sites in the CNS was mainly determined by immunodetection of the proteins and autoradiographic detection of their corresponding mRNA. These techniques enabled high anatomical resolution and specificity. Occasional mismatch between the anatomical localization of protein and its mRNA could be interpreted by region-specific transcriptional or translational regulation of mRNA formation and protein synthesis, respectively. The most complete mapping of the brain CRF system was performed in the rat (Table 1). In general, CRF is widely distributed in cells and fibers throughout the rat CNS, from the neocortex to the spinal cord (Olschowka et al. 1982; Cummings et al. 1983). The detection of CRF mRNA (Thompson et al. 1990) mostly corresponds to the presence of CRF immunoreactivity (CRF-ir). The most prominent sites of CRF production are the region of the anterior commissura, central amygdala, hypothalamus, and several brain stem nuclei. Projection pathways containing CRF-ir neurons have been observed from the central amygdala to several pontine nuclei, from the PVN to the median eminence, the brain stem and spinal cord, from the dorsal lateral tegmentum to the thalamus, septum and cortex, from

Table 1. Distribution of the CRF-like peptides and their binding sites in rat brain

Anatomical region	CRF-ir[a,b] cells (fibres)	Ucn-ir[c] cells (fibres)	CRFR1[d,e] mRNA (protein)	CRFR2[f] mRNA (protein)	CRF-BP[g] mRNA (protein)
Telencephalon					
Olfactory bulb	(++)		++++ (++++)		++ (++)
Neocortex	+		++++ (++++)		++ (++)
N. accumbens septi	++				
Dorsal lateral septum			+ (+)		
Intermediate lateral septum	++	+ (+++)		++++	– (–)
Ventral lateral septum	++ (+++)			++++	± (±)
Medial septum	(+)		+++ (++)	±	±
Bed nucleus of the stria terminalis	+++ (++)		++ (++)	++	++
Nucleus of the diagonal band	+++ (+)		+++ (+++)		
CA1 hippocampal area	+	+	++ (+)	++	+
Dentate gyrus	+	– (++)	++ (+)	++	++ (++)
Entorhinal cortex	±		+++ (+++)	++	+
Subiculum	+		++ (+++)		
Central amygdala	+++				++ (++)
Basolateral nucleus of the amygdala			++ (++)	±	++ (++)
Medial nucleus of the amygdala			++ (+)	++	
Posterior cortical nucleus of the amygdala				+++	++ (++)
Medial preoptic area	++		+ (+)		± (±)

Table 1. Continued

Anatomical region	CRF-ir[a,b] cells (fibres)	Ucn-ir[c] cells (fibres)	CRFR1[d,e] mRNA (protein)	CRFR2[f] mRNA (protein)	CRF-BP[g] mRNA (protein)
Diencephalon					
Anterior hypothalamic area	±		+ (+)		± (±)
Periventricular hypothalamic nucleus	+++ (++)	+++ (+)	+ (+)		± (±)
Paraventricular hypothalamic nucleus	+++ (++)	+++ (+)	+ (+++)	++	± (±)
Supraoptic nucleus	+ (++)	++ (+)	+ (++)	+++	±
Dorsomedial hypothalamic nucleus	+		+++		±
Ventromedial hypothalamic nucleus	+	+++	+	++++	±
Median eminence	(+++)				
Mammillary nuclei	++ (±)				
Retcular thalamic nucleus				(++)	
Ventral thalamic nucleus			+ (+)		
Mesencephalon					
Superior colliculus	+		+++ (++)	+	
Substantia nigra		++	++ (++)		
Edinger-Westphal nucleus		(++)	++++ (+)		
Nucleus raphe dorsalis	++ (++)	++	+ (+)	++	++ (++)
Central gray	+ (++)		++ (++)		
Darkschewitsch nucleus		++	(+++)		
Red nucleus		++	++++ (++++)		
Midbrain trigeminal nucleus	+		+++ (+++)		++ (++)
Pons/medulla					
Locus ceruleus	++				
Parabrachial nucleus	+++				
Inferior colliculus	±		++ (+)	++	
Olivary area	++ (+)		++ (++)		
Dorsal tegmental nucleus	++		++++ (+++)		
Sensory trigeminal nucleus				+++ (++)	±
Reticular pontine nucleus	++		++ (++)		++ (++)
Cuneate nucleus	++		++ (++)		
Interpeduncular nucleus			++ (++)	+++	
Peduncular nucleus			++ (+++)		
Cerebellum	+++	±	+++ (++++)		

[a] Olschowka et al. (1982); [b] Cummings et al. (1983); [c] Kozicz et al. (1998); [d] Potter et al. (1994); [e] Radulovic et al. (1998); [f] Chalmers et al. (1995); [g] Potter et al. (1992)
CRF-ir and Ucn-ir were expressed as: −, undetectable; ±, very weak; +, low; ++, moderate; and +++, strong cell staining or fiber staining (in parenthesis). The hybridization signal of CRFR1, CRFR2 and CRF-BP mRNA was expressed as: −, undetectable; +, weak; ++ moderate; +++, strong; and ++++, very strong. In parenthesis, the immunohistochemical staining of CRFR1 and CRF-BP was described as: +, undetectable; +, weak; ++ moderate; +++, strong; and ++++, very strong, according to the number of stained cells and density of staining within an area.

the spinal cord to the brain stem, central gray and thalamus and from the inferior olivary nucleus to the cerebellum (Cummings et al. 1983, 1988; Moga and
Gray, 1985).

Ucn immunoreactivity (Ucn-ir) appears to be more restricted to certain
brain areas, primarily to the lateral septum, hypothalamus and Edinger-
Westphal nucleus (Kozicz et al. 1998). Ucn mRNA has been detected in areas
which did not exhibit Ucn-ir, such as neocortex, olfactory bulb, amygdala and
cerebellum (Wong et al. 1996). A possibility that Ucn synthesis in these areas
occurs under specific conditions has not been investigated yet.

CRFR1 distribution in the CNS mostly parallels the distribution of CRF in
the cortex, area of the anterior commissurae, and most brain stem nuclei (Potter et al. 1994; Chalmers et al. 1995; Radulovic et al. 1998). CRFR1-ir was mainly visualized by membrane staining confined to the neuronal soma (Fig. 3). In
the hypothalamus, CRFR1 is present in areas containing both CRF and Ucn,

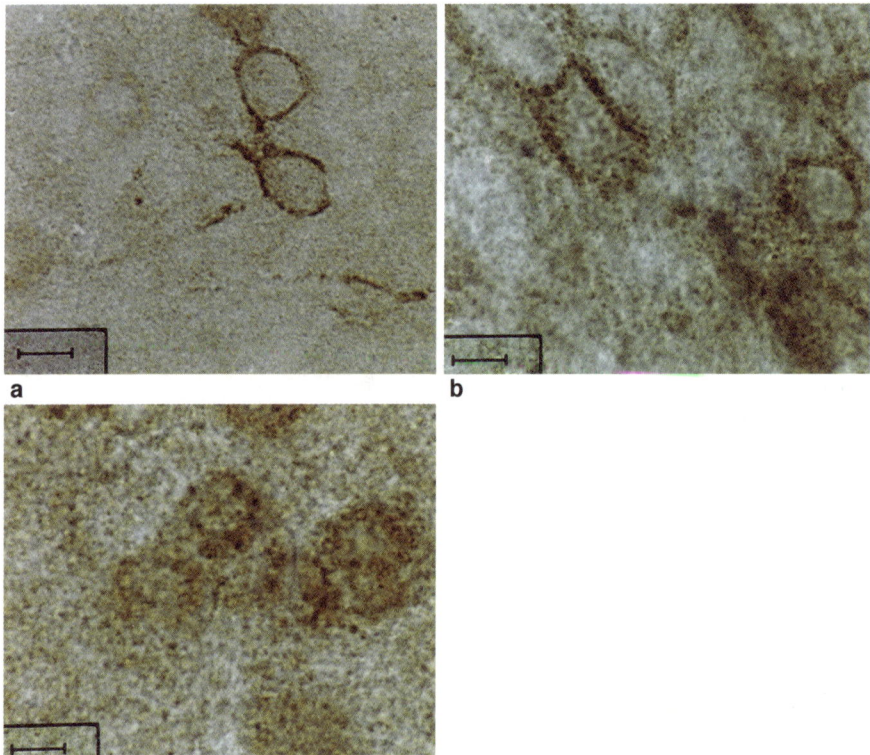

a   b

Fig. 3. CRFR1-positive cells in the cortex (**a**), reticular thalamic nucleus (**b**) and motor trigeminal nucleus (**c**) of a C57BL/6N mouse. Scale bar = 15.8 µm. Membrane staining was typically
observed in most of the brain areas. In the motor trigeminal nucleus and cerebellum (not shown)
cells were stained homogenously, suggesting high intracellular protein levels

whereas some mesencephalic regions, such as substantia nigra, the Dark-schewitsch nucleus and the red nucleus, contain CRFR1 and Ucn, but not CRF.

High CRFR2 mRNA formation seems to be limited to the lateral septum and hypothalamus, whereas moderate levels are detectable in hippocampal sub-fields and some nuclei of the amygdala (Chalmers et al. 1995). In general, these areas contain both CRF and Ucn. On the basis of the distribution patterns of CRF and Ucn in areas containing CRFR1, CRFR2 or both receptors, the initial assumption that Ucn may be the endogenous ligand for CRFR2 does not seem probable. The distribution of CRF-BP appears to be predominant in the neo-cortex, limbic system and several brain stem nuclei. In these areas, CRF-BP mainly does not exhibit cellular colocalization with CRF (Potter et al. 1992).

It is worthwhile mentioning that CRFR1, CRFR2 and CRF-BP could not be detected in brain areas such as the central amygdala, Edinger-Westphal nucle-us and locus ceruleus containing abundant CRF-ir or Ucn-ir. It appears that CRF-like peptides synthesized in these nuclei act through yet undefined recep-tor(s) or on remote areas, probably after axonal transport by projection path-ways (Fendt et al. 1996, 1997) to neuronal targets producing CRFR1 or CRFR2. However, it cannot be excluded that the expression of CRFR in these areas is too low for detection with the methods used. An opposite distribution mis-match, presence of CRFR1 in areas that do not contain CRF or Ucn, as in the basolateral and medial nuclei of the amygdala, may imply the existence of other, yet undiscovered CRF-like peptide(s).

CRF-like peptides exhibit different endocrine, autonomic and behavioral effects depending on the animal species investigated. This may be partly due to different molecular properties and distribution patterns of CRF-like pep-tides and their receptors. We have demonstrated recently that CRFR1 from rat and mouse brain differs in molecular size and regional density. The mouse produced significantly less CRFR1 in the neocortex but significantly more CRFR1 in the brain stem than the rat.

The widespread network of CRF-containing cells and fibers provides ana-tomical substrate for multiple regulatory roles of CRF, from neuroendocrine and autonomic to behavioral, ranging from instinctive behavior to highest cognitive functions. The complexity of the regulation of the interactions between CRF-like peptides and CRFR is further enhanced by the distinct dis-tribution pattern of the CRF-BP in areas associated with cognitive and emo-tional behaviors.

5
Distribution of CRF-like Peptides
and Their Binding Sites in the Pituitary

The first physiological role assigned to CRF was its ability to stimulate adre-nocorticotropic hormone (ACTH) secretion from the anterior lobe of the

Table 2. Distribution of CRF-like peptides and their binding sites in rat pituitary

Pituitary lobe (protein)	CRF[a] mRNA (protein) mRNA (protein)	Ucn[b]	CRFR1[c] mRNA (protein) mRNA (protein)	CRFR2[d]	CRF-BP[e] mRNA
Anterior (+++)	–	++	+++	±	+++
Intermediate	– (+)	+++	+++	±	– (–)
Posterior	–	–	–	–	– (–)

[a] Thompson et al. (1990); [b] Wong et al. (1996); [c] Potter et al. (1994); [d] Chalmers et al. (1995); [e] Potter et al. (1992)
The mRNA and protein levels of CRF, Ucn, CRFR1, CRFR2 and CRF-BP were expressed as: –, undetectable; +, weak; ++, moderate; and +++, strong.

pituitary (Vale et al. 1981), and thus initiate stress responses. Absence of CRF mRNA in the pituitary is consistent with the proposed endocrine role of CRF. CRF-ir has been observed in neurons of the intermediate lobe of the pituitary (Swanson et al. 1983), suggesting that CRF may reach this area by neuronal connections as well. Secretagogues other than CRF may also be involved in the release of ACTH, as demonstrated by ACTH secretion in female mice carrying a disrupted CRF gene (Muglia et al. 1995). Interestingly, high levels of Ucn mRNA were recently detected in the adenohypophysis (Wong et al. 1996). This observation, as well as the higher potency of Ucn to stimulate ACTH release from pituitary cells *in vitro* and *in vivo* (Vaughan et al. 1995) strongly supports a paracrine role of Ucn in the regulation of ACTH secretion. The corticotrophic actions of CRF and Ucn appear to be mediated by CRFR1 and regulated by CRF-BP (Table 2). High levels of CRFR1 in the anterior and intermediate lobes of the pituitary (Potter et al. 1994) support the role of CRF in the regulation of pro-opiomelanocortin (POMC)-derived peptide secretion. The rat pituitary exhibits higher density of CRF receptors in the anterior lobe, whereas bovine, porcine and mouse pituitaries contain higher density of CRF receptors in the intermediate lobe (De Souza. 1995). The hypophyseal expression of CRFR2 mRNA appears to be low to undetectable (Chalmers et al. 1995).

6
Immunomodulatory Effects of CRF

By activating the glucocorticoid and catecholamine secretion, CRF from the central nervous system mediates the suppressive effects of stress on the immune system (Irwin et al. 1988). Also, in inflammation, the immune system is activated and can stimulate the HPA axis by increased CRF levels in the hypophyseal portal circulation (Sapolsky et al. 1987). Mediators of such HPA axis stimulation by the immune system are mainly interleukin-1, tumor necrosis factor-alpha and interleukin-6 (Sapolsky et al. 1987; Besedovsky et al. 1991). These cytokines, are released, upon immune challenge, from monocytes/macrophages and neutrophils into the tissue and circulation (Dinarello 1994).

Taken together, injury and inflammation result in immediate secretion of cytokines from immune cells into the local tissue and circulation. This leads to mobilization of immune defense mechanisms to fight the intrusion of foreign material and, at the same time, to an activation of the hypothalamo-pituitary-adrenal axis which counteracts the immunologic activation (Besedovsky et al. 1991).

In addition to the effect of hypothalamic CRF on the immune system that is exerted indirectly, by activation of the HPA axis as described above, peripherally produced CRF may act directly by binding to CRF receptors on immune cells. Inflammation stimulates not only the expression of CRF within the hypothalamus but also at the site of inflammation (Hargreaves et al. 1989). It was indicated that mononuclear leukocytes, tissue fibroblasts and vascular endothelial cells at inflamed areas are able to produce CRF (Karalis et al. 1991). Consequently, CRF was found in inflamed synovial tissues at the same concentrations as detected in hypophyseal portal venous blood (Crofford et al. 1993). Binding sites for CRF were also found on immune cells (Audhya et al. 1991). CRF-binding sites of immune cells identified by binding of radioactively labeled CRF were localized to adherent splenocytes (Webster et al. 1990), monocytes, T lymphocytes (Singh and Fudenberg 1988) or monocyte-macrophages and T-helper cells. However, the studies mentioned did not determine the type(s) of CRF receptor on these cells. Binding sites for CRF have been mainly examined so far in the mononuclear cell fraction (Singh 1989; Singh and Christine-Leu 1990; Audhya et al. 1991; Salas et al. 1997), but not in polymorphonuclear cells (neutrophils) which represent the main cell type in acute inflammation. By employing polyclonal antibodies against CRFR1 (Radulovic et al. 1998) an immunohistochemical analysis of CRF receptor production of immune cells was recently performed for the first time. Thereby, CRF-R1+ mouse splenocytes were identified as mature and immature neutrophils (Radulovic and Spiess 1998). Interestingly, CRF-R1 mRNA was detected in the spleen of naive mice, but not in other major granulocyte pools such as bone marrow and peripheral blood leukocytes. This suggested that CRF-R1 was not produced constitutively by mature or immature neutrophils. Its production was rather induced by the splenic microenvironment (Radulovic and Spiess, 1999). A number of CRF-R1+ neutrophils in mouse spleen were shown to be transiently increased after psychological stress (Radulovic and Spiess 1998). Furthermore, an even more dramatic effect was shown after the induction of acute or chronic inflammation (Radulovic and Spiess, 1999). However, a physiological role of CRF-R1 on neutrophils still remains to be investigated.

A growing body of data now provides evidence for direct effects of CRF on other immune cell types (Stephanou et al. 1990). However, *in vitro* as well as *in vivo* experiments using different models to establish direct action of CRF in inflammatory processes have produced contradictory results. CRF was shown to induce the proliferation of rat splenocytes (McGillis et al. 1989) and to increase the IL-1 (Singh and Christine-Leu 1990) production of human monocytes. In contrast, inhibitory effects of CRF on IL-1 and IL-6 expression induced by LPS in the same cell type were observed (Hagan et al. 1992). More recent findings suggest that the ability of CRF to either enhance or inhibit the IL-1 production by monocytes depends on the level of monocyte activation (Pereda et al. 1995). CRF was thereby shown to increase the production of IL-1 and IL-1 receptor antagonist in resting cells while it acts as inhibitor when monocytes were activated.

Local paracrine actions of CRF at the site of inflammation appear to be mostly anti-inflammatory (Wei et al. 1993), while some studies also describe pro-inflammatory actions (Karalis et al. 1991; Correa et al. 1997).

In a rat model of acute inflammatory reaction induced by carrageenan, systemic pretreatment with anti-CRF antibodies reduced the inflammatory response (Karalis et al. 1991). On the other hand, CRF was reported to supress experimental autoimmune encephalomyelitis (Poliak et al. 1997). We have demonstrated that CRF exerts either a pro- or anti-inflammatory effect on carrageenan-induced rat paw oedema, depending on its concentration (Correa et al. 1997). In addition to its immunomodulatory effects, CRF was also reported to induce secretion of β-endorphin from immune cells in inflamed tissue, resulting in inhibition of pain (Schafer et al. 1994).

Thus, based on the experimental evidence obtained so far, the functional role of CRF and its receptors remains controversial, although CRF certainly possesses immunomodulatory activity. However, the upregulation of CRF and its receptor (Karalis et al. 1991; Mousa et al. 1996) in inflamed tissue points to a paracrine/autocrine role of peripherally expressed CRF in inflammation. More data are needed regarding the identification of immune cell types that secrete CRF or bear CRF receptors in different stressful situations. Also, further experiments focused on the time-course of appearance of CRF and CRF-R+ cells in inflamed tissue and main lymphoid organs, following the challenge of inflammation, are required in order to clarify our understanding of the physiological actions of CRF on immune cells.

7
Neurotransmitter-Like Actions of CRF

Electrophysiological studies (Siggins et al. 1985) indicate that CRF has predominantly excitatory actions in the locus ceruleus, the solitary tract complex, the hippocampus and some regions of the hypothalamus. Intracellular recordings with sharp microelectrodes have shown that CRF causes a significant increase in the number of spikes in a burst evoked by a depolarizing current pulse (Fig. 4). This increased excitability was shown to be mediated by a reduced magnitude and duration of the slow afterhyperpolarization (sAHP) following a burst of action potentials in CA1 and CA3 pyramidal cells (Aldenhoff et al. 1983; Smith and Dudek. 1994) and in Purkinje cells of the cerebellum (Fox and Gruol 1993). It has been suggested that the sAHP is generated by Ca^{2+}-dependent potassium channels which can be modulated by cAMP-dependent mechanisms (Madison and Nicoll 1982). An additional effect of CRF was observed in hippocampal slices at concentrations above 0.25 µM. At these high concentrations, CRF usually depolarized both CA1 and CA3 pyramidal cells and thereby increased the spontaneous firing rate (Aldenhoff et al. 1983; Siggins et al. 1985). Differential effects of CRF on neuronal activity were found in PVN neurons of the hypothalamus in rat brain slice preparations (Yamashita et al. 1991). From intracellular and extracellular recordings, excitation or inhibition was observed in PVN neurons following CRF application. However, after synaptic blockade, an increased firing rate and membrane depolarization by CRF were still observed, whereas no inhibition of the neuronal activity could be detected. Thus, the inhibitory responses to CRF may not be initiated by direct action of CRF on the PVN neurons but rather mediated through neighboring neurons that are sensitive to CRF.

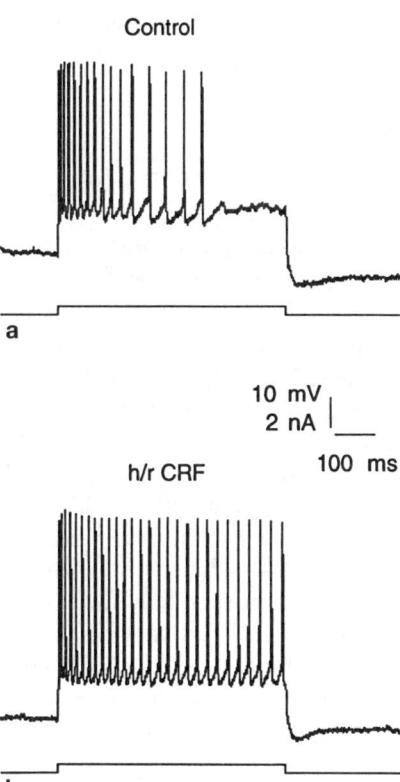

Fig. 4. Action potential firing pattern evoked by a constant depolarizing current pulse (600 ms, 800 pA; lower tracing) in the same mouse CA1 pyramidal cell. a) Control bathing solution; b) approximately 17 min after addition of 500 nM h/r CRF to the bathing solution. Holding potential was –70 mV

Similar differential effects of CRF were found in basolateral (BLA) and central (ACe) neurons of the rat amygdala *in vitro* (Rainnie et al. 1992). In BLA neurons CRF (12.5–250 nM) had no effect on either resting membrane potential or input resistance. In contrast, the majority of ACe neurons responded to CRF application with membrane hyperpolarization and decreased membrane input resistance at low nanomolar concentrations (10–50 nM). To examine whether CRF was acting via pre- or postsynaptic receptor activation, synaptic transmission was blocked by addition of the Na^+ channel blocker tetrodotoxin (TTX) to the superfusate. The reversal potential for the CRF effect of –84 mV in the presence of tetrodotoxin (0.5 µM) indicated that a postsynaptic increase in K+ conductance may underlie the CRF-mediated reduction of membrane excitability.

These direct actions of CRF on ACe neurons are unexpected because several groups using receptor autoradiography reported very low signals of CRF binding sites in the central amygdaloid nucleus (Wynn et al. 1984; De Souza et al. 1985). However, *in vivo* experiments clearly demonstrated that, for example, stressor exposure produced a sustained increase in the release of CRF at the central nucleus of the amygdala (Merali et al. 1998). Therefore, the observed inhibitory effects of CRF could reflect the autoinhibition of ACe neurons. It may be that endogenously released CRF acts on somatodendritic CRF autoreceptors to cause feedback inhibition of further peptide release.

8
CRF and Cardiovascular Regulation

It is generally accepted that CRF functions as a mediator of the response to stress. In this framework it is believed that CRF is involved in the cardiovascular regulation exercised by the autonomic nervous system (Brown et al. 1982; Fisher 1989). Both central (intracerebroventricular: icv) and peripheral (intravenous: iv) oCRF or h/rCRF administration produces a tachycardia in rats (Fisher 1989; Diamant and de Wied 1993; Richter and Mulvany 1995; Buwalda et al. 1998). The CRF-mediated tachycardia could be lowered by a high concentration of the CRF-antagonist α-helCRF(9-41) when injected before the administration of either oCRF or h/rCRF (Fisher 1989; Diamant and de Wied 1993; Richter and Mulvany 1995).

The CRF-mediated tachycardia in rats has been attributed to sympathetic activation independent from CRF-induced epinephrine release. However, a CRF-induced heart rate (HR) effect as a consequence of increased physical activity (Buwalda et al. 1998) was not excluded (Fisher 1989). In experiments using CRF injections into the central nucleus of the amygdala, the HR increase was explained by parasympathetic inhibition as no concomitant increase in the plasma catecholamine level was found (Wiersma et al. 1993). Differences in the oCRF- and h/rCRF-mediated tachycardia point to the involvement of both

CRFR. As CRF receptor 2β mRNA was isolated from heart tissue (Lovenberg et al. 1995b), the involvement of CRF receptor 2β in peripheral cardiovascular regulation is suggested. To date, a concomitant antagonistic effect of CRF on the sympathetic and parasympathetic system for tachycardia cannot be excluded. These cardiovascular results suggest that CRF acts as a stress mediator in rats. The role of the CRF receptor in the stress-induced regulation of behavioral, autonomic and endocrine function including the cardiovascular regulation in the mouse needs to be elucidated.

9
Role of the CRF System in Behavior

Intracerebroventricular (i.c.v.) administration of CRF in animals produces a variety of behavioral responses, which resemble behaviors elicited by stressful stimuli. The independence of CRF effects in the brain from the activation of the HPA axis could be demonstrated by i.c.v. application of CRF at doses that are ineffective after systemic administration of the peptide, and by the inability of dexamethasone and hypophysectomy to prevent these effects. Behavioral changes induced by CRF injection have been extensively reviewed (Dunn and Berridge 1990; Koob and Britton 1990; Owens and Nemeroff 1991). Although the data are not always consistent, depending on the doses of CRF, animal species and conditions employed among experiments, CRF is assumed to induce behavioral activation in a familiar environment (Sutton et al. 1982), to increase grooming (Britton et al. 1982), anxiety (Dunn and File 1987) and startle response (Swerdlow et al. 1986), and to suppress exploration of a novel environment (Berridge and Dunn 1989), ingestive behavior (Gosnell et al. 1983) and sexual behavior (Sirinathisinghji et al. 1983). In summary, CRF appears to increase arousal in a familiar environment and to enhance emotionality in a novel environment. Therefore, it has been suggested that CRF may mediate many physiological and behavioral effects observed during stress or fear (Koob and Bloom 1985; Fisher 1989). The role of endogenous CRF in stress-induced behaviors has been further supported by the ability of CRF receptor antagonists to prevent or reduce stress-induced decrease in feeding (Krahn et al. 1986) and investigative behavior (Berridge and Dunn 1987), shock-induced fighting (Tazi et al. 1987) and fear enhanced startle response (Swerdlow et al. 1989). The novel CRF-like peptide Ucn has also been reported to mediate some stress-related behaviors, such as appetite suppression (Spina et al. 1996) and anxiogenic-like behaviors (Moreau et al. 1997).

Widespread distribution of CRFR1 in the limbic system and neocortex, initially demonstrated by receptor autoradiography (De Souza et al. 1985) and subsequently confirmed by *in situ* hybridization (Potter et al. 1994) and immunohistochemistry (Radulovic et al. 1998), suggested that the CRF system may play a significant role in the modulation of emotional behaviors and cognition

under physiological and pathophysiological conditions. Although the anxiogenic effects of CRF and Ucn have been well documented in various animal
models of anxiety, the results generated by pharmacological approaches and
generation of mutant mice lacking a functional CRFR1 showed several discrepancies. In normal adult rats and mice, anxiogenic-like effects of CRF were
demonstrated by i.c.v. injection of the peptide and by the ability of the CRF
receptor antagonist α-helical CRF (9–41) to prevent stress-induced anxiogenic responses. The antagonist alone, however, did not produce anxiolytic effects
in several animal models of anxiety, such as the Geller Sefter conflict test
(Britton et al. 1986), social interaction test (Dunn and File 1987) and elevated
plus-maze test (Baldwin et al. 1991), suggesting that the endogenous tone of
the CRF system on anxiety levels may be very low in the absence of strong
prior stressful experience. In contrast, mice lacking a functional CRFR1 displayed reduced anxiety under baseline conditions (Smith et al. 1998; Timpl et
al. 1998). Taking into account that these mice suffered from atrophy of the
adrenal medulla (Timpl et al. 1998) or zona fasciculata of the adrenal cortex
(Smith et al. 1998) as well as other neuroendocrine abnormalities throughout
their entire development, conclusive evidence of the role of CRFR1 in baseline
and stress-induced anxiety is still lacking.

Another line of controversial results was obtained in an attempt to identify
the brain area involved in anxiogenic actions of CRF-like peptides. Several
studies employing CRF-receptor antagonists (Heinrichs et al. 1992; Rassnick et
al. 1993; Swiergiel et al. 1993) and CRFR1 antisense oligodeoxynucleotides
(Liebsch et al. 1995) indicated that stress-induced anxiogenic behaviors may
be mediated through the central amygdala, an area that does not contain
detectable levels of CRFR1 mRNA (Potter et al. 1994; Chalmers et al. 1995) or
CRFR1 protein (Radulovic et al. 1998). Consistent with the latter observations
is the finding that modulation of some anxiogenic-like behaviors, such as
potentiation of acoustic startle responses occurs in the pontine projection targets of the central amygdala (Fendt et al. 1997). Recently, an anatomical dissociation of CRF- and fear-enhanced startle was demonstrated (Lee and Davis
1997). CRF-enhanced startle was prevented by lesions of the bed nucleus of the
stria terminalis, whereas amygdala-lesioned animals showed blockade of fear-
but not CRF-potentiated startle. These findings suggested that some anxiogenic-like effects of CRF are mediated through the bed nucleus of the stria terminalis, a brain area containing CRFR1 and CRFR2.

Acute behavioral, autonomic and neuroendocrine responses to stressful
stimuli are assumed to play an important role in coping with stressful events
and in adaptation to environmental changes. Therefore, it is not surprising
that even a single brief acute stress induces long-term changes in neuronal
plasticity. Stressful experiences were reported to enhance or impair learning
and memory, depending on the stressor, learning task, animal species and a
number of other experimental conditions. It has been proposed that neurohormones released under stress may affect learning and memory (McGaugh

1983). The role of CRF in cognitive processes has been investigated by injections of exogenous CRF or displacement of endogenous CRF from the CRF binding protein. Intracerebral application of CRF shortly before or immediately after training enhanced learning of several tasks, such as visual discrimination learning (Koob and Bloom 1985), inhibitory avoidance learning (Liang and Lee 1988), spatial learning (Behan et al. 1995) and fear conditioning (Fig. 5a). Similarly, i.c.v. injection of CRF-BP ligands which displace CRF from CRF-BP [e.g. h/r CRF-(6-33)] enhances spatial learning, visual discrimination, inhibitory avoidance, active avoidance (Behan et al. 1995; Heinrichs et al. 1997) and fear conditioning (Fig. 5b).

An opposite effect, memory impairment, was observed when CRF was injected i.c.v. before the memory test of passive avoidance (Veldhuis and De Wied 1984; Diamant and De Wied 1993). A common conclusion drawn from these studies is that CRF modulates learning at much lower doses than the ones required for induction of anxiety (Behan et al. 1995) or activation of the HPA axis (Diamant and De Wied 1993). Interestingly, the behaviorally most potent CRF fragment (CRF-34-41) failed to affect ACTH release even at doses as high as 50 µg/rat (Diamant and De Wied 1993). Dissociation of cognitive from other behavioral effects of CRF was also observed after injection of CRF-BP ligands, which enhanced learning without affecting anxiety (Behan et al. 1995) and food intake (Heinrichs et al. 1997). These findings suggest that different behavioral actions of CRF are dose-dependent and may be mediated

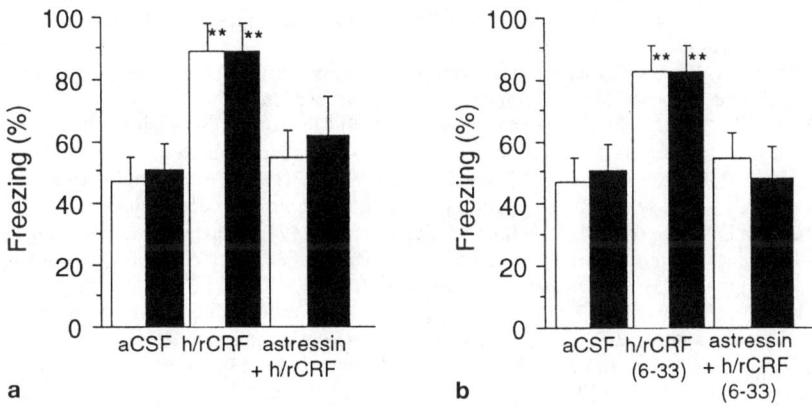

Fig. 5. Effect of i.c.v. injected h/rCRF (**a**) and h/rCRF (6-33) (**b**) on context- and tone-dependent fear conditioning. Balb/c mice (10–12/group) were injected with either peptide 5 min before training in a fear conditioning paradigm. Training consisted of exposure of mice to a context (3 min), tone (30 s) and footshock (2 s, 0.7 mA). Presented data are freezing scores recorded 24 h later during the memory test, which was performed by re-exposure of mice to the conditioning context (3 min, white bars) or to a tone (3 min, black bars) in a novel context. Astressin (300 ng/mouse) prevented the effects of both peptides (100 ng/mouse on context- and tone-dependent fear conditioning. Statistically significant differences: $**p < 0.001$ vs aCSF-injected control

through different brain areas and through different receptor types. Thus, the role of the CRF system in the plastic adaptation of the CNS to environmental stimuli may be complex and dynamic.

References

Aldenhoff JB, Gruol DL, Rivier J, Vale W, Siggins GR (1983) Corticotropin releasing factor decreases postburst hyperpolarization and excites hippocampal neurons. Science 221: 875–877

Audhya T, Rajeev J, Hollander C (1991) Receptor-mediated immunomodulation by corticotropin-releasing factor. Cell Immunol 134:77–84

Baldwin HA, Rassnick S, Rivier J, Koob GF, Britton KT (1991) CRF antagonist reverses the 'anxiogenic' response to ethanol withrawal in the rat. Psychopharmacology 103:227–232

Behan DP, Heinrichs SC, Troncoso JC, Liu X-J, Kawas CH, Ling N, De Souza EB (1995) Displacement of corticotropin releasing factor from its binding protein as a possible treatment for Alzheimer's disease. Nature 378:284–287

Behan D, Khongsaly O, Ling N, De Souza EB (1996) Urocortin interaction with corticotropin-releasing factor (CRF) binding protein (CRF-BP): a novel mechanism for elevating 'free' CRF levels in human brain. Brain Res 725: 263–267

Berridge CW, Dunn AJ (1987) A corticotropin-releasing factor antagonist reverses the stress-induced changes of exploratory behavior in mice. Horm Behav 21:393–401

Berridge CW, Dunn AJ (1989) CRF and restraint stress decrease exploratory behavior in hypophysectomized mice. Pharmacol Biochem Behav 34:517–519

Besedovsky HO, Del Rey A, Klusman I, Furukawa H, Monge Arditi G, Kabiersch A (1991) Cytokines as modulators of the hypothalamus-pituitary-adrenal axis. J Steroid Biochem Mol Biol 40:613–618

Beyermann M, Fechner K, Furkert J, Krause E, Bienert M (1996) A single-point slight alteration set as a tool for structure-activity relationship studies of ovine corticotropin releasing factor. J Med Chem 39:3324–3330

Britton DR, Koob GF, Rivier J, Vale W (1982) Intraventricular corticotropin-releasing factor enhances behavioral effects of novelty. Life Sci. 31:363–367

Britton KT, Lee G, Vale W, Rivier J, Koob GF (1986) Corticotropin-releasing factor antagonists block activating and 'anxiogenic' actions of CRF in the rat. Brain Res. 369:303–306

Brown MR, Fisher LA, Spiess J, Rivier C, Rivier J, Vale W (1982) Corticotropin-releasing factor: actions on the sympathetic nervous system and metabolism. Endocrinology 111:928–931

Buwalda B, Van Kalkeren AA, de Boer SF, Koolhaas JM (1998) Behavioral and physiological consequences of repeated daily intracerebroventricular injection of corticotropin-releasing factor in the rat. Psychoneuroendocrinology 23:205–218

Chalmers DT, Lovenberg TW, De Souza EB (1995) Localization of novel corticotropin-releasing factor receptor (CRF2) mRNA expression to specific subcortical nuclei in rat brain: comparison with CRF1 receptor mRNA expression. J Neurosci 15: 6340–6350

Chalmers DT, Lovenberg TW, Grigoriadis DE, Behan DP and De Souza EB (1996) Corticotropin-releasing factor receptors: from molecular biology to drug design. Trends Pharmacol Sci 17: 166–172

Chang CP, Pearse RV, O'Connell S, Rosenfeld MG (1993) Identification of a seven transmembrane helix receptor for corticotropin-releasing factor and sauvagine in mammalian brain. Neuron 11:1187–1195

Chen R, Lewis KA, Perrin MH, Vale WW (1993) Expression cloning of a human corticotropin-releasing-factor receptor. Proc Natl Acad Sci USA 90:8967–8971

Chou PY, Fasman GD (1978) Empirical prediction of protein conformation. Annu Rev Biochem 47:251–276

Correa, SG, Riera CM, Spiess J, Bianco ID (1997) Modulation of the inflammatory response by corticotropin-releasing factor. Eur J Pharmacol 319:85–90

Crofford LJ, Sano H, Karalis K, Friedman TC, Epps HR, Remmers EF, Mathern P, Chrousos GP, Wilder RL (1993) Corticotropin-releasing hormone in synovial fluids and tissues of patients with rheumatoid arthritis and osteoarthritis. J Immunol 151:1587–1596

Cummings S, Elde R, Ells J, Lindall A (1983) Corticotropin-releasing factor immunoreactivity is widely distributed within the central nervous system of the rat. J Neurosci 3:1355–1368

Cummings S, Sharp B, Elde R (1988) Corticotropin-releasing factor in cerebellar afferent systems: a combined immunohistochemistry and retrograde transport study. J Neurosci 8:543–554

Dathe M, Fabian H, Gast K, Zirwer D, Winter R, Beyermann M, Schumann M and Bienert M (1996) Conformational differences of ovine and human corticotropin releasing hormone – a CD, IR, NMR and dynamic light scattering study. Int J Pept Protein Res 47:383–393

Dautzenberg FM, Dietrich K, Palchaudhuri MR, Spiess J (1997) Identification of two corticotropin-releasing factor receptors from *Xenopus laevis* with high ligand selectivity: unusual pharmacology of the type 1 receptor. J Neurochem 69:1640–1649

Dautzenberg FM, Wille S, Lohmann R, Spiess J (1998) Mapping of the ligand-selective domain of *Xenopus laevis* CRF receptor: implications for the ligand-binding site. Proc Natl Acad Sci USA 95:4941–4946

De Souza EB (1995) Corticotropin-releasing factor receptors: physiology, pharmacology, biochemistry and role in central nervous system and immune disorders. Psychoneuroendocrinology 20:789–819

De Souza EB, Insel TR, Perrin MH, Rivier J, Vale W, Kuhar MJ (1985) Corticotropin releasing factor receptors are widely distributed within the rat central nervous system: an autoradiographic study. J Neurosci 5:3189–3203

Diamant M, de Wied D (1993) Structure-related effects of CRF and CRF-derived peptides: dissociation of behavioral, endocrine and autonomic activity. Neuroendocrinology 57:1071–1081

Dinarello CA (1994) The interleukin-1 family: 10 years of discovery. FASEB J 8:1314–1325

Donaldson C, Sutton SW, Perrin MH, Corrigan AZ, Lewis KA, Rivier JE, Vaughan JM, Vale WW (1996) Cloning and characterization of human urocortin. Endocrinology 137:2167–2170

Dunn AJ, File SE (1987) Corticotropin-releasing factor has an anxiogenic action in the social interaction test. Horm Behav 21:193–202

Dunn AJ, Berridge CW (1990) Physiological and behavioral responses to corticotropin-releasing factor administration: is CRF a mediator of anxiety or stress responses? Brain Res Rev 15:71–100

Esch F, Ling N, Bohlen P, Baird A, Benoit R, Guillemin R (1984) Isolation and characterization of the bovine hypothalamic corticotropin-releasing factor. Biochem Biophys Res Commun 122:899–905

Fendt M, Koch M, Schnitzler H-U (1996) Lesions of the central gray block conditioned fear as measured with the potentiated startle paradigm. Behav Brain Res 74:127–134

Fendt M, Koch M, Schnitzler H-U (1997) Corticotropin-releasing factor in the caudal pontine reticular nucleus mediates the expression of fear-potentiated startle in the rat. Eur J Neurosci 9:299–305

Fisher LA (1989) Corticotropin-releasing factor: endocrine and autonomic integration of responses to stress. Trends Pharmacol Sci 10:189–193

Fox EA, Gruol DL (1993) Corticotropin releasing factor suppresses the afterhyperpolarization in cerebellar Purkinje neurons. Neurosci Lett 149:103–107

Gosnell BA, Morley JE, Levine AS (1983) A comparison of the effects of corticotropin-releasing factor and sauvagine on food intake. Pharmacol Biochem Behav 19:771–775

Gulyas J, Rivier C, Perrin M, Koerber SC, Sutton S, Corrigan A, Lahrichi SL, Craig AG, Vale W, Rivier J (1995) Potent, structurally constrained agonists and competitive antagonists of corticotropin-releasing factor. Proc Natl Acad Sci USA 92:10575–10579

Hagan P, Poole S, Bristow A (1992) Immunosuppressive activity of CRF. Inhibition of IL-1 and IL-6 production by human mononuclear cells. Biochem J 281:251–254

Hargreaves KM, Costello AH, Joris JL (1989) Release from inflamed tissue of a substance with properties similar to corticotropin-releasing factor. Neuroendocrinology 49:476–482

Heinrichs SC, Pich EM, Miczek KA, Britton KT, Koob GF (1992) Corticotropin-releasing factor antagonist reduces emotionality in socially defeated rats via direct neurotropic action. Brain Res 581:190–197

Heinrichs SC, Vale EA, Lapsansky J, Behan DP, McClure LV, Ling N, De Souzza EB, Schulteis G (1997) Enhancement of performance in multiple learning tasks by corticotropin-releasing factor-binding protein ligand inhibitors. Peptides 18:711–716

Irwin M, Hauger RL, Brown M, Britton KT (1988) CRF activates autonomic nervous system and reduces natural killer cytotoxicity. Am J Physiol 255: R744–747

Jingami H, Mizuno N, Takahashi H, Shibahara S, Furutani Y, Imura H, and Numa S (1985) Cloning and sequence analysis of cDNA for rat corticotropin-releasing factor precursor. FEBS Lett 191:63–66

Karalis K, Sano H, Redwine J, Listwak S, Wilder RL Chrousos GP (1991) Autocrine or paracrine inflammatory actions of corticotropin-releasing factor in vivo. Science 254:241–423

Kishimoto T, Pearse RV, Lin CR, Rosenfeld MG (1995) A sauvagine/corticotropin-releasing factor receptor expressed in heart and skeletal muscle. Proc Natl Acad Sci USA 92:1108–1112

Koob GF, Bloom FE (1985) Corticotropin-releasing factor and behavior. Fed Proc 44:259–263

Koob GF, Britton KT (1990) Behavioral effects of corticotropin-releasing factor. In: De Souza EB, Nemeroff CB (eds) Corticotropin-releasing factor: basic and clinical studies of a neuropeptide. CRC Press, Boca Raton, pp 275–289

Kornreich WD, Galyean R, Hernandez JF, Craig AG, Donaldson CJ, Yamamoto G, Rivier C, Vale W, Rivier J (1992) Alanine series of ovine corticotropin releasing factor oCRF: a structure-activity relationship study. J Med Chem 35:1870–1876

Kozicz T, Yanaihara H, Arimura A (1998) Distribution of urocortin-like immunoreactivity in the central nervous system of the rat. J Comp Neurol 391:1–10

Krahn DD, Gosnell BA, Grace M, Levine AS (1986) CRF antagonist partially reverses CRF- and stress-induced effects on feeding. Brain Res Bull 17:285–289

Krause E, Beyermann M, Dathe M, Rothemund S, Bienert M (1995) Location of an amphipathic alpha-helix in peptides using reversed phase HPLC retention behavior of D-amino acid analogs. Anal Chem 67:252–258

Lederis K, Letter A, McMaster D, Moore G (1982) Complete amino acid sequence of urotensin-I, a hypotensive and corticotropin-releasing neuropeptide from Catostomus. Science 218: 162–164

Lee Y, Davis M (1997) Role oF hippocampus, the bed nucleus of the stria terminalis, and the amygdala in the excitatory effect of corticotropin-releasing hormone on the acoustic startle reflex. J Neurosci 17:6434–6446

Liang KC and Lee EHY (1988) Intra-amygdala injections of corticotropin releasing factor facilitate inhibitory avoidance learning and reduce exploratory behavior in rats. Psychopharmacology 96:232–236

Liaw CW, Lovenberg TW, Barry G, Oltersdorf T, Grigoriadis DE, De Souza EB (1996) Cloning and characterization of the human corticotropin-releasing factor-2 receptor complementary deoxyribonucleic acid. Endocrinology 137:72–77

Liaw CW, Grigoriadis DE, Lovenberg TW, De Souza EB, Maki RA (1997) Localization of ligand-binding domains of human corticotropin-releasing factor receptor: a chimeric receptor approach. Mol Endocrinol 11:980–985

Liebsch G, Landgraf R, Gerstberger R, Probst JC, Wotjak CT, Engelmann M, Holsboer F, Montkowski A (1995) Chronic infusion of a CRH-1 receptor antisense oligodeoxynucleotide into the central nucleus of the amygdala reducedanxiety-related behavior in socially defeated rats. Regulatory Peptides 59:229–239

Lovenberg TW, Liaw CW, Grigoriadis DE, Clevenger W, Chalmers DT, De Souza EB, Oltersdorf T (1995a) Cloning and characterization of a functionally disof the amygdala reduced anxiety-related behavior in socially defeated rats. Regulatory Peptides 59:229–239

Lovenberg TW, Liaw CW, Grigoriadis DE, Clevenger W, Chalmers DT, De Souza EB, Oltersdorf T (1995a) Cloning and characterization of a functionally distincy corticotropin-releasing factor receptor subtype from rat brain Proc Natl Acad Sci USA 92:836–840

Lovenberg TW, Chalmers T, Liu C, De Souza EB (1995b) CRF2α and CRF2β receptor mRNAs are diverentially distributed between the rat central nervous system and peripheral tissues. Endocrinology 136:4139–4142

Madison DV, Nicoll RA (1982) Noradrenaline blocks accommodation of pyramidal cell discharge in the hippocampus. Nature 299:636–638

McGaugh JL (1983) Hormonal influences on memory. Annu Rev Psychol 27:297–323

McGillis JP, Park A, Rubin-Fletter P, Turck C, Dallman MF Payan DG (1989) Stimulation of rat B-lymphocyte proliferation by corticotropin-releasing factor. J Neurosci Res 23:346–352

Merali Z, McIntosh J, Kent P, Michaud D, Anisman H (1998) Aversive and appetitive events evoke the release of corticotropin releasing hormone and bombesin-like peptides at the central nucleus of the amygdala. J Neurosci 18:4758–4766

Miranda A, Lahrichi SL, Gulyas J, Koerber SC, Craig AG, Corrigan A, Rivier C, Vale W, Rivier J (1997) Constrained corticotropin-releasing factor antagonists with I-(I+3) Glu-Lys bridges. J Med Chem 40:3651–3658

Mitchell AJ (1998) The role of corticotropin-releasing factor in depressive illness: a critical review. Neurosci Biobehav Rev 22:635–651

Moga MM, Gray TS (1985) Evidence for corticotropin-releasing factor, neurotensin and somatostatin in the neural pathway from the central nucleus of the amygdala to the parabrachial nucleus. J Comp Neurol 241:275–284

Montecucchi PC, Henschen A (1981) Amino acid composition and sequence analysis of sauvagine, a new active peptide from the skin of *Phyllomedusa sauvagei*. Int J Pept Protein Res 18:113–120

Moreau JL, Kilpatrick G, Jenck F (1997) Urocortin, a novel neuropeptide with anxiogenic-like properties. Neuroreport 8:1697–1701

Mousa SA, Schafer M, Mitchell WM, Hassan AHS, Stein C (1996) Local upregulation of corticotropin-releasing hormone and interleukin-1 receptors in rats with painful hindlimb inflammation. Eur J Pharmacol 311:221–231

Muglia L, Jacobson L, Dikkes P, Majzoub JA (1995) Corticotropin-releasing hormone deficiency reveals major fetal but not adult glucocorticoid need. Nature 373:427–432

Olschowka JA, O'Donohue TL, Mueller GP, Jacobowitz DM (1982) The distribution of corticotropin releasing factor-like immunoreactive neurons in rat brain. Peptides 3:995–1015

Owens MJ, Nemeroff CB (1991) Physiology and pharmacology of corticotropin-releasing factor. Pharmacol Rev 43:425–473

Patty M, Horvath J, Mason-Garcia M, Szoke B, Schlesinger DH, Schally AV (1985) Isolation and amino acid sequence of corticotropin-releasing factor from pig hypothalami. Proc Natl Acad Sci USA 82:8762–8766

Pereda MP, Sauer J, Castro CP, Finkeielman S, Stalla GK, Holsboer F, Arzt E (1995) Corticotropin-releasing hormone differentially modulates the interleukin-1 system according to the level of monocyte activation by endotoxin. Endocrinology 136:5504–5510

Perrin M, Haas Y, Rivier J, Vale W (1986) Corticotropin-releasing factor binding to the anterior pituitary receptor is modulated by divalent cations and guanine nucleotides. Endocrinology 118:1171–1179

Perrin M, Donaldson C, Chen R, Blount A, Berggren T, Bilezikjian L, Sawchenko P, Vale W (1995) Identification of a second corticotropin-releasing factor receptor gene and characterization of a cDNA expressed in heart. Proc Natl Acad Sci USA 92:2969–2973

Perrin MH, Sutton S, Bain DL, Berggren WT, Vale WW (1998) The first extracellular domain of corticotropin releasing factor-R1 contains major binding determinants for urocortin and astressin. Endocrinology 129:566–570

Poliak S, Mor F, Conlon P, Wong T, Ling N, Rivier J, Vale W, Steinman L (1997) The neuropeptides corticotropin-releasing factor and urocortin suppress encephalomyelitis via effects on both the hypothalamic-pituitary-adrenal axis and the immune system. J Immunol 158: 5751–5756

Potter E, Behan DP, Fischer WH, Linton EA, Lowry PJ, Vale WW (1991) Cloning and characterization of the cDNAs for human and rat corticotropin releasing factor-binding proteins. Nature 349:423–426

Potter E, Behan DP, Linton EA, Lowry PJ, Sawchenko PE, Vale WW (1992) The central distribution of a corticotropin releasing factor (CRF)-binding protein predicts multiple sites and modes of action. Proc Natl Acad Sci USA 89:4192–4196

Potter E, Sutton S, Donaldson C, Chen R, Perrin M, Lewis K, Sawchenko PE, Vale W (1994) Distribution of corticotropin-releasing factor receptor mRNA expression in the rat brain and pituitary. Proc Natl Acad Sci USA USA 91:8777–8781

Qi LJ, Leung AT, Xiong Y, Marx KA, Abou-Samra AB (1997) Extracellular cysteines of the corticotropin-releasing factor receptor are critical for ligand interaction. Biochemistry 36: 12442–12448

Radulovic J, Sydow S, Spiess J (1998) Characterization of native corticotropin-releasing factor receptor type 1 (CRF-R1) in the rat and mouse central nervous system. J Neurosci Res 54: 507–521

Radulovic M, Spiess J (1998) Effect of immobilization stress on the corticotropin releasing factor receptor 1 (CRF-R1) production by splenic neutrophils. J Neuroimmunol 90:43

Radulovic M, Dautzenberg F M, Sydow S, Radulovic J, Spiess J (1999) Corticotropin-releasing factor receptor 1 in mouse spleen: expression after immune stimulation and identification of receptor-bearing cells. J Immunol 162:3013–3021

Rainnie DG, Fernhout BJH, Shinnick Gallagher P (1992) Differential actions of corticotropin releasing factor on basolateral and central amygdaloid neurons *in vitro*. J Pharmacol Exp Ther 263:846–858

Rassnick S, Heinrichs SC, Britton KT, Koob GF (1993) Microinjection of a corticotropin-releasing factor antagonist into the central nucleus of the amygdale reverses anxiogenic-like effects of ethanol withrawal. Brain Res 605:25–32

Richter RM, Mulvany MJ (1995) Compoarison of hCRF and oCRF effects on cardiovascular responses after central, peripheral, and *in vivo* application. Peptides 16:843–849

Rivier J, Spiess J, Vale WW (1983) Characterization of rat hypothalamic corticotropin-releasing factor. Proc Natl Acad Sci USA 80:4851–4855

Rivier J, Rivier C, Galyean R, Miranda A, Miller C, Craig AG, Yamamoto G, Brown M, Vale WW (1993) Single point D-substituted corticotropin-releasing factors analogues effects on potency and physicochemical characteristics. J Med Chem 36:2851–2859

Rivier J, Lahrichi SL, Gulyas J, Erchegyi J, Koerber SC, Craig AG, Corrigan A, Rivier C, Vale WW (1998) Minimal-size, constrained corticotropin-releasing factor agonists with I-(I+3) Glu-Lys and Lys-Gln bridges. J Med Chem 41:2614–2620

Rothemund S, Krause E, Beyermann M, Bienert M (1997) Hydrophobically induced conformation in ovine corticotropin-releasing hormone. Int J Pept Protein Res 50:184–192

Rühmann A, Köpke AKE, Dautzenberg FM, Spiess J (1996) Synthesis and characterization of a photoactivatable analog of corticotropin-releasing factor for specific receptor labeling. Proc Natl Acad Sci USA 93:10609–10613

Rühmann A, Bonk I, Lin CR, Rosenfeld MG, Spiess J (1998) Structural requirements for peptidic antagonists of the corticotropin-releasing factor receptor (CRFR): development of CRFR2β selective antisauvagine-30. Proc Natl Acad Sci USA 95:15264–15269

Salas MA, Brown OA, Perone MJ, Castro MG, Goya RG (1997) Effect of corticotrophin releasing hormone precursor on interleukin-6 release by human mononuclear cells. Clin Immunol Immunopathol 85:35–39

Sapolsky R, Rivier C, Yamamoto G, Plotsky P, Vale W (1987) Interleukin-1 stimulates the secretion of hypothalamic corticotropin-releasing factor. Science 238:522–524

Schafer M, Carter L, Stein C (1994) Interleukin-1ß and corticotropin-releasing factor inhibit pain by releasing opioids from immune cells in inflamed tissue. Proc Natl Acad Sci USA 91:4219–4223

Siggins GR, Gruol DL, Aldenhoff JB, Pittman Q (1985) Electrophysiological actions of corticotropin releasing factor in the central nervous system. Fed Proc 44:237–242

Singh VK (1989) Stimulatory effect of corticotropin-releasing neurohormone on human lymphocyte proliferation and interleukin-2 receptor expression. J Neuroimmunol 23:257–262

Singh VK, Christine-Leu SJ (1990) Enhancing effect of corticotropin-releasing neurohormone on the production of interleukin-1 and interleukin-2. Neurosci Lett 120:151–154

Singh VK, Fudenberg HH (1988) Binding of [125I]corticotropin releasing factor to blood immunocytes and its reduction in Alzheimer's disease. Immunol Lett 18:5–8

Sirinathisinghji DJS, Rees LH, Rivier J, Vale W (1983) Corticotropin-releasing factor is a potent inhibitor of sexual receptivity in the female rat. Nature 305:232–235

Smith BN, Dudek FE (1994) Age-related effects of corticotropin releasing hormone in the isolated CA1 region of rat hippocampal slices. J Neurophysiol 72:2328–2333

Smith GW, Aubry J-M, Dellu F, Contarino A, Bilezikjian LM, Gold LH, Chen R, Marchuk Y, Hauser C, Bentley CA, Sawchenko PE, Koob GF, Vale W, Lee K-F (1998) Corticotropin releasing factor receptor 1-deficient mice display decreased anxiety, impaired stress response, and aberrant neuroendocrine development. Neuron 20:1093–1102

Sperle K, Chen A, Kostich W, Largent BL (1997) CRH-2γ: a novel CRH2 isoform found in human brain. Proc Neurosci Abstr 23, Part 2:1765

Spiess J, Rivier J, Rivier C, Vale WW (1981) Primary structure of corticotropin-releasing factor from ovine hypothalamus. Proc Natl Acad Sci USA 78:6517–6521

Spiess J, Rivier J, Vale WW (1983) Sequence analysis of rat hypothalamic corticotropin-releasing factor with the orthophthalaldehyde strategy. Biochemistry 22:4341–4346

Spiess J, Dautzenberg FM, Sydow S, Hauger RL, Rühmann A, Blank T, Radulovic J (1998) Molecular properties of the CRF receptor. Trends Endocrinol Metab 9:140–145

Spina M, Merlo-Pich E, Chan RK, Basso AM, Rivier J, Vale W, Koob GF (1996) Appetite-suppressing effects of urocortin, a CRF-related neuropeptide. Science 273:1561–1564

Stenzel-Poore MP, Heldwein KA, Stenzel P, Vale WW (1992) Characterization of the genomic corticotropin-releasing factor (CRF) gene form Xenopus laevis: two members of the CRF family exist in amphibians. Mol Endocrinol 6:1716–1724

Stenzel P, Kesterson R, Yeung W, Cone RD, Rittenberg MB, Stenzel-Poore MP (1995) Identification of a novel murine receptor for corticotropin-releasing hormone expressed in heart. Mol Endocrinol 9:637–645

Stephanou A, Jessop DS, Knight RA, Lightman SL (1990) Corticotropin-releasing factor-like immunoreactivity and mRNA in human leukocytes. Brain Behav Immun 4:67–73

Sutton RE, Koob GF, Le Moal M, Rivier J, Vale W (1982) Corticotropin releasing factor produces behavioral activation in rats. Nature 297:331–333

Sutton SW, Behan DP, Lahrichi, SL, Kaiser R, Corrigan A, Lowry P, Potter E, Perrin MH, Rivier J, Vale WW (1995) Ligand requirements of the human corticotropin-releasing factor binding protein. Endocrinology 136:1097–1102

Swanson LW, Sawchenko PE, Rivier J, Vale WW (1983) Organization of ovine corticotropin-releasing factor immunoreactive cells and fibers in the rat brain: an immunohistochemical study. Neuroendocrinology 36:165–186

Swerdlow NR, Geyer MA, Vale WW, Koob GF (1986) Corticotropin-releasing factor potentiates acoustic startle in rats: blockade by chlordiazepoxide. Psychopharmacology 88:147–152

Swerdlow NR, Britton KT, Koob GF (1989) Potentiation of acoustic startle by corticotropin-releasing factor (CRF) and by fear are both reversed by α-helical CRF (9-41). Neuropsychopharmacology 2:285–292

Swiergiel AH, Takahashi LK, Kalin NH (1993) Attenuation of stress-induced behavior by antagonism of corticotropin-releasing factor receptors in the central amygdala in the rat. Brain Res 623:229–234

Sydow S, Radulovic J, Dautzenberg FM, Spiess J (1997) Structure – function relationship of different domains of the rat corticotropin-releasing factor receptor. Mol Brain Res 52:182–193

Sydow S, Flaccus A, Fischer A, Spiess J (1997) The role of the fourth extracellular domain of the rat corticotropin-releasing factor receptor type 1 in ligand binding. Eur J Biochem 138 (in press)

Tazi A, Dantzer R, Le Moal M, Rivier J, Vale W, Koob GF (1987) Corticotropin-releasing factor antagonist blocks stress-induced fighting in rats. Regul Pept 18:37–42

Thompson RC, Seasholtz AF, Douglass JO, Herbert E (1990) Cloning and distribution of the rat corticotropin-releasing factor (CRF) gene. In: De Souza EB, Nemeroff CB (eds) Corticotropin-releasing factor: basic and clinical studies of the neuropeptide. CRC Press, Boca Raton, pp 1–12

Timpl P, Spanagel R, Sillaber I, Kresse A, Reul JMHM, Stalla GK, Blanquet V, Steckler T, Holsboer F, Wurst W (1998) Impaired stress response and reduced anxiety in mice lacking a functional corticotropin-releasing hormone receptor 1. Nature Genetics 19:162–166

Valdenaire O, Giller T, Breu V, Gottowik J, Kilpatric G (1997) A new functional isoform of the human CRF2 receptor for corticotropin-releasing factor. Biochim Biophys Acta 1352: 129–132

Vale WW, Spiess J, Rivier, C, Rivier J (1981) Characterization of a 41-residue ovine hypothalamic peptide that stimulates secretion of corticotropin and ß-endorphin. Science 213:1394–1397

Vaughan J, Donaldson C, Bittencourt J, Perrin MH, Lewis K, Sutton S, Chan R, Turnbull AV, Lovejoy D, Rivier C, Rivier J, Sawchenko PE, Vale W (1995) Urocortin, a mammalian neuropeptide related to fish urotensin I and to corticotropin-releasing factor. Nature 378:287–292

Veldhuis HD, De Wied D (1984) Differential behavior actions of corticotropin-releasing factor (CRF). Pharmacol Biochem Behav 21:707–713

Vita N, Laurent P, Lefort S, Chalon P, Lelias JM,Kaghad M, Le Fur G, Caput D, Ferrara P (1993) Primary structure and functional expression of mouse pituitary and human brain corticotrophin releasing factor receptors. FEBS Letters 335:1–5

Webster EL, Tracey DE, Jutila MA, Wolfe SA Jr, De Souza EB (1990) Corticotropin-releasing factor receptors in mouse spleen: identification of receptor-bearing cells as resident macrophages. Endocrinology 127:440–452

Wei ET, Gao GC, Thomas HA (1993) Peripheral anti-inflammatory actions of corticotropin-releasing factor. Ciba Found Symp 172:258–268

Wiersma A, Bohus B, Koolhaas JM (1993) Corticotropin-releasing hormone microinfusion in the central amygdala diminishes a cardiac parasympathetic outflow under stress-free conditions. Brain Res 625:219–227

Wong M-L, Al-Shekhleem A, Bongiorno PB, Esposito A, Khatri P, Sternberg EM, Gold PW, Licinio J (1996) Localization of urocortin messenger RNA in rat brain and pituitary. Mol Psychiat 1:307–312

Wynn PC, Hauger RL, Holmes MC, Millan MA, Catt KJ, Aguilera G (1984) Brain and pituitary receptors for corticotropin releasing factor: localization and differential regulation after adrenalectomy. Peptides 5:1077–1084

Yamashita H, Kasai M, Inenaga K (1991) Effects of corticotropin releasing factor on neurons in the hypothalamus paraventricular nucleus in vitro. Brain Res Bull 27:321–325

Yu J, Xie LY, Abou-Samra A-B (1996) Molecular cloning of a type A chicken corticotropin-releasing factor receptor with high affinity for urotensin I. Endocrinology 137:192–197

Neural Oxytocinergic Systems as Genomic Targets for Hormones and as Modulators of Hormone-Dependent Behaviors

Donald W. Pfaff, Sonoko Ogawa, and Lee-Ming Kow

The neuropeptide oxytocin has a wide spectrum of physiological actions, including important autonomic effects in both genders, endocrine effects on pituitary hormones, and behavioral effects. Regarding the latter, oxytocin contributes both directly and indirectly to the neurobiological mechanisms of mating behavior by the female, including the attendant elevation of sexual motivation (reviewed in Pfaff, The MIT Press, 1999). Relying heavily on this reference and on a voluminous recent literature on the molecular biology of oxytocin, its receptor and its actions, we review some aspects of the contributions of oxytocin to reproductive behavior (**Section I**), hormonal controls over the genes for oxytocin and its receptor (**Section II**), and the signal transduction subsequent to oxytocin binding, as subserved by G-proteins (**Section III**).

1
Behavioral Biology of Oxytocin

Oxytocin and vasopressin are neuropeptides which evolved from an ancient precursor, vasotocin, common in fish. Chemically they retain great similarity, seven of their nine amino acids remaining the same, and both neuropeptides retaining the same disulfide bond. Indeed, some of their functions also remain similar, related to the stimulation of smooth muscles, water balance, and major biological events such as giving birth, and responses to stress. If we follow Hans Selye (1974) in defining stress as a non-specific response of the body to any demand made upon it, then we can begin to distinguish the functions of vasopressin from those of oxytocin (Carter and Altemus, 1997). That is, vasopressin (AVP) defends the body against severe stress, by supporting aggressive responses and by enforcing restoration of water balance. Oxytocin protects normal social responses, especially in the female (reviewed in Carter et al., 1997), from interruption by low level, mild stress.

For two peptides which are chemically so similar, the functional differences are surprising. Vasopressin not only facilitates offensive aggression toward an intruder (Ferris and Delville, 1994) in a manner which depends on testosterone (Delville et al., 1996) but it also greatly stimulates aggressive social sig-

The Rockefeller University, 1230 York Avenue, New York, NY 10021

nals such as flank marks (Ferris, 1992). Within the hypothalamus AVP is released in response to emotional stress (Wotjak et al., 1996). Aggressive responses to stress may have derived from the ancient functions of arginine vasotocin, which determined masculine aggressiveness in fish.

In dramatic contrast, oxytocin is the neurohormone of affiliation and love – maternal behavior and positive social responses (Insel and Shapiro, 1992a, 1992b). Cort Pedersen and his colleagues have pointed out that after years of endocrine research on the secretion of oxytocin from the posterior pituitary and its effects in the uterus and mammary gland, oxytocin has now, in a wide variety of forms, been proven to contribute to the facilitation of maternal, sexual and other social behaviors (reviewed in Pedersen et al., 1992). Indeed, referring to "affiliation" as the set of social behaviors which serve to bring individuals closer together, Carter and her co-authors (1997) are almost able to suggest that the very prominence of oxytocin in promoting such behaviors could lead to a significant increase in research on this behavioral topic. The evidence from several zoologic forms and many laboratories on this point has been so overwhelming that entire books have been devoted to its exposition (Pedersen, Caldwell, Jirikowski and Insell, 1992; Carter, Lederhendler, and Kirkpatrick, 1997). For example, estrogen-primed female voles receiving oxytocin in the cerebral ventricles showed more friendly social contacts and reduced aggressive responses when tested with a male partner (reviewed in Carter, DeVries et al., 1997). Emotionally neutral or positive social contacts and investigative responses in both sexes, but especially in the female, depend on oxytocin (Witt, 1997). In fact, oxytocin exerts an anxiolytic action which depends on contemporaneous estrogen treatment (McCarthy et al., 1996, 1997). In doing so, oxytocin protects instinctive behaviors such as maternal and sexual responses from disruption by mild stress (McCarthy et al., 1991).

The protection by oxytocin of adaptive reproductive behaviors may be especially important, since some biologists feel that sexual behaviors set the paradigm for a broad variety of social behaviors. The dependence of certain maternal and reproductive responses on oxytocin carries through to human beings as well. Likewise, vasopressinergic functions are just as important in human biology as they are for water balance and other autonomic functions in experimental animals (Gross, Richter and Robertson, 1993).

The evidence for the involvement of oxytocin in sexual behavior is very strong. Not only did oxytocin infused ICV significantly increase sexual receptivity, but also the independence of this behavioral effect from intermediary changes in the adrenals was demonstrated, as was the requirement for classical oxytocin receptor occupation (Caldwell et al., 1984, 1986). Moreover, Arletti, Bertolini and their colleagues (1985) clearly showed that ICV administration of oxytocin would significantly increase the frequency of lordosis behavior. An oxytocin antagonist would not only block that effect of exogenous oxytocin but also reduced lordosis behavior by rendering endogenous oxytocin ineffective (from Arletti et al., Annals of the New York Academy of

Sciences, 1992). At least two sets of studies, one by Schumacher et al. (PNAS, 1989; Science, 1990), and the other by Gorzalka and Lester (1987), have indicated that the increase in lordosis behavior caused by oxytocin is much more impressive if the female is given not only estradiol, but also a supplementary

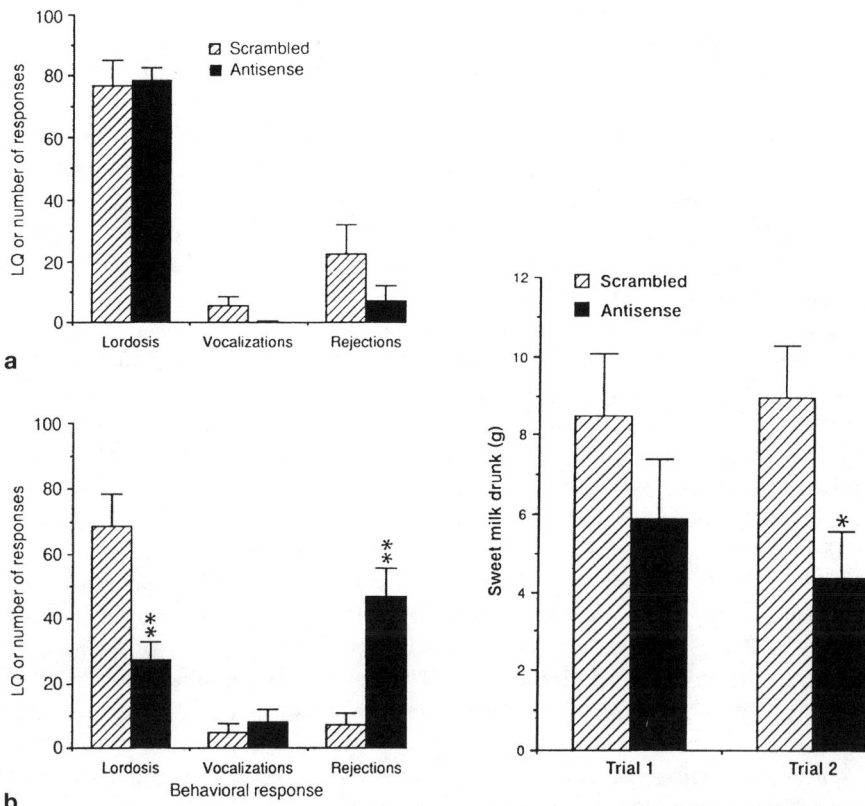

Fig. 1. Effect of antisense DNA directed against oxytocin receptor mRNA microinjected into the ventromedial hypothalamus upon lordosis behavior. *Top left panel* (**a**): When ovariectomized female rats were pretreated not only with estrogen but also with progesterone, to maximize female reproductive behavior, there were no significant effects of the antisense on any behavioral parameter. Thus, the antisense DNA does not simply destroy the neurons responsible for the behavior. *Bottom left panel* (**b**): Two weeks later, the same females were treated with estrogen only. There was a significant reduction in the frequency of lordosis behavior and a corresponding increase in rejection behaviors exhibited by the females infused with antisense DNA (compared to scrambled sequence controls) (From McCarthy et al., 1994). *Right panel:* As a control behavioral measurement, the same female rats which yielded the results in the left panels were tested for carbohydrate appetite by measuring the amount of sweetened condensed milk drank during a 30-minute period. If ventromedial hypothalamic neurons were damaged by the antisense DNA treatment, the amount of sweet milk consumed should go up. That did not occur – instead, there was a downward trend. Therefore, it is unlikely that the reproductive behavioral results were subsequent to neuronal damage. (From McCarthy et al., 1994)

injection of progesterone. Furthermore, female rats are much more sensitive to the facilitating effects of oxytocin on lordosis behavior during the dark phase of the daily light cycle, at a time when normally cycling females would show maximal levels of receptivity due to the amplification of the endogenous estrogen effect by endogenous progesterone (Schumacher et al., Behavioral Neuroscience, 1991). The clear role for progesterone in facilitating oxytocinergic action in the brain on mating behavior contrasts markedly with the ability of progesterone acting in the uterus to inhibit oxytocinergic signaling (Grazzini et al., 1998). A likely explanation is that differential coupling of the oxytocin receptor to different G-proteins in uterus (See Section II), compared to brain, allows for a differential action of progesterone on oxytocinergic signaling. Precedent for this suggestion is found in the work of Zou et al. (Circulation Research, 1998) in which cardiac myocytes and fibroblasts respond differently to a single ligand because of the activation of different G-protein pathways (see Section III, below).

Given the obvious behavioral interest of oxytocinergic actions in the brain, two major questions are raised. How are the expression of the genes for oxytocin and the oxytocin receptor controlled? With respect to neuroendocrine controls, this question is addressed in Section II. Finally, how does oxytocin execute its actions in the brain to achieve these behavioral effects? This question is approached below at the cellular (Section III A) and the neural systems (Section III B) levels, below.

Molecular evidence for the necessity of oxytocin receptor gene expression in the support of female reproductive behavior is especially clear. Antisense DNA against oxytocin receptor messenger RNA significantly reduced lordosis behavior (Fig. 1) if it was delivered to the ventromedial hypothalamus (McCarthy et al., Neuroendocrinology, 1994). The reversibility of the effect and the inclusion of other control behavioral responses showed that the antisense DNA effect was not due to simple damage of hypothalamic neurons. In agreement with the antisense DNA results, an oxytocin receptor antagonist decreases estrogen-stimulated lordosis behavior (Witt and Insel, 1991, 1992).

2
Hormonal Controls over Genes for Oxytocin and its Receptor

Both oxytocin and vasopressin genes were cloned by Professor Dietmar Richter and his colleagues at the University of Hamburg. Gene expression for oxytocin and for the oxytocin receptor are both elevated in specific subsets of neurons by estradiol (For references see Dellovade et al., 1998; Bale and Dorsa, 1997; Quinones-Jenab et al., 1997). Since both the ligand and its receptor are increased, the two estrogen effects should *multiply* each other for a powerful behavioral effect.

Oxytocin. Even though large-scale dissections of hypothalamic tissue failed to show a large hormone effect on oxytocin mRNA, a more painstaking cell-by-cell approach using *in situ* hybridization revealed a significant estrogen effect in a subset of anterior hypothalamic and paraventricular nucleus (PVN) neurons (Chung, et al., 1991 and manuscript submitted). In particular, estrogen increased oxytocin mRNA in the portion of PVN with cells projecting to the lower brain stem and spinal cord, that is, the cell group most likely to be relevant for the control of behavior. Moreover, work with thyroidectomized rats demonstrated an important additional condition upon the estrogen effect. High thyroid hormone levels militated against an estrogen effect, whereas it was robust in thyroidectomized females (Fig. 2) (Dellovade, et al. Journal of Neuroendocrinology, 1998). The estrogen effect is likely to depend on classical nuclear binding in the nerve cell nucleus, since isotopically-labeled estradiol is retained by certain oxytocinergic neurons (Rhodes et al., 1981a, 1981b).

PVN: OVX Rats

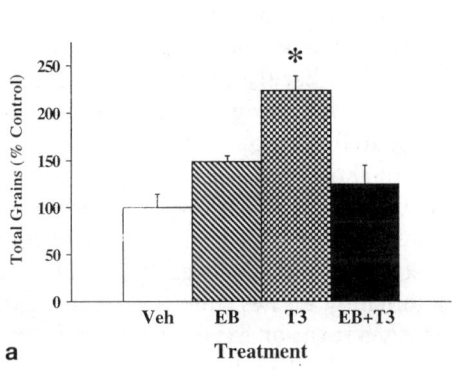

a

Fig. 2. Measurements of oxytocin messenger RNA, using in situ hybridization histochemistry. Estradiol benzoate treatment (EB) of ovariectomized (OVX) female rats was able to significantly increase oxytocin mRNA when thyroid hormone levels had been reduced by thyroidectomy (TX) *(Panel B)*, but not when normal, high thyroid hormone levels were present *(Panel A)*. Thus, endogenous thyroid hormones, presumably working through liganded thyroid hormone receptors, can interfere with the estrogen receptor effect. Likewise, experimentally administered thyroid hormone (T3) can virtually abolish the estrogen effect (Black bar in Panel B). These mRNA results find their parallel in reproductive behavioral results. (Dellovade et al., Journal of Neuroendocrinology, 1998)

PVN: TX/OVX Rats

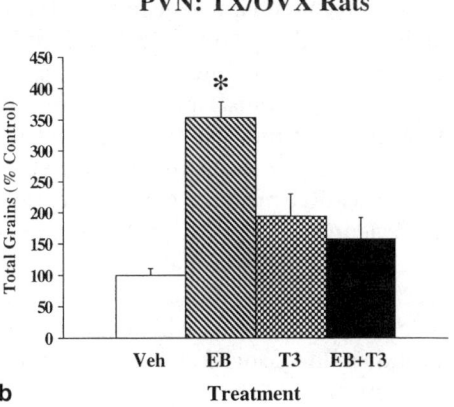

b

According to recent histochemical results, the effect would depend on ER-β (Shughrue et al., 1997).

Oxytocin receptor. A long series of neurochemical experiments showed that estrogen could increase oxytocin binding in the ventromedial hypothalamus, precisely where it could act to facilitate oxytocinergic promotion of lordosis behavior. The pioneering work of Jack Elands with Professor Ron DeKloet (DeKloet et al., 1986) demonstrating this point was extended by H. Coirini, M. Schumacher and Al Johnson in the McEwen lab at Rockefeller (Coirini, et al, 1992, 1989; Schumacher, 1989, 1990; Johnson et al., 1989a, 1989b). Especially since the promoter of the rat oxytocin receptor gene revealed a functional estrogen response element (Bale and Dorsa, 1997), it was reasonable to look for a hormone effect on oxytocin receptor gene expression in hypothalamic neurons. Vanya Quiñones-Jenab et al. (Neuroendocrinology, 1997) used *in situ* hybridization to do so, and found a significant estrogen stimulation of this gene in the VMH. By RT-PCR assay, as well, estradiol treatment, but not progesterone, led to large increases in oxytocin receptor mRNA in the ventromedial hypothalamus (Breton and Zingg, 1997). The cloning of the oxytocin receptor by Kimura and his colleagues (this volume, Chapter X) opens the way toward more detailed molecular studies of the oxytocin receptor in brain as well as in other tissues, for example, the uterus (see below).

Therefore, estrogen effects on the ligand, oxytocin, and on its receptors, (measured both by molecular (Quiñones-Jenab, 1997) and by electrophysiological (Kow et al., 1991, (see Fig. 3)) methods would *multiply* each other, in theory, to achieve an even greater behavioral impact (Pfaff, 1988). Extending that notion, the potential for an series of multiplicative mechanisms in the oxytocin neuronal system has been pointed out (Pfaff, 1988). Oxytocin predominantly excites other oxytocin neurons (Yamashita et al., 1987). Because estrogen turns on oxytocinergic neuronal electrical activity (but not vasopressinergic cells) (Akaishi and Sakuma, 1985), the hormone should actually set off a self-reexciting neuronal mechanism of considerable quantitative import.

Multiplicative hormone effects are not limited to oxytocin. Massive estrogen effects on enkephalin gene expression (Romano, et al., 1988; Lauber et al., 1990; cf. Priest et al., 1996) would be multiplied by estrogen effects on gene expression for the delta opioid receptor, through which enkephalins act (Quiñones-Jenab et al., manuscript in preparation). Likewise, long and strong estrogen treatment can increase GnRH mRNA levels (Rothfeld et al, 1989; Roberts et al, 1989; Rosie, Thomson and Fink, 1990), at the same time as estrogen stimulation of GnRH receptor gene expression has been reported (Quiñones-Jenab, 1996; Jennes, 1997). All of these systems should serve to provide molecular mechanisms for stimulating estrogen-dependent reproductive behaviors. In line with Section I, above, the behavioral actions of oxytocin should be subject to two qualifications, which mirror each other: (1) that the oxytocin effect would facilitate social encounters by reducing the anxiety inherent in meeting the opposite sex (McCarthy et al., 1991; McCarthy et al., 1996); and, converse-

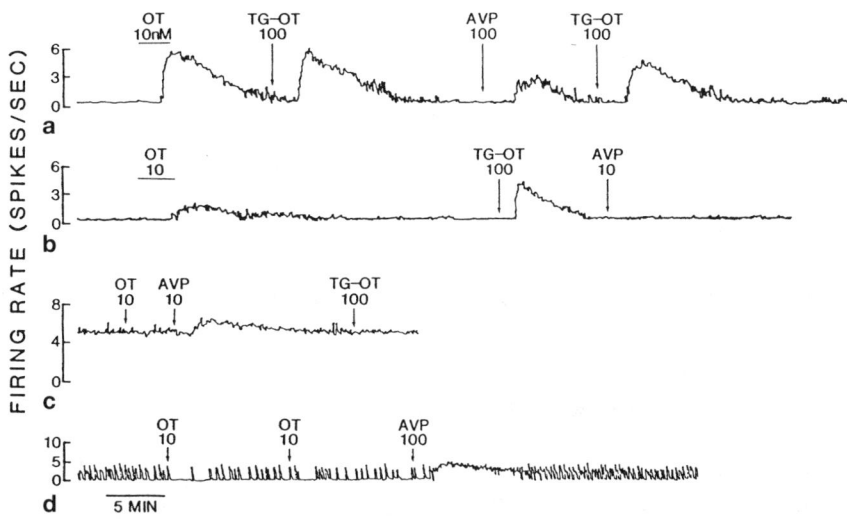

Fig. 3. Electrophysiological responses of ventromedial hypothalamic neurons to oxytocin and other agents. The results allow comparisons between neuronal responses to OT, TG-OT and AVP, for the purpose of inferring that the electrophysiological results are mediated by the classical oxytocin receptor. While OT and TG-OT always affected the same neuron in the same way (**A** and **B** with excitation, and **C** with no action), AVP effects could be similar (**A**), different (**B** and **C**), or even opposite (**D**) to those of OT and TG-OT. (From Kow et al., 1991)

ly, (2) that the oxytocin effect is limited to circumstances which provoke such anxiety (deriving from Fahrbach et al., 1986).

Do these neurobiological findings in experimental animals have any relation to human biology? The most primitive biological roots of human sex drive depend on the forebrain mechanisms, as argued in a current book (Pfaff, 1999) and previously. It is most interesting, then, that cellular and neurochemical features of these oxytocin neurons seem to have been conserved in the human brain. Their anatomical organization is comparable in the human brain to that in non-human species (Sukhov et al., 1993), and the estrogenic stimulation of oxytocin release remains in place in the human brain, as well (Bossmar et al., 1995).

In summary, estrogen turns on a potentially multiplicative oxytocinergic system which drives female reproductive behavior. Moreover, at least in mice, the oxytocin receptor system is strong in the female but not the male (Chritin, M. et al., 1996). In dramatic contrast to the positive estrogen effect, severe stress such as releases opioids, can, under some circumstances, inhibit oxytocin neuronal activity (John Russell et al., 1992). For example, naloxone, an opioid antagonist, impressively raised the electrical activity of oxytocin neurons, relieving them from their morphine-caused inhibition of electrical discharge, as one instance of how stress-caused opioid inputs could reduce oxytocin release (Bicknell et al., 1988). Under other circumstances, in rats, mild stress can, instead, cause oxytocin release. Overall, the pattern of results with estrogen, oxytocin and its receptor shows a pleasing isomorphism between molecular biology and reproductive behavior. And to date, many results fit the theory that oxytocin release serves to protect a broad variety of social responses from disruption due to mild stress.

3
Means of Execution of Oxytocinergic Actions

3.1
Oxytocin Mechanisms at the Cell Membrane Level

The signal transduction pathways used by oxytocin in its effects on behavior, particularly the lordosis response, are of great interest. This is in part because virtually all neurochemical agents which operate upon hypothalamic neurons with the effect of increasing lordosis behavior have been shown to work through phospholipase C, with a consequent increase in inositol triphosphate, in sharp contrast to other agents which work through adenylate cyclase (- review by Lee-Ming Kow in Neuroscience and Biobehavioral Reviews, 1994). Signal transduction systems implicated in oxytocinergic signaling, with an emphasis on the G-proteins involved, comprise a relatively well-developed and quickly growing area of research (reviewed in Sanborn et al., 1995). Strakova and Soloff (1997) have used pregnant rat myometrial membranes incubated with iodinated oxytocin antagonist, followed by detergent solubilization and size selection to follow the coupling of the oxytocin receptor. Using antibodies to various possible partners of the oxytocin receptor in the membrane, they showed an association with phospholipase C beta-1, as well as with significant amounts of the proteins G alpha (q/11) and G alpha (i/3). Their overall pattern of results was consistent with interactions between the oxytocin receptor and both Gq and Gi. In a different type of cell, human granulosa cells, Lopez Bernal et al. (1995) used immunoblotting with specific antibodies to show the expression of several G-proteins: G alpha-s, alpha (i/3), alpha (i/1), and alpha (q/11). Since these cells responded to hCG with a dose-dependent increase in cyclic AMP formation, a functional activation of G alpha-s, can be inferred. Most important, oxytocin stimulated the formation of inositol phosphates and increased intracellular calcium concentration which was partly resistant to treatment by pertussis toxin (PTX), providing evidence of G alpha (q/11) activation. However, part of the response to oxytocin could be blocked by pertussis toxin, indicating G-i-mediated phospholipase C activation as well.

With respect to the dynamics of oxytocin binding itself, Monga et al. (1996) used smooth muscle cells isolated from the myometrium of a pregnant woman for whole cell binding assays whose results were best summarized with a one-site binding model. These cells responded to oxytocin with an increase in intracellular free calcium and an increase in phosphatidylinositol turnover. Oxytocin-stimulated PI turnover was inhibited by a cyclic AMP blocker. Oxytocinergic signal transduction can be sharply distinguished from that of other signals such as epidermal growth factor (Anwer et al., 1996). In human myometrial cells, the pattern of pharmacological blocking agents which were effective on these two pathways were distinctly different.

Despite the Monga et al. (1996) result indicating one binding site on the receptor, in oxytocinergic signal transduction there is plenty of room for multiple mechanisms which might offer important opportunities for tissue specificity in oxytocin signaling. With respect to multiple opportunities for integration at the membrane level, the phenomenon of desensitization clearly involves a down-regulation of oxytocin receptor messenger RNA (Phaneuf et al., 1997) as well as down-regulation at post-receptor levels. Likewise, several intracellular domains in the oxytocin receptor are involved in its coupling to G alpha (q/11) (Qian et al., 1998). That is, examining the effects of co-expression of oxytocin receptor intracellular domains on oxytocin-stimulated phosphoinositide turnover in COS cells, several intracellular segments, the OTR 3i loop and others, were effective. There is also plenty of room for tissue specificity due to the multiple G-proteins and phospholipase C isoforms implicated in oxytocin action, in the results of Phaneuf et al. (1996). That is, by western blotting, G alpha-q, alpha-11, alpha-i1, alpha-i2, alpha-i3, alpha-z and certain splice variants were found expressed in human myometrial cells; and no less than five phospholipase C isoforms were discovered. A pattern of pharmacological blocking results suggested, for example, that in these cells oxytocin can activate phospholipase C beta by interacting with at least two types of G-proteins: a member of the Gq family which is PTX resistant, and a member of the Gi family, which is PTX sensitive (Phaneuf et al., 1996).

The effects of estrogen, of primary interest for behavioral functions mediated in brain, likewise can be expressed in more than one way. Clearly, estrogen leads to increased oxytocin receptor messenger RNA levels and oxytocin binding in both uterus and brain (See II). In addition, however, Phaneuf and his colleagues (1995) followed the effects of estradiol on the oxytocin-enhanced formation of inositol phosphates in human myometrial cells. Surprisingly, when phospholipase- C was activated with PGF 2 alpha or fluoroaluminate instead of oxytocin, estradiol and tamoxifen had the same stimulatory and inhibitory effects, respectively, as if oxytocin itself had been the primary stimulating agent. These results suggested to the authors that estradiol effects on oxytocinergic signaling and the signaling of other agents in the myometrium could in part involve phospholipase C activation at a post-receptor level.

Of special interest, the emphasis on G-protein activation, and subsequent phospholipase C stimulation by G alpha (q/11) (Ku et al., 1995) in myometrium is consonant with our neuropharmacological results in hypothalamus (Kow et al., 1997).

3.2
Mechanisms at the Systems Level

In Section I, above, we saw that oxytocin working through the oxytocin receptor, could foster reproductive behavior by facilitating lordosis. At the systems level, is that the only mechanism involved? The answer is a resounding "no", because data collected over the past fifteen years have made it clear that oxytocin action in the CNS can protect the instinctive behaviors associated with reproduction from the disruptive effects of mild stress.

During certain forms of mild stress, oxytocin is released. Endogenous oxytocin facilitates maternal behavior, as shown by the use of oxytocin antagonists (Fahrbach et al., Neuroendocrinology, 1985). The effect of oxytocin deposition is to maintain normal behavior responses under conditions that, without oxytocin, would be susceptible to stress disturbance. That is, during maternal behavior tested in the familiar environment, the home cage, oxytocin is not necessary. At the other extreme, under quite stressful conditions, oxytocin is not strong enough to protect maternal behavior responses. But in between, with a fixed period of adaptation to a novel environment, the female to be tested may be stressed, but under these circumstances oxytocin maintains maternal behavior (Fahrbach et al., Physiology and Behavior, 1986). In fact, oxytocin can act as an anxiolytic agent in female mice, but only in the presence of estradiol (which presumably facilitates synthesis of the oxytocin receptor) (McCarthy et al., Physiology and Behavior, 1996).

Mice with a "knockout" of the oxytocin gene shed light upon this aspect of oxytocinergic systems. In these experiments, parental behaviors were much more severely affected by stress in male mice than in female mice. Parental behaviors were not different between genotypes in the absence of stress, but they were greatly reduced in both the oxytocin gene knockout animals and in the heterozygotes (Pfaff et al., Soc. for Neuroscience, 1998). Males with the oxytocin gene knocked out also showed significantly higher anxiety in the light/dark transition test. As well, after one hour of restraint stress, activity in the light portion of the chamber was significantly reduced in both sexes of oxytocin gene knockout mice, but not wild-type mice. All of the data suggested that oxytocin regulates instinctive behavioral responses in stressful situations (Pfaff et al., Soc. for Neuroscience, 1998).

It has long been hypothesized, since the paper of Meisenberg (1981), that the behavioral effect of oxytocin administration might be dependent upon the environment in which the behavior was tested. Following on from that idea, McCarthy et al. (1991) hypothesized that the effect of oxytocin would vary according to the level and type of stress at the time of testing: that the level of stress would be an underlying factor which would govern the effect of exogenously applied oxytocin. As McCarthy et al. (1991) stated, lower animals and humans face varying levels of stress almost all the time. Consistent with the data above, therefore, one can state that the systematic role of oxytocin is to overcome the "fear" associated with mild environmental disturbance. The effect of oxytocin, in turn, would depend upon the stimulus environment: in the presence of pups, maternal behavior would be protected from the effects of stress. In the presence of a mating partner, sexual behavior would be protected from the effect of stress. According to this notion in its most general form, the most fundamental neural function of oxytocin would be to mediate responses to mild stress. It initiates a neural state which allows appropriate, specific behavioral responses to particular environmental stimuli.

4
Summary

At the molecular level, estradiol turns on the gene for oxytocin in a subset of paraventricular hypothalamic neurons and turns on the gene for the oxytocin receptor in other limbic and hypothalamic cell groups. As a result, oxytocin deposition, whose signal is transduced both through G alpha (q/11) and Gi to stimulate phosphatidylinositol turnover, facilitates electrical activity in certain hypothalamic neurons. Consequently, affiliative behaviors including those closely associated with reproduction – mating behaviors and parental behaviors – are promoted. One important aspect of this effect is the preservation of instinctive behaviors associated with reproduction, in the face of disturbances due to mild stress.

References

Akaishi T, Sakuma Y (1985) Oestrogen excites oxytocinergic but not vasopressinergic cells in the paraventricular nucleus of the female rat. Brain Res 335:302–305

Anwer K, Monga M, Sanborn BM (1996) Epidermal growth factor increases phosphoinositide turnover and intracellular free calcium in an immortalized human myometrial cell line independent of the arachidonic acid metabolic pathway. American Journal of Obstetrics and Gynecology 174(2):676–681

Arletti R, Bertolini A (1985) Oxytocin stimulates lordosis behavior in female rats. Neuropeptides 6:247–253

Arletti R, Benelli A, Bertolini A (1992) Oxytocin involvement in male and female sexual behavior. Annals, N.Y. Acad Sci 652:180–193

Bale T, Dorsa D (1997) Cloning, Novel promoter sequence, and estrogen regulation of a rat oxytocin receptor gene. Endocrinology 138:1151–1158

Bicknell RJ, Leng G, Lincoln DW, Russell JK (1988) Naloxone excites oxytocin neurons in the supraoptic nucleus of lactating rats after chronic morphine treatment. Journal of Physiology 396:297–31

Bossmar T, Forsling M, Akerlund M (1995) Circulating oxytocin and vasopressin is influenced by ovarian steroid replacement in women. Acta Obstetricia et Gynecolgica Scandinavica 74:544–548

Breton C, Zingg HH (1997) Expression and region-specific regulation of the oxytocin receptor gene in rat brain. Endocrinology 138:1857–1862

Caldwell JD, Prange AJ, Pedersen CA (1986) Oxytocin facilitates sexual behavior in estrogen treated ovariectomized rats. J Steroid Biochem 20:1510

Caldwell JD, Prange AJ, Pedersen CA (1986) Oxytocin facilitates the sexual receptivity of estrogen-treated female rats. Neuropeptides 7:175–189

Carter CS, Lederhendler II, Kirkpatrick B (Eds) (1997) The integrative neurobiology of affiliation. Annals, New York Acad Sci Vol 807

Carter CS, Altemus M (1997) Integrative functions of lactational hormones in social behavior and stress management. Annals, New York Acad Sci 807:164–174

Carter CS, DeVries AC, Taymans SE, Roberts RS, Williams JR, Getz LL (1997) Peptides, steroids, and pair bonding. Annals, New York Acad Sci 807:260–272

Chritin M, Ueta Y, Yamashita K, Dreifuss JJ, Tribollet (1996) Vasopressin, oxytocin and angiotensin II receptors in the brain of inbred polydipsic mice. Society for Neuroscience, 22: (abstract #246.15) p 618

Chung SR, McCabe JT, Pfaff DW (1991) Estrogen influences on oxytocin mRNA expression in preoptic and anterior hypothalamic regions studied by in situ hybridization. J Comp Neurol 307:281–295

Chung SK, Haldar J, Pfaff DW Estrogen effect on oxytocin mRNA-expressing neurons which project to spinal cord studied by combination of in situ hybridization and retrograde marker. J Comp Neurol (Submitted)

Coirini H, Johnson AE, McEwen BS (1989) Estradiol modulation of oxytocin binding in the ventromedial hypothalamic nucleus of male and female rats. Neuroendocrinology 50:193–198

Coirini H, Johnson AE, Schumacher M, McEwen B (1992) Sex differences in the regulation of oxytocin receptors by ovarian steroids in the ventromedial hypothalamus of the rat. Neuroendocrinology 55:269–275

DeKloet ER, Voorhuis DAM, Boschma Y, Elands J (1986) Estradiol modulates density of putative 'oxytocin receptors' in discrete rat brain regions. Neuroendocrinology 44:415–421

Dellovade T, Kia K, Zhu YS, Pfaff DW (1998) Thyroid hormone administration reduces estrogenic facilitation of oxytocin gene expression. J Neuroendo (in press)

Delville W, Mansour KM, Ferris CF (1996) Testosterone facilitates aggression by modulating vasopressin receptors in the hypothalamus. Physiology and Behavior. 60:25–29

Fahrbach SE, Morrell JI Pfaff DW (1985) Possible role for endogenous oxytocin in estrogen-facilitated maternal behavior in rats. Neuroendocrinology, 40:526–532

Fahrbach SE, Morrell JI, Pfaff DW (1986) Effect of varying the duration of pre-test cage habituation on oxytocin induction of short-latency maternal behavior. Physiology and Behavior 37:135–139

Ferris CF (1992) Role of vasopressin in Aggressive and Dominant/Subordinate Behaviors. Annals of the New York Academy of Sciences 652:212–226

Ferris CF, Delville Y (1994) Vasopressin and serotonin interactions in the control of agonistic behavior. Psychoneuroendocrinology 19:593–601

Grazzini E, Guillon G, Mouillac B, Zingg H (1998) Inhibition of oxytocin receptor function by direct binding of progesterone. Nature 392:509–512,

Gross P, Richter D, Robertson GL (1993) Vasopressin. Paris: John Libby Eurotext

Insel TS, Shapiro LE (1992) Oxytocin receptor dsitribution reflects social organization in monogamous and polygamous voles. PNAS 89:5981–5985

Insel TR, Shapiro LE (1992a) Oxytocin receptors and maternal behavior. In: Pedersen C, Caldwell J, Jirikowski G, Insel T (eds) Oxytocin in Maternal, Sexual and Social Behaviors. Ann. of the NY Acad of Sci 652:122–141

Jennes L, McShane TM, Brame B, Centers AJ (1996) J Neuroendocrinol 8:275–281

Jennes L, Eyigor O, Janovick J, Conn M. Brain gonadotropin releasing hormone receptors: Localization and regulation. In: Conn M (ed) Recent Progress in Hormone Research, Vol. 52. Bethesda, The Endocrine Society, pp 475–491

Johnson AE, Ball GF. Coirini H, Harbaugh CR, McEwen BS, Insel TR (1989) Time course of the estradiol-dependent induction of oxytocin receptor binding in the ventromedial hypothalamic nucleus of the rat. Endocrinology 125:1414–1419

Johnson AE, Coirini H, Ball GF, McEwen BS (1989) Anatomical localization of the effects of 17β-estradiol on oxytocin receptor binding in the ventromedial hypothalamic nucleus. Endocrinology 124:207–211

Kow L-M, McEwen BS, Pfaff DW, Weiland NG (1997) G-protein activation in hypothalamic ventromedial nucleus by phenylephrine, and its potentiation by estrogen. Soc for Neurosci 23:1852 (Abstr. #721.4)

Kow L-M, Mobbs CV, Pfaff DW (1994) Roles of second-messenger systems and neuronal activity in the regulation of lordosis by neurotransmitters, neuropeptides and estrogen: a review. Neuroscience & Biobehavioral Reviews 18:251–268

Kow L-M, Johnson AE, Ogawa S, Pfaff DW (1991) Electrophysiological actions of oxytocin on hypothalamic neurons, in vitro: neuropharmacological characterization and effects of ovarian steroids. Neuroendocrinology 54:526–535

Ku CY, Qian A, Wen Y, Anwer K, Sanborn BM (1995) Oxytocin stimulates myometrial guanosine triphosphatase and phospholipase C activities coupling to G alpha q/11. Endocrinology 136 (4):1509–1515

Lauber AH, Romano GJ, Mobbs CV, Howells RD, Pfaff DW (1990) Estradiol induction of proenkephalin messenger RNA in hypothalamus: dose-response and relation to reproductive behavior in the female rat. Molecular Brain Research 8:47–54

Lopez Bernal A, Bellinger J, Marshall JM, Phaneuf S, Europe-Finner GN, Asboth G, Barlow DH (1995) G protien expression and second messenger formation in human granulosa cells. Journal of Reproduction & Fertility 104:77–83

McCarthy MM, Chung SR, Ogawa S, Kow L-M, Pfaff DW (1991) Behavioral effects of oxytocin: Is there a unifying principle? In: Jard S, Jamison R (eds) Vasopressin. Colloque INSERM/ John Libbey Eurotext Ltd. vol. 208, pp 195–212

McCarthy MM, Schwartz-Giblin S, Wang SM (1997) Does estrogen facilitate social behavior by reducing anxiety? Annals, New York Acad Sci 807:541–542

McCarthy MM, Kleopoulos SP, Mobbs CV, Pfaff DW (1994) Infusion of antisense oligodeoxynucleotides to the oxytocin receptor in the ventromedial hypothalamus reduces estrogen-induced sexual receptivity and oxytocin receptor binding in the female rat. Neuroendocrinology 59:432–440

McCarthy MM, McDonald C, Phillip B, Goldman D (1996) An anxiolytic action of oxytocin is enhanced by estrogen in the mouse. Physiology and Behavior 60:1209–1215

Meisenberg,G (1981) Short-term behavioral effects of posterior pituitary peptides in mice. Peptides 2:1–8

Monga M, Ku CY, Dodge K, Sanborn BM (1996) Oxytocin-stimulated responses in a pregnant human immortalized myometrial cell line. Biology of Reproduction 55(2):427–432

Pedersen C, Caldwell J, Jirikowski G, Insel T (1992) Oxytocin in Maternal, Sexual, and Social Behaviors. Annals of the New York Academy of Sciences, Vol 652. New York: The New York Academy of Sciences

Pfaff DW (1988) Multiplicative responses to hormones by hypothalamic neurons. In Recent Progress in Posterior Pituitary Hormones. In: Yoshida S, Share L (eds), Elsevier Science Publishers B. V. (Biomedical Division), Excerpta Medica, Amsterdam, International Congress Series No. 797, pp 257–267

Pfaff DW (1999) Drive: Neural and Molecular Mechanisms of Sexual Motivation. Cambridge, The MIT Press

Pfaff DW, Galindo K, Larcher S, Luedke C, Muglia L, Ogawa S (1998) Role of oxytocin gene in behavioral responses to stress. Society for Neuroscience Abstracts

Phaneuf S, Europe-Finner GN, MacKenzie IZ, Watson SP, Lopez Bernal A (1995) Effects of oestradiol and tamoxifen on oxytocin-induced phospholipase C activation in human myometrial cells. Journal of Reproduction & Fertility 103:121–126

Phaneuf S, Carrasco MP, Europe-Finner GN, Hamilton CH, Lopez Bernal A (1996) Multiple G proteins and phospholipase C isoforms in human myometrial cells: implication for oxytocin action. Journal of Clinical Endocrinology & Metabolism 81:2098–2103

Phaneuf S, Asboth G, Carrasco MP, Europe-Finner GN, Saji F, Kimura T, Harris A, Lopez Bernal A (1997) The desensitization of oxytocin receptors in human myometrial cells is accompanied by down-regulation of oxytocin receptor messenger RNA. Journal of Endocrinology 154:7–18

Priest CA, Borsook D, Pfaff DW (1996) Estrogen and stress interact to regulate the hypothalamic expression of a human proenkephalin promoter-b-galactosidase fusion gene in a site-specific and sex-specific manner. J Neuroendocrinology 9:317–326

Qian A, Wang W, Sanborn BM (1998) Evidence for the involvement of several intracellular domains in the coupling of oxytocin receptor to G alpha (q/11). Cellular Signalling 10(2):101–105

Quiñones-Jenab V, Jenab S, Ogawa S, Adan RAM, Burbach PH, Pfaff DW (1997) Effects of estrogen on oxytocin receptor messenger ribonucleic acid expression in the uterus, pituitary and forebrain of the female rat. Neuroendocrinology 65:9–17

Quiñones-Jenab V, Jenab S, Ogawa S, Funabashi T, Weesner GD, Pfaff DW (1996) Estrogen regulation of gonadotropin-releasing hormone receptor messenger RNA in female rat pituitary tissue. Molecular Brain Research 38:243–250

Rhodes CH, Morrell JI, Pfaff DW (1981a) Immunohistochemical analysis of magnocellular elements in rat hypothalamus: Distribution and numbers of cells containing neurophysin, oxytocin, and vasopressin. Journal of Comparative Neurology 198:45–64

Rhodes CH, Morrell JI, Pfaff DW (1981b) Estrogen- Neurophysin-Containing Hypothalamic Magnocellular Neurons in the Vasopressin-Deficient (Brattleboro) Rat: A Study Combining Steroid Autoradiography and Immunocytochemistry. Journal of Neuroscience 2:1718–1724

Roberts JL, Dutlow CM, Jakubowski M, Blum M, Miller RP (1989) Estradiol stimulates preoptic area-anterior hypothalamic proGnRH-GAP gene expression in ovariectomized rats. Mol Brain Res 6:127–134

Romano GJ, Harlan RE, Shivers BD, Howells RD, Pfaff DW (1988) Estrogen increases proenkephalin messenger ribonucleic acid levels in the ventromedial hypothalamus of the rat. Molecular Endocrinology 2:1320–1328

Rosie R, Thomson E, Fink G (1990) Oestrogen positive feedback stimulates the synthesis of LHRH mRNA in neurones of the rostral diencephalon of the rat. J Neuroendocrinol 8:185–191

Rothfeld J, Hejtmancik JF, Conn PM, Pfaff DW (1989) In situ hybridization for LHRH mRNA following estrogen treatment. Molecular Brain Research 6:121–125

Russell JA, Douglas AJ, Bull P, Pumford KM, Bicknell RJ, Leng,G (1992) Pregnancy and opioid interactions with the anterior peri-third ventricular input to magnocellular oxytocin neurones. Prog Brain Res 91:41–53

Sanborn BM, Qian A, Ku CY, Wen Y, Anwer K, Monga M, Singh SP (1995) Mechanisms regulating oxytocin receptor coupling to phospholipase C in rat and human myometrium (Review). Advances in Experimental Medicine and Biology 395:469–479

Schumacher M, Coirini H, Frankfurt M, McEwen BS (1989) Localized actions of progesterone in hypothalamus involve oxytocin. Proc Natl Acad Sci USA 86:6798–6801

Schumacher M, Coirini H, Pfaff DW, McEwen BS (1991) Light-dark differences in behavioral sensitivity to oxytocin. Behavioral Neuroscience 105(3):487–492

Schumacher M, Coirini H, Pfaff DW, McEwen BS (1990) Behavioral effects of progesterone associated with rapid modulation of oxytocin receptors. Science 250:691–694

Selye H (1974) Stress Without Distress. Philadelphia: Lippincott,

Shughrue PJ, Lane MV, Merchenthaler IJ (1997) J Comp Neurol 388:507–525

Strakova Z, Soloff MS (1997) Coupling of oxytocin receptor to G proteins in rat myometrium during labor: Gi receptor interaction. American Journal of Physiology 272 (5 Pt 1): E870–E876

Sukhov R, Walker W, Rance N, Price D, Young W (1993) Vasopressin and Oxytocin Gene Expression in the Human Hypothalamus. The Journal of Comparative Neurology 337: 295–306

Witt DM, Ensel TR (1991) A selective oxytocin antagonist attenuates gonadal steroid facilitation of female sexual behavior. Endocrinology 128:3269–3276

Witt DM, Insel TR (1992) Central oxytocin antagonism decreases female reproductive behavior. In: Pedersen C, Caldwell J, Jirikowski G, Insel T (eds) Oxytocin in Maternal, Sexual and Social Behaviors. Annals, NY Acad Sci 652:445–447

Witt DM (1997) Regulatory mechanisms of oxytocin-mediated sociosexual behavior. Annals, New York Acad Sci 807:287–301

Wotjak CT, Masaharu K, Liebsch G, Montkowski A, Holsboer F, Neumann I, Landgraf R (1996) Release of vasopressin within the rat paraventricular nucleus in response to emotional stress: a novel mechanism of regulating adrenocorticotropic hormone secretion. J Neurosci 16:7725–7732

Yamashita H, Okuya S, Inenaga K, Kasai M, Uesugi S, Kannan H, Kaneko T (1987) Oxytocin predominately excites putative oxytocin neurons in the rat supraoptic nucleus in vitro. Brain Res 416:364–368

Zou Y, Komuro I, Yamazaki T, Kudoh S, Aikawa R, Zhu W, Shiojima I, Hiroi Y, Kazuyuki T, Kadowaki T, Yazaki Y (1998) Cell type-specific angiotensin II-evoked signal transduction pathways: Critical roles of Gbg subunit, Src family, and Ras in cardiac fibroblasts. Circ Res 82:337–345

Vasopressin Receptors: Structural Functional Relationships and Role in Neural and Endocrine Regulation

Oscar Schoots[1], Fernando Hernando[2], Nine V. Knoers[2], and J. Peter H. Burbach[1]

1
Introduction

Vasopressin (VP) is the biologically active peptide whose endocrine and neural actions, chemical structure, biosynthesis and receptors were described before any other peptide hormone. In this respect, VP has fulfilled an important role as a prototype peptide in the fields of endocrinology and neurosciences, and has always been on the brink of novel advances and concepts. VP was described originally as an entity in posterior pituitary extracts that led to antidiuresis and vasoconstriction, hence its name. In the 1950s it was purified from this gland and its chemical structure determined, an achievement by Vincent du Vigneaud that was honored by the Nobel prize in 1955 (Du Vigneaud 1956). It turned out to be a nonapeptide with an internal disulfide bridge and amidated C terminus, *Cys-Tyr-Phe-Gln-Asn-Cys*-Pro-Arg-GlyNH$_2$, and to be closely related to oxytocin (OT). VP has been the leading component in the concepts of neurosecretion and neuropeptides (Bargmann 1966; De Wied 1997). The VP/OT signaling system is evolutionarily old as is illustrated by the presence of highly similar peptides and receptors in diverse metazoans such as mollusca, arthropoda, annelida and chordata.

Although diversity in VP receptors (VPRs) was described biochemically and pharmacologically early on, much insight was gained by cloning their genes and cDNAs. In particular, the understanding of receptor diversity in relation to the physiological functions and pharmacology of VP has been enhanced by recent studies. Various molecular cloning strategies have resulted in the cloning of four established mammalian receptors for VP or OT, the V1a, V1b (or V3), V2 and OT receptors (V1aR, V1bR, V2R and OTR). These receptors form the VP/OT receptor family (VP/OT-R) and belong to the super-family of seven transmembrane G protein-coupled receptors (GPCRs). This chapter focuses largely on the structure, function, expression and pathophysiology of the three

[1] Rudolf Magnus Institute for Neurosciences, Department of Medical Pharmacology, Medical Faculty, Utrecht University, Universiteitsweg 100, 3584 CG Utrecht, The Netherlands
[2] Department of Human Genetics, University Hospital, P.O. Box 9101, 6500 HB Nijmegen, The Netherlands

VP-selective receptors. The OTR, which binds VP with high affinity as well and can be considered a non-selective VP receptor, is the subject of a separate chapter by Kimura and Ivell (this Vol.). A synopsis of the main physiological functions of VP is provided in Section 2. Detailed and extensive reviews on separate aspects of VP are available (Gross et al. 1993; Urban et al. 1998). Two additional putative vasopressin receptors that have been cloned from kidney are not members of the family of GPCRs. One, the vasopressin-activated calcium mobilizing protein 1, VACM1 (Burnatowska-Hledin et al. 1995), is a member of the *cullin* gene family of cell-cycle regulators (Kipreos et al. 1996). The other was identified as a dual angiotensin II and VP receptor (Ruiz-Opazo et al. 1995) and homology searches of protein databases indicate that it is not a member of any known protein family. However, some sequence similarity with RNAse inhibitors is evident. Neither of these two putative receptors will be reviewed in this chapter because the pharmacological data are limited and at present it is unclear whether they represent true VPR.

2
Physiological Functions of Vasopressin Systems

Depending on the source of VP and localization of VP receptors several, functionally different VP systems can be distinguished in mammals. VP was discovered originally as a hormone serving a number of endocrine functions. The major endocrine target tissues are rich in expression of specific VP receptors, e.g. the V2R in kidney and V1aR in vascular epithelium and liver. VP is also endogenous to the brain and serves as a neuropeptide controlling several nervous system functions. Multiple receptor subtypes have been implicated in the central actions of VP. At the interface of brain and periphery, a specialized set of VP neurons are destined for neuroendocrine regulation of anterior pituitary functions. Also, VP is produced locally in several organs together with receptors, suggesting paracrine action. Thus, VP participates in cell regulatory control at various levels.

2.1
Endocrine Functions

The origin of circulating VP is the hypothalamo-neurohypophysial system in which VP is produced by magnocellular neurons in the hypothalamus, transported to and released from nerve terminals in the posterior lobe of the pituitary gland. In quantitative terms, this system is the most important neurosecretory output system of the brain. VP released from the posterior lobe serves hormonal actions on peripheral target tissues. The predominant endocrine function of VP is the retention of water in the kidney through the increase of water permeability in the collecting ducts. This function is controlled primarily by the release of VP from the posterior pituitary gland in response to plasma osmolality, and primarily mediated by the V2R present in the basolateral membrane of the collecting ducts.

Proof of principle has come from human and animal pathology. Diabetes insipidus (DI) is a syndrome characterized by excessive diuresis, and is caused at the level of either the hypothalamo-neurohypophysial system or the kidney. In familial neurohypophysial DI (FNDI), an autosomal dominant disorder, the cause of pathology is insufficient VP synthesis due to mutations in the VP prohormone (Rittig et al. 1996). In the X-linked form of congenital nephrogenic DI (CNDI) the V2R receptor protein is mutated (see Sect. 5). In rare cases, autosomal dominant and recessive forms of CNDI involving mutations of the aquaporin-2 gene, encoding the channel for water reabsorbtion in kidney collecting ducts, have been described.

A unique animal model in which physiological functions of VP were derived is the Brattleboro rat, a mutant rat strain that suffers from autosomal recessive FNDI (Valtin 1967; Ivell et al. 1990). Due to the virtually complete absence of VP biosynthesis, this animal suffers from severe DI: its 24-h-water excretion is about equal to its body weight. This animal also shows that other well-known actions of VP are less dominant in physiology. Well-established endocrine actions of VP include the contraction of smooth muscle cells in the vascular wall, the stimulation of glycogenolysis and gluconeogenesis in the liver, platelet aggregation and factor VIII release, all of which are mediated by the V1aR.

In addition, VP plays a role in the regulation of secretion of pro-opiomelanocortin (POMC) peptides [adrenocorticotrophic hormone (ACTH), melano-stimulating hormone (MSH), lipotropins, endorphins] from the anterior pituitary gland. It synergizes with corticotropin releasing hormone (CRH) for this action, while each peptide employs its own receptor, the V1bR for VP and the CRH-R1 for CRH. This neuroendocrine VP originates from CRH-producing neurons in the parvicellular PVN. This neuronal sub-population is critically regulated by glucocorticoids and stressors (Aguilera 1994). In this way VP fine-tunes the release of ACTH as an essential step in the hypothalamo-pituitary-adrenal (HPA) axis and its responses to stress. In particular, VP predominates in the setting of HPA activity during chronic stress, and is thought to participate in neuroendocrine alterations in depression and anxiety disorders (Holsboer 1989).

2.2
Functions in the Brain

Part of the VP-producing neurons of the hypothalamus and additional cell groups in the suprachiasmatic nucleus, bed nucleus and the amygdala send axonal projections to other brain structures, and are thus the anatomical basis for the central functions of VP. The central effects of VP are diverse (De Wied et al. 1993; Urban 1998). VP exerts different functions depending on origin and projection area. Generally, VP serves adaptive functions, as is evident from its effects on behavior related to memory processes, rewarded stimuli, and social interactions. Also, its role in central regulation of cardiovascular activities and body temperature and the regulation of diurnal rhythms point to adaptive functions (Urban 1998).

For many of the central effects of VP the complete nonapeptide is required. The V1aR is widely involved in these effects in rodents, although it has been suggested that a receptor with V2R-like pharmacology may also be implicated (Diaz Brinton and Brownson 1993; Croiset and De Wied 1997). Significantly, other effects, particular those on social, memory- and reward-associated behaviors can be elicited by C-terminal fragments of VP which lack affinity to the known V1a, V1b and V2R receptors or OT receptor. C-terminal VP fragments are intermediate metabolites of the major metabolic conversion route of extracellular VP in the brain (Burbach and Lebouille 1983). The peptide [pGlu4,Cyt6]VP 4–9, which is a predominating metabolite, appears to be more potent than VP in several of the behavioral paradigms (Burbach et al. 1983). Using this labeled VP metabolite, or anti-idiotypic antibodies mimicking the C terminus of VP, binding sites have been found in the brain with an anatomical distribution different from the classical receptors. Furthermore, electrophysiological and calcium imaging studies support the notion that a different type of receptor responding to VP metabolites exists in the brain (reviewed in Burbach et al. 1998). The identity of this receptor is as yet totally obscure, although there are experiments underway towards its cloning.

2.3
Paracrine Functions

VP and its mRNA have been found in several tissues and organs, indicating local production of the peptide for paracrine action. These include the adrenal gland, testis, ovarium, thyroid gland, thymus, spleen, pancreas and intestine (Nicholson 1996). In most of these tissues VP receptors have been identified (see Sect. 3). It should be noted that the expression of VP and its receptors displays species-dependency to a considerable degree. Furthermore, VP receptors have been found in organs in which no VP production has been demonstrated (yet), like the cochlea (Kitano et al. 1997), where VP production may be limited to a restricted number of cells. It cannot be excluded that such VPR bearing tissues respond to VP in the circulation. The data suggest that several organs have adopted an endogenous VP system, including peptide production and VP receptors. One may speculate that these systems serve paracrine functions in the local communication between cells. Paracrine effects may include stimulation of cell proliferation and differentiation. It has been suggested that ectopically produced VP by tumors, as in small cell lung carcinoma, serves growth promotion (North et al. 1998).

3
Vasopressin Receptor Distribution

The distribution of the VPRs in different mammals, including human, rat and mouse has been studied in relation to the physiological functions of VP (for a more complete review see (Tribollet 1992; Barberis and Tribollet 1996). The most extensively used technique has been *in vitro* receptor-binding autoradiography using radiolabelled VP or VP ligands, in particular iodinated antagonists. More recently, *in situ* hybridization and RT-PCR studies have provided additional insights in this field. Using these techniques, all VPRs have been found in peripheral tissues. The V2R is almost exclusively located in renal collecting ducts, being responsible for the antidiuretic effect of VP (Jard et al. 1987; Tribollet et al. 1988; Birnbaumer et al. 1992; Lolait et al. 1992). The V1aR has been mainly found in blood vessels and liver, although it is also present in kidney (Gerstberger and Fahrenholz 1989; Phillips et al. 1990; Ostrowski et al. 1993). The V1bR is mainly restricted to the anterior pituitary corticotropes and is involved in the ACTH secretion (Jard et al. 1986; Antoni 1993), although it seems to be located in other pituitary cells, such as thyrotropes (Lolait et al. 1995).

Until recently, only one of the VPRs, the V1aR, has been well characterized in the rat central nervous system. Its distribution has been described in detail, using in particular V1aR selective antagonists (Tribollet 1992; Johnson et al. 1993; Barberis et al. 1995). The highest binding has been found in the olfactory system, nucleus accumbens, lateral septum, ventral hippocampus, hypothalamic stigmoid nucleus and lateral hypothalamus, locus coeruleus, interpeduncular nucleus, nucleus solitary tract, inferior olive and choroid plexus (see Table 1 for more detail). In addition, large superficial arteries supplying the brain, such as the anterior cerebral artery, the internal carotid artery and basilar arteries were also labelled with a radioiodinated linear VP antagonist, suggesting a role for VP in the regulation of cerebral blood flow (Barberis et al. 1995). In the human brain VPRs are present at much lower levels compared to the rat brain (Loup et al. 1991). V1aR mRNA in the rat brain, as shown by *in situ* hybridization, showed a high degree of overlap between the V1aR mRNA distribution and the binding of radioiodinated V1a antagonists (Ostrowski et al. 1994). However, a different pattern distribution has been reported by Szot et al. (1994), who could not find V1aR expression in some areas intensely labelled by selective V1a antagonists, such as the lateral septum. The reason for this discrepancy has not been clarified yet.

With the cloning of the rat and human V1bRs (De Keyzer et al. 1994; Sugimoto et al. 1994; Lolait et al. 1995; Saito et al. 1995), V1bR mRNA distribution was determined by RT-PCR. A number of peripheral tissues, including pituitary, thymus, heart, lung, spleen, and kidney, were found positive although expression in the pituitary gland was by far the most notable. Strikingly, most rat brain areas studied expressed V1bR mRNA: olfactory bulb, caudate putamen, septum, cerebral cortex, hippocampus, hypothalamus, and cerebellum (Lolait et al. 1995; Saito et al. 1995). This surprising finding suggests that some of the central effects of VP previously thought to be mediated by the V1aR or OTR could be mediated by the V1bR.

In our laboratory we have characterized an antibody, raised against a synthetic peptide fragment derived from the carboxy terminal of the V1bR protein, that recognizes the V1bR with a high specific affinity (antiserum provided by Dr. S. J. Lolait, University of Bristol, Bristol). This antibody displayed an intense immunohystochemical staining in different brain areas, including the olfactory bulb, taenia tecta, caudate putamen, medial habenula, hippocampus, hypothalamus, and cerebellum (for more details see Table 1). The distribution

Table 1. Distribution of V1a and V1b receptors in rat brain

Region	V1a receptor	V1b receptor-like immunoreactivity		
	[^{125}I] V1a antagonists binding	V1a receptor mRNA	Pericardia and dendrites	Axon terminals
Telencephalon				
Olfactory system				
Main olfactory bulb				
Internal plexiform layer	+	+		
Internal glomerulosa layer	+			
Internal granule cell layer		+		
Anterior olfactory nucleus		++		
Posterior olfactory nucleus		+		
Olfactory cortical area	++			
Piriform cortex	++	+/++	++	
Olfactory tubercle	++	+	++	
Cerebral cortex				
Layer IV	+/++			
Parietal cortex (layer VI, area 2)	+			
Cingulate cortex (layer II)			++	
Frontal cortex (layer II)			++	
Basal forebrain				
Bed nucleus of stria terminalis				
Anterior division	++/+++	++		
Posterior division		+		+
Diagonal band of Broca (hor limb)	+	+		
Fundus striati	++	+		
Substantia innominata		+		
Medial forebrain bundle		++		
Basal nucleus of Meynert		++		
Organum vasculosum lamina term				++
Basal ganglia				
Nucleus accumbens	+++	++		
Caudate putamen			++	
Ventral pallidum		+		

Table 1. Continued

Region	V1a receptor		V1b receptor-like immunoreactivity	
	[^{125}I] V1a antagonists binding	V1a receptor mRNA	Pericardia and dendrites	Axon terminals
Lateral septum				
Dorsal part	+++/++++	+++/++++		
Intermediate part	+++/++++	+++/++++		
Ventral part	++	++/+++		
Septofimbrial nucleus		+		
Taenia tecta			++	
Hippocampal formation				
Dentate gyrus	++	++	++	
Fields of Ammon's horn	+/++	+/++	+++	
Ventral subiculum and presubiculum	+++			
Amygdala				
Anterior amygdaloid area		++		
Basal nucleus		++		+
Medial nucleus				+
Amygdalostriatal transition area	++/+++	++		
Amygdalo-hippoc trans area		+		
Diencephalon				
Subfornical organ	+	+		+
Lateral habenula		++		
Medial habenula			++	
Thalamus				
Anteroventral nucleus	++	+		
Ventromedial nucleus	++			
Ventrolateral nucleus	++			
Paraventricular nucleus		+		
Posterior nuclear group	+	+		
Mediodorsal nucleus	+			
Suprafascicular nucleus	+			
Periventricular nucleus		++		
Parafascicular nucleus		++		
Reticular nucleus		+		
Hypothalamus				
Stigmoid nucleus	+++	+++		
Zona incerta	+++	+		
Tuber cinereum	+++			
Suprachiasmatic nucleus	++	+++		
Arcuate nucleus	++	+++		+
Lateral hypothalamic area	++	++		+
Magnocellular preoptic nucleus	+			
Retrochiasmatic area		+++		
Periventricular nucleus		++		+
Medial preoptic area		+		
Anterior, dorsal and posterior areas		+		
Parvocellular PVN		+		
Mammillary nuclei		+		++

Table 1. Continued

	V1a receptor	V1b receptor-like immunoreactivity		
Region	[^{125}I] V1a antagonists binding	V1a receptor mRNA	Pericardia and dendrites	Axon terminals
Dorsomedial nucleus		+++		
Median eminence				++++
Mesencephalon				
Pineal gland	++	+++		
Central gray	+/+++			
Periaqueductal central gray	+	+		
Edinger-Westphal nucleus	++			
Nucleus of Darkschewitsch	++	++/+++		
Linear raphe nucleus	++	++		
Dorsal nucleus of the raphe	++/+++	++		
Superior colliculus				
Zonal superficial	++/+++			
Optic layers	++/+++	++		
Interpeduncular nucleus	+++	++		
Substantia nigra	+	++		
Ventral tegmental area	+	++		
Pons and medulla				
Area postrema	++			
Locus coeruleus	++	++		
Nucleus of the solitary tract	++++	++		
Inferior olive (medial part)	++++	+++/++++		
Pontine and medullary raphe nuclei		++		
Intermediate reticular nucleus		+		
Paragigantocellular reticular nucleus		+		
Spinal trigeminal nucleus, interpolar	+++			
Parvocellular reticular nucleus	+			
Hypoglossal nucleus	+			
Nucleus ambiguus		++		
Ventral cochlear nucleus			+	
Nucleus trapezoid body			+	
Oculomotor nucleus		+		
Facial nucleus		+		
Accessory facial nucleus		+		
Motorhypoglossal nucleus		+		
Prepositus hypoglossal nucleus		+		
Trigeminal ganglion		+		
Supratrigeminal nucleus		+		
Dorsal motor nucleus of the vagus		+		
Cerebellum				
Granule cell layers		+	++	
White matter				++

Data from V1a receptor distribution was taken from Barberis et al. (1995), Barberis and Tribollet (1996) and Ostrowski et al. (1994). +, low density; ++, moderate density; +++, high density; ++++, very high density.

revealed by this antibody is consistent with that of the RT-PCR expression studies. At the cellular level, we found the V1bR-like immunoreactivity to be located exclusively in neuronal cell bodies and dendrites as well as in axon terminals (Fig. 1). This indicates that VP may exert both presynaptic and postsynaptic effects throughout the V1bR in the central nervous system.

Double labeling experiments using the V1bR antibody together with a monoclonal antibody raised against β-endorphin demonstrated that the V1bR is present not only in about 60–70% of corticotrope cells, but also in as yet unidentified other subpopulation(s) of pituitary cells. In addition, we could reveal expression of the V1bR in the neuronal lobe of the pituitary (Fig. 1). This may explain the labeling of this gland using [³H]VP, a labeling previously suggested to be due to the binding of VP to neurophysin.

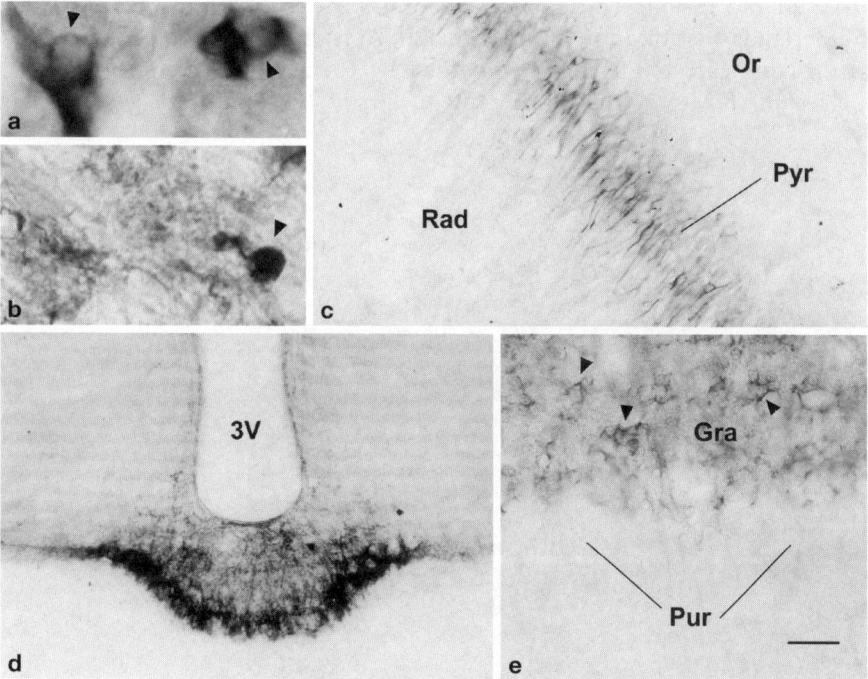

Fig. 1. V1bR-like immunoreactivity in rat brain and pituitary. **A** Immunoreactivity in the anterior pituitary gland. Cells were mainly stained in the cytoplasmic membrane (*arrowheads*). **B** Detail of a pituicyte (*arrowhead*) stained in the pituitary neuronal lobe. **C** Dendrite and pericardia immunostaining in the pyramidal (*Pyr*) cell layer from the hippocampal field CA1 of Ammon's horn. **D** Intense fiber staining in the median eminence, mainly concentrated in the zona externa. Fibers running along the periventricular nucleus are also positive. **E** Immunoreaction in cells located in the granular cell layer (*Gra*) from the cerebellum (*arrowheads*). *Rad* stratum radiatum; *Or* oriens layer; *3V* third ventricle; *Pur* cerebellar Purkinje cells. Bar: 50 μm in A and B; 200 μm in C, D and E

Immunohistochemistry for VPRs using antibodies raised against synthetic peptides is an important tool to specifically locate receptor subtypes at the cellular level, although it appears to have a serious limitation. We consider it conceivable that antibodies raised against peptides derived from intracellular or extracellular domains of the receptor may not recognize the entire population of a receptor, as we reported before for the OTR (Adan et al. 1995). These domains are involved in binding of agonists, G protein-coupling, are subject to modifications such as glycosylation, palmitoylation and phosphorylation, and may interact with other proteins. We propose that antibodies raised against such peptide fragments derived from G protein-coupled receptors may recognize the protein depending on its conformational and/or activation state, and would thus only detect a subpopulation of a receptor.

Our data indicate that this may be the case for three V1aR-specific antisera raised against different peptide fragments. Two were raised against different parts of the i3 loop (raised in collaboration with F.W. van Leeuwen, The Netherlands Institute for Brain Research, Amsterdam), and one was raised against the carboxy terminal (provided by Dr. Soloff, University of Texas, Galveston, Texas) of the V1aR protein. Firstly, these antisera gave similar

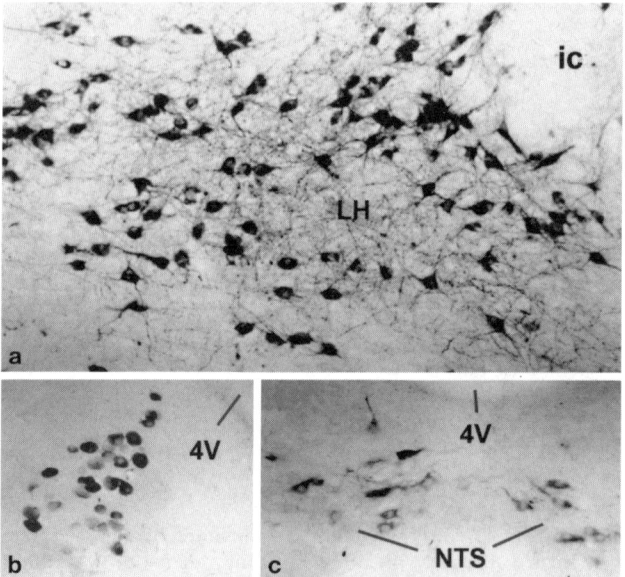

Fig. 2. V1aR-like immunoreactivity in rat brain. Three antisera raised against different peptide fragments derived from V1aR protein (two from the third intracellular loop of the receptor and one from the carboxyl terminal) gave a similar staining pattern in the lateral hypothalamus (*LH*; **a**), locus coeruleus (**b**) and nucleus of the solitary tract (*NTS*; **c**). These areas displayed an intense binding for iodinated V1aR antagonists and V1aR mRNA is very abundant. *ic* Internal capsule; *4V* fourth ventricle

immunohystochemical staining patterns in many different rat brain areas known to express the V1aR, including hippocampus, hypothalamus, thalamus, cortex, amygdala, different nuclei in the brainstem and pineal gland, indicating that they bind the V1aR (an example is shown in Fig. 2). Remarkably, in some areas devoid of VP binding and V1aR mRNA expression, like the supraoptic nucleus (Ostrowski et al. 1994, Barberis et al. 1995), all antisera showed strong labeling. The immunostaining observed with these three antisera did not show complete overlap, indicating that they may recognize different conformations or subpopulations of the V1aR. In line with this hypothesis we could not find a clear immunostaining with either of these antisera in the lateral septum, indicating that the antibodies used do not recognize the conformational state of the receptor in this area. Secondly, it appears impossible with several VPR antisera to obtain satisfying results from "positive" control experiments, like Western blotting, staining in transfected cells. Such difficulties render immunocytochemistry of VPRs still inappropriate.

4
Structural and Functional Relationships of Vasopressin Receptors

4.1
Structure

The structural hallmark of the GPCR superfamily is the configuration of seven hydrophobic transmembrane α-helices (TM-I to TM-VII) joined by alternating extracellular (e1, e2 and e3) and intracellular (i1, i2, i3) domains (Fig. 3). Within the family of VP/OT-R the highest degree of sequence similarity and identity is found in the transmembrane domains and in the extracellular loops connecting the transmembrane domains (Fig. 4). Of the conserved residues, some are more or less exclusive for the VP/OT-R family (Fig. 4, black circles) while others are present in many other GPCRs (Fig. 4, shaded circles). The latter presumably play an important role in general structural and functional characteristics of GPCRs, while the former may be responsible for VP/OT-R specific features such as ligand specificity. The majority of the residues conserved among VP/OT-Rs reside in the extracellular domains and parts of the transmembrane α-helices which are closest to the extracellular face of the membrane. Many of these residues have indeed been implicated in ligand binding (see Sect. 4.3). In contrast, residues generally conserved in GPCRs do not show this asymmetric distribution of homology. As shown in Figs. 3 and 4, part of the i3 loop just before TM-VI is also conserved between VP/OT-Rs, although potential functional or structural implications remain enigmatic. Among the different VP/OT-Rs, the extracellular amino terminus, i3, and the carboxy terminus are least conserved and diverge substantially in both length

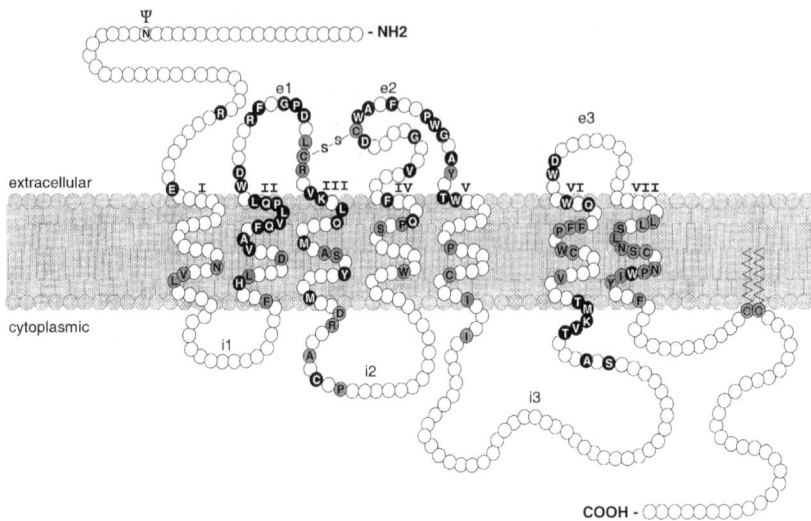

Fig. 3. Two-dimensional diagram of the membrane topology of VP/OT-Rs. *Shaded circles* represent residues that are conserved among many GPCR families. *Black circles* represent amino acids that are conserved in the VP/OT-R family. Extra- and intracellular loops are marked *e1, e2, e3* and *i1, i2, i3* respectively. Transmembrane domains are marked in *roman numbers*. Conserved Cys residues in e1 and e2 probably form a disulfide bridge. Two conserved Cys residues in the carboxyl terminus are palmitoylated and anchor the cytoplasmic tail to the plasma membrane. All VP/OT-Rs, have a consensus sequence for N-linked glycosylation in the amino terminus but additional putative glycosylation sites may be present in individual members of the VP/OT-Rs

and amino acid composition. Two conserved cysteine residues present in e1 and e2 likely form a disulfide bridge and play a role in the tertiary structure of the receptors (Gopalakrishnan et al. 1988; Pavo and Fahrenholz 1990; Thibonnier et al. 1993).

For the human V2R, the amino terminal part of the protein including the first transmembrane domain and the positively charged i1 loop is important for proper insertion and orientation in the membrane (Schulein et al. 1996).

──▶

Fig. 4a. Amino acid sequence alignment of VP/OT-like receptors. The alignment includes sequences from mammalian (V1aR, V1bR, V2R and OTR), fish (VTR and ITR), amphibian (mesotocin receptor, MTR), and mollusc (conopressin receptors 1 and 2, CPR1, CPR2) origin. Residues in *black* are conserved among VP/OT-like receptors while *shaded* residues are conserved among many GPCR families. Transmembrane regions are indicated by *arrows*. Sequences were obtained from the Genbank Database. Accession numbers are: V1aR sheep, L41502; V1aR rat, P30560; V1aR mouse, D49730; V1aR human, L24615; V1bR rat, U27322; V1bR human, L37112; V2R rat, Z22758; V2R mouse, AJ00691; V2R human, L22206; V2R cow, X83741; V2R pig, X71795; OTR rat, U15280; OTR mouse, D86599; OTR sheep, X87986; OTR pig, X71796; OTR monkey, U82440; OTR cow, S80965; OTR human, X64878; ITR fish, X87783; MTR toad, X93313; VTR fish, X76321; CPR1 snail, U27464; CPR2 snail, U40491

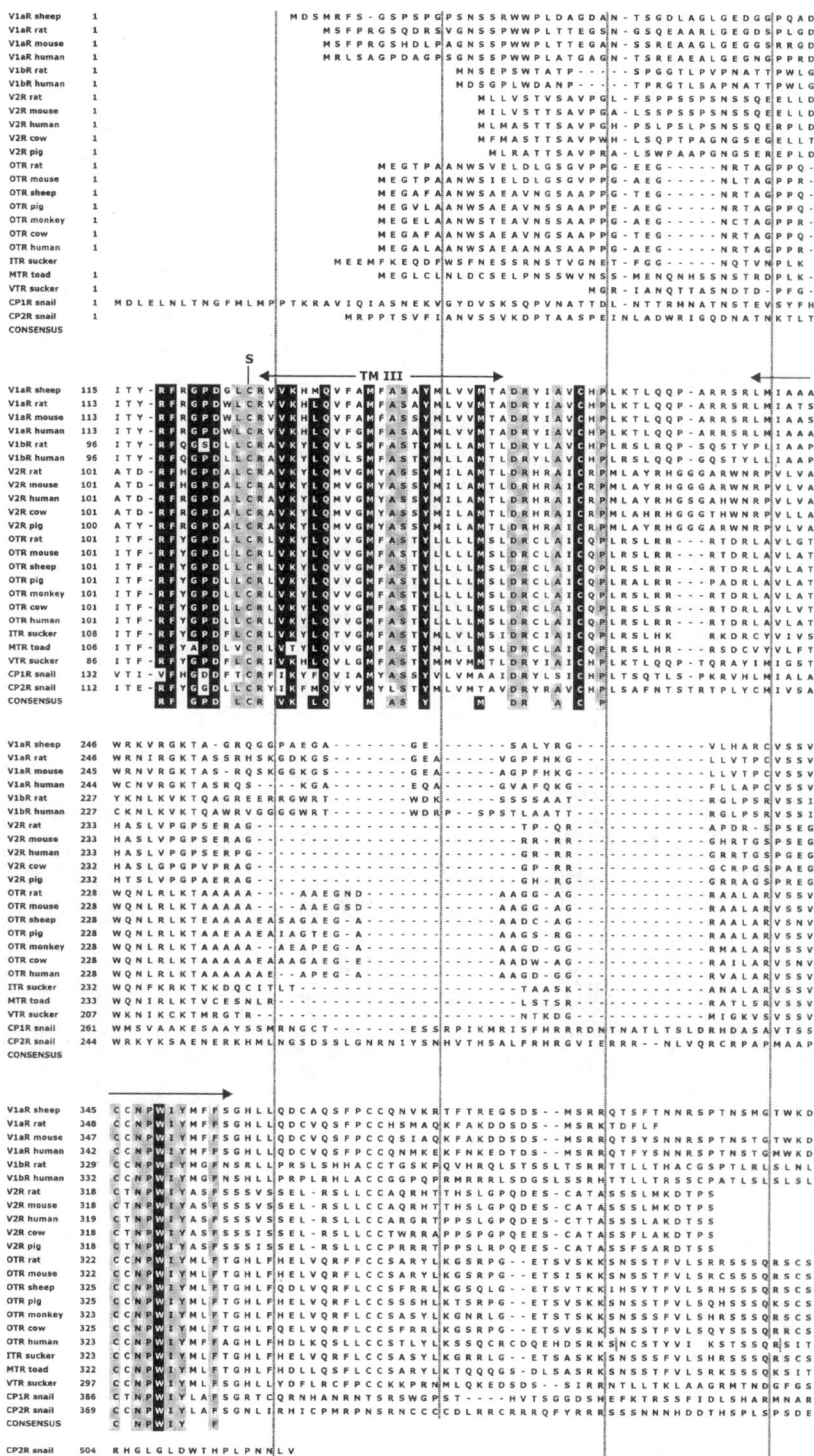

a CP2R snail 504 R H G L G L D W T H P L P N N L V

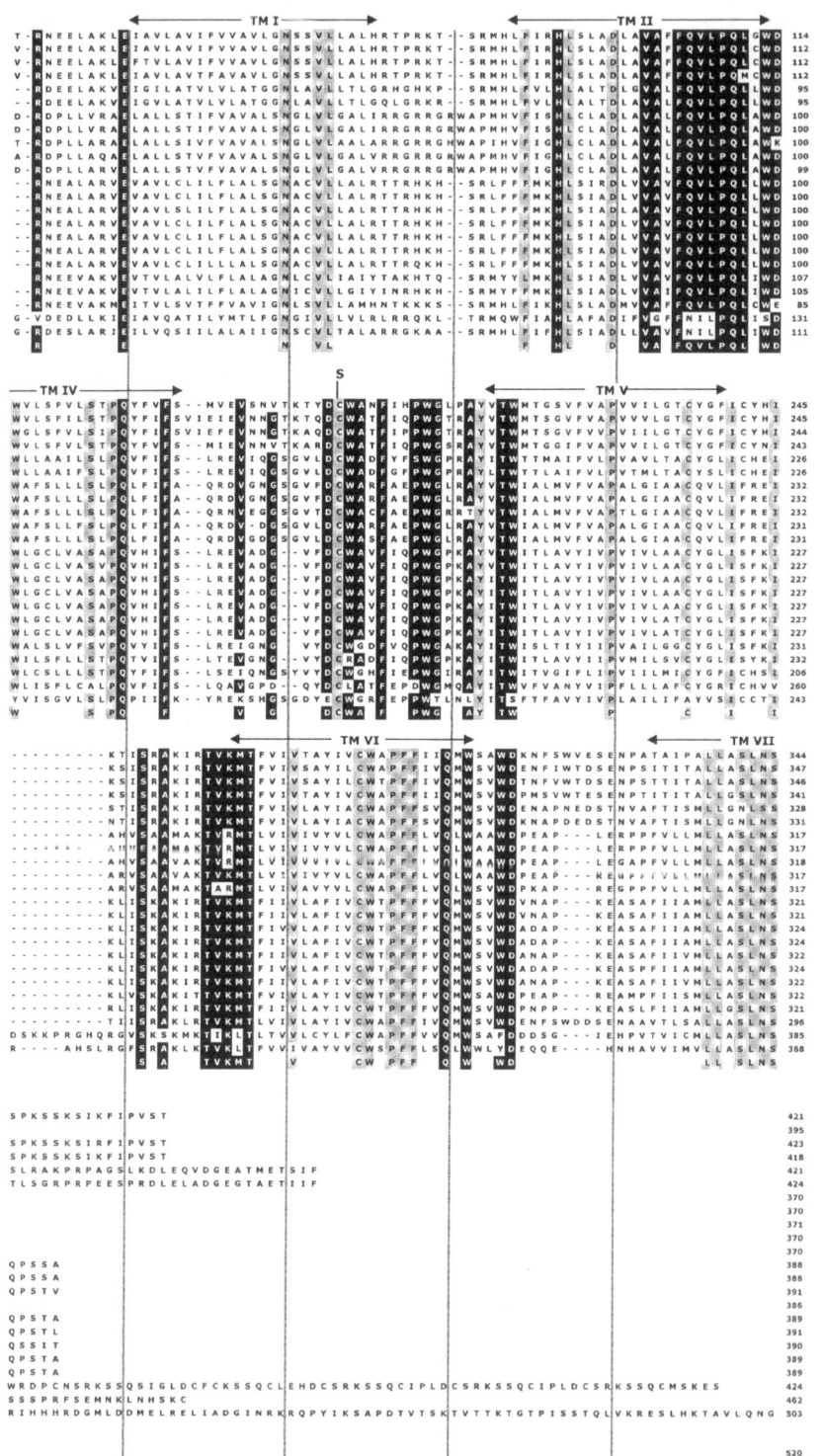

b

The topology of the VPR is in agreement with the "positive-inside" rule of Von Heijne. This rule says that regions of multi-spanning membrane proteins facing the cytoplasm are generally enriched in arginine and lysine residues. These positively charged amino acids, which are also found in other VP/OT-R, may be a major determinant of the transmembrane topology (Hartmann et al. 1989; von Heijne 1994).

Competition binding experiments of VP with a peptide corresponding to the e2 loop of the V1a receptor demonstrate that this e2 fragment inhibits VP binding (Mendre et al. 1997). The interference is receptor-subtype specific (Ki for a fragment of V1a receptor on the V1a, V1b and V2 receptors of 3.7, 14.6 and 64.5 µM respectively) and is via an interaction of the V1a fragment with the receptor and not with the hormone. The mechanism and site of interaction remain unknown and may involve intra- and/or intermolecular protein protein interactions. For rhodopsin it has been shown that intramolecular interactions play a role in protein structure (Ridge et al. 1996). Alternatively, intermolecular protein protein interactions may participate in the formation of a VP receptor dimer. Receptor dimerization has been described for several other GPCRs, but the structural and functional consequences remain elusive (Hebert and Bouvier 1998).

4.2
Post-Translational Modifications

As in many GPCRs, VP/OT-Rs accommodate a number of sites subject to post-translational modifications of the receptor protein. Such modifications generally can alter specific properties of proteins. For GPCRs they have a particular role in the consecutive events that take place after ligandbinding, e.g. G protein-coupling, desensitization and internalization. All VP/OT-Rs have consensus sequences for N-linked glycosylation in the extracellular amino terminus. For the human V2 receptor glycosylation appears not to affect ligand binding, Gs coupling, desensitization and internalization (Innamorati et al. 1996). For the porcine V2 receptor it was demonstrated that glycosylation plays a minor role in biosynthesis and transport of the receptor (Jans et al. 1992).

Fig. 4b. Amino acid sequence alignment of VP/OT-like receptors. The alignment includes sequences from mammalian (V1aR, V1bR, V2R and OTR), fish (VTR and ITR), amphibian (mesotocin receptor, MTR), and mollusc (conopressin receptors 1 and 2, CPR1, CPR2) origin. Residues in *black* are conserved among VP/OT-like receptors while *shaded* residues are conserved among many GPCR families. Transmembrane regions are indicated by *arrows*. Sequences were obtained from the Genbank Database. Accession numbers are: V1aR sheep, L41502; V1aR rat, P30560; V1aR mouse, D49730; V1aR human, L24615; V1bR rat, U27322; V1bR human, L37112; V2R rat, Z22758; V2R mouse, AJ00691; V2R human, L22206; V2R cow, X83741; V2R pig, X71795; OTR rat, U15280; OTR mouse, D86599; OTR sheep, X87986; OTR pig, X71796; OTR monkey, U82440; OTR cow, S80965; OTR human, X64878; ITR fish, X87783; MTR toad, X93313; VTR fish, X76321; CPR1 snail, U27464; CPR2 snail, U40491

A number of putative phosphorylation sites are present in all VP/OT-R, and, as is the case for other GPCRs, binding of agonist to the V1aR or V2R causes phosphorylation of the receptor. The V1aR can be phosphorylated by both protein kinase C (PKC) and G protein-coupled receptor kinases (GRK). Agonist-induced phosphorylation by GRK is fast and quickly reversed in the continuous presence of ligand (Innamorati et al. 1998a). Moreover, GRK-promoted phosphorylation and sequestration of the agonist-occupied receptor are concurrent and appear to participate in desensitization. Truncation of the V1aR at S374 shows that the carboxy terminus, which includes multiple potential PKC phosphorylation sites, is not essential for either desensitization or signal transduction (Ancellin et al. 1997). Indeed, phosphorylation by PKC alone does not promote sequestration or alter desensitization of the V1aR. The V2R is also rapidly phosphorylated by GRK upon binding of VP, but, unlike the V1aR, V2R phosphorylation of a carboxy terminal serine cluster is long-lasting (Innamorati et al. 1997; Innamorati et al. 1998b). A truncated V2R lacking the last 14 amino acids including this serine cluster was able to bind VP and activate Gs, but could not be phosphorylated and desensitized and was sequestered less efficiently. Internalized V2R does not recycle to the plasma membrane for many hours after removal of the ligand, and this slow recycling may be due to the inability of cytoplasmic phosphatases to dephosphorylate the serine cluster (Innamorati et al. 1998b). Generally, it appears that the conformation that the receptor acquires after ligand binding is necessary for phosphorylation by GRK and that desensitization and sequestration are enhanced by GRK phosphorylation

In the C-terminal domain of all vertebrate VP/OT-Rs, as well as in many other GPCRs, two conserved adjacent cysteines are found. For the V2R and many other GPCRs it has been demonstrated that both are palmitoylated (Sadeghi et al. 1997b) thereby anchoring the carboxyl tail to the plasma membrane (Fig. 3). Except for the fact that the palmitoylated V2R displayed a higher cell surface expression, no effects of palmitoylation were found on ligand binding affinity, adenylyl cyclase activation, receptor internalization, and desensitization.

4.3
Ligand Receptor Interactions

Recently, combined mutagenesis and pharmacological studies have significantly increased the understanding of structural requirements for ligand receptor interactions in the VP/OT-R family. Most progress has been made in resolving agonist binding although some data are available for antagonist receptor interactions.

Mutagenesis studies focusing on VP/OT-R-specific residues have been used to a large extent to try to identify amino acids essential for ligand binding. Replacement of conserved Gln residues in TM-II, -III, -IV and -VI of the V1aR by Ala resulted in reduced affinities for the agonists VP, OT and [Phe2,Orn8]-vasotocin (VT; Mouillac et al. 1995). Similarly, when a conserved Lys in TM-III was substituted by Met or Ala in either the rat V1aR or fish VT receptor (VTR), VT binding was reduced. Substitution Q104L of the VTR also resulted in loss of VT binding (Hausmann et al. 1996). The conserved Gln residue corresponds to an Asp residue at the same position in TM-III of catecholamine receptors. In these receptors the Asp residue plays a key role in binding of the amine group of dopamine or norepinephrine.

Using chimeric V2R-OTR protein, it was shown that e1, e2 and e3 contribute to OT binding at the OTR, but not to binding of the OT antagonist d(CH2)5[Tyr(Me)2,Thr4,Orn8,Tyr9]-VT (Postina et al. 1996). Experiments with chimeras of the VTR and teleost isotocin (IT) receptor (ITR) indicate that in the VTR, the N terminus, TM-IV to TM-V, and e2 determine ligand potency and affinity probably by recognizing amino acids that are conserved in the nonapeptides. TM-VI and e3 likely recognize nonapeptide-specific structures, thereby determining peptide selectivity (Hausmann et al. 1996). Replacement of Tyr115 in e1 of the rat V1aR by either Asp or Phe, as naturally present in the human V2R or OTR, increased the affinity for V2 or OT agonists respectively (Chini et al. 1995), suggesting an essential role for high affinity binding and receptor selectivity. Involvement of the e2 loop in hormone binding was also demonstrated by using a photoactivatable [Lys8]-VP analogue (Kojro et al. 1993). Some of the naturally occurring V2R mutations causing CNDI in humans also affect ligand binding and provide additional information on ligand receptor interactions (see Sect. 5).

Antagonist binding domains of VP/OT-R are as yet still poorly defined. Especially the site(s) of interaction of the large side chains and hydrophobic ring structures present in many cyclic or linear peptide antagonists is unclear. Often mutations affecting agonist binding have little effect on antagonist binding (Mouillac et al. 1995; Postina et al. 1996). This is surprising since many peptide antagonists are based on the backbone of the natural hormone, and overlap in structural requirements in the receptor may be expected. However, this is not the case. By using a photoactivatable linear V1aR antagonist and molecular modeling, Phalipou et al. (1997) defined a cluster of aromatic residues in TM-VI (W304, F307, and F308) involved in binding of antagonists with an aromatic group at position 1. This cluster is highly conserved among GPCRs and maybe interaction at this site is required for antagonism. Mutation of F307V dramatically reduced antagonist affinity while agonist affinity was minimally affected. However, it is anticipated that other residues must add specificity.

Molecular modeling supports many of the experimental data on receptor ligand interactions (Mouillac et al. 1995; Hausmann et al. 1996) but has its limitations. Perhaps the most serious one is the absence of the extracellular domains in modeling studies. Both amino acid sequence alignment (Fig. 4) and experimental data indicate that the extracellular loops are important for ligand binding. In general, ligand binding to VPRs appears to involve many points of interaction and likely occurs in a pocket formed by the transmembrane domains, as has been found for other peptide-binding GPCRs. Residues specifically conserved among VP/OT-Rs are likely involved in formation of the ring-like binding pocket and some appear to be direct points of interaction.

4.4
G Protein-Coupling

Traditionally, the different VP/OT-Rs have been known to couple to two distinct second messenger systems. The V2R activates, via Gs, adenylyl cyclase, which in turn results in increased intracellular cAMP concentrations. The V1aR, V1bR and OTR activate, via Gq/G11, phospholipase C. This leads to hydrolysis of phosphatidyl-inositol 4,5-biphosphate into inositol 1,4,5-triphosphate and diacylglycerol and causes an increase in intracellular Ca^{2+}. However, in recent years, activation of several other signal transduction systems has been demonstrated in tissues or cell culture. The V1aR was shown to activate phospholipases A2 and D, and stimulate cell acidification through an Na^+/H^+ exchanger (Briley et al. 1994). The V1bR was also shown to couple to these phospholipases and in addition can activate the mitogen-activated protein kinase (MAPK) signaling pathway (Thibonnier et al. 1997).

Functional studies aiming to identify the domains of the vasopressin receptors that confer G protein-coupling selectivity have been undertaken. Using chimaeric V1a-V2 receptors, Liu and Wess (1996) determined that a single intracellular domain can confer G protein-coupling on a receptor. When the i2 loop of the V2R was replaced by V1aR-i2, the hybrid gained the ability to couple efficiently to Gq/G11 while the capability to activate Gs was preserved. Analogously, replacement of the V1aR-i3 loop by V2R-i3 generated a hybrid that acquired the capacity to activate Gs while coupling to Gi/G11 was maintained. Evidently, the V1aR-i2 loop plays an essential role in activation of Gi/G11, while V2R-i3 is imperative for Gs coupling. The carboxy terminus, which for some GPCRs is involved in G protein-coupling, does not appear to serve this role for either the V1a or V2 receptor (Liu and Wess, 1996; Innamorati et al. 1997; Innamorati et al. 1998a)

A number of individual amino acids important for G protein activation have also been identified. At the interface of TM-II and i2 of most GPCRs, including the VP/OT-Rs, the triplet motif D-R-Y/H/C is highly conserved (Figs. 3 and 4) and for the V2R it is required for efficient G protein-coupling (Savarese and Fraser 1992). Also, a highly conserved proline present in TM-VII (Figs. 3 and 4) has been found to play a key role in activation of GPCR (Wess et al. 1993). This proline is part of a motif of 18 amino acids that forms the bulk of TM-VII and is highly conserved in GPCRs. Furthermore, site-directed mutagenesis of the V1a receptor indicates that D97, present in TM-II, is also involved in activation (Mouillac et al. 1995) as is the case for many other GPCRs (Savarese and Fraser 1992).

5
V2 Receptor Mutations
in Congenital Nephrogenic Diabetes Insipidus

The physiological importance of the V2R in the VP-mediated renal concentrating mechanism became more apparent when mutations in the V2R gene were identified in patients with congenital nephrogenic diabetes insipidus (CNDI). This is a rare inherited disorder, characterized by insensitivity of the renal distal nephron to the antidiuretic effect of VP. Consequently, large volumes of hypotonic urine are excreted which may lead to severe dehydration. Patients present in their first year of life with aspecific symptoms such as anorexia, vomiting, fever, growth retardation and developmental delay. After infanthood, the clinical picture is dominated by the less alarming symptoms of polyuria and polydipsia (Knoers and Monnens 1992). The most severe complication of the disorder is mental retardation, which is assumed to be a sequel of severe brain dehydration. Nowadays this complication is rare due to earlier recognition and treatment of CNDI (Hoekstra et al. 1996).

In most cases (about 90%) CNDI is transmitted as an X-linked recessive trait caused by mutations in the V2R gene. A minority of patients (about 10%) show an autosomal recessive inheritance as a result of mutations in the aquaporin-2 (AQP2) water channel gene (Deen et al. 1994; Van Lieburg et al. 1994a; Hochberg et al. 1997; Mulders et al. 1997). Two families have been described with clear autosomal dominant inheritance of CNDI. Recently, it has been shown that AQP2 mutations are responsible also for the autosomal dominant type of this disorder (Mulders et al. 1998).

To date, more than 80 distinct putative disease-causing mutations in the V2R gene have been detected in families with X-linked CNDI (Cheong et al. 1997; and reviewed in Knoers and van Os 1996; Bichet et al. 1997; Oksche and Rosenthal 1998; Fig. 5). The mutations are not clustered in one domain of the V2R but are scattered throughout the protein, except for the part coding for the tail of the receptor. The finding of so many diverse mutations in the V2R gene in CNDI patients demonstrates that the disease is highly heterogeneous at the molecular level, as expected for a typical X-linked disorder with reduced reproductive fitness in males. The mutations consist of nucleotide deletions and insertions which might be attributed to slipped mispairing during DNA replication, and nucleotide substitutions, some of which could be the result of 5-methylcytosine deamination at a CpG dinucleotide (Bichet et al. 1994; Faux and Scott 1996; Cheong et al. 1997).

As yet, only a minority of the documented mutations in the V2R gene have been subjected to *in vitro* expression studies to determine the molecular cause of CNDI. For several mutations defective folding, processing and/or intracellular trafficking of the receptor have been demonstrated. These mutants are retained in intracellular compartments [endoplasmic reticulum (ER), Golgi

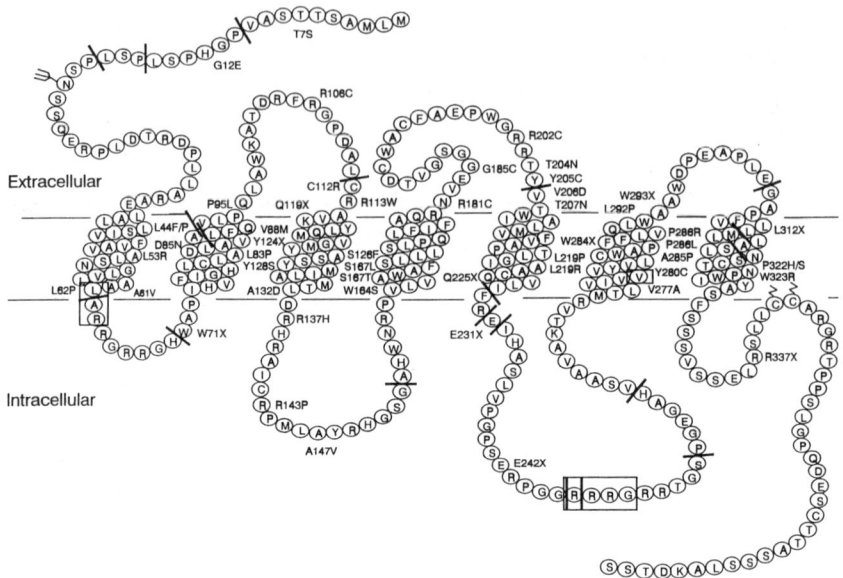

Fig. 5. Schematic model of V2R mutations in CNDI. Mutations identified in patients with X-linked nephrogenic diabetes insipidus are indicated by *text* (missense/nonsense), *boxes* (deletions of one or more amino acids), or *bars* (nucleotide insertions or deletions resulting in a frameshift). (Data from Knoers and van Os 1996; Bichet et al. 1997; Cheong et al. 1997; Oksche and Rosenthal 1998

apparatus] and degraded, and therefore their expression at the cell surface is either absent or substantially decreased. Intracellular retention and lack of surface expression especially holds true for truncation mutants. For example, the nonsense mutation R337X is functionally ineffective due to impaired export from the ER (Wenkert et al. 1996; Sadeghi et al. 1997a). Using an exofacial epitope-tagging technique, Tsukaguchi et al. (1995a,c) elegantly demonstrated that the 804insG mutant is not expressed at the cellular surface due to ineffective biosynthesis of the receptor protein. In several V2R mutants with missense or *in frame* deletions (L44P, R113W, R143P, W164S, S167T, S167L, del278/279, Y280C, L292P), misfolding/misrouteing was identified to be the molecular cause of CNDI also (Birnbaumer et al. 1994; Tsukaguchi et al. 1995b,c; Oksche et al. 1996; Wenkert et al. 1996). Some of the amino acids involved in these missense mutations are well-conserved among G protein-coupled receptors (W164, S167), or within the vasopressin subfamily (Y280), which is an indication of their importance for the structure and/or function of the protein. The introduction of prolines in amino acids found in transmembrane domains (L44, L292) is believed to disrupt the α-helical structure and therefore likely to have a strong impact on receptor folding (Faux and Scott 1996; Oksche et al. 1996).

For ten missense and several nonsense and frame-shift mutations in the V2R gene, reduced or absent binding of VP appeared to underlie CNDI. The missense mutations abolishing affinity for VP are the L44F, Y128S, R202C, V206D and P286D mutations (Pan et al. 1994; Van Lieburg et al. 1994b; Tsukaguchi et al. 1995a,b,c; Oksche et al. 1996). Reduced binding of VP was demonstrated for the R113W, R181C, T204N, Y205C, and V277A missense mutations (Birnbaumer et al. 1994; Pan et al. 1994; Van Lieburg et al. 1994b; Wenkert et al. 1996; Yokoyama et al. 1996). It has been proposed that some of the mutations that are found in the first two extracellular loops (R113W, T204N, V206D) disrupt the disulfide-bond between C112 and C192 which is believed to be involved in forming the ligand binding site, whereas others, which introduce a cysteine (R181C, Y205C), form a new bond with one of the important cysteines. Mutations affecting the affinity for AVP are generally found in the TM domains or e-loops and are enriched in areas highly conserved among the VP/OT-R or GPCR. As proposed for V277 (Wenkert et al. 1996), many of these mutations might be involved in the conformation of the receptor for agonist binding. Interestingly, for one of the mutants, T204N, the reduced but not absent vasopressin binding capacity of the mutant receptor, as shown in the expression system, was reflected *in vivo* by a substantial increase in urine osmolality after administration of a double dose of DDAVP to the patient carrying this mutation (Van Lieburg et al. 1994b).

As yet, only one V2R mutation has been identified, which abolishes stimulation of the adenylate cyclase system (Rosenthal et al. 1993). This mutation (R137H) is located in the second intracellular loop and stresses the importance of this loop for G protein-coupling, which is also known for other G protein-coupled receptors. Mutations of P322 in the human V2R, as found in several families suffering from CNDI, results in receptors with reduced or disrupted activation of Gs (Tajima et al. 1996). Mutation P322S, which causes a mild form of CNDI, yields a receptor with a reduced binding affinity and capacity to stimulate adenylyl cyclase. However, the P322H mutation completely abolishes signal transduction and, consequently, patients with this mutation suffer more severely from CNDI. The R113W mutation is of special interest since it causes a combination of functional defects: a decrease in ligand binding (about 20-fold), a reduced ability to stimulate adenylate cyclase (about three-fold), and lowered expression at the cell surface (about ten-fold) (Birnbaumer et al. 1994). In contrast to the above mentioned mutations, no consequences for ligandbinding or adenylate cyclase stimulation were found for the G12E and A61V mutations nor for the deletion of four amino acids in the third cytoplasmic loop, suggesting that these mutations represent rare polymorphisms and thus are not causal for CNDI (Pan et al. 1994; Wenkert et al. 1996).

6
Concluding Remarks

As one of the first known regulatory peptides, VP is still at the forefront of research on peptides and their cognate receptors. The cloning of the receptors has contributed significantly to insights into the physiological and endocrine systems in which VP participates. Furthermore, the broad knowledge of the pharmacology of VPRs and the extensive collection of agonists and antagonists appear important in probing for functional and structural aspects of the VP/OT-R family, such as the identification of the ligand-binding pocket and the residues determining ligand specificity. It may be expected that novel ligands with improved selectivity will be designed on the basis of this work.

In particular, the VP field is still in need of selective V1bR agonists and antagonists to work out the physiological involvement of this receptor in the tissues where it has been encountered only very recently, like the brain. A major future challenge will be to understand the physiological significance of the diversity in VPRs. Why are multiple receptors for the same ligand required? What is the contribution of each receptor in organs that contain more than one receptor, and what is the role of the OTR in VP action? Answers to such questions, together with the identity of still other VPRs and receptors for metabolites will be an essential contribution to the VP field.

Acknowledgments. We thank R. van Vlaardingen for secretarial assistance. O. Schoots is supported by NWO grant GB-MW-970-10-015 and F. Hernando by TMR-grant ERBFMBI CT-96-1315 from the EU.

References

Adan RAH, Van Leeuwen FW, Sonnemans MA, Brouns M, Hoffman G, Verbalis JG, Burbach JPH (1995) Rat oxytocin receptor in brain, pituitary, mammary gland, and uterus: partial sequence and immunocytochemical localization. Endocrinology 136:4022–2028

Aguilera G (1994) Regulation of pituitary ACTH secretion during chronic stress. Front Neuroendocrinol 15:321–350

Ancellin N, Preisser L, Corman B, Morel A (1997) Role of protein kinase C and carboxyl-terminal region in acute desensitization of vasopressin V1a receptor. FEBS Lett 413:323–326

Antoni FA (1993) Vasopressinergic control of pituitary adrenocorticotropin secretion comes of age. Front Neuroendocrinol 14:76–122

Barberis C, Tribollet E (1996) Vasopressin and oxytocin receptors in the central nervous system. Crit Rev Neurobiol 10:119–154

Barberis C, Balestre MN, Jard S, Tribollet E, Arsenijevic Y, Dreifuss JJ, Bankowski K, Manning M, Chan WY, Schlosser SS (1995) Characterization of a novel, linear radioiodinated vasopressin antagonist: an excellent radioligand for vasopressin V1a receptors. Neuroendocrinology 62:135–146

Bargmann W (1966) Neurosecretion. Int Rev Cytol 19:183–201

Bichet DG, Birnbaumer M, Lonergan M, Arthus MF, Rosenthal W, Goodyer P, Nivet H, Benoit S, Giampietro P, Simonetti S (1994) Nature and recurrence of AVPR2 mutations in X-linked nephrogenic diabetes insipidus. Am J Hum Genet 55:278–286

Bichet DG, Oksche A, Rosenthal W (1997) Congenital nephrogenic diabetes insipidus. J Am Soc Nephrol 8:1951–1958

Birnbaumer M, Seibold A, Gilbert S, Ishido M, Barberis C, Antaramian A, Brabet P, Rosenthal W (1992) Molecular cloning of the receptor for human antidiuretic hormone. Nature 357: 333–335

Birnbaumer M, Gilbert S, Rosenthal W (1994) An extracellular congenital nephrogenic diabetes insipidus mutation of the vasopressin receptor reduces cell surface expression, affinity for ligand, and coupling to the Gs/adenylyl cyclase system. Mol Endocrinol 8:886–894

Briley EM, Lolait SJ, Axelrod J, Felder CC (1994) The cloned vasopressin V1a receptor stimulates phospholipase A2, phospholipase C, and phospholipase D through activation of receptor-operated calcium channels. Neuropeptides 27:63–74

Burbach JP, Lebouille JL (1983) Proteolytic conversion of arginine-vasopressin and oxytocin by brainsynaptic membranes. Characterization of formed peptides and mechanisms of proteolysis. J Biol Chem 258:1487–1494

Burbach JP, Kovacs GL, De Wied D, van Nispen JW, Greven HM (1983) A major metabolite of arginine vasopressin in the brain is a highly potent neuropeptide. Science 221:1310–1312

Burbach JPH, Schoots O, Hernando F (1998) Biochemistry of vasopressin fragments. In: Urban IJA, Burbach JPH, De Wied D (eds) Progr Brain Res 119:127–136

Burnatowska-Hledin MA, Spielman WS, Smith WL, Shi P, Meyer JM, Dewitt DL (1995) Expression cloning of an AVP-activated, calcium-mobilizing receptor from rabbit kidney medulla. Am J Physiol 268:F1198–210

Cheong HI, Park HW, Ha IS, Moon HN, Choi Y, Ko KW, Jun JK (1997) Six novel mutations in the vasopressin V2 receptor gene causing nephrogenic diabetes insipidus. Nephron 75: 431–437

Chini B, Mouillac B, Ala Y, Balestre MN, Trumpp-Kallmeyer S, Hoflack J, Elands J, Hibert M, Manning M, Jard S (1995) Tyr115 is the key residue for determining agonist selectivity in the V1a vasopressin receptor. EMBO J 14:2176–2182

Croiset G, De Wied D (1997) Proconvulsive effect of vasopressin; mediation by a putative V_2 receptor subtype in the central nervous system. Brain Res 759:18–23

Deen PM, Verdijk MA, Knoers NV, Wieringa B, Monnens LA, van Os CH, van Oost BA (1994) Requirement of human renal water channel aquaporin-2 for vasopressin-dependent concentration of urine. Science 264:92–95

De Keyzer Y, Auzan C, Lenne F, Beldjord C, Thibonnier M, Bertagna X, Clauser E (1994) Cloning and characterization of the human V3 pituitary vasopressin receptor. FEBS Lett 356:215–220

De Wied D (1997) The neuropeptide story. Geoffrey Harris Lecture, Budapest, Hungary, July 1994. Front Neuroendocrinol 18:101–113

De Wied D, Diamant M, Fodor M (1993) Central nervous system effects of the neurohypophyseal hormones and related peptides. Front Neuroendocrinol 14:251–302

Diaz Brinton R, Brownson EA (1993) Vasopressin-induction of cyclic AMP in cultured hippocampal neurons. Brain Res Dev Brain Res 71:101–105

Du Vigneaud V (1956) The isolation and proof of structure of the vasopressin and the synthesis of octapeptide amides with pressor-antidiuretic activity. Proc 3rd Int Congr Biochem Brussels, 1955, Academic Press, New York, USA, pp 49–58

Faux MC, Scott JD (1996) More on target with protein phosphorylation: conferring specificity by location. Trends Biochem Sci 21:312–315

Gerstberger R, Fahrenholz F (1989) Autoradiographic localization of V1 vasopressin binding sites in rat brain and kidney. Eur J Pharmacol 167:105–116

Gopalakrishnan V, McNeill JR, Sulakhe PV, Triggle CR (1988) Hepatic vasopressin receptor: differential effects of divalent cations, guanine nucleotides, and N-ethylmaleimide on agonist and antagonist interactions with the V1 subtype receptor. Endocrinology 123:922–931

Gross P, Richter D, Robertson GL (eds) (1993) Vasopressin. John Libbey Eurotext

Hartmann E, Rapoport TA, Lodish HF (1989) Predicting the orientation of eukaryotic mem-brane-spanning proteins. Proc Natl Acad Sci USA 86:5786–5790

Hausmann H, Richters A, Kreienkamp HJ, Meyerhof W, Mattes H, Lederis K, Zwiers H, Richter D (1996) Mutational analysis and molecular modeling of the nonapeptide hormone binding domains of the [Arg8]vasotocin receptor. Proc Natl Acad Sci USA 93:6907–6912

Hebert TE, Bouvier M (1998) Structural and functional aspects of G protein-coupled receptor oligomerization. Biochem Cell Biol 76:1–10

Hochberg Z, Van Lieburg A, Even L, Brenner B, Lanir N, van Oost BA, Knoers NV (1997) Autosomal recessive nephrogenic diabetes insipidus caused by an aquaporin-2 mutation. J Clin Endocrinol Metab 82:686–689

Hoekstra JA, van Lieburg AF, Monnens LA, Hulstijn-Dirkmaat GM, Knoers VV (1996) Cognitive and psychosocial functioning of patients with congenital nephrogenic diabetes insipidus. Am J Med Genet 61:81–88

Holsboer F (1989) Psychiatric implications of altered limbic-hypothalamic-pituitary-adrenocor-tical activity. Eur Arch Psychiatry Neurol Sci 238:302–322

Innamorati G, Sadeghi H, Birnbaumer M (1996) A fully active nonglycosylated V2 vasopressin receptor. Mol Pharmacol 50:467–473

Innamorati G, Sadeghi H, Eberle AN, Birnbaumer M (1997) Phosphorylation of the V2 vasopres-sin receptor. J Biol Chem 272:2486–2492

Innamorati G, Sadeghi H, Birnbaumer M (1998a) Transient phosphorylation of the V1a vaso-pressin receptor. J Biol Chem 273:7155–7161

Innamorati G, Sadeghi HM, Tran NT, Birnbaumer M (1998b) A serine cluster prevents recycling of the V2 vasopressin receptor. Proc Natl Acad Sci USA 95:2222–2226

Ivell R, Burbach JPH, van Leeuwen FW (1990) The molecular biology of the Brattleboro rat. Front Neuroendocrinol 11:313–338

Jans DA, Jans P, Luzius H, Fahrenholz F (1992) N-glycosylation plays a role in biosynthesis and internalization of the adenylate cyclase stimulating vasopressin V2-receptor of LLC-PK1 renal epithelial cells: an effect of concanavalin A on binding and expression. Arch Biochem Biophys 294:64–69

Jard S, Gaillard RC, Guillon G, Marie J, Schoenenberg P, Muller AF, Manning M, Sawyer WH (1986) Vasopressin antagonists allow demonstration of a novel type of vasopressin receptor in the rat adenohypophysis. Mol Pharmacol 30:171–177

Jard S, Barberis C, Audigier S, Tribollet E (1987) Neurohypophyseal hormone receptor systems in brain and periphery. Prog Brain Res 72:173–187

Johnson AE, Audigier S, Rossi F, Jard S, Tribollet E, Barberis C (1993) Localization and charac-terization of vasopressin binding sites in the rat brain using an iodinated linear AVP antago-nist. Brain Res 622:9–16

Kipreos ET, Lander LE, Wing JP, He WW, Hedgecock EM (1996) cul-1 is required for cell cycle exit in C. elegans and identifies a novel gene family. Cell 85:829–839

Kitano H, Takeda T, Suzuki M, Kitanishi T, Yazawa Y, Kitajima K, Kimura H, Tooyama I (1997) Vasopressin and oxytocin receptor mRNAs are expressed in the rat inner ear. Neuroreport 8:2289–2292

Knoers N, Monnens LA (1992) Nephrogenic diabetes insipidus: clinical symptoms, pathogene-sis, genetics and treatment. Pediatr Nephrol 6:476–482

Knoers NV, van Os CH (1996) Molecular and cellular defects in nephrogenic diabetes insipidus. Curr Opin Nephrol Hypertens 5:353–358

Kojro E, Eich P, Gimpl G, Fahrenholz F (1993) Direct identification of an extracellular agonist binding site in the renal V2 vasopressin receptor. Biochemistry 32:13537–13544

Liu J, Wess J (1996) Different single receptor domains determine the distinct G protein coupling profiles of members of the vasopressin receptor family. J Biol Chem 271:8772–8778

Lolait SJ, O'Carroll AM, McBride OW, Konig M, Morel A, Brownstein MJ (1992) Cloning and characterization of a vasopressin V2 receptor and possible link to nephrogenic diabetes insip-idus. Nature 357:336–339

Lolait SJ, O'Carroll AM, Mahan LC, Felder CC, Button DC, Young WS 3rd, Mezey E, Brownstein MJ (1995) Extrapituitary expression of the rat V1b vasopressin receptor gene. Proc Natl Acad Sci USA 92:6783–6787

Loup F, Tribollet E, Dubois-Dauphin M, Dreifuss JJ (1991) Localization of high-affinity binding sites for oxytocin and vasopressin in the human brain. An autoradiographic study. Brain Res 555:220–232

Mendre C, Dufour MN, Le Roux S, Seyer R, Guillou L, Calas B, Guillon G (1997) Synthetic rat V_{1a} vasopressin receptor fragments interfere with vasopressin binding via specific interaction with the receptor. J Biol Chem 272:21027–21036

Mouillac B, Chini B, Balestre MN, Elands J, Trumpp-Kallmeyer S, Hoflack J, Hibert M, Jard S, Barberis C (1995) The binding site of neuropeptide vasopressin V1a receptor, evidence for a major localization within transmembrane regions. J Biol Chem 270:25771–25777

Mulders SM, Knoers NV, van Lieburg AF, Monnens LA, Leumann E, Wuhl E, Schober E, Rijss JP, van Os CH, Deen PM (1997) New mutations in the AQP2 gene in nephrogenic diabetes insipidus resulting in functional but misrouted water channels. J Am Soc Nephrol 8:242–248

Mulders SM, Bichet DG, Rijss JP, Kamsteeg EJ, Arthus MF, Lonergan M, Fujiwara M, Morgan K, Leijendekker R, van der Sluijs P, van Os CH, Deen PM (1998) An aquaporin-2 water channel mutant which causes autosomal dominant nephrogenic diabetes insipidus is retained in the Golgi complex. J Clin Invest 102:57–66

Nicholson HD (1996) Oxytocin: a paracrine regulator of prostatic function. Rev Reprod 1:69–72

North WG, Fay MJ, Longo KA, Du JL (1998) Expression of all known vasopressin receptor subtypes by small cell tumors implies a multifaceted role for this neuropeptide. Cancer Res 58:1866–1871

Oksche A, Rosenthal W (1998) The molecular basis of nephrogenic diabetes insipidus. J Mol Med 76:326–337

Oksche A, Schulein R, Rutz C, Liebenhoff U, Dickson J, Muller H, Birnbaumer M, Rosenthal W (1996) Vasopressin V2 receptor mutants that cause X-linked nephrogenic diabetes insipidus: analysis of expression, processing, and function. Mol Pharmacol 50:820–828

Ostrowski NL, Young WS, 3d, Knepper MA, Lolait SJ (1993) Expression of vasopressin V1a and V2 receptor messenger ribonucleic acid in the liver and kidney of embryonic, developing, and adult rats. Endocrinology 133:1849–1859

Ostrowski NL, Lolait SJ, Young WS 3rd (1994) Cellular localization of vasopressin V1a receptor messenger ribonucleic acid in adult male rat brain, pineal, and brain vasculature. Endocrinology 135:1511–1528

Pan Y, Wilson P, Gitschier J (1994) The effect of eight V2 vasopressin receptor mutations on stimulation of adenylyl cyclase and binding to vasopressin. J Biol Chem 269:31933–31937

Pavo I, Fahrenholz F (1990) Differential inactivation of vasopressin receptor subtypes in isolated membranes and intact cells by N-ethylmaleimide. FEBS Lett 272:205–208

Phalipou S, Cotte N, Carnazzi E, Seyer R, Mahe E, Jard S, Barberis C, Mouillac B (1997) Mapping peptide-binding domains of the human V1a vasopressin receptor with a photoactivatable linear peptide antagonist. J Biol Chem 272:26536–26544

Phillips PA, Abrahams JM, Kelly JM, Mooser V, Trinder D, Johnston CI (1990) Localization of vasopressin binding sites in rat tissues using specific V1 and V2 selective ligands. Endocrinology 126:1478–1484

Postina R, Kojro E, Fahrenholz F (1996) Separate agonist and peptide antagonist binding sites of the oxytocin receptor defined by their transfer into the V2 vasopressin receptor. J Biol Chem 271:31593–31601

Ridge KD, Lee SS, Abdulaev NG (1996) Examining rhodopsin folding and assembly through expression of polypeptide fragments. J Biol Chem 271:7860–7867

Rittig S, Robertson GL, Siggaard C, Kovacs L, Gregersen N, Nyborg J, Pedersen EB (1996) Identification of 13 new mutations in the vasopressin-neurophysin II gene in 17 kindreds with familial autosomal dominant neurohypophyseal diabetes insipidus. Am J Hum Genet 58:107–117

Rosenthal W, Antaramian A, Gilbert S, Birnbaumer M (1993) Nephrogenic diabetes insipidus. A V2 vasopressin receptor unable to stimulate adenylyl cyclase. J Biol Chem 268: 13030–13033

Ruiz-Opazo N, Akimoto K, Herrera VL (1995) Identification of a novel dual angiotensin II/vasopressin receptor on the basis of molecular recognition theory. Nat Med 1: 1074–1081

Sadeghi HM, Innamorati G, Birnbaumer M (1997a) An X-linked NDI mutation reveals a requirement for cell surface V2R expression. Mol Endocrinol 11: 706–713

Sadeghi HM, Innamorati G, Dagarag M, Birnbaumer M (1997b) Palmitoylation of the V2 vasopressin receptor. Mol Pharmacol 52: 21–29

Saito M, Sugimoto T, Tahara A, Kawashima H (1995) Molecular cloning and characterization of rat V1b vasopressin receptor: evidence for its expression in extra-pituitary tissues. Biochem Biophys Res Commun 212: 751–757

Savarese TM, Fraser CM (1992) In vitro mutagenesis and the search for structure function relationships among G protein-coupled receptors. Biochem J 283: 1–19

Schulein R, Rutz C, Rosenthal W (1996) Membrane targeting and determination of transmembrane topology of the human vasopressin V2 receptor. J Biol Chem 271: 28844–28852

Sugimoto T, Saito M, Mochizuki S, Watanabe Y, Hashimoto S, Kawashima H (1994) Molecular cloning and functional expression of a cDNA encoding the human V1b vasopressin receptor. J Biol Chem 269: 27088–27092

Szot P, Bale TL, Dorsa DM (1994) Distribution of messenger RNA for the vasopressin V1a receptor in the CNS of male and female rats. Brain Res Mol Brain Res 24: 1–10

Tajima T, Nakae J, Takekoshi Y, Takahashi Y, Yuri K, Nagashima T, Fujieda K (1996) Three novel AVPR2 mutations in three Japanese families with X-linked nephrogenic diabetes insipidus. Pediatr Res 39: 522–526

Thibonnier M, Goraya T, Berti-Mattera L (1993) G protein coupling of human platelet V1 vascular vasopressin receptors. Am J Physiol 264: C1336–C1344

Thibonnier M, Preston JA, Dulin N, Wilkins PL, Berti-Mattera LN, Mattera R (1997) The human V_3 pituitary vasopressin receptor: ligand binding profile and density-dependent signaling pathways. Endocrinology 138: 4109–4122

Tribollet E (1992) Vasopressin and oxytocin receptors in the rat brain. In: Björklund A, Hökfelt T, Kuhar MJ (eds) Handbook of chemical neuroanatomy. Elsevier Amsterdam, 289 pp

Tribollet E, Barberis C, Dreifuss JJ, Jard S (1988) Autoradiographic localization of vasopressin and oxytocin binding sites in rat kidney. Kidney Int 33: 959–965

Tsukaguchi H, Matsubara H, Inada M (1995a) Expression studies of two vasopressin V2 receptor gene mutations, R202C and 804insG, in nephrogenic diabetes insipidus. Kidney Int 48: 554–562.

Tsukaguchi H, Matsubara H, Mori Y, Yoshimasa Y, Yoshimasa T, Nakao K, Inada M (1995b) Two vasopressin type 2 receptor gene mutations R143P and delta V278 in patients with nephrogenic diabetes insipidus impair ligand binding of the receptor. Biochem Biophys Res Commun 211: 967–977

Tsukaguchi H, Matsubara H, Taketani S, Mori Y, Seido T, Inada M (1995c) Binding-, intracellular transport-, and biosynthesis-defective mutants of vasopressin type 2 receptor in patients with X-linked nephrogenic diabetes insipidus. J Clin Invest 96: 2043–2050

Urban IJA (1998) Effects of vasopressin and related peptides on neurons of the rat lateral septum and ventral hippocampus. In: Urban IJA, Burbach JPH, De Wied D (eds) Progress in brain research. Elsevier, Amsterdam (in press)

Valtin H (1967) Hereditary hypothalamic diabetes insipidus in rats (Brattleboro strain). A useful experimental model. Am J Med 42: 814–827

Van Lieburg AF, Verdijk MA, Knoers NVAM, Van Essen AJ, Proesmans W, Mallmann R, Monnens LA, Van Oost BA, Van Os CH, Deen PM (1994a) Patients with autosomal nephrogenic diabetes insipidus homozygous for mutations in the aquaporin 2 water-channel gene. Am J Hum Genet 55: 648–652

Van Lieburg AF, Verdijk MA, Knoers NVAM, Afer E, Pastina R, Fahrenholz F, van Oost BA (1994b) In vitro expression of mutations in the V2 receptor gene confirmation of their role in the pathogenesis of X-linked nephrogenic diabetes insipidus. Pediatr Nephrol 8: C75

von Heijne G (1994) Membrane proteins: from sequence to structure. Annu Rev Biophys Biomol Struct 23:167–192

Wenkert D, Schoneberg T, Merendino JJ Jr, Rodriguez Pena MS, Vinitsky R, Goldsmith PK, Wess J, Spiegel AM (1996) Functional characterization of five V2 vasopressin receptor gene mutations. Mol Cell Endocrinol 124:43–50

Wess J, Nanavati S, Vogel Z, Maggio R (1993) Functional role of proline and tryptophan residues highly conserved among G protein-coupled receptors studied by mutational analysis of the m3 muscarinic receptor. EMBO J 12:331–338

Yokoyama K, Yamauchi A, Izumi M, Itoh T, Ando A, Imai E, Kamada T, Ueda N (1996) A low-affinity vasopressin V2-receptor gene in a kindred with X-linked nephrogenic diabetes insipidus. J Am Soc Nephrol 7:410–414

The Oxytocin Receptor

Tadashi Kimura[1] and Richard Ivell[2]

1
Introduction

The oxytocin receptor was first identified using a pharmacological ligand-binding assay in the rat myometrium (Soloff and Swartz 1973). Its peptide ligand, oxytocin, belongs to the nonapeptide hormone family comprising both oxytocin-like (mesotocin, isotocin, etc.) and vasopressin-like (vasotocin, phenypressin, etc.) cyclic peptides. Amongst eutherian mammals, arginine[8]-vasopressin (AVP), though having its own specific receptors, was shown to bind to the oxytocin receptor with almost as high affinity as oxytocin itself. Oxytocin was the first peptide hormone whose amino acid sequence was completely elucidated, and the first to be chemically synthesized (Du Vigneaud et al. 1953a,b). Because of the relatively simple structure of these peptide hormones, a large number of synthetic agonists and antagonists have been developed and tested (reviewed by Manning et al. 1995). Pharmacological studies using such peptides did allow a degree of speculation on structural aspects of the receptor-ligand binding interaction, though only upon the cloning of the receptor some 40 years after the elaboration of the peptide hormone has it been possible to pursue such studies in detail.

Initially oxytocin was considered solely as the hormone implicated in the contraction of uterine muscle at parturition (labour) and of the myoepithelial cells of the mammary gland during milk ejection. From more recent experiments using a variety of agonists and antagonists, as well as from studies on manipulating gene expression, we know that oxytocin is involved at paracrine as well as endocrine levels in a wide range of physiological functions including sexual and maternal behavior, feeding and satiety, memory, follicular and oocyte regulation in the ovary, as a modulator of luteolysis and hence the estrous cycle, penile erection, ejaculation, kidney function and sodium homeostasis (reviewed in Ivell and Russell 1995). These actions all appear to be mediated by a single type of specific oxytocin receptor (OTR) expressed

[1] Department of Obstetrics and Gynecology, Osaka University Medical School, Osaka 5650871, Japan
[2] Department of Reproductive Science, IHF Institute for Hormone and Fertility Research, University of Hamburg, 22529 Hamburg, Germany

from a single copy gene in different organs and tissues of the body. Because we are dealing with a single type of ligand and a single type of receptor, yet mediating a wide variety of discrete physiological responses, an essential part in understanding oxytocin physiology is to understand how the ligand and its receptor are differentially regulated in a defined time- and cell-specific manner in the different tissues.

In the uterus, the first known target tissue for oxytocin (Dale 1906), the level of receptor protein indicated by a ligand-binding assay increased dramatically at the time of parturition (Soloff et al. 1979; Fuchs et al. 1984) within a smaller time-frame even than the increasing levels of oxytocin being released at this time from the pituitary. OTR levels in the brain were shown using similar assays to be upregulated by *in vivo* administration of estrogens. However, there is a limitation to being able to analyze these phenomena using such classical ligand-binding assays.

The first molecular cloning of members of the nonapeptide hormone receptor family (oxytocin receptor, Kimura et al. 1992b; vasopressin V1 receptor, Morel et al. 1992; V2 receptor, Birnbaumer et al. 1992; Lolait et al. 1992) not only provided the primary amino acid sequence of the OTR, but also opened up a whole new set of tools and methods for analyzing the regulation of oxytocin-dependent functions. Comparison of the structures from different species for the various oxytocin and vasopressin receptors has allowed a detailed analysis of the receptor structure-binding activity relationship. Also, the genes for oxytocin receptors have been cloned from several different species (Inoue et al. 1994; Bathgate et al. 1995; Rozen et al. 1995; Kubota et al. 1996; Young et al. 1996), providing further valuable information on the molecular mechanisms of receptor expression and transcription. The present chapter reviews current progress being made to investigate the structure of the oxytocin receptor, its expression and transcriptional regulation.

2
Pharmacology of the Oxytocin Receptor and its Signal Transduction

Molecular cloning of the oxytocin receptor (OTR) from a variety of eutherian species, as well as of the closely related mesotocin receptor from marsupials, birds and amphibea, has shown that it belongs to the G-protein coupled receptor family with seven transmembrane domains. A comparison of the encoded protein structure with that of vasopressin receptors and other nonapeptide hormone receptors allows intelligent speculation as to which amino acid residues or domains of the receptor are important for ligand binding as well as for ligand selectivity. The signal transduction system of the OTR has also been intensively analyzed in several different tissues. In the following, we summarize a number of studies employing mutation analysis and pharmacological assays, as well as investigations on the receptor signal transduction system, in order to provide an overview of our current understanding of OTR structure-function relationships.

2.1
Extracellular N-Terminal Domain

Comparing the OTR from six eutherian species (human, Kimura et al. 1992b; pig, Gorbulev et al. 1993; sheep, Riley et al. 1995; cow, Bathgate et al. 1995; rat, Rozen et al. 1995; mouse, Kubota et al. 1996) indicates that the amino acid sequence of the N-terminal domain is relatively poorly conserved (Fig. 1). As Kojro et al. (1991) have shown, the extracellular domain of the OTR molecule is highly glycosylated *in vivo*. There are either two (rat and mouse) or three (human, pig, bovine, sheep) putative N-glycosylation sites (Asn-X-Ser/Thr) within this N-terminal domain. It was thus of interest to determine whether alteration of the glycosylation pattern could affect ligand-binding or ligand selectivity. However, a mutational analysis of the three putative N-glycosylation sites in the human OTR, converting the asparagine (Asn) residues at positions 8, 15, or 26 to aspartate (Asp), showed that loss of one or two of these three N-glycosylation sites in the human OTR failed to influence either ligand-binding or ligand selectivity (Kimura et al. 1997). The non-glycosylated vasopressin V2 receptor (V2R) also exhibited full activity in terms of ligand-binding and signal transduction (Innamorati et al. 1996). A chimeric V2 receptor possessing the N-terminal domain of the OTR was shown to have an approximately six-times higher affinity for oxytocin than the original V2 receptor, although the affinity for AVP and arginine vasotocin (AVT) was not altered by this chimeric mutation (Postina et al. 1996). A similar phenomenon was observed for the vasotocin receptor (VTR) of the fish *Catostomus commersoni*; when the N-terminal domain of the teleost isotocin receptor (ITR) was substituted for the equivalent domain of the VTR, the binding characteristics towards vasotocin and isotocin were similar to those of the original VTR. Moreover, a VTR with a truncated N-terminal domain still retained a relatively high K_d for vasotocin (0.06 nM for wild type VTR vs 1.46 nM for the N-terminally truncated VTR; Hausmann et al. 1996). Together, these observations suggest that the N-terminal domain of the OTR, including the glycosylation sites, only plays a partial role in nonapeptide ligand selectivity, and is not essential for ligand binding.

2.2
Extracellular Loop Domains

There are three extracellular loops in the OTR, all of which are well conserved among species. By applying photoaffinity labeling using a radioactive agonist and sequencing of the linked peptide fragments, amino acid residues within

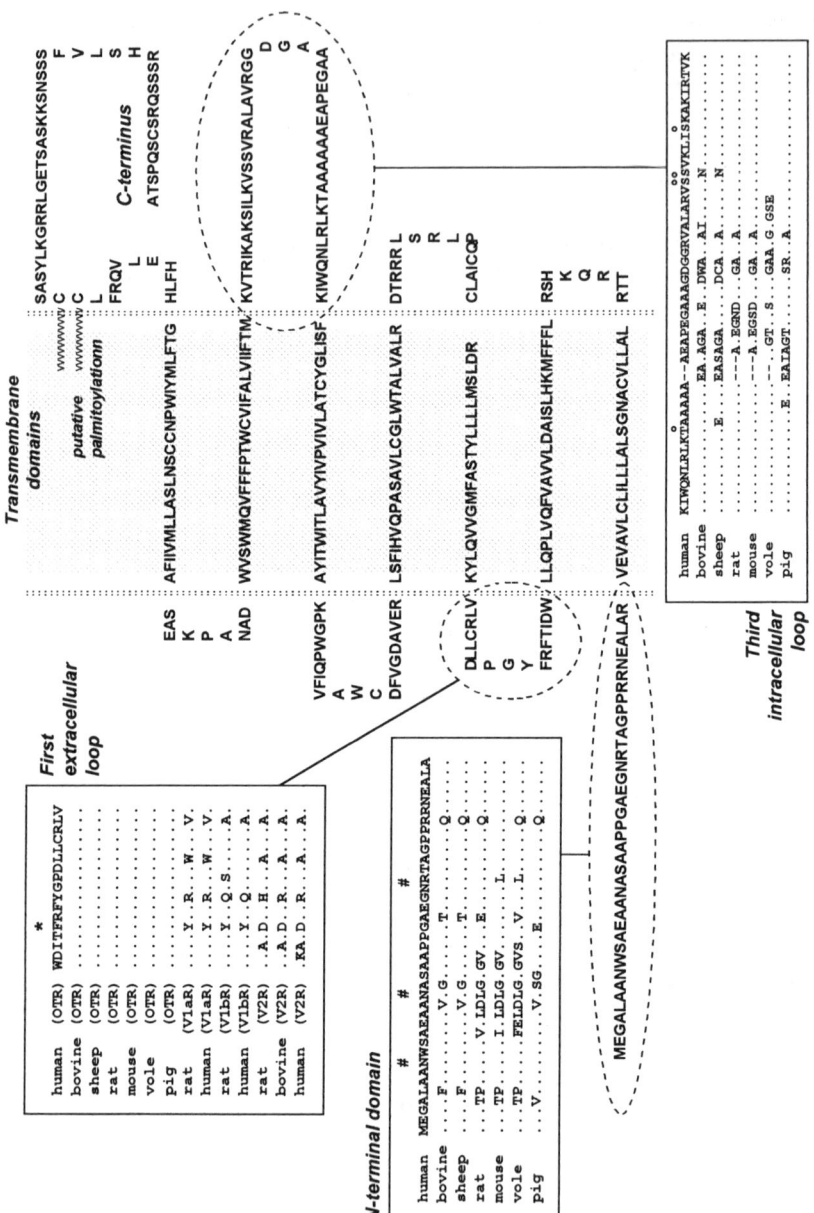

Fig. 1. Human oxytocin receptor primary and secondary structure, indicating extracellular, transmembrane and intracellular domains. Regions of special interest are *boxed*, where primary sequence of human OTR is compared with equivalent sequences from other species (see text for details). * indicates the hydrophobic residue in the first extracytoplasmic loop equivalent to Tyr[115] of the rat V1a receptor; # indicates sites of potential N-glycosylation in the N-terminal domain; ° indicates potential sites of phosphorylation in the third intracytoplasmic loop

the first extracellular loop of the bovine V2 receptor were identified as a part of the hormone binding site (Kojro et al. 1993). Furthermore, synthetic peptide mimetics of the first extracellular loop of the rat V1aR were able to inhibit effectively the binding of AVP, whereas peptide mimetics of the N-terminal domain or the second extracellular loop were less effective, though still showed some interference (Howl and Wheatley 1996). In the rat V1a receptor, substitution of Tyr[115] (conserved in V1a receptors) by Phe (conserved in OTR and VTR) increased the affinity of the mutant V1aR towards oxytocin by approximately 19-fold. The alteration of Tyr[115] to Asp (conserved in V2 receptors) had no effect on the affinity of the mutant receptor toward AVP, though this mutant receptor did have a much higher affinity toward DAVP or dDAVP, which are specific V2 receptor agonists (Chini et al. 1995). In the bovine V2R, substitution of Asp by Tyr also caused a loss of affinity toward V2 agonists (Ufer et al. 1995). These results all suggest that the first extracellular domain plays a crucial role for ligand binding and selectivity. However, when a chimeric porcine V2R was constructed with the first, second, or third extracellular domains of the OTR, the new molecule had a similar low affinity toward oxytocin as the original V2R. If the entire extracellular portion of the V2R (i.e. the N-termimal and all three extracellular loops) were replaced by the equivalent domains of the OTR, the affinity for oxytocin was only ten-fold higher than in the original V2R, still well below that of the OTR (Postina et al. 1996). These results imply that oxytocin is not captured only by the extracellular loops of the receptor, but requires interaction with other parts of the receptor molecule.

2.3
Transmembrane Domains

Although the transmembrane domains (TM) are predicted to localize within the membrane lipid bilayer, they nevertheless appear to be important for ligand binding and selectivity. By docking AVP into a three-dimensional model of the V1aR, constructed by comparison with the low resolution map of bovine rhodopsin, it has been speculated that the hormone might be buried into a 15 to 20 Å deep cleft defined by the transmembrane helices of the receptor. Experiments in which point mutations were introduced into the second, third, or fourth TMs of V1aR showed that such alterations strongly interfered with AVP binding. Interestingly, these mutated receptors still had a high affinity for peptide or nonpeptide AVP antagonists (d(CH$_2$)$_5$[Tyr(Me)^2]AVP and SR 49059) compared to the original V1aR (Mouillac et al. 1995). This suggests that there are different contact sites within the receptor molecule between the receptor and the natural ligand and for ligand-specific antagonists. A point mutation of Asp[85] in the second TM of the OTR to Asn caused a complete loss of binding activity and physiological responsiveness to oxytocin (Yarwood et al. 1997). This result also underlines the structural importance of the second

TM. However, exchanging the first and second TM of a chimeric V2R to OTR sequences led only to a low affinity for oxytocin, and still a high affinity for AVP (Postina et al. 1996). As the amino acid sequences of the first and second TMs are highly homologous between the OTR and the V2R, the exchange of these domains might not be expected to affect ligand selectivity.

AVP appears to act as a partial agonist of the human OTR. When measured by rat uterine muscle contraction, or by the electrophysiological response in OTR-expressing *Xenopus laevis* oocytes following OTR-cRNA injection, the EC_{50} values for AVP to give the same physiological response were 25- to 100-fold higher than for oxytocin (Kimura et al. 1994; Chan et al. 1996). Substitution of aromatic amino acid Tyr^{209} to Phe in the fifth TM or Phe^{284} to Tyr in the sixth TM of the human OTR, to mimic the equivalent residues in the V1aR, improved the physiological responsiveness to AVP as measured by the production of inositol phosphates. The dose-response curve for oxytocin was nevertheless still very similar to that of wild-type OTR. This observation suggests that the TMs may also influence the ligand selectivity of the signal transduction of the OTR (Chini et al. 1996).

Also, the spatial alignment of the TMs seems to be important for ligand binding and selectivity. When the human OTR tagged by a partial c-myc sequence was expressed in insect Sf9 cells using a baculovirus expression vector, the OTR protein could be shown to be successfully expressed by immunoblotting; however, the receptor affinity toward oxytocin was very low ($K_d = 215$ nM). The cell membranes of insect cells are known to have a lower cholesterol content than mammalian cells, thus in terms of membrane fluidity appearing more rigid. When cholesterol was supplied to the membrane fraction of Sf9 cells transfected with OTR using β-cyclodextrin as a cholesterol carrier, the affinity to oxytocin was dramatically improved ($K_d = 0.96$ nM) (Gimpl et al. 1995). When these authors applied the same procedure to the natural OTR in rat myometrium, they were able to alter the K_d for oxytocin about 87-fold (1.5 to 131 nM) by depleting or increasing the cholesterol content of the membrane lipids. This effect was specific for cholesterol; the substitution of cholesterol by related sterols (stigmasterol, 5-cholesten-3-one, 5-cholesten, and pregnenolone) only succeeded in worsening the OTR binding affinity for oxytocin (Klein et al. 1995). These results highlight the importance of the spatial distribution and/or dynamics of the TMs of the receptor, and may offer a possible explanation for the existence of different oxytocin receptor subtypes as postulated by several authors (Maggi et al. 1990; Pliska and Kohnhauf Albertin 1991).

2.4
Intracellular Loops, C-Terminus and Signal Transduction

Ligand-dependent information signalling by the OT-OTR system is mediated by G-protein-receptor coupling. Since oxytocin stimulation causes the intracellular release of inositol triphoshophate and Ca^{2+} (Marc et al. 1986), the major signal transducer of the OTR was considered to be $G_{\alpha q}$ (Phaneuf et al. 1993). However, when OTR was solubilized and fractionated by size-exclusion chromatography, oxytocin binding activity co-eluted with both $G_{\alpha q/11}$ and $G_{\alpha i3}$ (Strakova and Soloff 1997). The interaction between the G-protein and the OTR should theoretically be mediated by the intracellular loops or the C-terminus of the receptor, although to date there is still no structural evidence for this. The amino acid sequences of the first and second intracellular loops in the OTR are well conserved among species, suggesting that they have a significant role in signal transduction. Indeed, in the human V2R, a point mutation of Arg^{137} to His abolished coupling to G-proteins, and caused a receptor-dependent diabetes insipidus (Rosenthal et al. 1993). The amino acid sequence of the third intracellular loop appears to be highly variable between different species (Kaluz et al. 1996). In contrast, the OTR signal transduction system appears to be very compatible among species. For example, the human OTR can be functionally reconstituted in *Xenopus laevis* oocytes (Morley et al. 1988; Kimura et al. 1992b), green monkey COS cells, mouse LTK^- cells or hamster COS7 cells (Chini et al. 1996). Although it is possible to speculate about possible important amino acid residues in the third intracellular loop on the basis of their conservation in different species (Fig. 1), no mutation studies of the OTR have yet been reported for this region of the receptor.

Oxytocin can also cause activation of mitogen-activated protein (MAP) kinase activity by phosphorylation. This phosphorylation can be attenuated by islet-activating protein (IAP), suggesting that the activation of the MAP kinase was indeed mediated by an OTR-coupled G-protein (Ohmichi et al. 1995). The phosphorylation of the MAP kinase is mediated by MAP kinase kinase (MEK), and an inhibitor of MEK can partially attenuate the uterine contraction elicited by oxytocin (Nohara et al. 1996). Therefore a protein kinase cascade can mediate one aspect of OTR-specific signal transduction, although the G-protein involved in the physical interaction between the receptor and the kinase pathway is still uncharacterized.

Another structural feature of the OTR is the putative palmitoylation at the paired Cys residues (Cys^{346}–Cys^{347}) in the C-terminus. The sequence around these cysteines is well conserved among species. In the V2R, it has been shown that the equivalent Cys^{341}–Cys^{342} residues of the C-terminus are indeed palmitoylated. When these two Cys residues were substituted by Ser, palmitoylation was totally prevented. This mutant receptor had the same affinity for AVP, and the same physiological response in terms of cAMP production; however, the number of cell surface receptors was significantly decreased (Sadeghi et al. 1997). As the Cys-Cys residues exist at almost the same position in the OTR, we can probably assume that, also for this receptor, palmitoylation could act as an anchoring stabilizer of the receptor to the cell membrane.

The OTR appears to be subject to desensitization upon continuous oxytocin stimulation *in vitro* or in actual clinical cases *in vivo* (Phaneuf et al. 1997,

1998). This process, however, is very slow, and prominent receptor desensitization is only evident after 12 h of ligand stimulation; in comparison, the pituitary GnRH receptor is desensitized within 2 min (Weiss et al. 1995). The binding capacity on the membrane surface, intracellular Ca^{2+} elevation after stimulation, and the expression level of OTR-mRNA were all markedly suppressed after 12 h. However, the receptor protein level detected by an anti-OTR antibody was the same before as after the desensitization process (Phaneuf et al. 1997). These results suggest that there may be modification of the receptor protein causing the decreased affinity and inactivation of the OTR. There are several putative amino acid targets for phosphorylation in the third intracellular loop, which might be addressed during the desensitization-related modification of the receptor, and which could be studied by site-specific mutagenesis.

2.5
Summary

Ligand-receptor interaction of the OTR seems to involve mostly the first extracellular loop and the transmembrane domains. N-glycosylation in the N-terminal domain is not critical for ligand-binding. Signal transduction is partly regulated by the transmembrane domains; however, the structural basis for the interaction between the OTR and G-proteins or the MAP kinase system still requires detailed investigation. Such studies could encourage the development of new substances for a safe and effective OTR blockade, which obstetricians are keenly demanding.

3
Expression of the Oxytocin Receptor *in Vivo* and its Functional Differentiation

The oxytocin receptor is expressed in a wide variety of different tissues and organs (Table 1), and has been studied using several independent techniques (ligand-binding to tissue extracts, receptor autoradiography, immunohistochemistry, RNA measurements, *in situ* hybridization) in a broad range of species (see Ivell and Russell 1995 for a recent overview). In the following sections, we shall focus on those biological systems where the regulation of the receptor is well documented, and where we can use such systems to learn about the molecular mechanisms involved in that regulation.

3.1
The Uterus of Pregnancy

For historical reasons, the target tissue which has been subjected to most intensive study is the uterus at term of pregnancy, where oxytocin is involved

Table 1. Oxytocin receptor concentrations in different tissues and cells calculated as Bmax values from Scatchard analyses. All concentrations unless otherwise stated are expressed as fmol OTR per mg protein in membrane preparations, and represent basal, unstimulated levels. For conversion purposes, 1 mg DNA is equivalent to 1.67×10^5 cells, and 1 fmol is equal to 6×10^8 receptor molecules

Species	Tissue/cell type	OTR concentration (fmol/mg protein)	Alternative values	Reference
Bovine	cervix mucosa (estrus)	2500–7000		Fuchs et al. (1996b)
	(mid-cycle)	100– 800		Fuchs et al. (1996b)
	cervix mucosa (labour)	2000–6000		Fuchs et al. (1996a)
	cervix muscle (labour)	100– 400		Fuchs et al. (1996a)
	endometrium (prepubertal)	1000–2500		Fuchs et al. (1998)
	endometrium (early pregnancy)	40		Fuchs et al. (1992a)
	endometrium (labour)	7300		Fuchs et al. (1992a)
	myometrium (estrus)	344		Fuchs et al. (1992a)
	myometrium (early pregnancy)	180		Fuchs et al. (1992a)
	myometrium (labour)	1850		Fuchs et al. (1992a)
	fetal membranes, etc. (non-term)	<20		Fuchs et al. (1992b)
	cotyledons (labour)	163		Fuchs et al. (1992b)
	chorioallantois	270		Fuchs et al. (1992b)
	amnion	311		Fuchs et al. (1992b)
	endometrium (cycle av.)	563		Jenner et al. (1991)
	endometrium (estrus)	1600		Jenner et al. (1991)
	myometrium (cycle av.)	82		Jenner et al. (1991)
	myometrium (estrus)	1500		Jenner et al. (1991)
	endometrium (estrus)	1600		Bathgate et al. (1995)
	myometrium (mid cycle)	100– 400		Soloff and Fields (1989)
	myometrium (estrus)	800–1200		Soloff and Fields (1989)
	ovarian stroma		56– 292 fmol/mg DNA	Fuchs et al. (1990)
	corpus luteum		42–3740 fmol/mg DNA	Fuchs et al. (1990)
	large follicles		76– 475 fmol/mg DNA	Fuchs et al. (1990)
	granulosa cells (small follicles)		<10 fmol/µg DNA	Okuda et al. (1997)
	luteal cells		<10 fmol/µg DNA	Okuda et al. (1997)
	cultured endometrial epithelial cells	1500–2000		Horn et al. (1998)
	adrenal medulla	90		Nussey et al. (1987)
Ovine	uterus	588		Sernia et al. (1989)
	corpus luteum (pregnancy)	9–40		Sernia et al. (1989)
	endometrium (ovex)	1346		Wathes et al., (1996a)

Table 1. Continued

Species	Tissue/cell type	OTR concentration (fmol/mg protein)	Alternative values	Reference
Ovine	endometrium (ovex + E2)	140		Wathes et al., (1996a)
	endometrium (ovex + P)	0–306		Wathes et al., (1996a)
	myometrium (ovex)	277		Wathes et al., (1996a)
	myometrium (ovex + E2)	255		Wathes et al., (1996a)
	myometrium (ovex + P)	61		Wathes et al., (1996a)
	cultured endometrium	400–500		Sheldrick and Flick-Smith (1993)
	pineal gland	68		Rahmani et al. (1997)
Rat	uterus	800		Grazzini et al. (1998a)
	myometrium (term)	20–34		Alexandrova and Soloff (1980)
	myometrium (term)		56–230 fmol/mg DNA	Fuchs et al. (1983)
	mammary gland (lactation)	2000		Soloff and Wieder, (1983)
	mammary gland (non-lactating)	100–300		
	thymus	42		Elands et al. (1990)
Human	myometrium (term)	70–223		Akerlund et al. (1995)
	myometrium (non-pregnant)		28 fmol/mg DNA	Fuchs et al. (1982)
	myometrium (labour)		2300–3500 fmol/mg DNA	Fuchs et al. (1982)
	myometrium (before labour)	10		Maggi et al. (1990)
	myometrium (early labour)	400		Maggi et al. (1990)
	myometrium (max. labour)	1000		Maggi et al. (1990)
	total ovary		72–157 fmol/mg DNA	Fuchs et al. (1990)
	corpus luteum (cycle)	25–125		Behrens et al. (1995)
	ovarian stroma	30–56		Behrens et al. (1995)
	mammary tissue (basal)	28–60		Taylor et al. (1990)
	mammary cell-line (HS5787)	58		Copland et al. (1997)
	mammary cell-line (MCF7)		87000 receptors/cell	Taylor et al. (1990)
	primary myometrial cells	ca. 300	20000 receptors/cell	Phaneuf et al. (1997)
	primary myometrial cells		500 fmol/106 cells	Maggi et al. (1994)
	cloned myometrial cell-line (D6)		2940 fmol/106 cells	Maggi et al. (1994)

Table 1. Continued

Species	Tissue/cell type	OTR concentration (fmol/mg protein)	Alternative values	Reference
Human	smooth muscle vascular epithelial cells	122		Yazawa et al. (1996)
	prostate carcinoma cells (DU145)	55		Bathgate R (unpubl.)
Porcine	endometrium (pregnancy)	200–1200		Lundin-Schiller et al. (1996)
	myometrium (pregnancy)	250–600		Lundin-Schiller et al. (1996)
	mammary gland (pregnancy)	500–2400		Lundin-Schiller et al. (1996)
	corpus luteum (cycle)	65– 116		Pitzel et al. (1993)
	testis tunica albuginea	45		Maggi et al. (1987)
	epididymis	26		Maggi et al. (1987)
	vas deferens	20		Maggi et al. (1987)
	porcine kidney cells (LLC-PK1)	100		Cantau et al. (1990)
Rabbit	myometrium (non-pregnant)	20		Maggi et al. (1988)
	myometrium (labour)	2962		Maggi et al. (1988)
	vagina	not detectable		Maggi et al. (1988)
	oviduct	not detectable		Maggi et al. (1988)
Wallaby	myometrium (non-pregnant)	100– 400		Parry et al. (1997)
	myometrium (term)	600–1200		Parry et al. (1997)
Chicken	kidney	42– 129		Takahashi et al. (1996)
Transfected cells	rat OTR in CHO cells	960		Grazzini et al. (1998a)
	porcine OTR in COS7 cells		80 fmol/106 cells	Gorbulev et al. (1993)
	human OTR in 293 cells		820	Jasper et al. (1995)
	bovine OTR in COS7 cells	3200		Bathgate et al. (1995)

in labour and the expulsion of the fetus. The uterus in pregnancy comprises two principal compartments, the myometrium and the immediately juxtaposed complex of endometrium, decidua, trophoblast and fetal membranes. The physiological uterine response to exogenous oxytocin is primarily contraction of the smooth muscle cells of the myometrium. In the uterus at term of pregnancy the sensitivity of the myometrium to oxytocin is 200 to 1000-fold greater than in the uterus of the cycle (Caldeyro-Barcia and Theobald 1968).

This reflects the massive upregulation of oxytocin receptors seen, for example, in human term myometrium (Table 1). At a molecular level, the human term myometrium appears to express the highest level of OTR-mRNA of any species so far examined (Kimura et al. 1992a,b). In general, the level of OTR-mRNA and protein within the myometrium increases during the course of pregnancy, reaching dramatic peak values just prior to the onset of labour (Fuchs et al. 1984; Kimura et al. 1996). Similar results have been obtained also for the cow (Ivell et al. 1995). However, in rats (Larcher et al. 1995) and mice (Kubota et al. 1996), there appears to be a more disjointed pattern of expression, with upregulation of the OTR observed only on the day of parturition itself. That this is not restricted to eutherians is shown by a very similar pattern of expression for the mesotocin receptor in the uterus of the pregnant tammar wallaby (Parry et al. 1997). It should be pointed out, however, that it is still not clear whether the relatively high level of OTR expression (about one-third of term myometrium) obseved in the late second or early third trimester observed in the human uterus may not be an artifact due to operative manipulation.

In the human term myometrium, OTR-mRNA and protein appear not to be distributed homogeneously in the myometrial layer even at the time of maximum upregulation (Kimura et al. 1996). During labour, there is increased expression in the myometrium of connexin 43 which facilitates cell-cell communication by gap junction formation (Garfield et al. 1977; Balducci et al. 1993). It is therefore not essential that all of the myometrial cells respond to oxytocin stimuli and contract independently; the OTR-positive cells could respond to oxytocin stimulation, raise intracellular calcium ion concentration and then activate neighbouring cells via the gap junctions. This hypothesis would explain many of the observations made concerning uterine contraction at labour; however, electrophysiological studies are required to explore this idea further.

In the ruminant uterus of pregnancy, the highest expression of OTR appears to be within the endometrium, in the epithelial cells of the uterine lumen and endometrial glands, rather than in the myometrium (Bathgate et al. 1995; Ivell et al. 1995; Wu et al. 1996; Table 1). In the human, endometrial epithelial cells are very reduced during pregnancy; instead, OTR expression was observed in decidual cells, extravillous trophoblast cells (Takemura et al. 1994) and amnion cells (Moore et al. 1988). In these non-myometrial tissues, the OTR-mRNA levels increase during the course of parturition, there being up to five-fold more at parturition compared with before the onset of labour, also implicating these receptors in uterine contraction. It is known that oxytocin stimulation of both endometrial cells and fetal membranes causes secretion of $PGF_{2\alpha}$ (Fuchs et al. 1981), which itself is an important contractile agent. During successful labour induction by a continuous infusion of oxytocin, serum levels of PGF metabolites increase during the course of parturition (Husslein et al. 1981). Oxytocin can also induce PGE_2 secretion in human amnion cells, which also express OTR (Moore et al. 1988). Although it is assumed that the principal uterotonic oxytocin derives from the posterior pituitary, it should not be forgotten that there is also extraneurohypophyseal *de novo* oxytocin synthesis in the rat endometrium (Lefebvre et al. 1992) and in human fetal membranes (Chibbar et al. 1993) at term of pregnancy. In the human, oxytocin-mRNA has been localized in the decidua, extravillous trophoblast and in the amnion, where oxytocin receptors are also detected. This oxytocin gene expression was also upregulated about three- to four-fold around parturition. Together, these results suggest that in addition to the systemic oxytocin released from the pituitary, there is also a local autocrine/paracrine oxytocin-OTR circuit within the fetal membranes and uterus which might be controlling or modulating the progress of parturition.

3.2
The Uterus of the Estrous Cycle

The uterus of the estrous cycle is also an important target organ for oxytocin, although the physiological importance of this action seems to vary between species. It is particularly evident in the ruminant, where OTR expression is mainly observed in the endometrial epithelium (Bathgate et al. 1995; Ivell et al. 1995). Both in the cow and sheep, the maximum OTR expression was observed on the day of ovulation, marking also the end of the cycle (Jenner et al. 1991; Stewart et al. 1993; Ivell et al. 1995). The endometrial OTR plays a key role in the determination of cycle length in ruminants and in the persistence of luteal function upon maternal recognition of pregnancy. Circulating oxytocin, mostly secreted from the corpus luteum in ruminants, stimulates the endometrial OTR, leading to an upregulation of prostaglandin (PG)$F_{2\alpha}$ synthesis and secretion. This $PGF_{2\alpha}$ then itself interacts with ovarian prostaglandin receptors causing further secretion of luteal oxytocin, but at the same time inducing luteoloysis and the termination of the cycle (reviewed by Flint et al. 1994). This creates a positive feedback loop between the corpus luteum and the uterus which is only interrupted in the event of conception occurring. Then the early blastocyst produces an interferon-like molecule (interferon-τ) which interacts with specific receptors on the endometrial epithelial cells, and prevents the upregulation of the OTR on these cells. The key element in this system is the very restricted expression of the OTR on the endometrial epithelial cells. For most of the cycle these receptors are suppressed, and only at the end of the cycle are they massively upregulated (unless conception occurs) and allow the luteolytic feedback loop to become established. Thus, in the sheep, hysterectomy by also interrupting the feedback loop markedly prolongs the estrous cycle (Dobrowlski and Hafez, 1971). The OTR is chiefly found in the endometrial glandular epithelial cells (Ayad et al. 1991) and in the luminal epithelium (Bathgate et al. 1995).

In the human, the luteolytic process is considered to be independent of the uterus, although the OTR concentration in the endometrium is higher in the follicular phase than in the luteal phase of the cycle (Fuchs et al. 1985).

Furthermore, *in situ* hybridization and semi-quantitative PCR revealed the expression level of OTR mRNA to be highest at the time of ovulation (Takemura et al. 1993), implying that in humans as in ruminants there appears to be a similar underlying regulation for the OTR in the endometrium of the cycle. The function of endometrial OTR at ovulation in the human is unclear. Endometrial samples of subfertile patients showed fewer oxytocin-binding sites than those from normal women (Baker et al. 1990). Since it has also been shown that mouse oocyte-cumulus complexes also produce oxytocin, and that oxytocin can influence the implantation rate in this species (Furuya et al. 1995), this result suggests there might be an important role for oxytocin in the fertilization or implantation process.

The function of myometrial OTR in the uterus of the cycle is also unclear. In women, the receptor concentration was higher in the luteal phase than in the follicular phase (Fuchs et al. 1985), whereas in cattle myometrial OTR-mRNA does not greatly change through the cycle (Ivell et al. 1995). It has been suggested that, in the human, oxytocin might be involved in modulating spontaneous myometrial contractions whose pattern shows a clear cyclic dependence (Kunz et al. 1997). Furthermore, it has been shown than dysmenorrhea can be treated by administration of the oxytocin receptor antagonist Atosiban (Akerlund et al. 1995), though whether this acts by modulating uterine contraction or by influencing the endometrium is not clear. Moreover, this antagonist also appears to interact at V1a receptors.

3.3
The Cervix

Although the cervix and vagina appear to be an anatomical extension of the uterus with apparently similar divisions into myoid and mucosal layers, in terms of regulation there are some distinct differences. As in the uterus the secretory function of the mucosal layer varies in a cyclic fashion, and in pregnancy. Especially at term, the cervix undergoes major morphological changes and connective tissue remodelling allowing sufficient distension to tolerate the expulsion of the foetus. In both humans and ruminants the mucosal layer of the cervix is equipped with oxytocin receptors (Matthews and Ayad 1994; Fuchs et al. 1996a, b; Wathes et al. 1996b). At term these increase dramatically at both mRNA and protein levels, reflecting the changes also occuring within the uterus. In fact, it has been demonstrated that reverse-transcription polymerase chain reaction analysis of OTR-mRNA in cervical scrapings can provide a useful test for cervical maturation or for the ability of the uterus to respond to exogenous oxytocin during birth induction (Kubota et al. 1994).

Also during the estrous cycle, in the ruminant the cervix, particularly the mucosa, exhibits high levels of OTR with, as in the uterus, maximum levels of OTR-mRNA and protein at estrus itself (Matthews and Ayad 1994; Fuchs et al. 1996b; Table 1). Since it has been shown that oxytocin in the cervix can cause local release of prostaglandins, particularly PGE_2, and it is known that PGE_2 is an active cervical softening agent, it has been speculated that oxytocin may act to modulate the physical properties of the cervix. There is, however, one marked distinction in the expression of the OTR in the cervix compared to that in the uterus, at least in the pregnant cow. Whereas in the uterus there is a gradual increase in OTR at the protein and mRNA levels through the second and third trimester, with a further dramatic upregulation at term, in the cervix the OTR appears to be suppressed to basal levels throughout pregnancy; only immediately before birth does one observe the dramatic upregulation of receptors associated with parturition (Fuchs et al. 1996a). This would be in agreement with the changing function of the cervix: throughout pregnancy, and particularly in the third trimester, its job is to keep the foetus in; only at parturition is this function reversed.

3.4
The Mammary Gland

Besides the induction of labour, the milk letdown reflex represents possibly the best known physiological function for oxytocin. Oxytocin is released from the nerve endings of the hypothalamic magnocellular neurones in the posterior pituitary upon a suckling stimulus. These oxytocin pulses pass via the blood stream to the mammary gland where they cause alveolar contraction and milk letdown. The only major pathological consequence observed in the oxytocin knock-out mouse was an inability to show oxytocin-induced milk letdown, with the consequence that newborn pups died because they were unable to receive milk from their knock-out mothers (reviewed in Russell and Leng 1998). In accord with this physiology in rodents it has been shown for a long time that the mammary gland is rich in OTR, and that these are upregulated in late pregnancy and during lactation (Soloff and Wieder 1983). More recently, immunohistochemistry with an anti-receptor polyclonal antibody localized OTR to myoid cells in the rat mammary gland (Adan et al. 1995). However, this might not be as simple for non-rodent species. In a study on the human breast using tissues obtained by surgery following the diagnosis of breast cancer, Kimura et al. (1998) were unable to locate any OTR on myoid cells. Instead, they were clearly present on the luminal epithelial cells of the alveoli, and appeared to be upregulated there in one sample from a breast during lactation. These observations were essentially confirmed for another primate, the marmoset monkey, though in this species a low level of OTR could additionally be detected sporadically on some myoid cells (Kimura et al. 1998). Supporting these findings are studies which have shown that breast cancer cell-lines (e.g. MCF7 cells) which are believed to be epithelioid in origin also have OTR (Taylor et al. 1990; Bussolati et al. 1995; Bale and Dorsa 1998). For the human, therefore, the milk ejection reflex may be less simple than in rodents, pituitary oxytocin interacting with receptors on epithelial cells, which in turn probably secrete some myotonic agent, possibly a prostaglandin (Cobo et al. 1974).

3.5
The Ovary

Much attention has been paid to the ability of the ovarian corpus luteum to produce oxytocin, particularly in ruminants, where it is assumed that the hormone is addressing systemic targets, for example, the endometrium. In other species, where oxytocin levels are lower, it is presumed that these are operating at a local paracrine level. Oxytocin receptors or specific effects of oxytocin peptide have been detected in the corpus luteum of the cow (Fuchs et al. 1990; Okuda et al. 1995; Sakumoto et al. 1996), pig (Pitzel et al. 1988, 1993), human (Fuchs et al. 1990; Maas et al. 1992) and baboon (Khan-Dawood et al. 1993, 1996), and there it has been postulated that luteal oxytocin may play a local role in luteolysis at the end of the cycle (Wuttke et al. 1995). However, recent studies in the cow (Okuda et al. 1997; Uenoyama and Okuda 1997) and particularly the marmoset monkey (Einspanier et al. 1997) indicate that oxytocin is intimately involved in the differentiation of the granulosa cells of the preovulatory follicle. Following the LH surge, OTR, which previously were restricted to a few basal cells, is upregulated in all granulosa cells, and together with an upregulation of oxytocin synthesis, establishes a functional local circuit, which increases progesterone production in these cells. Given also an LH-dependent increase in progesterone receptors and progesterone synthesis, oxytocin, together with progesterone, appears to orchestrate a positive feedback system leading to the irreversible differentiation (luteinization) of the follicle cells (Einspanier et al. 1997; Ivell 1998).

3.6
The Brain

It was discovered relatively early on that oxytocin receptors were present in the brain. Most research has been carried out in the rat, though other rodents (e.g. voles, Insel et al. 1995; Young et al. 1996; mice, Insel et al. 1993) and the human brain (Loup et al. 1991) have also been examined. The role of oxytocin receptors in brain function has been reviewed in depth by several authors recently (Bale et al. 1995; Freund-Mercier and Stoeckel 1995; Neumann et al. 1995; Russell and Leng 1998) and will not be discussed in detail here. It would appear from studies in the rat that we can distinguish several discrete modes of action of central oxytocin. Firstly, it appears to act in an autocrine/paracrine fashion within the magnocellular nuclei to orchestrate the expression and firing of the oxytocinergic neurones during specific stimulation (Neumann et al. 1996). At the same time projections from these nuclei and elsewhere convey oxytocin to other regions of the brain where it may influence behavioural paradigms, such as sexual, maternal and affiliative behaviour (McCarthy 1995; Insel et al. 1995, 1997). Thirdly, central oxytocin is also involved in the control of natriuresis and appetite (Verbalis et al. 1995), and has been suggested to influence memory (De Wied 1997). One of the richest sites of OTR expression appears to be within the hypothalamus (Yoshimura et al. 1993; Breton and Zingg 1997), in the areas of the magnocellular nuclei, as well as in the preoptic area and in the ventromedial hypothalamus, and may be related to the auto-crine/paracrine control of the magnocellular nuclei as well as to reproductive behaviour. Of interest in the context of the present review are the observations on the regulation of these receptors. It has been shown that in castrated rats *in vivo* application of both testosterone (in males) and estradiol (in females) is able to induce a significant upregulation of OTR in the ventromedial hypothalamus (Bale and Dorsa 1995a, b). However, in similar studies in the mouse, receptors were firstly distributed differently within the brain, and some were in fact inhibited by testosterone (Insel et al. 1993). Also progesterone and glucocorticoids have been shown to modulate OTR expression in the rat ventromedial hypothalamus (Patchev et al. 1993; Schumacher et al. 1993). In the rat hippocampus OTR appears also to be under glucocorticoid control (Liberzon et al. 1994).

3.7
The Kidney

Oxytocin receptors have been characterized at the protein and mRNA levels also in the kidney, particularly of the rat (Conrad et al. 1993; Ostrowski and Lolait 1995; Breton et al. 1996), and have been cloned from a porcine renal epithelial cell-line (Gorbulev et al. 1993). Recent results have also shown the equivalent mesotocin receptors to be present in the kidney of the chicken (Takahashi et al. 1996, 1997). They are localized on epithelial cells both in the macula densa of the glomerulus as well as in the proximal tubules, where they appear to be responsible for regulating natriuresis (Conrad et al. 1993; Windle et al. 1997). Of particular interest in the context of this review is their regulation by steroids. Like OTR in the brain, the kidney receptors also appear to be upregulated by estradiol administration. However, this effect needs to be differentiated, since whereas the receptors in the proximal tubules are downregulated by the anti-estrogen tamoxifen, those in the macula densa are not (Ostrowski and Lolait 1995). Furthermore, unlike for the receptors in the brain and uterus, renal OTR-mRNA is not upregulated at parturition, but instead is downregulated (Breton et al. 1996).

3.8
The Male Reproductive System

Just as oxytocin is shown to be an acute regulator of different organs of the female reproductive system, so also in the male can oxytocin influence diverse organs and accessory glands (reviewed in Ivell et al. 1997). Receptors for oxytocin have been localized in the testes of several species particularly on the Leydig cells, where oxytocin appears to influence the production of testosterone (Frayne and Nicholson 1995). Weak immunohistochemical staining for OTR is also evident, at least in the primate, on Sertoli cells (Einspanier and Ivell 1997; Ivell et al. 1997). Receptors have not been identified at the molecular level on the peritubular myoid cells, although there is physiological evidence that nanomolar concentrations of oxytocin can affect the contractility of the seminiferous tubules (Harris and Nicholson 1998). OTR is also present on some myoid cells of the epididymis, prostate, and bulbourethral glands in primates (Einspanier and Ivell 1997; Ivell et al. 1997), as well as on smooth muscle cells of the porcine testis, epididymis and vas deferens (Maggi et al. 1987). These receptors probably respond to the systemic pulse of oxytocin that accompanies ejaculation (reviewed in Ivell et al. 1997), though within the prostate local production of oxytocin may be involved in the 5α-reduction of testosterone to dihydrotestosterone (Nicholson and Jenkin 1995). Here, OTR has also been identified on some stromal cells (Einspanier and Ivell 1997).

3.9
Oxytocin-Responsive Cell-Lines

With the cloning of the gene for the OTR, it was logical to look for a cell-line in which the receptor is expressed sufficiently in order to carry out studies on the regulation of the receptor at the molecular level. Unfortunately, to date, although a number of cell-lines have been shown to express low levels of the receptor, none of these express it at a level sufficiently high to be appropriate for studies of gene regulation. Of the cells found to be expressing OTR at both protein and/or mRNA levels, the majority are derived from tumors of the reproductive system; for example, OTR-mRNA is expressed in the human leiomyosarcoma cell-line SKN (T. Kimura, unpubl.; Bale and Dorsa 1998), the mouse Leydig tumour cell-line MA10 (R. Bathgate, pers. comm.), the prostate carcinoma cell-line DU145 (K. Augustin, N. Abend, R. Bathgate and R. Ivell, unpubl.), as well as the breast carcinoma cell-lines MCF7 (Bussolati et al. 1995; Bale and Dorsa 1998), HS578T (Copland et al. 1997), MDA-MB231 and T47D (Bussolati et al. 1995). Additionally, an immortalized human myometrial cell-line has been shown to retain low levels of OTR (Monga et al. 1996), as has the porcine kidney cell-line LLC-PK$_1$ (Gorbulev et al. 1993).

However, the majority of pharmacological studies have made use of primary cell cultures or cells of very early passages derived from these, where there are much higher levels of OTR expression. The most useful in this context have been rat or human myometrial cells (e.g. Maggi et al. 1996; Phaneuf et al. 1997), rabbit amnion cells (Hinko and Soloff 1993; Jeng et al. 1995), bovine endometrial epithelial cells (Horn et al. 1998), and ovarian granulosa cells from several different species (e.g. Einspanier et al. 1997; Uenoyama and Okuda 1997).

It is important to note that evidence of OTR expression should be obtained not only at the mRNA level, but also at the level of ligand-binding or ligand-dependent cell activation. We have been able to show using a sensitive quantitative PCR assay that both the HeLa cervical carcinoma cell-line and the prostate carcinoma cell-line LNCaP express OTR-mRNA, but we have failed to detect receptor expression at the cell surface (R. Bathgate, pers. comm.).

4
Regulation of Oxytocin Receptor Expression

Expression of the oxytocin receptor shows a marked time- and tissue-specific pattern of expression, a pattern that correlates especially with the circulating levels of sex steroids. As noted above, there is an association between expression of the OTR and estradiol in several different tissues. In the uterus of the ovariectomized rat, estrogen can induce the expression of OTR and further administration of progesterone is able to antagonise this effect of estrogen (Larcher et al. 1995). However, also testosterone can return brain OTR levels to normal in castrated male rats. Additionally, both glucocorticoids and progesterone can influence the pattern of expression of OTR in the brain. However, attempts to influence OTR expression by sex steroids in cell culture systems, as opposed to *in vivo*, have consistently failed. Even the application of so-called progesterone-withdrawal protocols, designed to mimic the steroidal milieu at birth, or at estrus, where progesterone is sinking rapidly and estradiol levels are high, fail to reproduce in culture the *in vivo* upregulation of the OTR (e.g. Horn et al. 1998). Interestingly, in a recent exhaustive study looking at the influence of sex steroids on OTR in the uterus of the ovariectomized sheep, Wathes and coworkers (1996a) were also unable to find evidence for a direct effect of such steroids on the expression of the OTR.

Some of the most interesting experiments have been carried out using the ruminant endometrial model. OTR-mRNA and protein are massively upregulated *in vivo* during the estrous cycle at around ovulation (estrus), at a time when circulating progesterone is very low and estradiol produced by the new growing follicles is high. However, if one removes endometrial tissue from the mid-cycle uterus, when the endogenous receptor is low, and places it in culture, either as discrete endometrial fragments (Sheldrick and Flick-Smith 1993) or as dispersed endometrial epithelial cell culture (Horn et al. 1998), then the OTR spontaneously upregulates to reach moderately high levels in the absence of exogenous steroids. It is concluded that progesterone *in vivo* suppresses the expression both of the OTR gene and its protein. Measurement of OTR-mRNA shows clearly that this affect is at the transcriptional level (though see below). Progesterone *in vitro* does not influence the levels of OTR-mRNA and protein in endometrial epithelial cell cultures (Horn et al. 1998). A related observation is that in the uteri of prepubertal calves, where progesterone levels are also low, the endogenous OTR is expressed at high levels equivalent to those at estrous in the sexually mature animal (Fuchs et al. 1998). These levels are reduced to the low basal mid-cycle level of the adult upon attainment of puberty.

Recently, it could be shown for the cloned rat OTR transfected into CHO cells, as well as in the native uterus, that progesterone or some of its metabolites can directly influence the receptor protein in a non-genomic manner (Grazzini et al. 1998a). The steroid appears to bind with high affinity to the receptor so as to influence ligand binding and signal transduction. However, this effect appears to be species-specific, progesterone being without effect on the human receptor, although the metabolite 5β-dihydroprogesterone was able to influence the human OTR. Medroxyprogesterone acetate, as a conventional analogue of progesterone, was without effect on the bovine receptor (Horn et al. 1998). Although this specific inhibition of OTR by gestagens appears to offer a possible explanation for the cyclicity of OTR responsiveness, the observed non-genomic effects do not influence gene expression at the transcriptional level; thus other mechanisms must still be sought to explain the hormone-dependent expression of the OTR gene. The idea that gestagens might influence the OTR at the cell membrane is not new. A similar hypothesis was put forward for the brain receptors by Schumacher and colleagues (1993).

Other factors that have been shown to influence OTR expression are the classic signal transduction agents cAMP and phorbol esters, presumably acting though protein kinase C. Both compounds are able to upregulate the endogenous receptor in MCF7 breast carcinoma cells (Bale et al. 1998), as well as induce reporter gene expression of transfected rat OTR promoter-reporter constructs (Bale et al. 1998; Grazzini et al. 1998b). Soloff and colleagues have shown also that cAMP can induce the upregulation of the OTR gene and receptors in rabbit amnion cells (Hinko and Soloff 1993; Jeng et al. 1995). We were also able to show an increase in specific OTR-mRNA upon forskolin treatment to activate protein kinase A in the DU145 prostate carcinoma cell-line (K. Augustin, N. Abend, R. Bathgate and R. Ivell, unpubl.).

An interesting observation has also been made in the context of homologous desensitization of the receptor. Prolonged treatment of human myometrial cells in dispersed culture with the ligand oxytocin led to a marked reduction in the levels of OTR-mRNA in these cells (Phaneuf et al. 1997). However, oxytocin at different concentrations had no effect on the levels of OTR-mRNA in the DU145 prostate carcinoma cells (K. Augustin, N. Abend, R. Bathgate and R. Ivell, unpubl.).

The only other factor that has been shown to have a marked effect on OTR gene expression is the ruminant-specific cytokine, interferon-τ. This molecule belongs to the interferon-α family and is produced in large amounts by the implanting bovine blastocyst in early pregnancy. The interferon-τ interacts with specific receptors on the endometrial epithelial cells of the uterus and prevents the estrous-related upregulation of the OTR at the transcriptional level. This interrupts the positive feedback loop between luteal oxytocin and the oxytocin-dependent secretion of prostaglandin-$F_{2\alpha}$ by the endometrium which leads to luteolysis and the termination of the cycle (reviewed in Flint et al. 1994). It is not clear how interferon-τ works. It interacts directly with endometrial epithelial cells, as shown by its effect on OTR expression in pure dispersed cell cultures (Horn et al. 1998). On the one hand, it appears to induce a downregulation of the gene for the estradiol receptor(ER)α (Spencer et al. 1995, 1996; Spencer and Bazer 1996), which, as outlined above, is implicated in OTR expression. On the other hand, we have recently been able to show that the bovine OTR gene promoter possesses a functional interferon-responsive *cis* element which might offer a direct link to the interferon-τ induced signal transduction pathway (Bathgate et al. 1998).

Summarizing the many data from *in vivo* and *in vitro* studies leads to the following general synopsis for OTR gene expresion. It would appear to be expressed at a low basal level in both myoid and epithelial cell types in many different tissues of the reproductive system, as well as in other tissues. This low basal expression is probably dictated by a combination of cell-specific and constitutive factors. In addition, there appears to be a specific, hormonally dependent upregulation, possibly mediated again by cell-specific factors in combination with the activation of protein kinase A or phorbol ester-dependent pathways. In addition to this, there is a clear role for specific inhibition as part of the control system for regulating the OTR gene. This is witnessed by the progesterone-dependent suppresion of the OTR in the mid-cycle ruminant uterus, and by the clear effect of interferon-τ in the same system. A further interesting example supporting this view is provided by the tammar wallaby. This marsupial has separated twin uteri, one of which only becomes gravid during pregnancy. Just as in eutherian mammals there is a massive upregulation of mesotocin (oxytocin) receptors immediately prior to parturition, but only in the gravid uterus. In the non-gravid uterus, not only is there no upregulation of the receptors, but also these are even further suppressed below the normal basal level (Parry et al. 1997). These observations preclude the possibility of any systemic effectors, such as steroids, influencing OTR gene expression. This must be being regulated at a local level, possibly by some product of the foetus itself.

5
Structure and Regulation of the OTR Gene

The structure of the oxytocin receptor gene itself has been described for the human (Inoue et al. 1994), bovine (Bathgate et al. 1995), rat (Zingg et al. 1995; Bale and Dorsa 1997), mouse (Kubota et al. 1996), and vole (Young et al. 1996). In most species there are three exons with two intervening sequences (Fig. 2). Only in the human and the mouse is there an extra intron separating the region of the first exon in other species into two discrete parts. The first exon (in the human and mouse, exons 1 and 2) includes only the 5' non-coding region (5'UTR) downstream of the transcription start site. Interestingly, in the bovine OTR gene, RT-PCR analysis indicates variability in the splice donor and acceptor sites of intron 1 (Bathgate et al. 1995; Fig. 2, dashed box). Exon 2 (in the human, exon 3) encompasses most of the protein coding region including the N-terminal extracellular domain and the first five transmembrane domains. There then follows a long intron in all species. Exon 3 (in the human, exon 4) encodes the remaining transmembrane domains 6 and 7, the cytoplasmic C-terminus and the entire 3'UTR.

The region of genomic DNA upstream of the transcription start site, which is fairly similar in location in all species studied, has been sequenced to varying degrees in the different species. For the human approximately 7 kb is known, for the bovine approximately 4 kb, and for rats, mice and voles approximately 3 kb each. This is the region that is supposedly responsible for interaction with the majority of nuclear proteins (transcription factors) specifically involved in regulating the gene. Comparing the promoter sequences from the different species firstly shows that there is a high degree of homology, indicating functional conservation, within the approximately 700 bp immediately upstream of the transcription start site (Fig. 2, stippled region). However, there are no recognizable motifs within this region that can be readily associated with well-known and relevant transcription factors. A computer search for known motifs would of course provide an ample list of possibilities based on the random association of nucleotides.

For this reason, our approach has been to apply differential nuclear protein binding assays to the promoter regions of both the human and bovine genes. As sources of the specific nuclear proteins we have used myometrial or endometrial samples from different times in the estrous cycle or pregnancy or from cultured cells, and applied these in electrophoretic mobility shift assays (EMSA) using subfragments of the promoter regions as radiolabelled probes. Considering first the bovine OTR gene, although there are a number of DNA-protein complexes specifically formed with uterine nuclear proteins, as opposed to a control tissue, such as liver, none of these appear to be differentially expressed comparing uterine tissues where the endogenous OTR is up- or downregulated. A detailed analysis of the *cis* elements responsible for the formation of some of these complexes shows that all are novel motifs, and thus

──►

Fig. 2. Structure and organization of the oxytocin receptor (OTR) genes from rat, mouse, bovine and human. The vole gene structure (Young et al. 1996) is essentially identical to that of rat. The *stippled box* in the 5' promoter region of the genes indicates sequence of high homology between species. The *dashed box* in exon 1 of the bovine gene indicates the region of variable splicing (Bathgate et al. 1995). *Below* are shown the corresponding human mRNA structure and encoded protein with positions of the transmembrane domains (*TM1–TM7*)

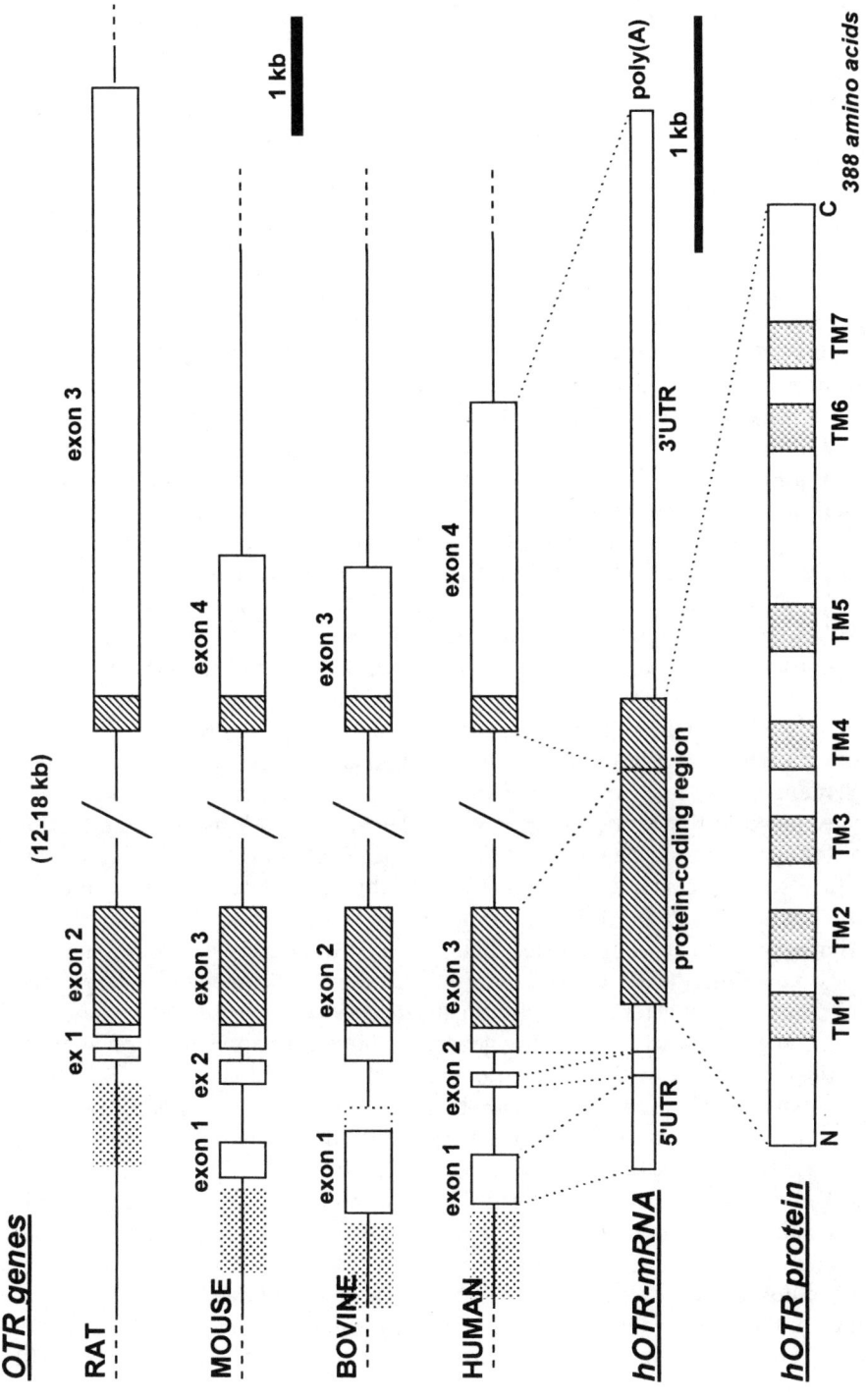

not at this stage readily identifiable. The same general finding is also true for the human gene promoter, except that in this case two *cis* motifs can be identified in a relatively distal region, which appear to form differentially expressed EMSA complexes comparing non-pregnant with term-pregnant human myometrium. However, both these elements have not been previously described in the literature, and neither appear to have a homologous counterpart in the promoter regions of the OTR genes from other species. Also the other EMSA complexes formed with the fragments of the human promoter do not appear to have their correlates in the bovine gene promoter. At this stage we are left with the impression that there may indeed be specific protein-DNA complexes formed, but these appear to be less defined by precise *cis* motifs than by their general positioning within the promoter region, or possibly by some secondary structural feature which we cannot yet ascertain. These characteristics are typical for genes being constitutively expressed at a basal level.

Because of the attention given to the role of estrogens in terms of the *in vivo* regulation of OTR expression (see above), we and others have searched for and analysed DNA sequences showing similarity to the classic palindromic estrogen responsive element (ERE, GGTCANNNTGACC). In the rat OTR gene promoter, besides two ERE half-sites (TGACC), there appears to be a near perfect ERE palindrome at position –2400 upstream of the transcription start site. Although the excised element appears to be functional in a truncated reporter construct, it is not functional in the natural context of the OTR promoter (Bale and Dorsa 1997). In both the bovine and human genes there are three so-called ERE half-sites within the first 3 kb upstream of the transcription start site, though in non-homologous positions. Also for these genes, none of these elements represents functional EREs using transfected reporter constructs in the context of the natural promoters (Ivell et al. 1998). Moreover, examination of the EMSA results showed that none of these half-sites was involved in specific DNA-nuclear protein interactions. Within the protein-coding region of the bovine OTR gene downstream from the transcription start site, there is what appears to be a near-perfect palindromic ERE. However, also this appears to be non-functional even when excised and used as a cassette in an appropriate promoter-reporter context (Bathgate et al. 1998).

Another approach to studying gene promoters is to apply deletion analysis in the context of transfected promoter-reporter expression constructs. Here it is important that the constructs are being transfected into an appropriate cell type which has an endogenous expression of the receptor gene. Using this approach, Bale and Dorsa (1998) have investigated the rat promoter by transfecting into the human MCF7 cell-line. They were able to show that there is basal activity within the first 500 bp, but no further increase upon including longer stretches of upstream promoter. This basal activity could be increased when the cells were stimulated with cAMP and phorbol esters, suggesting that these second messengers were acting via *cis* elements present within this proximal promoter region. A similar study was also performed for the human OTR promoter in hepG2 hepatoma cells lacking endogenous OTR expression and in the Hs578T breast tumor cell-line with basal OTR expression (Hoare et al. 1997), which also delimited the basal promoter to a relatively small stretch of DNA close to the transcription start site. Also we have performed such studies for the human and bovine OTR genes using mainly the human HeLa cell-line, which we have shown to express OTR transcripts, though not protein, to a moderate level (Ivell et al. 1998). All our results support the notion that the basal activity of the OTR gene resides within the proximal 500 bp, and that this is sufficient for constitutive basal expression of the gene. Whether or not this basal region is directly activated by the effects of protein kinase A or protein kinase C *in vivo* remains, however, an open question. Against the notion are the general observations that *in vivo* reagents known to cause an upregulation of cAMP in uterine smooth muscle cells, such as classic β2-adrenergic stimulants (ritodoline, terbutaline, etc.), theophylline and coffeine, all tend to be tocolytic, and do not appear to lead to an upregulation of OTR.

Both studies of endogenous OTR gene expression, as well as transfection data (e.g. Bale and Dorsa 1997) imply that not only activation of the gene is important, but also suppression or inhibition of gene expression may play a significant role. For the bovine it is known that members of the interferon-α family may act via some inhibitory interferon-responsive transcription factors to suppress OTR gene expression (Spencer et al. 1996; Horn et al. 1998). Inhibiting transcription factors not only may interact with the upstream promoter of genes, but also are known to act downstream of the transcription start site, for example within introns. In order to investigate this possibility we have analysed the large third intron of the human OTR gene using differential EMSA analysis and have been able to map a 20-bp *cis* element which appears to bind a myometrial nuclear protein complex correlating with the downregulation of the gene in the non-pregnant myometrium (Mizumoto et al. 1997). The element identified represented a novel sequence, so that it has not yet been possible to characterize the molecular components involved in this interaction.

6
General Conclusions

The oxytocin receptor occupies a central position in the control of labour and birth. It is also the key molecule mediating the release of milk from the breast during lactation. It is implicated in the regulation of various reproductive and feeding-associated behaviour patterns in the brain. Furthermore, it plays an active role in a large number of local paracrine control systems in most organs of the body.

The molecular cloning of the OTR-cDNA and gene has led to the development of a large number of important methodological systems, which have extended our knowledge of the receptor and its physiology beyond what can be obtained from classic ligand-binding studies. Primary sequence information has led to an intimate knowledge of the structure of the receptor protein and of structure-function relationships in regard to ligand specficity, ligand affinity and signal transduction. The construction of cell-lines transfected with the wild type or variously mutated receptor genes has provided important tools to extend our understanding of OTR pharmacology, and hopefully will lead on to the development of a new generation of pharmacological effectors of benefit in the clinic.

Although steroids appear to influence OTR expression *in vivo*, we have no evidence to date of any direct action of steroids on the OTR gene in cell culture *in vitro*, though indirect effects cannot be absolutely excluded. There may, however, be a non-genomic effect at least of some progesterone-like compounds on the binding properties of the membrane receptor protein. All studies on the OTR gene promoter in several species concur that expression is governed by a complex interaction of different types of transcription factors. There appears to be a constitutive gene activation allowing moderate expression in a wide variety of cells (though particularly epithelial cells) from reproductive and other tissues. There may additionally be a specific hormone-dependent further upregulation to account, for example, for the high levels of expression at estrus in the ruminant endometrium or at term in the uterus and cervix of most mammals. This might involve cAMP-dependent gene activation, though direct effects of protein kinase A activation appear unlikely because of the generally tocolytic effects of cAMP-inducing agents on uterine smooth muscle *in vivo*. Finally, specific inhibition of transcriptional activity appears to play an important role in determining the pattern of OTR expression. For the ruminant, both progesterone and interferons appear to be specific inhibitory effectors, though the mechanisms by which such inhibition occurs are still unclear.

Acknowledgements. We gratefully acknowledge the help and advice of many colleagues in both the Osaka and Hamburg laboratories. In particular, we should like to thank Dr. Ross Bathgate for critically reading this manuscript and offering many helpful comments, and also for allowing us to cite a number of unpublished observations. Thanks are also due to the Deutsche Forschungsgemeinschaft, the Alexander-von-Humboldt Foundation, the CIBA-GEIGY Foundation of Japan, and the Japanese Ministry of Education, Science and Culture.

References

Adan RA, Van Leeuwen FW, Sonnemans MA, Brouns M, Hoffman G, Verbalis JG, Burbach JP (1995) Rat oxytocin receptor in brain, pituitary, mammary gland, and uterus: partial sequence and immunocytochemical localization. Endocrinology 136:4022–4028

Akerlund M, Melin P, Maggi M (1995) Potential use of oxytocin and vasopressin V1a antagonists in the treatment of preterm labour and primary dysmenorrhea. Adv Exp Med Biol 395:595–600

Alexandrova M, Soloff MS (1980) Oxytocin receptors and parturition. I. Control of oxytocin receptor concentration in the rat myometrium at term. Endocrinology 106:730–735

Ayad VJ, Guldenaar SEF, Wathes DC (1991) Characterization and localization of oxytocin receptors in the uterus and oviduct of the non-pregnant ewe using iodinated receptor antagonist. J Endocrinol 128:187–195

Baker PN, Peat ML, Symonds EM, Maynard V (1990) Endometrial oxytocin binding sites in normal women and in subfertile patients. Postgrad Med J 166:195–199

Balducci J, Risek B, Gilula NB, Hand A, Egan JFX, Vintzileos AM (1993) Gap junction formation in human myometrium: a key role to preterm labor? Am J Obstet Gynecol 168:1609–1615

Bale TL, Dorsa DM (1995a) Sex differences in and effects of estrogen on oxytocin receptor messenger ribonucleic acid expression in the ventromedial hypothalamus. Endocrinology 136:27–32

Bale TL, Dorsa DM (1995b) Regulation of oxytocin receptor messenger ribonucleic acid in the ventromedial hypothalamus by testosterone and its metabolites. Endocrinology 136:5135–5138

Bale TL, Dorsa DM (1997) Cloning, novel promoter sequence, and estrogen regulation of a rat oxytocin receptor gene. Endocrinology 138:1151–1158

Bale TL, Dorsa DM (1998) NGF, cyclic AMP, and phorbol esters regulate oxytocin receptor gene transcription in SK-N-SH and MCF7 cells. Mol Brain Res 53:130–137

Bale TL, Pedersen CA, Dorsa DM (1995) CNS oxytocin receptor mRNA expression and regulation by gonadal steroids. Adv Exp Med Biol 395:269–280

Bathgate R, Rust W, Balvers M, Hartung S, Morley S, Ivell R (1995) Structure and expression of the bovine oxytocin receptor gene. DNA Cell Biol 14:1037–1048

Bathgate RAD, Tillmann G, Ivell R (1998) Molecular mechanisms of bovine oxytocin receptor gene regulation. Biol Reprod 59:(Suppl 1)

Behrens O, Maschek H, Kupsch E, Fuchs AR (1995) Oxytocin receptors in human ovaries during the menstrual cycle. Adv Exp Med Biol 395:485–486

Birnbaumer M, Seibold A, Gilbert S, Ishido M, Barberis C, Antaramian A, Brabet P, Rosenthal W (1992) Molecular cloning of the receptor for human antidiuretic hormone. Nature 357:333–335

Breton C, Zingg HH (1997) Expression and region-specific regulation of the oxytocin receptor gene in rat brain. Endocrinology 138:1857–1862

Breton C, Neculcea J, Kingg HH (1996) Renal oxytocin receptor messenger ribonucleic acid: characterization and regulation during pregnancy and in response to ovarian steroid treatment. Endocrinology 137:2711–2717

Bussolati G, Cassoni P, Negro F, Stella A, Sapino A (1995) Effect of oxytocin on breast carcinoma cell growth. Adv Exp Med Biol 395:553–554

Caldeyro-Barcia R, Theobald GW (1968) Sensitivity of the pregnant human myometrium to oxytocin. Am J Obstet Gynecol 102:1181

Cantau B, Barjon JN, Chicot D, Baskevitch PP, Jard S (1990) Oxytocin receptors from LLC-PK1 cells: expression in *Xenopus* oocytes. Am J Physiol 258:F963–972

Chan WY, Wo N-C, Manning M (1996) The role of oxytocin receptors and vasopressin V1a receptors in uterine contractions in rats: implications for tocolytic therapy with oxytocin antagonists. Am J Obstet Gynecol 175:1331–1335

Chibbar R, Miller FD, Michell BF (1993) Synthesis of oxytocin in amnion, chorion, and decidua may influence the timing of human parturition. J Clin Invest 91:185–192

Chini B, Mouillac B, Ala Y, Balestre MN, Trumpp-Kallmeyer S, Hoflack J, Elands J, Hibert M, Manning M, Jard S, Barberis C (1995) Tyr115 is the key residue for determining agonist selectivity in the V1a vasopressin receptor. EMBO J 14:2176–2182

Chini B, Mouillac B, Balestre MN, Trumpp-Kallmeyer S, Hoflack J, Hibert M, Andriolo M, Pupier S, Jard S, Barberis C (1996) Two aromatic residues regulate the response of the human oxytocin receptor to the partial agonist arginine vasopressin. FEBS Letters 397:201–206

Cobo E, Rodriguez A, Villamizar M (1974) Milk-ejecting activity induced by prostaglandin F2α. Am J Obstet Gynecol 118:831–836

Conrad KP, Gellai M, North WG, Valtin H (1993) Influence of oxytocin on renal hemodynamics and sodium excretion. Ann NY Acad Sci 689:346-362

Copland JA, Jeng YJ, Strakova Z, Ives K, Hellmich MR, Soloff MS (1997) Regulation and function of the oxytocin receptor (OTR) in human mammary HS578T cells. Endocrine Society Annual Meeting, Mineapolis, Abs P2-389

Dale HH (1906) On some physiological actions of ergot. J Physiol (Lond) 34:163-206

De Wied D (1997) The neuropeptide story. Front Neuroendocrinol 18:101-113

Dobrowlski W, Hafez ESE (1971) The uterus and control of ovarian function. Acta Obstet Gynecol Scand, Suppl 12:5-126

Du Vigneaud V, Resser C, Trippett S (1953a) The sequence of amino acids in oxytocin, with a proposal for the structure of oxytocin. J Biol Chem 205:949-957

Du Vigneaud V, Resser C, Swan JM, Roberts CW, Katsoyannis PG, Gordon S (1953b) The synthesis of an octapeptide amide with the hormonal activity of oxytocin. J Am Chem Soc 75:4879-4880

Einspanier A, Ivell R (1997) Oxytocin and oxytocvin receptor expression in reproductive tissues of the male marmoset monkey. Biol Reprod 56:416-422

Einspanier A, Jurzinski A, Hodges JK (1997) A local oxytocin system is part of the luteinization process in the preovulatory follicle of the marmoset monkey (Callithrix jacchus). Biol Reprod 57:16-26

Elands J, Resink A, De Kloet ER (1990) Neurohypophyseal hormone receptors in the rat thymus, spleen and lymphocytes. Endocrinology 126:2703-2710

Flint APF, Lamming GE, Stewart HJ, Abayasekara DRE (1994) The role of the endometrial oxytocin receptor in determining the length of the sterile oestrous cycle and ensuring maintenance of luteal function in early pregnancy in ruminants. Phil OS Trans R Soc Lond 344:291-304

Frayne J, Nicholson HD (1995) Effect of oxytocin on testosterone production by isolated rat Leydig cells is mediated via a specific oxytocin receptor. Biol Reprod 52:1268-1273

Freund-Mercier MJ, Stoeckel ME (1995) Somatodendritic autoreceptors on oxytocin neurones. Adv Exp Med Biol 395:185-194

Fuchs AR, Husslein P, Fuchs F (1981) Oxytocin and the initiation of human parturition. II. Stimulation of prostaglandin production in human decidua by oxytocin. Am J Obstet Gynecol 141:694-697

Fuchs AR, Fuchs F, Husslein P, Soloff MS, Fernström MJ (1982) Oxytocin receptors and human parturition: a dual role for oxytocin in the initiation of labor. Science 215:1396-1398

Fuchs AR, Periyasamy S, Alexandrova M, Soloff MS (1983) Correlation between oxytocin receptor concentration and responsiveness to oxytocin in pregnant rat myometrium: effects of ovarian steroids. Endocrinology 113:742-749

Fuchs AR, Fuchs F, Husslein P, Soloff MS (1984) Oxytocin receptors in the human uterus during pregnancy and parturition. Am J Obstet Gynecol 150:734-741

Fuchs AR, Fuchs F, Soloff MS (1985) Oxytocin receptors in nonpregnant human uterus. J Clin Endocrinol Metab 60:37-41

Fuchs AR, Behrens O, Helmer H, Vangsted A, Ivanisevic M, Grifo J, Barros C, Fields M (1990) Oxytocin and vasopressin binding sites in human and bovine ovaries. Am J Obstet Gynecol 163:1961-1967

Fuchs AR, Helmer H, Behrens O, Liu HC, Antonian L, Chang SM, Fields MJ (1992a) Oxytocin and bovine parturition: a steep rise in endometrial oxytocin receptors precedes onset of labor. Biol Reprod 47:937-944

Fuchs AR, Helmer H, Chang SM, Fields MJ (1992b) Concentration of oxytocin receptors in the placenta and fetal membranes of cows during pregnancy and labour. J Reprod Fertil 96:775-783

Fuchs AR, Ivell R, Balvers M, Chang SM, Fields MJ (1996a) Oxytocin receptors in bovine cervix during pregnancy and parturition: gene expression and cellular localization. Am J Obstet Gynecol 175:1654-1660

Fuchs AR, Ivell R, Fields PA, Chang SM, Fields MJ (1996b) Oxytocin receptors in bovine cervix: distribution and gene expression during the estrous cycle. Biol Reprod 54:700–708

Fuchs AR, Drolet P, Fortier MA, Balvers M, Fields MJ (1998) Ontogeny of oxytocin receptors and oxytocin-induced stimulation of prostaglandin synthesis in prepubetal heifers. Endocrinology 139:2755–2764

Furuya K, Mizumoto Y, Makimura N, Mitsui C, Murakami M, Tokuoka S, Ishikawa N, Nagata I, Kimura T, Ivell R (1995) A novel biological aspect of ovarian oxytocin: gene expression of oxytocin and oxytocin receptor in cumulus/luteal cells and the effect of oxytocin on embryogenesis in fertilized oocytes. Adv Exp Med Biol 395:523–528

Garfield RE, Sims S, Daniel EE (1977) Gap junctions: their presence and necessity in myometrium during parturition. Science 198:958–960

Gimpl G, Klein U, Reilander H, Fahrenholz F (1995) Expression of the human oxytocin receptor in baculovirus-infected cells: high affinity binding is induced by a cholesterol-cyclodextrin complex. Biochemistry 34:13794–13801

Gorbulev V, Buchner H, Akhundova A, Fahrenholz F (1993) Molecular cloning and functional characterization of V2 [8-lysine] vasopressin and oxytocin receptors from a pig kidney cell-line. Eur J Biochem 215:1–7

Grazzini E, Guillon G, Mouillac B, Zingg HH (1998a) Inhibition of oxytocin receptor function by direct binding of progesterone. Nature 392:509–512

Grazzini E, Russo C, Zingg HH (1998b) Oxytocin receptor gene promoter: induction by protein kinase C. Proc Endocrine Society Annual Meeting. New Orleans, Abstr P3–29

Harris GC, Nicholson HD (1998) Characterization of the biological effects of neurohypophyseal peptides on seminiferous tubules. J Endocrinol 156:35–42

Hausmann H, Richters A, Kreienkamp HJ, Meyerhof W, Mattes H, Lederis K, Zwiers H, Richter D (1996) Mutational analysis and molecular modeling of the nonapeptide hormone binding domains of the [Arg8]vasotocin receptor. Proc Natl Acad Sci USA 93:6907–6912

Hinko A, Soloff MS (1993) Up-regulation of oxytocin receptors in rabbit amnion by adenosine 3′, 5′-monophosphate. Endocrinology 132:126–132

Hoare S, Wood TG, Copland JA, Acosta M, Jeng YJ, Izban MG, Soloff MS (1997) Characterization of the human oxytocin receptor gene promoter. Endocrine Society Annual Meeting, Mineapolis, Abstr P3–551

Horn S, Bathgate R, Lioutas C, Bracken K, Ivell R (1998) Bovine endometrial epithelial cells as a model system to study oxytocin receptor regulation. Human Reprod Update 4:605–614

Howl J, Wheatley M (1996) Molecular recognition of peptide and non-peptide ligands by the extracellular domains of neurohypophysial hormone receptors. Biochem J 317:577–582

Husslein P, Fuchs AR, Fuchs F (1981) Oxytocin and the initiation of human parturition. I. Prostaglandin release during induction of labor by oxytocin. Am J Obstet Gynecol 141:688–693

Innamorati G, Sadeghi H, Brinbaumer M (1996) A fully active nonglycosylated V2 vasopressin receptor. Mol Pharmacol 50:467–473

Inoue T, Kimura T, Azuma C, Inazawa J, Takemura M, Kikuchi T, Kubota Y, Ogita K, Saji F (1994) Structural organization of the human oxytocin receptor gene. J Biol Chem 269:32451–32456

Insel TR, Young L, Witt DM, Crews D (1993) Gonadal steroids have paradoxical effects on brain oxytocin receptors. J Neuroendocrinol 5:619–628

Insel TR, Winslow JT, Wang ZX, Young L, Hulihan TJ (1995) Oxytocin and the molecular basis of monogamy. Adv Exp Med Biol 395:227–234

Insel TR, Young L, Wang Z (1997) Central oxytocin and reproductive behaviors. Rev Reprod 2:28–37

Ivell R (1998) The physiology of ovarian oxytocin. Reprod Med Rev (in press)

Ivell R, Russell JA (eds) (1995) Oxytocin: Cellular and Molecular approaches in medicine and research. Plenum Press, New York

Ivell R, Rust W, Einspanier A, Hartung S, Fields M, Fuchs AR (1995) Oxytocin and oxytocin receptor gene expression in the reproductive tract of the pregnant cow: rescue of luteal oxytocin production at term. Biol Reprod 53:553–560

Ivell R, Balvers M, Rust W, Bathgate R, Einspanier A (1997) Oxytocin and male reproductive function. Adv Exp Med Biol 424:253–264

Ivell R, Bathgate RA, Walther N, Kimura T (1998) The molecular basis of oxytocin and oxytocin receptor gene expression in reproductive tissues. Adv Exp Med Biol 449:297–306

Jasper JR, Harrell CM, O'Brien JA, Pettibone DJ (1995) Characterization of the human oxytocin receptor stably expressed in 293 human embryonic kidney cells. Life Sci 57:2253–2261

Jeng YJ, Hinko A, Soloff MS (1995) Effectors of cyclic adenosine 5'-monophosphate up-regulating oxytocin receptors in rabbit amnion cells: isoproterenol, parathyroid related protein, and potentiation by cortisol. Biol Reprod 53:1051–1056

Jenner LJ, Parkinson TJ, Lamming GE (1991) Uterine oxytocin receptors in cyclic and pregnant cows. J Reprod Fertil 91:49–58

Kaluz S, Kaluzova M, Flint APF (1996) Heterogeneity in the third intracytoplasmic region of the oxytocin receptor-encoding gene. Gene 172:313–314

Khan-Dawood FS, Kanu EJ, Dawood MY (1993) Baboon corpus luteum: presence of oxytocin receptors. Biol Reprod 49:262–266

Khan-Dawood FS, Chellaram R, Dawood MY (1996) In vitro microdialysis of baboon corpus luteum: effects of oxytocin on total and pulsatile progesterone secretion. Regul Pept 66: 137–147

Kimura T, Azuma C, Saji F, Takemura M, Tokugawa Y, Miki M, Ono M, Mori K, Tanizawa O (1992a) Estimation by an electrophysiological method of the expression of oxytocin receptor mRNA in human myometrium during pregnancy. J Steroid Biochem Mol Biol 42: 253–258

Kimura T, Tanizawa O, Mori K, Brownstein MJ, Okayama H (1992b) Structure and expression of a human oxytocin receptor. Nature 356:526–529

Kimura T, Makino Y, Saji F, Takemura M, Inoue T, Kikuchi T, Kubota Y, Azuma C, Nobunaga T, Tokugawa Y, Tanizawa O (1994) Molecular characterization of a cloned human oxytocin receptor. Eur J Endocrinol 131:385–390

Kimura T, Takemura M, Nomura S, Nobunaga T, Kubota Y, Inoue T, Hashimoto K, Kumazawa I, Ito Y, Ohashi K, Koyama M, Azuma C, Kitamura Y, Saji F (1996) Expression of oxytocin receptor in human pregnant myometrium. Endocrinology 137:780–785

Kimura T, Makino Y, Bathgate R, Ivell R, Nobunaga T, Kubota Y, Kumazawa I, Saji F, Murata Y, Nishihara T, Hashimoto M, Kinoshita M (1997) The role of N-terminal glycosylation in the human oxytocin receptor. Mol Hum Reprod 3:957–963

Kimura T, Ito Y, Einspanier A, Toya K, Nobunaga Y, Takemura M, Kubota Y, Ivell R, Matsuura N, Saji F, Murata Y (1998) Expression and immunolocalization of the oxytocin receptor in human lactating and non-lactating mammary glands. Hum Reprod 13:2645–2653

Klein U, Gimpl G, Fahrenholz F (1995) Alteration of the myometrial plasma membrane cholestrol content with β-cyclodextrin modulates the binding affinity of the oxytocin receptor. Biochemistry 34:13784–13793

Kojro E, Hackenberg M, Zsigo J, Fahrenholz F (1991) Identification and enzymatic deglycosylation of the myometrial oxytocin receptor using a radioligand photoreactive antagonist. J Biol Chem 266:21416–21421

Kojro E, Eich P, Gimpl G, Fahrenholz F (1993) Direct identification of an extracellular agonist binding site in the renal V2 vasopressin receptor. Biochemistry 32:13537–13544

Kubota Y, Kimura T, Takemura M, Inoue T, Kikuchi T, Ono M, Kanai T, Azuma C, Saji F, Tanizawa O (1994) A novel method for detection of uterine oxytocin receptor in preparation for delivery using scraped endocervical cells. Horm Metab Res 26:442–443

Kubota Y, Kimura T, Hashimoto K, Tokugawa Y, Nobunaga T, Azuma C, Saji F, Murata Y (1996) Structure and expression of the mouse oxytocin receptor gene. Mol Cell Endocrinol 124:25–32

Kunz G, Beil H, Deiniger H, Einspanier A, Mall G, Leyendecker G (1997) The uterine peristaltic pump: normal and impeded sperm transport within the female genital tract. Adv Exp Med Biol 424:267–278

Larcher A, Neculcea J, Breton C, Arslan A, Rozen F, Russo C, Zingg HH (1995) Oxytocin receptor gene expression in the rat uterus during pregnancy and the estrous cycle and in response to gonadal steroid treatment. Endocrinology 136:5350–5356

Lefebvre DL, Giaid A, Bennett H, Lariviere R, Zingg HH (1992) Oxytocin gene expression in rat uterus. Science 256:1553–1555

Liberzon I, Chalmers DT, Mansour A, Lopez JF, Watson SJ, Young EA (1994) Glucocorticoid regulation of hippocampal oxytocin receptor binding. Brain Res 650:317–322

Lolait SJ, O'Carroll AM, McBride OW, König M, Morel A, Brownstein MJ (1992) Cloning and characterization of a vasopressin V2 receptor and a possible link to nephrogenic diabetes insipidus. Nature 357:336–339

Loup F, Tribollet E, Dubois-Dauphin M, Dreifuss JJ (1991) Localization of high-affinity binding sites for oxytocin and vasopressin in the human brain. An autoradiographic study. Brain Res 555:220–232

Lundin-Schiller S, Kreider DL, Rorie RW, Hardesty D, Mitchell MD, Koike TI (1996) Characterization of porcine endometrial, myometrial, and mammary oxytocin binding sites during gestation and labor. Biol Reprod 55:575–581

Maas S, Jarry H, Teichmann A, Rath W, Kuhn W, Wuttke W (1992) Paracrine actions of oxytocin, prostaglandin F_{2a}, and estradiol within the human corpus luteum. J Clin Endocrinol Metab 74:306–312

Maggi M, Malozowski S, Kassis S, Guardabasso V, Rodbard D (1987) Identification and characterization of two classes of receptors for oxytocin and vasopressin in porcine tunica albuginea, epididymis and vas deferens. Endocrinology 120:986–994

Maggi M, Genazzani AD, Giannini S, Torrisi C, Baldi E, Di Tomasso M, Munson PJ, Rodbard D, Serio M (1988) Vasopressin and oxytocin receptors in vagina, myometrium and oviduct of rabbits. Endocrinology 122:2970–2980

Maggi M, Del Carlo P, Fantoni G, Giannini S, Torrisi C, Casparis D, Ulassi G, Serio M (1990) Human myometrium during pregnancy contains and responds to V1 vasopressin receptors as well as oxytocin receptors. J Clin Endocrinol Metab 70:1142–1154

Maggi M, Fantoni G, Baldi E, Cioni A, Rossi S, Vannelli GB, Melin P, Akerlund M, Serio M (1994) Antagonists for the human oxytocin receptor: an in vitro study. J Reprod Fertil 101:345–352

Maggi M, Peri A, Baldi E, Mancina R, Granchi S, Fantoni G, Finetti G, Forti G, Raggi CC, Serio M (1996) Interferon-alpha downregulates expression of the oxytocin receptor in cultured human myometrial cells. Am J Physiol 271:E840–846

Manning M, Cheng L, Klis WA, Stoev S, Przybylski J, Bankowski K, Sawyer WH, Barberis C, Chan WY (1995) Advances in the design of selective antagonists, potential tocolytics, and radioiodinated ligands for oxytocin receptors. Adv Exp Med Biol 395:559–583

Marc S, Leiber D, Harbon S (1986) Carbachol and oxytocin stimulate the generation of inositol phosphates in the guinea pig myometrium. FEBS Lett 201:9–14

Matthews EL, Ayad VJ (1994) Characterization and localization of a putative oxytocin receptor in the cervix of the oestrous ewe. J Endocrinol 142:397–405

McCarthy MM (1995) Estrogen modulation of oxytocin and its relation to behavior. Adv Exp Med Biol 395:235–246

Mizumoto Y, Kimura T, Ivell R (1997) A genomic element within the third intron of the human oxytocin receptor gene may be involved in transcriptional suppression. Mol Cell Endocrinol 135:129–138

Monga M, Ku CY, Dodge K, Sanborn BM (1996) Oxytocin-stimulated responses in a pregnant human immortalized myometrial cell-line. Biol Reprod 55:427–432

Moore JJ, Dubyak GR, Moore RM, Kooy DV (1988) Oxytocin activates the inositol-phospholipid-protein kinase-C system and stimulates prostaglandin production in human amnion cells. Endocrinology 123:1771–1777

Morel A, O'Carroll AM, Brownstein MJ, Lolait SJ (1992) Molecular cloning and expression of a rat V1a arginine vasopressin receptor. Nature 356:523–526

Morley SD, Meyerhof W, Schwarz J, Richter D (1988) Functional expression of the oxytocin receptor in *Xenopus laevis* oocytes primed with mRNA from bovine endometrium. J Mol Endocrinol 1:77–81

Mouillac B, Chini B, Balestre MN, Elands J, Trumpp-Kallmeyer S, Hoflack J, Hibert M, Jard S, Barberis C (1995) The binding site of neuropeptide vasopressin V1a receptor. Evidence for a major localization within transmembrane regions. J Biol Chem 270:25771–25777

Neumann I, Pittman QJ, Landgraf R (1995) Release of oxytocin within the supraoptic nucleus: mechanisms, physiological significance and antisense targetting. Adv Exp Med Biol 395: 173–184

Neumann I, Douglas AJ, Pittman QJ, Russell JA, Landgraf R (1996) Oxytocin released within the supraoptic nucleus of the rat brain by positive feedback action is involved in parturition-related events. J Neuroendocrinol 8:227–233

Nicholson HD, Jenkin L (1995) Oxytocin and prostatic function. Adv Exp Med Biol 395:529–538

Nohara A, Ohmichi M, Koike K, Masumoto N, Kobayashi M, Akahane M, Ikegami H, Hirota K, Miyake A, Murata Y (1996) The role of mitogen-activated protein kinase in oxytocin-induced contraction of uterine smooth muscle in pregnant rat. Biochem Biophys Res Commun 229:938–944

Nussey SS, Prysor-Jones RA, Taylor A, Ang VTY, Jenkins JS (1987) Arginine vasopressin and oxytocin in the bovine adrenal gland. J Endocrinol 115:141–149

Ohmichi M, Koike K, Nohara A, Kanda Y, Sakamoto Y, Zhang Z-X, Hirota K, Miyake A (1995) Oxytocin stimulates mitogen-activated protein kinase activity in cultured human puerperal uterine myometrial cells. Endocrinology 136:2082–2087

Okuda K, Uenoyama Y, Miyamoto A, Okano A, Schweigert FJ, Schams D (1995) Effects of prostaglandins and oestradiol-17β on oxytocin binding in cultured bovine luteal cells. Reprod Fertil Dev 7:1045–1051

Okuda K, Uenoyama Y, Fujita Y, Iga K, Sakamoto K, Kimura T (1997) Functional oxytocin receptors in bovine granulosa cells. Biol Reprod 56:625–631

Ostrowski NL, Lolait SJ (1995) Oxytocin receptor gene expression in female rat kidney: the effect of estrogen. Adv Exp Med Biol 395:329–340

Parry LJ, Bathgate RAD, Shaw G, Renfree MB, Ivell R (1997) Evidence for a local fetal influence on myometrial oxytocin receptors during pregnancy in the tammar wallaby (*Macropus eugenii*). Biol Reprod 56:200–207

Patchev VK, Schlosser SF, Hassan AH, Almeida OF (1993) Oxytocin binding sites in rat limbic and hypothalamic structures: site-specific modulation by adrenal and gonadal steroids. Neuroscience 57:537–543

Phaneuf S, Europe-Finner GN, Varney M, Mackenzie Z, Watson SP, Lopez-Bernal A (1993) Oxytocin stimulated involvement of pertussis toxin-sensitive and insensitive G-proteins. J Endocrinol 136:497–509

Phaneuf S, Asboth G, Carrasco MP, Europe-Finner GN, Saji F, Kimura T, Harris A, Lopez-Bernal A (1997) The desensitization of oxytocin receptors in human myometrial cells is accompanied by down-regulation of oxytocin receptor messenger RNA. J Endocrinol 154:7–18

Phaneuf S, Asboth G, Carrasco MP, Linares BR, Kimura T, Harris A, Lopez-Bernal A (1998) Desensitization of oxytocin receptors in human myometrium. Human Reprod Update 4:625–633

Pitzel L, Probst I, Jarry H, Wuttke W (1988) Inhibitory effect of oxytocin and vasopressin on steroid release by cultured porcine luteal cells. Endocrinology 122:1780–1785

Pitzel L, Jarry H, Wuttke W (1993) Demonstration of oxytocin receptors in procine corpora lutea: effects of the cycle stage and the distribution on small and large luteal cells. Biol Reprod 48:640–646

Pliska V, Kohlhauf, Albertin H (1991) Effect of Mg^{2+} on the binding of oxytocin to sheep myometrial cells. Biochem J 277:97–101

Postina R, Kojro E, Fahrenholz F (1996) Separate agonist and peptide antagonist binding sites of the oxytocin receptor defined by their transfer into the V2 vasopressin receptor. J Biol Chem 271:31593–31601

Rahmani HR, Muge DK, Ingram CD (1997) Pharmacological characterization of oxytocin bind-
    ing sites in the ovine pineal gland. Regul Pept 70:23–27
Riley PR, Flint AP, Abayasekara RE (1995) Structure and expression of an ovine oxytocin recep-
    tor cDNA. J Mol Endocrinol 15:195–202
Rosenthal W, Antaramian A, Gilbert S, Birnbaumer M (1993) Nephrogenic diabetes insipidus: a
    V2 vasopressin receptor unable to stimulate adenyl cyclase. J Biol Chem 268:13030–13033
Rozen F, Russo C, Banville D, Zingg HH (1995) Structure, characterization, and expression of the
    rat oxytocin receptor gene. Proc Natl Acad Sci USA 92:200–204
Russell JA, Leng G (1998) Sex, parturition and motherhood without oxytocin? J Endocrinol
    157:343–359
Sadeghi HM, Innamorati G, Dagarag M, Birnbaumer M (1997) Palmitoylation of the V2 vaso-
    pressin receptor. Mol Pharmacol 52:21–29
Sakumoto R, Ando Y, Okuda K (1996) Progesterone release of bovine corpus luteum in response
    to oxytocin in different culture systems. J Reprod Dev 42:199–204
Schumacher M, Coirini H, Johnson AE, Flanagan LM, Frankfurt M, Pfaff DW, McEwen BS (1993)
    The oxytocin receptor: a target for steroid hormones. Regul Pept 45:115–119
Sernia C, Gemmell RT, Thomas WG (1989) Oxytocin receptors in the ovine corpus luteum.
    J Endocrinol 121:117–123
Sheldrick EL, Flick-Smith HC (1993) Effect of ovarian hormones on oxytocin receptor concen-
    trations in explants of uterus from ovariectomized ewes. J Reprod Fertil 97:241–245
Soloff MS, Fields MJ (1989) Changes in uterine oxytocin receptor concentration throughout the
    estrous cycle of the cow. Biol Reprod 40:283–287
Soloff MS, Swartz TL (1973) Characterization of a proposed oxytocin receptor in rat mammary
    gland. J Biol Chem 248:6471–6478
Soloff MS, Wieder MH (1983) Oxytocin receptors in rat involuting mammary gland. Can J
    Biochem Cell Biol 61:631–635
Soloff MS, Alexandrova M, Fernstrom MJ (1979) Oxytocin receptors: triggers for parturition and
    lactation. Science 204:1313–1315
Spencer TE, Bazer FW (1996) Ovine interferon-tau suppresses transcription of the estrogen recep-
    tor and oxytocin receptor genes in the ovine endometrium. Endocrinology 137: 1144–1147
Spencer TE, Becker WC, George P, Mirando MA, Ogle TF, Bazer FW (1995) Ovine interferon-tau
    regulates expression of endometrial receptors for estrogen and oxytocin but not progeste-
    rone. Biol Reprod 53:732–745
Spencer TE, Mirando MA, Mayes JS, Watson GH, Ott TL, Bazer FW (1996) Effects of interferon-
    tau and progesterone on oestrogen-stimulated expression of receptors for oestrogen, proges-
    terone and oxytocin in the endometrium of ovariectomized ewes. Reprod Fertil Dev
    8:843–853
Strakova Z, Soloff MS (1997) Coupling of oxytocin receptor to G proteins in rat myometrium
    during labor; Gi receptor interaction. Am J Physiol 272 (Endocrinol Metab 35):E870–E876
Stewart HJ, Stevenson KR, Flint APF (1993) Isolation and structure of a partial sheep oxytocin
    receptor cDNA and its use as a probe for Northern analysis of endometrial RNA. J Mol
    Endocrinol 10:359–361
Takahashi T, Kawashima M, Yasuoka T, Kamiyoshi M, Tanaka K (1996) Mesotocin binding to
    receptors in hen kidney plasma membranes. Poult Sci 75:910–914
Takahashi T, Kawashima M, Yasuoka T, Kamiyoshi M, Tanaka K (1997) Mesotocin receptor
    binding of cortical and medullary kidney tissues of the hen. Poult Sci 76:1302–1306
Takemura M, Nomura S, Kimura T, Inoue T, Onoue H, Azuma C, Saji F, Kitamura Y, Tanizawa
    O (1993) Expression and localization of oxytocin receptor gene in human uterine endometri-
    um in relation to the menstrual cycle. Endocrinology 132:1830–1835
Takemura M, Kimura T, Nomura S, Makino Y, Inoue T, Kikuchi T, Kubota Y, Tokugawa Y,
    Nobunaga T, Kamiura S, Onoue H, Azuma C, Saji F, Kitamura Y, Tanizawa O (1994) Ex-
    pression and localization of human oxytocin receptor mRNA and its protein in chorion and
    decidua during parturition. J Clin Invest 93:2319–2323

Taylor AH, Ang VTY, Jenkins JS, Silverlight JJ, Coombes RC, Luqmani YA (1990) Interaction of vasopressin and oxytocin with human breast carcinoma cells. Cancer Res 50:7882–7886

Uenoyama Y, Okuda K (1997) Regulation of oxytocin receptors in bovine granulosa cells. Biol Reprod 57:569–574

Ufer E, Postina R, Gorbulev V, Fahrenholz F (1995) An extracellular residue determines the agonist specificity of V2 vasopressin receptors. FEBS Lett 362:19–23

Verbalis JG, Blackburn RE, Hoffman GE, Stricker EM (1995) Establishing behavioral and physiological functions of central oxytocin: insights from studies of oxytocin and ingestive behaviors. Adv Exp Med Biol 395:209–226

Wathes DC, Mann GE, Payne JH, Riley PR, Stevenson KR, Lamming GE (1996a) Regulation of oxytocin, oestradiol and progesterone receptor concentrations in different uterine regions by oestradiol, progesterone and oxytocin in ovariectomized ewes. J Endocrinol 151:375–393

Wathes DC, Smith HF, Leung ST, Stevenson KR, Meier S, Jenkin G (1996b) Oxytocin receptor development in ovine uterus and cervix throughout pregnancy and at parturition as determined by in situ hybridization analysis. J Reprod Fertil 106:23–31

Weiss J, Cote CR, Jameson JL, Crowley WF (1995) Homologous desensitization of gonadotropin-releasing hormone (GnRH)-stimulated luteinizing hormone secretion in vitro occurs within the duration of an endogenous GnRH pulse. Endocrinology 136:138–143

Windle RJ, Judah JM, Forsling ML (1997) Effect of oxytocin receptor antagonists on the renal actions of oxytocin and vasopressin in the rat. J Endocrinol 152:257–264

Wu WX, Verbalis JG, Hoffman GE, Derks JB, Nathanielsz PW (1996) Characterization of oxytocin receptor expression and distribution in the pregnant sheep uterus. Endocrinology 137:722–728

Wuttke W, Jarry H, Knoke I, Pitzel L, Spiess S (1995) Luteotropic and luteolytic effects of oxytocin in the porcine corpus luteum. Adv Exp Med Biol 395:495–506

Yarwood NJ, Howl J, Wheatley M (1997) Characterization of the human oxytocin receptor ligand binding site. Biochem Soc Trans 25:436S

Yazawa H, Hirasawa A, Horie K, Saita Y, Iida E, Honda K, Tsujimoto G (1996) Oxytocin receptors expressed and coupled to Ca^{2+} signalling in a human vascular smooth muscle cell-line. Br J Pharmacol 117:799–804

Yoshimura R, Kiyama H, Kimura T, Araki T, Maeno H, Tanizawa O, Tohyama M. (1993) Localization of oxytocin receptor messenger ribonucleic acid in the rat brain. Endocrinology 133:1239–1246

Young LJ, Huot B, Nilson R, Wang Z, Insel TR (1996) Species differences in central oxytocin receptor gene expression: comparative analysis of promoter sequences. J Neuroendocrinol 8:777–783

Zingg HH, Rozen F, Breton C, Larcher A, Neculcea J, Chu K, Russo C, Arslan A (1995) Gonadal steroid regulation of oxytocin and oxytocin receptor gene expression. Adv Exp Med Biol 395:395–404

Targeted Mutagenesis of the Murine Opioid System

Michael D. Hayward and Malcolm J. Low

1
Introduction

The opioid system is a powerful modulator of pain and pleasure. Membrane-bound receptors that bind opioid drugs have been identified and characterized pharmacologically, and the genes that encode these receptors have been cloned and mapped to their chromosomal loci. A family of endogenous peptides that produce opioid responses bind these receptors with affinities specific to individual receptors. Currently, experiments using genetically manipulated mice are unambiguously identifying the physiological processes that opioid peptides and their cognate receptors control. A major strength of this technique is that experiments are no longer subject to the limitations of pharmacological specificity (e.g., an agonist's affinity for more than one receptor). Technology for transgenesis and targeted mutagenesis in mice has progressed to a point where we can now examine single gene changes *in vivo*, using biologically relevant paradigms. Genetically modified mice, mostly those harboring a null allele for an opioid gene, behave consistently with several hypotheses based on earlier pharmacological data, but unexpected results have been observed as well. We review here what those experiments have found and identify important experiments that have not been done. In addition, we review the use of genetically manipulated mice, pointing out caveats and pitfalls of this technology, particularly as it applies to the opioid-mediated behaviors of analgesia and addiction.

2
A Brief History of the Opioid System

The three opioid receptor genes that have been identified and cloned are the μ (MOR, OPR-3), κ (KOR, OPR-2), and δ (DOR, OPR-1) (Evans et al. 1992; Keiffer et al. 1992; Chen et al. 1993; Meng et al. 1993; Thompson et al. 1993). These receptors are 60% similar at the amino acid level, contain seven membrane

Vollum Institute, L474, Oregon Health Sciences University, 3181 S.W. Sam Jackson Park Rd., Portland, Oregon 97201, USA spanning domains and share a similar coupling to intracellular signaling pathways (reviewed in Christie et al. 1987; Brownstein 1993; Reisine and Bell 1993; Mansour et al. 1995). The three opioid receptors are negatively coupled to adenylyl cyclase via a pertussis toxin-sensitive GTP-binding regulatory (G) protein, likely G_i and/or G_o. Thus, the opioid receptors are rightfully members of the seven-transmembrane, G-protein coupled receptor family, which also includes many neurotransmitter and neurohormone receptors including all of the dopamine receptors, the substance P/neurokinin-A receptor and the β-adrenergic receptor. All three opioid receptors can also be coupled to a calcium current and an inwardly rectifying potassium current, resulting in neuronal hyperpolarization. In most cells, the opioid receptors inhibit both action potentials and cAMP formation, which counters other neurotransmitters from generating signals via cAMP-dependent protein kinase. Another similarity among the opioid receptors is that their genes each contain three coding exons with conserved splice sites and at least one large intron between corresponding exons. The physical and functional conservation between the opioid receptors suggests that they may have evolved from a common receptor. The highly related OFQ/nociceptin receptor (ORL-1) is the subject of Reinscheid et al. (this Vol.).

The corresponding endogenous opioid peptides are β-endorphin and endomorphins 1 and 2, which bind to the μ receptor; enkephalin, which binds to the δ receptor; and dynorphin, which binds to the κ receptor. This distinction may not be absolute, however, as β-endorphin only has a slightly higher affinity for the μ receptor than for the δ receptor (Reisine and Pasternak 1996). Presumably, the colocalization of peptide and receptor also provides specificity in the endogenous opioid system. For example, experiments using electrically evoked release of β-endorphin in brain slices suggest that, functionally, β-endorphin uses presynaptic μ receptors in neocortical slices (Schoffelmeer et al. 1991). The most recently discovered of the opioid peptides are endomorphins 1 and 2, two naturally occurring peptides isolated from bovine cortex but, as of yet, uncloned (Zadina et al. 1997). Endomorphins 1 and 2 have a higher affinity for the μ receptor than β-endorphin (K_i = 0.36 and 0.69 nM for endomorphins 1 and 2, respectively, while K_i = 4.4 nM for β-endorphin). Consequently, there are at least two different endogenous peptides that bind to the μ receptor with high affinity.

Similar to the opioid receptors, several structural features are also shared among the endogenous opioid peptides. The peptides for β-endorphin, enkephalin, and dynorphin all contain the identical core penta-peptide sequence (Tyr-Gly-Gly-Phe-Met/Leu) and all arise from the proteolytic cleavage of a precursor molecule. The pro-opioimelanocortin (POMC) protein, the first of the three opioid peptide precursors to be sequenced (Bradbury et al. 1976; Roberts and Herbert 1977; Nakanishi et al. 1979), produces β-endorphin from the 31 C-terminal amino acids of POMC. The POMC precursor gives rise to three primary peptides, melanocyte-stimulating hormone (α MSH), ACTH, and β-lipotropin (β-LPH); β-endorphin is derived from β-LPH. β-endorphin contains the sequence of Met-enkephalin at its amino terminus, but in higher species Met-enkephalin is derived from the processing of proenkephalin. Proenkephalin was first isolated from bovine adrenal medulla and each pro-hormone molecule gives rise to one copy of Leu-enkephalin, four copies of Met-enkephalin and one copy each of a Met-enkephalin C-terminal-extended heptapeptide and octapeptide (Lewis et al. 1980; Comb et al. 1982; Noda et al. 1982). Prodynorphin gives rise to α- and β-neoendorphin, dynorphin A (1–17), dynorphin A (1–8) and dynorphin B (Kakidani et al. 1982; Horikawa et al. 1983). Among the shared features, all three precursor proteins contain almost the same number of amino acids and all three possess multiple sequences of opioid peptides that are contained in the C-terminal half of the precursor. Another similarity that exists is sequence conservation between proenkephalin and prodynorphin, which share more that 50% amino acid identity. Organization of the exon and intron structures are also similar between POMC, proenkephalin, and prodynorphin. Consequently, it has been suggested that these genes also evolved from a common ancestor gene (Simon and Hiller 1994).

The endogenous opioid system is usually considered a regulator of nociception, but it can also be considered part of a general physiological stress regulatory system working in parallel with the autonomic nervous system. Since the endogenous opioids can function as neurotransmitters, modulators of neurotransmission, or neurohormones, they are capable of influencing a diverse array of behaviors. Evidence suggests that the endogenous opioids can modulate endocrine, cardiovascular, gastrointestinal and immune functions in addition to their effects through the central and peripheral nervous systems on nociception, behavioral reinforcement, feeding and locomoter activity (Illes 1989; Mansour et al. 1995; Olson et al. 1995; Yamada and Nabeshima 1995). Most of what we know of the opioid system is based on pharmacological experiments with synthetic drugs and peptides. For example, we know that morphine can produce analgesia, decrease body temperature, depress respiration, dilate peripheral blood vessels and decrease intestinal motility (Reisine and Pasternak 1996). Unfortunately, it is difficult to dissect out relevant physiological processes from pharmacological actions when administering exogenous drugs. Elimination of the endogenous peptides stands as a definitive test for their specific roles in many of these physiological processes.

3
Techniques in Murine Mutagenesis

Gene-targeting and transgenic techniques are a relatively new technology that is rapidly developing and will be of enormous value in future studies of both opioid receptors and their respective ligands. Induced mutations of the mouse can be produced by three different methods: random mutagenesis by a chemical agent, transgenesis, or targeted mutagenesis. Transgenesis involves the introduction of a free-standing construct that is randomly integrated within the genome, often in multiple copies. This technique is most useful for studies using reporter genes fused to the promoter of a gene of interest in order to precisely examine expression patterns (Low 1992). It is also useful for experiments involving dominant-negative proteins, antisense molecules, and overexpression or ectopic expression of a particular protein. Use of the transgenic technique does not lend itself easily to introducing mutations into specific genetic loci.

Targeted genetic manipulations of mice use transfection of embryonic stem cells (ES) to produce a null allele (a "knock-out") or replacement of a gene with a mutated version or completely different gene (a "knock-in") (reviewed in Soriano 1995). Most of the work using targeted mutagenesis has generated null mutations, most commonly done by inserting a drug resistance marker into an exon or deleting exons necessary for the gene's function (for methods see Hogan et al. 1994). However, mutations can also be created with temporal and/or anatomical restrictions (a "conditional knock-out") by using site-specific recombinases such as Cre from bacteriophage P1 and FLP from yeast (Meyers et al. 1998; reviewed in Kilby et al. 1993). The DNA recognition sites for these recombinases are introduced into a specific gene by homologous recombination in ES cells and then genomic sequences flanked by two of these sites can be excised *in vivo* by expression of the recombinase. The recombinase is typically expressed by a transgene that can have a tissue specific promoter or an inducible (or repressible) promoter such as the tetracycline controlled system (Furth et al. 1994; Kistner et al. 1996; St-Onge et al. 1996). Small or subtle mutations can be created by gene knock-in using the hit-and-run/in-and-out method (Hasty et al. 1991; Valancius and Smithies 1991), the tag and exchange method (Askew et al. 1993), or the double replacement method (Wu et al. 1994). These techniques are especially useful for generating point mutations of amino acids or replacing functional domains *in vivo*. A plethora of techniques is now available for generating mutant mice and the opioid system is an excellent target for the use of such techniques.

4
Gene Deletion Studies in the Opioid System

Null mutations of all of the cloned members of the opioid system have been produced. Mice null for the μ opioid receptor have been generated independently in at least five labs (Matthes et al. 1996; Bolan et al. 1997; Schuller et al. 1997; Sora et al. 1997b; Tian et al. 1997) Mice lacking the κ opiate receptor (Simonin et al. 1998) and the δ receptor (Zhu et al. 1997) have also been reported. β-Endorphin was the first member of the opioid system to be knocked-out

(Rubinstein et al. 1996) and a null mutant of enkephalin (König et al. 1996) has also been reported. Gene targeting of the prodynorphin gene has resulted in viable and fertile mice in the absence of dynorphin peptides throughout development and additional studies are ongoing (U. Hochgeschwender, Oklahoma Medical Research Foundation, pers. comm.). In addition to the single gene mutations, double knock-out mice that lack both β-endorphin and enkephalin (M.J. Low and J. Pintar, unpubl. data) and double knock-outs for each pair of μ, κ, and δ receptors (J. Pintar, U.M.D.N.J.-Robert Wood Johnson Medical School, pers. comm.) are viable and healthy.

4.1
Stress-Induced Analgesia

One way to infer a gene's involvement in a specific physiological process is to examine that process in the absence of the gene product. A remarkable feature of the opioid system is the apparent redundancy of multiple pathways that regulate antinociception. This quandary is particularly amenable to the use of gene deletion studies. Drugs and peptides that are specific to any of the three different opioid receptors are capable of producing antinociception to varying degrees. Additionally, environmental stressors can also produce an antinociception, mediated in part by endogenous opioid pain-inhibition mechanisms, which is called stress-induced analgesia (SIA). The relative importance of opioid and nonopioid pathways in SIA depends on the type and magnitude of the stressor (Mogil et al. 1996). For example, mice that are forced to swim in room temperature water for 3 min take longer to respond in a test for nociception such as the hot-plate assay or show a reduced writhing response in the abdominal constriction test. These measures of analgesia can be reversed by naloxone, indicating the involvement of an opioid pathway. A nonopioid mediated SIA, which by definition is not reversible by naloxone, is produced by a cold water swim. Thus, SIA is a physiological process that can be examined in the absence of a gene product believed to be involved in that action.

Mice lacking β-endorphin were produced by a unique point-mutation strategy that introduced a premature stop codon at the N terminus of the β-endorphin coding region within the POMC gene (Rubinstein et al. 1993). These mice no longer demonstrated the opioid-mediated SIA produced by a room temperature swim and measured using the abdominal constriction assay (Rubinstein et al. 1996). The nonopioid form of SIA was actually potentiated in these mice. Contrary to some expectations, mice null for enkephalin had intact opioid and nonopioid forms of SIA (König et al. 1996). It remains to be shown whether mice lacking the μ receptor no longer produce opioid SIA. This experiment would be of considerable value as it is still not certain whether all β-endorphin mediated behaviors are transduced through the μ receptor because β-endorphin has nearly equivalent affinities for both the μ and δ receptors (Reisine and Bell 1993).

The finding that mutant mice lacking ß-endorphin, but not enkephalin, no longer expressed opioid SIA is one example where previous data from pharmacological experiments were not as definitive as the data produced by gene deletion studies. However, classical pharmacological experiments were critical for distinguishing between opioid and nonopioid SIA in the mutant mice. This distinction held true following additional experiments in mice null for the substance P/neurokinin-A receptor (NK-1) (De Felipe et al. 1998). NK-1 knockout mice, in contrast to both β-endorphin and enkephalin deficient mice, had decreased non-opioid but intact opioid SIA. Mice lacking the pre-protachykinin A gene which encodes the tachykinins substance P and neurokinin A reportedly had no change in SIA (Zimmer et al. 1998) or were not tested for SIA (Cao et al. 1998). These differences in SIA between strains lacking the NK-1 receptor or the preprotachykinin A peptides are likely due to cross-talk between the remaining receptors and peptides encoded by distinct, but highly related, genes in the tachykinin family.

4.2
Nociception

Although SIA was preserved in mice lacking enkephalin, a basal nociceptive phenotype was discovered. Mice lacking enkephalin had normal tail-flick responses but decreased jump latencies on a 55 °C hot-plate test (i.e., they were hyperalgesic) (König et al. 1996). This finding is in contrast to the β-endorphin deficient mice which had no or minimal alteration in their basal response to thermal stimuli (J. Mogil and M.J. Low, unpubl. data). The enkephalin deficient mice were reported to demonstrate increased apprehension or anxiety in the open field test and the elevated "zero-maze" (Shepherd et al. 1994). Additionally, male enkephalin knock-out mice were more aggressive in the "resident-intruder" test during the initial exposure to an intruder. After repeated exposure to an intruder the mutant mice were similar to wild-type mice, suggesting a dysregulation in aggression rather than a constitutive increase in aggressive behavior. Since spinal nociception was not affected by the proenkephalin mutation, the authors have suggested that a lack of enkephalin resulted in an exaggerated response to painful or threatening environmental stimuli at a supraspinal level (König et al. 1996). Thus, supraspinal processing of fearful stimuli may be aberrant in these mutant mice. It remains to be seen whether similar phenotypes will be demonstrated in δ receptor mutant mice.

The tail-immersion/withdrawal assay tests thermal nociception at the spinal level with supraspinal modulation. As with many other tests for nociception, the response time of the mouse, in this case withdrawing its tail from hot water, is a measurement of nociception and is labeled the latency period. The hot-plate assay tests nociception at supraspinal levels. This assay consists of placing the animal on a hot plate and measuring the time it takes for the animal to respond to the pain by hind-paw shaking, paw licking, or jumping (latency period). Results from these nociception assays using μ receptor deficient mice generated by different laboratories were inconsistent. In one study there were significant basal nociceptive differences between wild-type mice and μ receptor mutants, depending on the test parameters. Homozygote null mice responded quickest (shortest latency) to the painful stimulus in the tail-withdrawal assay when the water temperature was at 50 or 53°C (Sora et al. 1997b). There was also a gene-dosage effect such that the response latencies of heterozygote mutants were halfway between wild-type mice and homozygous null mutants. The μ receptor mutants also took less time to respond in the 55°C hot-plate assay, with a similar non-significant trend at 52°C, but only the homozygote mutants differed from wild-type (Sora et al. 1997b). When the stimuli were stronger (56°C for tail withdrawal and 58°C for hot-plate) there were no differences in latency periods among the three genotypes. Results from this study suggest that the μ receptor influences basal nociception to a modest degree, because hyperalgesia was measured only in response to less intense stimuli. However, in another study using an independently derived strain of μ receptor knock-out mice, basal nociception was measured by both tail withdrawal (54°C) and hot-plate (52°C) methods and it was not altered in the μ receptor mutants (Matthes et al. 1996). Unfortunately, the stimulus temperatures used in this study were not exactly the same as Sora et al. 1997b so comparison between these two studies is hampered.

Sora et al. 1997b suggest that μ receptor signaling may have greater functional significance in spinal nociception than in supraspinal nociception to explain the differing results between the tail-withdrawal and hot-plate assays with their μ receptor knock-out mice. Alternatively, the exceptional difference measured only in the tail-withdrawal assay may be due to synergistic alterations in both spinal and supraspinal pathways resulting from the absence of μ receptors. One way of resolving this debate would be to construct a region-specific knock-out of the μ receptor by using the Cre-loxP system to produce mice with selective deletion of μ receptors from the spinal cord or the CNS.

Changes in basal nociception in μ knock-out mice are consistent with a "tonic" analgesic state, which has been suggested previously based on clinical data using naloxone treatment (Levine et al. 1979). However, basal nociception was not altered in the β-endorphin-deficient mice (Rubinstein et al. 1996). The newly identified endogenous opioids, the endomorphins, might therefore play a role in basal nociception since intracerebral-ventricular (I.C.V.) and intrathecal (I.T.) injections of the peptides produced spinal analgesia (Stone et al. 1997; Zadina et al. 1997). Endomorphin modulation of basal nociception would also be consistent with the limited and specific alterations in nociception found in the μ-receptor knock-out. Endomorphin mutant mice will contribute significantly to answering this question.

Specific thermal nociceptive alterations in μ receptor knock-out mice were detected, but loss of functional κ receptors produced changes in nociception that were exquisitely selective to non-thermal stimuli. Mice null for the κ receptor were identical to wild-type mice in assays using thermal stimuli (tail-withdrawal and hot-plate tests), mechanical stimuli (tail-pressure test) and local tissue damage (the formalin test) (Simonin et al. 1998). However, mice lacking the κ receptor reacted twice as strongly (i.e., hyperalgesia) to the writhing test, which uses an injection of dilute acetic acid into the peritoneal cavity. Previous pharmacological studies have shown that κ receptor agonists can influence chemically induced visceral pain (Millan 1990), but the null mutant κ receptor study suggests that endogenous κ agonists (i.e., dynorphin) play an important physiological role in mediating basal nociception to noxious visceral stimuli. δ-Receptor knock-out mice were reported to have no changes in baseline tail-withdrawal latencies (Zhu et al. 1997), but a more extensive study of nociception from these mice should be undertaken.

4.3
Opioid Pharmacology of Opioid System Knock-out Mice

Much of our understanding of the endogenous opioid system comes from research using the large number of selective agonists and antagonists. Using these tools, opioid binding sites were first identified as distinct receptors. Pharmacological agents have also delineated receptor subtypes, which have still not been molecularly characterized. Experiments that combine exogenous opioid ligands with the mutant mouse strains have confirmed several old predictions, led to new pharmacological insights, but have also raised a number of intriguing and unanswered questions.

4.3.1
μ Receptor Pharmacology in the μ Receptor Knock-out Mice

The prototypical opiate, morphine, is a relatively selective agonist for μ receptors. In fact, μ receptors were defined by their affinity for morphine (Reisine and Pasternak 1996). However, at high enough concentrations morphine will interact with κ and δ receptors. There have been suggestions that some analgesic effects of morphine are, in fact, mediated through non-μ opioid receptors (Jiang et al. 1990; Sutters et al. 1990). The μ receptor mutant mice had an extraordinary loss of morphine-induced analgesia by both the hot-plate and the tail-withdrawal assays (Matthes et al. 1996; Sora et al. 1997b). At doses below 56 mg/kg there was essentially no analgesic action of morphine and at greater doses there was only a trend toward modest analgesia. However, at such heroic doses it is likely that other opioid receptors are involved (Sora et al. 1997b). The μ receptor mutant mice confirmed that the μ receptor is the only receptor directly responsible for the analgesic actions of morphine to thermal stimuli at clinically relevant doses.

Intriguingly, an independently derived strain of μ receptor mutant mice, which also had no analgesia in response to morphine, responded normally to other μ receptor selective ligands. The opioids that produced analgesia in these μ receptor null mice include fentanyl, heroin, 6-acetylmorphine and morphine-6-glucuronide (Schuller et al. 1997). [^{3}H]-morphine-6-glucuronide binding could also be detected in these μ receptor knock-out mice (Bolan et al. 1997). One possible explanation for these unexpected findings is the existence of an uncloned receptor that binds morphine-6-glucuronide and heroin. Alternatively, there may be a μ receptor splice variant that was not eliminated by the construct used to generate the mutant mice. It appears that there are as many as three forms of the μ receptor in humans as well as other forms found in rats and mice, although only one μ receptor gene has been identified in each species (Bare et al. 1994; Mayer et al. 1996; Du et al. 1997). In fact, morphine-6-glucoronide has been proposed as a tool for detecting a new μ receptor splice variant (Rossi et al. 1995,1997). However, an important point to make is that many of the spliced forms have not been shown to bind opioid ligands.

4.3.2
δ and κ Pharmacology in the μ Receptor Knock-out Mice

μ Receptor mutant mice were analyzed by saturation radioligand binding on brain membranes and, predictably, there was nearly no detectable μ receptor binding by [^{3}H]DAMGO ([D-Ala^2, N-McPhe^4, Gly^5-ol]enkephalin) in homozygous mutants (Matthes et al. 1996; Sora et al. 1997b). These studies also analyzed δ and κ binding in the μ receptor mutant mice by saturation binding and autoradiography and gene expression by *in situ* hybridization. The μ receptor mutant mice did not have any alteration in the binding or expression of either δ or κ receptors. Surprisingly, these studies found that δ receptor mediated analgesia was nearly absent in the μ receptor mutant mice. The δ selective agonist BuBu [Tyr-D-Ser(O-tert-butyl)-Gly-Phe-Leu-Thr(O-tert-butyl)] did not induce any analgesia measured by the tail-flick test, but δ analgesia on the hot-plate test was normal (Matthes et al. 1996). Another δ selective agonist, DPDPE ([D-Pen, D-Pen]enkephalin) injected I.C.V. produced nearly no analgesic effect assessed by both the hot-plate and tail-withdrawal assays (Funada et al. 1997; Sora et al. 1997a). In addition, DPDPE analgesia in heterozygote mice was intermediate between wild-type and knock-out mice (Sora et al. 1997a). These results were intriguing because substantial evidence indicates that there are interactions between the μ receptor and other opioid receptor types. Morphine analgesia can be modulated by specific δ agonists (Jiang et al. 1990; Traynor and Elliott 1993) and analgesia produced by μ agonists can also be modulated by κ agonists (Sutters et al. 1990).

Similar to studies on basal nociception, there were also conflicting data using δ-specific ligands between the different μ receptor knock-out mice produced in separate laboratories. One report claimed that analgesia produced by DPDPE and DELT (D-Ala^2-deltorphin II) was only reduced in the μ receptor mutant mice when tested by tail flick but not affected by the hot-plate assay

(Matthes et al. 1997) whereas another report claimed that DPDPE was analgesic by both hot plate and tail withdrawal (Sora et al. 1997b). There are also data that κ receptor analgesia is compromised in the μ receptor mutant mice when measured by the hot-plate assay but not the tail-flick assay (Funada et al. 1997). However, Matthes et al. (1997) found no decrease in either tail-withdrawal or hot-plate assays following the κ agonist U50,488 (Matthes et al. 1997). Nonetheless, previous pharmacological data have indicated that some mechanism exists by which receptor cross-talk occurs, possibly even by heterodimerization of the receptors themselves. Experiments using μ receptor mutant mice support the theory of receptor cross-talk, but, more importantly, the mutant mice provide a new tool to further investigate the mechanism of this communication.

4.3.3
Pharmacology in the δ and κ Receptor Knock-out Mice

A functional μ receptor may be required for some analgesic properties of δ receptor agonists but it does not appear that the reverse is true. δ Receptor mutant mice were recently demonstrated to have normal morphine analgesia as measured by the tail-flick assay (Zhu et al. 1997). The δ receptor mutant mice, as expected, did not exhibit DPDPE or DELT spinal analgesia. Unexpectedly, supraspinal analgesia produced by the same δ receptor specific compounds remained intact (Zhu et al. 1997). The supraspinal analgesia produced by DPDPE and DELT may be mediated by μ receptors in the δ receptor knock-out mice. In fact, there have even been suggestions that all thermal, supraspinal opioid analgesia in the mouse is mediated solely via μ receptors (Fang et al. 1986; Baamonde et al. 1991).

κ Receptor mutant mice, as expected, lost analgesia produced by U50,488 when the animals were tested by both the hot-plate and tail-withdrawal assays (Simonin et al. 1998). Analgesia produced by morphine was not altered when tested by either assay, demonstrating that the κ receptor is not required for μ receptor analgesia. Taken together, these studies suggest that although κ and δ mediated analgesia may require a functional μ receptor, morphine analgesia does not require either a functional δ or κ receptor.

4.3.4
Opioid Pharmacology in the β-Endorphin Knock-out Mice

Loss of an endogenous ligand may have consequences on the expression or function of its receptor. We are currently examining both the expression and function of all three opioid receptors in the β-endorphin mutant mice and have found no apparent change in the distribution or amount of opioid receptor expression in brain and spinal cord (M.D. Hayward and M.J. Low, unpubl. data). Analgesic dose-response curves to morphine administered I.P. and

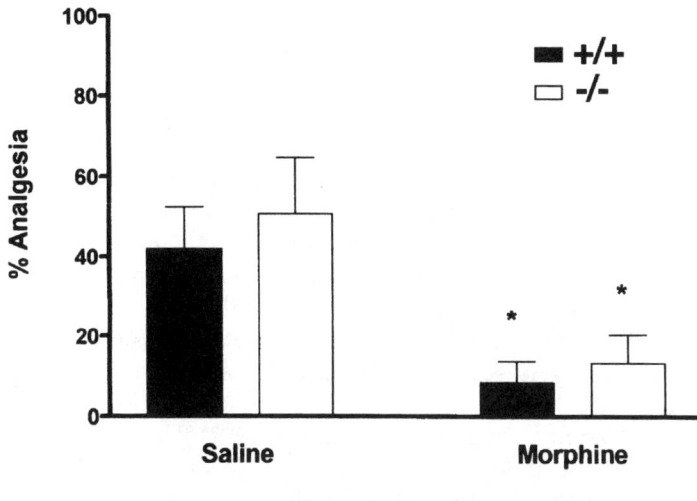

Fig. 1. Tolerance to the analgesic effect of an acute test dose of morphine following chronic morphine administration. Tolerance was studied in adult male wild-type (+/+) and homozygous mutant (−/−) β-endorphin-deficient mice (Rubinstein et al. 1996) on an F_2 (129/SvEv, C57BL/6) genetic background. Separate groups of mice were injected subcutaneously twice daily for 7 days with either saline vehicle (10 ml/kg) or morphine (16 mg/kg). On day 8, all mice were given an acute, intraperitoneal injection of morphine (16 mg/kg) and analgesia was measured for the first time. Mice were scored on a 55 ± 1° C hot-plate for latency (sec) to exhibit hind-paw licking, hind-paw flutter, jumping, or a cut-off time of 60 sec as the end-point immediately before (pre) and 20 min after (post) the test morphine injection. Data are mean ± SEM of maximal possible analgesic effect calculated as MPE = [(post − pre)/(60 − pre)] × 100%. n 10 mice per group; * p < 0.05 by t-test compared to respective saline control

measured by abdominal constriction and hot-plate tests were normal in the mutant mice (Rubinstein et al. 1996). We have also found that analgesic tolerance to morphine I.P. develops normally in β-endorphin-deficient mice (Fig. 1). However, further tests for analgesia have suggested that a differential change in sensitivity to μ agonists occurs in the β-endorphin-deficient mice dependent on whether the agonists are administered I.P, I.T. or I.C.V. (J. Mogil et al., in prep.). Thus, it appears that the loss of β-endorphin has specifically affected the μ receptor but that change appears to be at a functional level and not at the level of expression or ligand binding. The change in μ receptor function is apparently quite specific since the development of tolerance was not affected.

4.4
Behavioral Reinforcement and Reward

The opioid system is equally well known for its roles in antinociception and operant behavioral reinforcement. μ Agonists produce reinforcing effects that can lead to the development of sensitization, drug craving, and conditioned withdrawal. Conversely, κ agonists appear to be aversive (Mucha and Herz 1985; Nabeshima et al. 1988; Bals-Kubik et al. 1989; Székely 1994). The meso-limbic dopamine pathway to the nucleus accumbens (NA) originates from the ventral tegmental area (VTA) and this pathway appears to be involved in the rewarding properties of several drugs of abuse. Experiments have shown that dopamine is released in the NA following administration of both opioids and psychostimulants (Di Chiara and Imperato 1988a; Di Chiara and North 1992; Koob 1992; Koob et al. 1992; Stolerman 1992). The μ agonist DAMGO increas-es the release of dopamine in the NA while κ agonists decrease the amount of mesolimbic dopamine release (Di Chiara and Imperato 1988b; Spanagel et al. 1992; Devine et al. 1993). These data support the hypothesis that μ agonists are reinforcing because they increase dopamine release in the NA while κ agonists are aversive because they inhibit dopamine release in the NA. Consistent with these behavioral and biochemical data, dense binding of μ ligands and μ receptor mRNA expression are detected in the VTA while κ ligand binding and κ receptor mRNA expression are substantially less in the VTA (Mansour et al. 1995). Additional support for the idea that the μ receptor is necessary for rein-forcement comes from the μ receptor null mice, which did not prefer mor-phine in conditioned place preference tests (Matthes et al. 1997). Thus, the reinforcing property of μ agonists may be due largely to the expression of μ receptors in the VTA.

Experiments with the κ receptor mutant mice strongly suggest that there is little role for κ receptors in the rewarding potential of morphine, since the κ receptor knock-outs still have normal morphine reward (Simonin et al. 1998). This result was somewhat unexpected since κ agonists can block the reward-ing properites of morphine (Funada et al. 1993). However, signs of physical withdrawal in morphine-dependent mice were reduced in the absence of the κ opiate receptor (Simonin et al. 1998), even though the κ peptide agonist dynor-phin A-(1–17) also attenuates physical signs of morphine withdrawal in wild-type mice (Suzuki et al. 1992). In fact, the κ receptor antagonist nor-binaltor-phimine potentiated the weight loss that occurred during withdrawal, although other signs of opiate withdrawal were not significantly altered (Suzuki et al. 1992). Thus, it appears that κ receptor expression does not affect the rewarding properties of morphine, but it can modulate physical withdraw-al from morphine. It would be helpful, however, to know if κ receptor ligands are still aversive to κ receptor knock-out mice.

5
Genetic Background of Mutant Mice

The complicated behaviors that are modulated by the opioid system are undoubtedly polygenic. Thus, the genetic context in which opioid pathways are studied is likely to have a significant influence on these behaviors and, consequently, experimental results. Different strains of mice diverge for their basal sensitivities in nociceptive assays, their analgesic sensitivities to opioid drugs, and their propensity to self-administer opioid drugs. Since these behaviors are influenced by multiple neurotransmitter systems, it is expected that the source of strain variation will not be found in only one gene. Experiments using targeted genetic mutations must be carefully designed so that a genetically equivalent wild-type control is used and interpretations of experimental results must be in the context of the particular mouse strain used. With the advent of studies using mutant mice it is important for researchers who plan on continuing with these experiments to be aware of the role of genetic background and some of the idiosyncrasies specific mouse strains can play in determining phenotypes. Of utmost importance is the adoption of standards for maintaining mutant strains of mice.

A recent consensus report stressed that the genetic background of induced mutant mice should be known and reproducible (Banbury Conference 1997). A flawed but commonly used technique is to maintain a line of mutant mice by inbreeding F_2 homozygous littermates. Superficially, this is an enticing method because it eliminates the need for genotyping. However, this method is flawed because the genetic background of such inbred mice is not known and cannot be reproduced. In fact, after 20 generations of brother-sister matings a new inbred line is generated. This procedure generally maintains a parallel wild-type control line that is a separate population generated from the original F_2 generation. The result of these matings is two inbred lines whose genetic makeup, in addition to the targeted locus, can differ drastically. These differences arise because of random segregation and fixation of polymorphic alleles, including genes linked to the targeted locus, that are unique to the wild-type or the targeted line. Thus, it is preferable by far to avoid this scheme at all costs. If this method has been used, then the targeted mutation should be transferred to standard inbred backgrounds, which can be done in ten generations of standard mating or as little as three to four generations using "speed congenics" (Lander and Schork 1994; Markel et al. 1997).

The method recommended by the Banbury Conference delegates to maintain a mutant mouse line on a known genetic background is to initially mate chimeras to both a 129 strain (origin of the ES cells and mutant locus) and a C57BL/6 strain (origin of the blastocyst) (Banbury Conference 1997). Heterozygous mice are then repeatedly backcrossed in parallel on the inbred C57BL/6 and 129 lines, such that congenic heterozygotes are created on both backgrounds. Ten backcrosses are required to produce 99.9% congenic mice. However, after only five generations the genome is nearly 97% congenic to the background strains and these mice could be used, at least for pilot studies, although the backcrossing procedure should continue indefinitely. Heterozygotes from the two congenic strains are then crossed to produce F_1 hybrid mice in Mendelian proportions of 1 wild-type:2 heterozygotes:1 homozygote for experiments. F1 hybrid mice have the advantages of reproductive vigor and the absence of deleterious recessive traits from both of the background strains. Unfortunately, this breeding procedure is often not possible because some behaviors to be studied are only present in certain inbred strains and the procedure is prohibitively expensive for most laboratories. For example, studying a double mutant by this method would require massive numbers of mating pairs because double mutants would only be 1/16 of the F_1 progeny of double heterozygous parents.

A less expensive alternative is to backcross the chimeric mice onto only one inbred strain and use littermates from heterozygote matings as the experimental subjects. However, this method requires some knowledge of the strain to be used and should not be done without some comprehensive literature review or pilot studies. Many studies of null mutations have reported learning deficits, but many inbred lines have phenotypic abnormalities in learning themselves. For example, several 129 lines have spatial learning deficits (Gerlai 1996) and most 129 and DBA strains show poor hippocampal-dependent learning (Upchurch and Wehner 1988; Wolfer et al. 1997). Mice from the C57BL/6 line become deaf to certain frequencies at an early age (Willot 1986) and they are poor avoidance learners (Schwegler and Lipp 1983). Studies that are especially relevant to experiments on the opioid system have shown that the C57BL/6 strain is considerably more tolerant to the analgesic effects of μ-selective opioids such as morphine and DAMGO compared to other strains (Belknap et al. 1995; Mogil and Wilson 1997). These strain differences are not limited to performances on behavioral tests. For example, kainic acid induced seizures occur in the absence of hippocampal neuronal degeneration in C57BL/6 and Balb/C mice (Choi 1997; Schauwecker and Steward 1997) and the corpus callosum is defective in many mice from BALB and 129 substrains (Livy and Wahlsten 1991; Wahlsten and Bulman-Fleming 1994; Wahlsten and Schalomon 1994). An F_1 hybrid cross of inbred lines often eliminates some of these abnormalities because homozygous recessive genes are often present in the inbred strains. For example, of all inbred lines tested, the C57BL/6 is the best performer on the Morris water maze yet all F_1 hybrid lines performed better than the C57BL/6 mice (Upchurch and Wehner 1988). If the Morris water maze is an intended assay, then an F_1 hybrid might be the preferred genetic background for studying the effects of single gene deletions that are expected to decrease spatial learning.

A fundamental problem of each of the breeding strategies described is the phenomenon of "hitchhiking genes" closely linked to the targeted genetic locus (Crusio 1996). These alleles are derived only from the genetic background of the ES cells and even after 12 generations of backcrossing to produce a congenic line they may represent as much as16 cM of the ES genome, or 1% of the total containing 100 to 300 genes (Banbury Conference 1997). Virtually all available and reliable ES cells are derived from 129 substrain mice, so these closely linked "hitchhiking" genes are 129 alleles. One important aspect of this reality is that 129 substrains are actually quite diverse genetically and many different substrains have been used for the derivation of stem cells (reviewed in Simpson et al. 1997). Standardized behavioral testing of these substrains is incomplete and many are no longer commercially available. Moreover, certain substrains are notorious for inherited neurological disorders (Gerlai 1996) as well as low fecundity. There are at least two instances in which phenotypic differences between induced mutant mice and wild-type control mice have been attributed to closely linked gene alleles originating from the 129 background strain, rather than the targeted mutation itself (Choi 1997; Schauwecker and Steward 1997; Kelly et al. 1998). Remedies for this problem are not easy. The most desirable, but still not available, solution is the development of a new series of ES cells derived from a panel of different inbred strains that can be used for gene targeting.

6
Differences Between Knock-out Lines

Although the generation of mutant mice was intended to clarify the physiologic role of individual molecular components of the opioid system, some of the data are apparently inconsistent between laboratories for independently produced strains of mice with targeting of identical genes. Phenotypic inconsistencies have similarly arisen between pairs of nonopioid system mutant mice produced in separate laboratories. There are many possible explanations for these discrepancies. The use of different targeting construct strategies may be the source of some of the differences observed and, inadvertently, provide us with useful information to identify new genes or alternatively spliced forms of the μ opioid receptor. Alternatively, differences between data from two laboratories may be due to genetic background of mutant mice and the effect of epistasis, i.e., the mutated protein normally interacts with a number of other proteins, any of which could contribute to a new phenotype.

6.1
Construct Differences Between μ Receptor Knock-out Mice Lines

Two different strategies have been used to target the μ receptor locus to create a "null" mutant. The construct that was used by Sora et al. (1997b) replaced the first exon of the μ receptor with a gene cassette coding for neomycin resistance. It is possibile that an alternative transcriptional start site could exist somewhere within the very large (>26 kb) first intron of the μ receptor or other location in the gene, and a splice variant derived from exons 2, 3, and 4 could encode a receptor which cannot bind morphine but can bind other μ receptor agonists. In contrast, Matthes et al. (1996) targeted the second and third exon of the μ receptor. These two different targeting construct strategies may be the source of the phenotypic differences observed in basal nociception and in analgesic responses to δ agonists described earlier. In addition, the μ knock-out strain, which had intact analgesia and binding to morphine-6-glu-curonide (Bolan et al. 1997; Schuller et al. 1997; see Sect. 4.3.1), was produced by a strategy similar to Sora et al. (1997b) that deleted exon 1. An intriguing finding is that the μ receptor knock-out mice produced by targeting of the second and third exons (Matthes et al. 1996) apparently do not have an analgesic response to heroin and morphine-6-glucuronide (J. Pintar, pers. comm.). The multiple strategies used in designing null, or possibly hypomorphic, mutations of the μ receptor may have produced a serendipitous tool for identifying an uncloned μ-like receptor or splice variant.

6.2
Genetic Differences Between Background Strains Used for Producing Mutant Mice

The phenotypic differences among the μ receptor knock-out mice from different laboratories could also lie in the genetic background of these mice, and a close comparison of their pedigrees should be pursued. Rarely do papers using mutant mice give a sufficiently detailed pedigree of their mice, indicating origin of the ES cells (i.e., which 129 substrain they were derived from), backcrossing methods and the generation of mice used for the experiments. Indeed, of the six full publications on targeted null mutants of the opioid system reviewed here, one lacked a reference to the origin of the ES cells, four gave a reference, but only one clearly stated the 129 substrain. Descriptions of backcrossing in these papers was even less detailed; two papers gave partial information, one gave none and indicated that they used only inbred hybrid F_2 generation mice for their experiments. None of the papers indicated what generation of mice was used for experiments. β-Endorphin-deficient mice were an F_2 hybrid maintained as heterozygotes on the (129/SvEv, C57BL/6) background using wild-type sibling mice as controls from heterozygote mating pairs. Enkephalin knock-out mice were backcrossed onto the outbred CD-1 background, but no generation was given. Controls included CD-1 wild-types. This type of information, no matter how mundane it may appear, should be discussed more fully in the future. Experiments on the opioid system should especially be concerned with providing this data since there exist significant variations in the responses to opioid drugs among mice strains.

The two strains of mice most often used for homologous recombination techniques are, ironically, two strains that represent many extremes of opioid responses in mice. The 129/J and C57BL/6J strains exhibit different sensitivities on a variety of basal nociceptive assays and their sensitivity to opioid analgesia (Mogil and Wilson 1997; Mogil et al. 1998). In addition, C57BL/6J mice prefer orally available opiates more than many other strains (Belknap et al. 1993). On the other hand, 129/SvEvTac mice lack the development of tolerance to chronic morphine and DPDPE (but not U50,488H), a condition which may be due to a defective NMDA receptor (Kolesnikov et al. 1998). The decreased sensitivity of C57BL/6J mice to opiate analgesia has actually been exploited by the production of recombinant inbred strains. These studies have revealed phenotypic differences that exist among commonly used inbred strains manifested as deficiencies in opioid analgesia and receptor binding (Mogil and Wilson 1997). For example, one series of recombinant inbred mice derived from a BALB/c × C57BL/6 cross (CXB) have low opiate binding for several ligands and have a low analgesic response. Strains derived from a C57BL/6 × DBA/2 (BXD) cross display greater morphine analgesia (Belknap et al. 1995) and increased preference for oral consumption of morphine (Berrettini et al. 1994). Interestingly, the chromosomal region responsible for the increased analgesia was mapped to the same region as the µ receptor gene and the serotonin-1B gene (Kozak et al. 1994; Belknap et al. 1995). Nonetheless, significant support still exists for the C57BL/6 strain as the background strain of choice for several behavioral tests (Crawley et al. 1997). C57BL/6 mice are moderate learners, thus permitting the detection of single gene mutations that either improve or impair learning processes. These data support the need to carefully examine the contribution of polymorphic alleles from the background strains and stress the importance of using congenic mice for nociceptive/analgesic assays of single gene mutants.

7
Conclusions

The first wave of genetically engineered mice that contain targeted mutations in the genes encoding opioid peptides and opioid receptors has had important implications for our understanding of the physiological function of these molecules. These new models are complementary to pharmacological and antisense strategies, not a replacement. Converging data suggests that β-endorphin and not enkephalins are the critical endogenous opioid peptides involved in stress-induced analgesia. The µ opioid receptor has a fundamental role in the analgesic and rewarding actions of morphine and likely is also essential for supraspinal analgesia mediated by δ agonists. κ Receptors appear to be vital in establishing the tonic nociceptive threshold for visceral pain.

These studies have also demonstrated important limitations of the current technology and the need for improved experimental standards. To make valid comparisons among the mouse models generated by different laboratories, it is essential that better documentation of genetic background be provided in publications. Initial results using F_2 hybrid mice must be followed by more comprehensive analyses in congenic or inbred strains to truly evaluate the consequences of the single gene that has been deliberately mutated. Finally, there needs to be a continued effort to develop reliable techniques for introducing more subtle mutations in specific genes and to permit the spatial and temporal control of gene inactivation and reactivation *in vivo*.

References

Askew G, Doetschman T, Lingrel J (1993) Site-directed point mutations in embryonic stem cells: a gene-targeting tag-and-exchange strategy. Mol Cell Biol 13:4115–4124

Baamonde AI, Dauge V, Gacel G, Rozues BP (1991) Systemic administration of [Tyr-D-Ser(O-Tert-Butyl)-Gly-Phe-Leu-Thr(O-Tert-Butyl], a highly selective delta opioid agonist, induces mu receptor-mediated analgesia in mice. J Pharmacol Exp Ther 257:767–773

Bals-Kubik R, Herz A, Shippenberg TS (1989) Evidence that the aversive effects of opioid antagonists and κ-agonists are centrally mediated. Psychopharmacology 98:203–206

Banbury Conference (1997) Mutant mice and neuroscience: recommendations concering genetic background. Neuron 19:755–759

Bare L, Mannson E, Yang D (1994) Expression of two variants of the human mu opioid receptor mRNA in SK-N-SH cells and human brain. FEBS Lett 354:213–216

Belknap JK, Crabbe JC, Riggan J, O'Toole LA (1993) Voluntary consumption of morphine in 15 inbred mouse strains. Psychopharmacology 112:352–358

Belknap JK, Mogil JS, Helms ML, Richards SP, O'Toole LA, Bergeson SE, Buck KJ (1995) Localization to chromosome 10 of a locus influencing morphine analgesia in crosses derived from C57BL/6 and DBA/2 strains. Life Sci. 57:117–124

Berrettini WH, Ferraro TN, Alexander RC, Buchberg AM, Vogel WH (1994) Quantitative trait loci mapping of three loci controlling morphine preference using inbred mouse strains. Nat. Genet. 7:54–58

Bolan EA, Schuller A, Yang K, Brown GP, Pintar JE, Pasternak GW (1997). ^{3}H-morphine-6-glucoronide binding in MOR-1 knockout mice. In Soc Neurosci Abstr. Society for Neuroscience, Washington DC, pp. 584, Abst #235.5.

Bradbury AF, Smyth DG, Snell CR, Birdsall NJM, Hulme EC (1976) C fragment of lipotropin has an affinity for brain opiate receptor. Nature 260:793–795

Brownstein MJ (1993) A brief history of opiates, opioid peptides, and opioid receptors. Proc Natl Acad Sci USA 90:5391–5393

Cao YQ, Mantyh PW, Carlson EJ, Gillespie A-M, Epstein CJ, Basbaum AI (1998) Primary afferent tachykinins are required to experience moderate to intense pain. Science 392:390–394

Chen Y, Mestek A, Liu J, Hurley JA, Yu L (1993) Molecular cloning and functional expression of a μ-opioid receptor from rat brain. Mol Pharm 44:8–12

Choi DW (1997) Background genes: out of sight, but not out of brain. Trends Neurosci 20: 499–500

Christie MJ, Williams JT, North RA (1987) Cellular mechanisms of opioid tolerance: studies in single brain neurons. Mol Pharmacol 32:633–638

Comb M, Seeburg PH, Adelman J, Eiden L, Herbert E (1982) Primary structure of the human Met-and Leu-enkephalin precursor and its mRNA. Nature 295:663–666

Crawley JN, Belknap JK, Colins A, Crabbe JC, Frankel W, Henderson N, Hitzemann RJ, Maxson SC, Miner LL, Silva AJ, Wehner JM, Wynshaw-Boris A, Paylor R (1997) Behavioral phenotypes of inbred mouse strains: implications and recommendations for molecular studies. Psychopharmacology 132:107–124

Crusio WE (1996) Gene-targeting studies: new methods, old problems. Trends Neurol Sci 19:186–187

De Felipe C, Herrero JF, O'Brien JA, Palmer JA, Doyle CA, Smith AJH, Laird JMA, Belmonte C, Cervero F, Hunt SP (1998) Altered nociception, analgesia and aggression in mice lacking the receptor for substance P. Science 392:394–397

Devine DP, Leone P, Pocock D, Wise RA (1993) Differential involvement of ventral tegmental mu, delta and kappa opioid receptors in modulation of basal mesolimbic dopamine release: in vivo microdialysis studies. J Pharmacol Exp Ther 266:1236–1246

Di Chiara G, Imperato A (1988a) Drugs abused by humans preferentially increase synaptic dopamine concentrations in the mesolimbic system of freely moving rats. Proc Natl Acad Sci USA 85:5274–5278

Di Chiara G, Imperato A (1988b) Opposite effects of mu and kappa opiate agonists on dopamine release in the nucleus accumbens and in the dorsal caudate of freely moving rats. J Pharmacol Exp Ther 244:1067–1080

Di Chiara G, North RA (1992) Neurobiology of opiate abuse. Trends Pharmacol 13:185–192

Du Y-L, Elliott K, Pan Y-X, Pasternak GW, Inturrisi CE (1997). A splice variant of the the mu opioid receptor is present in human SHSY-5Y cells. In Soc Neurosci Abst. Society for Neuroscience, Washington DC, pp. 1206 Abst # 479.3.

Evans CJ, Keith DEJ, Morrison H, Magendzo K, Edwards RH (1992) Cloning of a delta opioid receptor by functional expression. Science 258:1952–1955

Fang FG, Fields HL, Lee NM (1986) Action at the mu receptor is sufficient to explain the supraspinal analgesic effect of opiates. J Pharmacol Exp Ther 238:1039–1044

Funada M, Suzuki T, Narita M, Misawa M, Nagase H (1993) Blockade of morphine reward through the activation of κ-opioid receptors in mice. Neuropharmacology 32:1315–1323

Funada M, Sora I, Goodman N, Li X, Kinsey S, Uhl GR (1997). DPDPE & U50,488 in μ receptor knockout mice: evidence for μ-dependent κ and δ effects on analgesia, tolerance and/or dependence. In Soc Neurosci Abstr. Society for Neuroscience, Washigton DC, pp. 584 Abstr. #235.3.

Furth PA, St. Onge L, Böger H, Gruss P, Gossen M, Kistner A, Bujard H, Hennighausen L (1994) Temporal control of gene expression in transgenic mice by a tetracycline-responsive promoter. Proc Natl Acad Sci USA 91:9302–9406

Gerlai R (1996) Gene-targeting studies of mammalian behavior: is it the mutation or the background genotype? Trends Neurosci 19:177–189

Hasty P, Ramifrez-Solis R, Krunlauf R, Bradley A (1991) Introduction of a subtle mutation into the Hox-2.6 locus in embryonic stem cells. Nature 350:243–246

Hogan B, Beddington R, Costantini F, Lacy E (1994) Manipulating the mouse embryo, 2nd edn. Cold Spring Harbor Laboratory Press, Plainview, New York

Horikawa S, Takai T, Toyosato M, Takahashi H, Noda M, Kakidani H, Kudo T, Hirose T, Inayama S, Hayashida H, Miyata T, Numa S (1983) Isolation and structural organization of the human preproenkephalin B gene. Nature 306:611–614

Illes P (1989) Modulation of transmitter and hormone release by multiple neuronal opioid receptors. Rev Physiol Biochem Pharmacol 112:140–233

Jiang Q, Mosberg HI, Porreca F (1990) Modulation of the potency and efficacy of mu-mediated antinociception by delta agonists in the mouse. J Pharmacol Exp 254:683–689

Kakidani H, Furutani Y, Takahashi H, Noda M, Morimoto Y, Hirose T, Asai M, Inayama S, Nakanishi D, Numa S (1982) Cloning and sequence analysis of cDNA for porcine β-neoendorphin/dynorphin precursor. Nature 298:245–249

Keiffer BL, Befort K, Baveriaux-Ruff C, Hirth CG (1992) The δ-opioid receptor: isolation of a cDNA by expression and pharmacological characterization. Proc Natl Acad Sci USA 89:12048–12052

Kelly MA, Rubinstein M, Phillips TJ, Lessov CN, Burkhart-Kasch S, Zhang G, Bunzow JR, Fang Y, Gerhardt GA, Grandy DK, Low MJ (1998) Locomoter activity in D2 dopamine receptor-deficient mice is determined by gene dosage, genetic background, and developmental adaptations. J Neurosci 18:3470–3479

Kilby N, Snaith M, Murray J (1993) Site-specific recombinases: tools for genome engineering. Trends Genet 9:413–421

Kistner A, Gossen M, Zimmermann F, Jerecic J, Ullmer C, Lübbert H, Bujard H (1996) Doxycycline-mediated quantitative and tissue-specific control of gene expression in transgenic mice. Proc Natl Acad Sci USA 93:10933–10938

Kolesnikov Y, Jain S, Wilson R, Pasternak GW (1998) Lack of morphine and enkephalin tolerance in 129/SvEv mice: evidence for a NMDA receptor defect. J. Pharmacol Exp Ther 284:455–459

König M, Zimmer AM, Steiner H, Holmes PV, Crawley JN, Brownstein MJ, Zimmer A (1996) Pain aggression in mice deficient in pre-proenkephalin. Nature 383:535–538

Koob GF (1992) Drugs of abuse: anatomy, pharmacology and function of reward pathways. Trends Pharmacol 13:177–184

Koob GF, Maldonado R, Stinus L (1992) Neural substrates of opiate withdrawal. Trends Neurosci 15:186–191

Kozak CA, Filie J, Adamson MC, Chen Y, Yu L (1994) Murine chromosomal location of the μ and κ opioid receptor genes. Genomics 21:659–661

Lander ES, Schork NJ (1994) Genetic dissection of complex traits. Science 265:2037–2048

Levine JD, Gordon NC, Fields HL (1979) Naloxone dose dependently produces analgeisa and hyperalgesia in postoperative pain. Nature 278:740–741

Lewis RV, Stern AS, Kimura S, Rossier J, Stein S, Udenfriend SA (1980) A 50000-dalton protein in adrenal medulla that may be a common precursor of [Met]- and [Leu] enkephalin. Science 208:1459–1461

Livy DJ, Wahlsten D (1991) Tests of genetic allelism between four inbred mouse strains with absent corpus callosum. J Hered 82:459–464

Low MJ (1992) The identification of neuropeptide gene regulatory elements in transgenic mice. In: Longstaff A, Revest P (eds) Methods in molecular biology, The Humana Press, Totowa, New Jersey, pp 181–204

Mansour A, Fox CA, Akil H, Watson SJ (1995) Opioid-receptor mRNA expression in the rat CNS: anatomical and functional implications. Trends Neurosci 18:22–29

Markel P, Shu P, Ebeling C, Carlson GA, Nagle DL, Smutko JS, Moore KJ (1997) Theoretical and empirical issues for marker-assisted breeding of congenic mouse strains. Nat Genet 17:280–284

Matthes HWD, Maldonado R, Simonin F, Valverde O, Slowe S, Kitchen I, Befort K, Dierich A, Le Meur M, Dollé P, Tzavara E, Hanoune J, Roques BP, Kieffer B (1996) Loss of morphine-induced analgesia, reward effect and withdrawal symptoms in mice lacking the μ-opioid-receptor gene. Nature 383:819–823

Matthes HWD, Maldonado R, Smadia C, Valverde O, Roques B, Kieffer BL (1997). Functional response of delta- and kapa-opioid receptors in mu-opioid receptor knock-out mice. In Soc Neurosci Abstr. Society for Neuroscience, Washington DC, pp 583

Mayer P, Schulzeck S, Kraus J, Zimprich A, Höllt V (1996) Promoter region and alternatively spliced exons of the rat μ-opioid receptor gene. J. Neurochem 66:2272–2278

Meng F, Xie G-X, Thompson RC, Mansour A, Goldstein A, Watson SJ, Akil H (1993) Cloning and pharmacologicial characterization of a rat κ opioid receptor. Proc Natl Acad Sci USA 90:9954–9958

Meyers E, Lewandoski M, Martin G (1998) An Fgf8 mutant allelic series generated by Cre- and Flp-mediated recombination. Nature Genetics 18:136–141

Millan MJ (1990) κ-Opioid receptors and analgesia. Trends Pharmacol Sci 11:70–76

Mogil JS, Sternberg WF, Balian H, Liebeskind J C, Sadowski B (1996) Opioid and nonopioid swim stress-induced analgesia: a parametric analysis in mice. Physiol Behav 59:123–132

Mogil JS, Wilson SG (1997) Nociceptive and morphine antinociceptive sensitivity of 129 and C57BL6 inbred mouse strains: implications for transgenic knock-out studies. Eur J Pain 1:293–297

Mogil JS, Lichtensteiger CA, Wilson SG (1998) The effect of genotype on sensitivity to inflammatory nociception: characterization of resistant (A/J) and sensitive (C57BL/6) inbred mouse strains. Pain 76:115–125

Mucha RF, Herz A (1985) Motivational properties of kappa and mu opioid receptor agonists studied with place and taste preference conditioning. Pysochopharmacology 86:274–280

Nabeshima T, Kamei H, Kameyama T (1988) Opioid κ receptors correlate with the development of conditioned suppression of motility in mice. Eur J Pharmacol 152:129–133

Nakanishi S, Inoue A, Kita T, Nakamura M, Chang AC, Cohen SN, Numa S (1979) Nucleotide sequence of cloned cDNA for bovine corticotropin-β-lipotropin precursor. Nature 278:423–427

Noda M, Furutani Y, Takahashi H, Toyosato M, Hirose T, Inayama S, Nakanishi S, Numa S (1982) Cloning and sequence analysis of cDNA for bovine preproenkephalin. Nature 295:1982

Olson GA, Olson RD, Kastin AJ (1995) Endogenous opiates: 1995. Peptides 17:1421–1466

Reisine T, Bell GI (1993) Molecular biology of opioid receptors. Trends Neurosci 16:506–510

Reisine T, Pasternak G (1996) Opioid analgesics and antagonists. In: Hardman JG, Gilman AG, Limbird LE (eds) Goodman and Gilman's The pharmacological basis of therapeutics, McGraw-Hill, New York, pp 521–555

Roberts JL, Herbert E (1977) Characterization of a common precursor to corticotropin and β-lipotropin: Cell-free synthesis of the precursor and identification of corticotropin peptides in the molecule. Proc Natl Acad Sci USA 74:4826–8430

Rossi GC, Pan Y-X, Brown GP, Pasternak GW (1995) Antisense mapping the MOR-1 opioid receptor: evidence for alternative splicing and a novel morphine-6β-glucuronide receptor. FEBS Lett 369:192–196

Rossi GC, Leventhal L, Pan Y-X, Cole J, Su W, Bodnar RJ, Pasternak GW (1997) Antisense mapping of MOR-1 in rats: distinguising between morphine and morphine-6β-glucoronide antinociception. J Pharmacol Exp Ther 281:109–114

Rubinstein M, Japón M, Low MJ (1993) High efficiency introduction of a point mutation into the mouse genome in embryonic stem cells by homologous recombination using a replacement type vector and selectable markers. Nucleic Acids Res 21:2613–2617

Rubinstein M, Mogil JS, Jápon M, Chan EC, Allen RG, Low MJ (1996) Absence of opioid stress-induced analgesia in mice lacking β-endorphin by site-directed mutagenesis. Proc Natl Acad Sci USA 93:2577–2582

Schauwecker PE, Steward O (1997) Genetic determinants of susceptibility to excitotoxic cell death: Implications for gene targeting approaches. Proc Natl Acad Sci USA 94:4103–4108

Schoffelmeer ANM, Wardeh G, Hogenboom F, Mulder AH (1991) β-Endorphin: a highly selective endogenous opioid agonist for presynaptic *mu* opioid receptors. J Pharmcol Exp Ther 258:237–242

Schuller AGP, King M, Zhang J, Czick M, Unterwald E, Pasternak GW, Pintar JE (1997). Heroin and M6G analgesia are retained in mu opioid receptor deficient mice. In Soc Neurosci Abstr. Society for Neuroscience Washington DC, pp. 584, Abst #235.6.

Schwegler J, Lipp H-P (1983) Hereditary covariations of neuronal circuitry and behavior: correlations between the proportions of hippocampal synaptic fields in the regio inferior and two-way avoidance in mice and rats. Behav Brain Res 7:1–39

Shepherd JK, Grewal SS, Fletcher A, Bill D, Dourish CT (1994) Behavioural and pharmacological characterization of the elevated "zero-maze" as an animal model of anxiety. Psychopharmacology 116:56–64

Simon EJ, Hiller JM (1994) Opioid peptides and opioid receptors. In: Siegel GJ, Agranoff BW, Albers RW, Molinoff PB (eds) Basic neurochemistry: molecular, cellular, and medical aspects, New York University Medical Center, New York, pp 321–339

Simonin F, Valverde O, Smadja C, Slowe S, Kithchen I, Dierich A, Le Meur M, Roques BP, Maldonado R, Kieffer BL (1998) Disruption of the κ-opioid receptor gene in mice enhances sensitivity to chemical visceral pain, impairs pharmacological actions of the selective κ-agonist U-50,488H and attenuates morphine withdrawal. EMBO J 17:886–897

Simpson E, Linder CC, Sargent EE, Davisson MT, Mobraaten LE, Sharp JJ (1997) Genetic variation among 129 substrains and its importance for targeted mutagenesis in mice. Nat Genet 16:19–27

Sora I, Funada M, Uhl G (1997a) The μ-opioid receptor is necessary for [D-Pen^2,D-Pen^5] enkephalin-induced analgesia. Eur J Pharmacol 256:281–286

Sora I, Takahashi N, Funada M, Ujiki H, Revay RS, Donovan DM, Miner LL, Uhl GR (1997b) Opiate receptor knockout mice define μ receptor roles in endogenous nociceptive respones and morphine-induced analgesia. Proc Natl Acad Sci USA 94:1544–1549

Soriano P (1995) Gene targeting in ES cells. Annu Rev Neurosci 18:1–18

Spanagel R, Herz A, Shippenberg TS (1992) Opposing tonically active endogenous opioid systems modulate the mesolimbic dopaminergic pathway. Proc Natl Acad Sci USA 89:2046–2050

Stolerman I (1992) Drugs of abuse: behavioural principles, methods and terms. Trends Pharmacol 13:170–176

Stone LS, Fairbanks CA, Laughlin TM, Ngyuen HO, Bushy TM, Wessendorf MW, Wilcox GL (1997) Spinal analgesic actions of the new endogenous opioid peptides endomorphin-1 and -2. NeuroReport 8:3131–3135

St-Onge L, Furth PA, Gruss P (1996) Temporal control of the Cre recombinase in transgenic mice by a tetracycline responsive promoter. Nucl Acids Res 24:3875–3877

Sutters KA, Miaskowski C, Taiwo YO, Levine JD (1990) Analgesic synergy and improved motor function produced by combinations of μ-δ and μ-κ-opioids. Brain Res 530:290–294

Suzuki T, Narita M, Takahashi Y, Misawa M, Nagase H (1992) Effects of nor-binaltorphimine on the development of analgesic tolerance to and physical dependence on morphine. Eur J Pharmacol 213:91–97

Székely JI (1994) μ-Agonist induced euphoria as opposed to dysphoria elicited by κ-agonists in humans and experimental animals In: Opioid peptides in substance abuse, CRC Press, Boca Raton, Florida, pp 55–80

Thompson RC, Mansour A, Akil H, Watson SJ (1993) Cloning and pharmacological characterization of a rat μ opioid receptor. Neuron 11:903–913

Tian M, Broxmeyer HE, Fan T, Lai Z, Zhang S, Aronica S, Cooper S, Bigsby RM, Steinmetz R, Engle SJ, Mestek A, Pollock J, Lehman MN, Jansen HT, Ying M, Stambrook PJ, Tischfield JA, Yu L (1997) Altered hematopoiesis, behavior, and sexual function in μ opioid receptor-deficient mice. J Exp Med 185:1517–1522

Traynor JR, Elliott J (1993) Delta-opioid receptor subtypes and cross-talk with mu-receptors. Trends Pharmacol Sci 14:84–86

Upchurch M, Wehner JM (1988) Differences between inbred strains of mice in Morris water maze performance. Behav Genet 18:55–68

Valancius V, Smithies O (1991) Testing an "in-out" targeting procedure for making subtle genomic modifications in mouse embryonic stem cells. Mol Cell Biol 11:1402–1408

Wahlsten D, Bulman-Fleming B (1994) Retarded growth of the medial septum: a major gene effect in acallosal mice. Dev Brain Res 77:203–214

Wahlsten D, Schalomon PM (1994) A new hybrid mouse model for agenesis of the corpus callosum. Behav Brain Res 64:111–117

Willot JF (1986) Effects of aging, hearing loss, and anatomical location on thresholds of inferior colliculus neurons in C57BL/6 and CBA mice. J Neurophysiol 56:391–408

Wolfer DP, Stagliar-Bozizevic M, Müller U, Lipp H-P (1997) Assessing the effects of the 129Sv genetic background on swimming navigation learning in transgenic mutants: a study using mice with a modified β-amyloid precursor gene. Brain Res 771:1–13

Wu H, Liu X, Jaenisch R (1994) Double replacement: strategy for efficient introduction of subtle mutations into the murine *Colla-1* gene by homologous recombination in embryonic stem cells. Proc Natl Acad Sci USA 91:2819–2823

Yamada K, Nabeshima T (1995) Stress-induced behavioral responses in multiple opioid systems in the brain. Behav Brain Res 67:133–145

Zadina JE, Hackler L, Ge L-J, Kastin AJ (1997) A potent and selective endogenous agonist for the ζ-opiate receptor. Nature 386:499–502

Zhu Y, King M, A. S, Unterwald E, Pasternak G, Pintar JE (1997). Genetic disruption of the mouse delta opioid gene. In Soc Neurosci Abstr: Society for Neuroscience, Washington DC, pp. 584, Abst #235.8.

Zimmer A, Zimmer AM, Baffi J, Usdin T, Reynolds K, Konig M, Palkovits M, Mezey E (1998) Hypoalgesia in mice with a targeted deletion of the tachykinin 1 gene. Proc Natl Acad Sci USA 95:2630–2635

Orphan Receptors and the Concept of Reverse Physiology: Discovery of the Novel Neuropeptide Orphanin FQ / Nociceptin

Rainer K. Reinscheid[1], Hans-Peter Nothacker, and Olivier Civelli[2]

1
Summary

The cloning of numerous orphan members from the supergene family of G protein-coupled receptors implies the existence of many as yet undiscovered neurotransmitters and neuropeptides. Recently, new technologies were developed to isolate natural ligands for orphan receptors, using the receptor as a biological sensor during the purification process. This manuscript will present the concept and technology of an approach which starts from a cloned receptor to ultimately describe the physiological functions of the transmitter system. This strategy inverts the classical order of biomedical research and was thus termed "reverse physiology".

The first natural ligand isolated by this strategy is a peptide with significant similarity to the opioid peptides and has been named orphanin FQ or nociceptin (OFQ/NOC). Evidence for characterizing OFQ/NOC as a genuine neuropeptide will be reviewed. OFQ/NOC is biosynthetically derived from a larger precursor protein which may encode additional bioactive peptides. Since its discovery, a large number of studies have described numerous physiological functions of OFQ/NOC. Because of its relation to the opioid system, much attention has been focused on the involvement of OFQ/NOC in nociception, sometimes with controversial results. However, the pharmacological profile of the OFQ/NOC system suggests a clear separation from the opioids. The discovery of OFQ/NOC and the subsequent analyses of its physiological functions is an example which has already been followed by the identification of two other novel neuropeptides. The orphan receptor strategy holds a lot of promises for the postgenomic era, helping to fill the vast amount of sequence data with life.

---

[1] Institute for Cell Biochemistry and Clinical Neurobiology, University Hospital Eppendorf, Süderfeldstr. 24, D-22529 Hamburg, Germany
[2] Department of Pharmacology, University of California, Irvine, Med Surge II, Irvine, CA 92697-4625

2
Orphan Receptors and the Concept of Reverse Physiology

Traditionally, the description of a physiological effect has been the starting point for investigation of biological systems and finally the molecular components, e.g. proteins, involved in it. The use of molecular biology has ultimately allowed for the identification of genes responsible for particular physiological responses. Not only could molecular biology prove the genetic entity of biological components previously described only physiologically, such as ion channels and neurotransmitter receptors. It also yielded spectacular insight in the multiplicity of these systems by the identification of a multitude of subtypes in protein families. Undoubtedly, the available genetic information will vastly increase by sequencing of the entire human genome but will also challenge scientists with thousands of protein sequences of unknown function (Rowen et al. 1997). This situation requires the development of new techniques and approaches, starting from the sequence of a gene to finally elucidate its function, approaches which have been called "reverse physiology" or "functional genomics".

A first step in the reverse physiology approach will be the identification of molecules that interact specifically with the novel protein. This will allow for investigation of the biochemical network in which any novel protein is participating. The elucidation of functional or signaling cascades will enable researchers to develop pharmacological tools, either chemical or genetical, that may finally help to reveal the biological functions of the novel protein.

As a result of extensive cloning experiments, the family of G protein-coupled receptors (GPCRs) turned out to be one of the largest gene families in the mammalian genome, comprising receptors for ligands as diverse as small molecules (e.g. acetylcholine, amino acids and the biogenic amines), neuropeptides, glycoproteins, lipids and odorant molecules (Horn et al. 1998; Marchese et al. 1998; Henikoff et al. 1997). They regulate a multitude of physiological processes and it is thus not surprising, that GPCRs represent the most successful group of biochemical targets for current drug therapy. The structural similarity of GPCRs made them very rewarding targets for further cloning approaches using low stringency screening or PCR. These experiments helped to identify a number of additional GPCRs whose endogenous ligands had already been well known and pharmacologically characterized before (Bunzow et al. 1988; Bunzow et al. 1992). As a first surprising – and then thoroughly studied – side-product, these approaches also produced an ever since increasing collection of sequences which are clearly members of the family but can not be activated by any of the known neurotransmitters (Libert et al. 1989). These "lonely" receptors were called "orphan GPCRs". (Using the same type of approach, similar observations were also made in the fields of receptor-tyrosine kinases (Lai and Lemke 1991), membrane transporter proteins (Liu et al. 1993) and nuclear DNA-binding proteins (Laudet 1997), all being "receptor-like" molecules as well.)

Since their genes have been conserved during evolution it can well be assumed that orphan GPCRs all serve a distinct biological function and thus *must* have an endogenous ligand. We have chosen these orphan receptors as examples for a reverse physiology approach, because identification of their natural ligands is a prerequisite of studying and understanding their roles in the organism.

3
Identification of the Natural Ligand of an Orphan GPCR: Orphanin FQ/Nociceptin

The orphan receptor itself can be used as a sensor to identify its natural ligand. Technologically this is achieved by exposing cells expressing the orphan receptor to tissue extracts expected to contain the ligand. Binding of the natural ligand to the GPCR will induce changes in intracellular second messenger concentrations, by which the purification process can be monitored.

However, there are two unknown variables in this scheme: the chemical nature of the natural ligand and the second messenger pathway which will be activated. For each chemical class of molecules (small hydrophilic molecules, peptides, proteins, lipids, etc.) an individual purification technique is necessary. Furthermore, GPCRs can activate a variety of second messenger responses including adenylyl cyclase, phospholipase C and A2, ion channels, phosphodiesterase and possibly others (Ross, 1992). Although homology to known GPCRs can be used for predictions about the physical nature of the expected endogenous ligand and the second messengers that will be activated, searching for the ligand of an orphan GPCR requires the application of a variety of technologies and a considerable amount of courage. The search for natural ligands of orphan GPCRs was first applied to an opioid-like orphan receptor.

The opioid-like receptor ORL-1 (also known under other names; Bunzow et al. 1994; Chen et al. 1994; Fukuda et al. 1994; Mollerau et al. 1994; Wang et al. 1994; Wick et al. 1994; Standifer et al. 1994; Lachowicz et al. 1995) had been cloned by the numerous laboratories in search for new subtypes of opioid receptors and showed significant sequence homology to the known opioid receptors, μ, δ, and κ. Surprisingly, ORL-1 did not bind any of the known opioid peptides or opiates. Because of its similarities to the opioid receptors, it could be assumed that the endogenous ligand for this receptor might also be a peptide and that after ligand binding it would negatively couple to adenylyl cyclase.

Two independent groups succeeded in purifying the natural ligand of the opioid-like orphan receptor, followed soon after by a third one (Meunier et al. 1995; Reinscheid et al. 1995; Okuda-Ashitaka et al. 1996). The opioid-like orphan receptor cDNA was stably expressed in cells and cAMP levels were monitored. As sources for the natural ligand porcine or bovine hypothalami or total rat brains were extracted and processed according to protocols for the isolation of neuropeptides. After several purification steps, an active compound was purified to homogeneity and proved to be a peptide with the primary structure FGGFTGARKSARKLANQ. This peptide was named orphanin FQ (OFQ), to mark its relationship to a formerly orphan receptor using the first and last amino acids of its sequence as hallmarks (Reinscheid et al. 1995), and nociceptin (NOC) according to one of its activities (see below) (Meunier et al. 1995).

4
Evidence for OFQ/NOC being a Neurotransmitter

By definition, a ligand molecule has to match certain criteria in order to be classified as a true (neuro-) transmitter. These criteria include: 1) saturable, high-affinity and reversible binding to its receptor, 2) a specific mode of synthesis and inactivation, 3) storage and release from specialized cells, e.g. neurons, 4) production of biochemical or electrochemical effects in the target cell carrying the receptor and 5) modulation of complex physiological effects in organotypic assays or intact organisms. Each of these questions have been analyzed in the recent past.

4.1
Receptor Binding of OFQ/NOC

To investigate the physical constants of binding of OFQ/NOC to its receptor, a radioligand was developed (Reinscheid et al. 1995). Because OFQ/NOC does not contain Tyr residues, a series of Tyr-substituted peptide analogs were synthesized. The Tyr^{14}-substituted OFQ/NOC was shown to be an agonist with equivalent potency in cAMP assays as the unsubstituted OFQ/NOC (EC_{50} values of 1.02 ± 0.11 nM). The $[^{125}I]$-labeled Tyr^{14}-substituted peptide displayed saturable, displaceable and reversible binding to membranes of opioid-like orphan receptor transfected cells with a K_d of 0.1 ± 0.02 nM (Reinscheid et al. 1995). Its binding constants are well in the range of affinities observed for other neuropeptides. This demonstrates that this novel peptide is a natural ligand of the opioid-like receptor (hereafter called OFQ/NOC receptor). Tyr^{14}-substituted OFQ/NOC has been used extensively as a radioligand to detect and quantify OFQ/NOC receptor levels, confirming OFQ/NOC as a true pharmacological agent to the OFQ/NOC receptor (Reinscheid et al. 1996; Dooley and

Houghton 1996; Shimoshigashi et al. 1996; Ardati et al. 1997; Butour et al. 1997; Guerrini et al. 1997).

4.2
Cellular Responses to OFQ/NOC

Activation of the OFQ/NOC receptor results in a variety of intracellular effects. First, the OFQ/NOC receptor was shown to inhibit adenylyl cyclase in CHO cells, as this effect was used to initially purify OFQ/NOC (Meunier et al. 1995; Reinscheid et al. 1995). Next, modulation of cellular excitability was detected when OFQ/NOC was found to increase inwardly rectifying K-conductance in dorsal raphe nucleus neurons (Vaughan and Christie 1996) and in the arcuate nucleus (Wagner et al. 1998), to increase K conductance in periaqueductal gray neurons (Morgan et al. 1997; Vaughan et al. 1997) and in locus coeruleus neurons (Connor et al. 1996a), to couple to G protein-activated K channels (Ikeda et al. 1997), to inhibit voltage-gated calcium channels in freshly dissociated CA3 hippocampal neurons (Knoflach et al. 1996), to inhibit T-type calcium channels in sensory neurons (Abdulla and Smith 1997), and to inhibit N-type Ca channels in SH-SY5Y cells (Connor et al. 1996b). Also, the OFQ/NOC receptor appears to couple to K channels in Xenopus oocytes (Matthes et al. 1996). Furthermore, OFQ/NOC has been shown to inhibit the release of glutamate and GABA from nerve terminals (Nicol et al. 1996; Faber et al. 1996), to block acetylcholine release from retina (Neal et al. 1997) and parasympathetic nerve terminals (Patel et al. 1997), to inhibit synaptic transmission and long-term potentiation in the hippocampus (Yu et al. 1997), to suppress dopamine release in the nucleus accumbens (Murphy et al. 1996) and to inhibit tachykinin and calcitonin gene-related peptide release from sensory nerves (Giuliani and Maggi 1996; Helyes et al. 1997). OFQ/NOC was demonstrated to activate mitogen-activated protein kinase in receptor transfected CHO cells (Fukuda et al. 1997; Lou et al. 1998). Together, these results show that OFQ/NOC is able to modulate the biochemical properties of cells, alter the electrophysiological properties of neurons and to affect their transmitter release. In organotypic assays, OFQ/NOC has been shown to inhibit electrically-induced contractions of the vas deferens, ileum and myenteric plexus preparations (Berzetei-Gurske et al. 1996; Calo et al. 1996; Zhang et al. 1997; Nicholson et al. 1998). Importantly, these activities were not inhibited by opiate antagonists as were none of the effects of OFQ/NOC described above, underscoring the pharmacological difference between the opioid and the OFQ/NOC systems.

4.3
Biosynthesis, Degradation and Release of OFQ/NOC

Like all bioactive peptides, OFQ/NOC is synthesized as part of a larger polypeptide (176 amino acids in human). The cDNA encoding the OFQ/NOC precursor has been cloned from mice (Saito et al. 1995), rat and human (Mollerau et al. 1996; Nothacker et al. 1996). The primary structure of the precursor protein exhibits all the features typical of a neuropeptide precursor, in particular an aminoterminal signal sequence necessary for its secretion (Fig. 1). The OFQ/NOC sequence is found in the C-terminal half of the precursor and is flanked by pairs of Lys-Arg residues indicating that its maturation requires trypsin-like cleavage as commonly found for the maturation of bioactive peptides. Inactivation of OFQ/NOC has been shown to involve cleavage of the aminoterminal Phe residue by aminopeptidase N, followed by further degradation through endopeptidases (Noble and Roques 1997; Montiel et al. 1997). The availability of an OFQ/NOC precursor cDNA allowed for a detailed inves-

OFQ/NOC gene

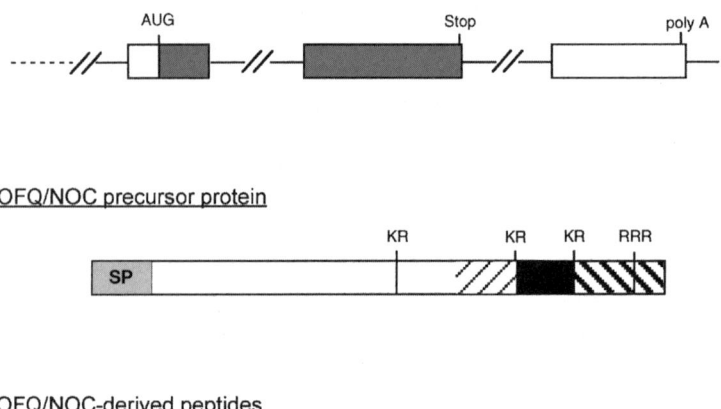

OFQ/NOC precursor protein

OFQ/NOC-derived peptides

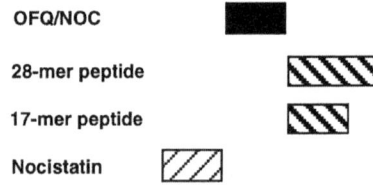

Fig. 1. OFQ/NOC gene, precursor and maturation products.
Organization of the OFQ/NOC gene (top), the preprohormone as deduced from the cDNA (middle) and peptides generated by enzymatic precursor processing (bottom). Top: Exons are indicated by boxes, introns appear as lines. The organization of the 5'-upstream region of the gene has not been defined but at least one additional exon encoding the 5'-untranslated region of the mRNA must exist. Middle: Structure of the human OFQ/NOC precursor polypeptide (not at the same scale as the gene). Location of pairs of basic amino acids are indicated, representing possible prohormone convertase cleavage sites. SP: signal peptide; K: Lys; R: Arg. Bottom: Putative peptide products after the precursor maturation (see text for discussion) (adapted from: Nothacker et al. 1996)

tigation of its site of synthesis (see below). These experiments in conjunction with immunohistochemical analyses clearly showed that OFQ/NOC is synthesized in neurons of the CNS. This is further supported by the discovery of its expression in neuroblastoma cells (Saito et al. 1996).

Unfortunately, no selective and high-affinity antagonist to the OFQ/NOC receptor is available thus far, despite a recent announcement (Guerrini et al. 1998; but see also: Butour et al. 1998; Xu et al. 1998). Therefore, experiments to show neuronal release of OFQ/NOC have not yet been possible. However in a recent study, a polyclonal anti-OFQ/NOC has been administered centrally to rats and was found to block development of tolerance to electroacupuncture as compared to vehicle injected animals (Tian et al. 1998). This report suggests, that the release of endogenous OFQ/NOC from neurons might occur physiologically. In conclusion, the effects of OFQ/NOC on neurons and its synthesis in the CNS strongly suggest that OFQ/NOC is a genuine neuropeptide.

4.4
Regional Expression of OFQ/NOC and its Receptor

Important suggestions about possible roles of a novel transmitter in the regulation of physiological and behavioral effects can be deduced from a characterization of its neuronal pathways, i.e. the site of synthesis and the projection areas of these neurons. Studies of the localization of OFQ/NOC immunoreactivity and of OFQ/NOC precursor and receptor by *in situ* hybridization studies have revealed the broad spectrum of the OFQ/NOC system (Bunzow et al. 1994; Mollerau et al. 1994; Lachowicz et al. 1995; Anton et al. 1996; Mollerau et al. 1996; Nothacker et al. 1996; Houtani et al. 1996; Pan et al. 1996; Riedl et al. 1996; Schulz et al. 1996; Sim et al. 1996; Florin et al. 1997b; Lai et al. 1997; Makman et al. 1997; Schuligoi et al. 1997; Neal et al. 1998). These data have been reviewed (Henderson and McKnight 1997; Meunier 1997) and will only be briefly discussed here.

In general, a good correlation between the distribution of the OFQ/NOC precursor mRNA and that of the OFQ/NOC peptide has been found. Both can be detected in the bed nucleus of the stria terminalis, amygdala, lateral septum, cortex, thalamus, striatum, substantia nigra and hypothalamus. Noteworthy, OFQ/NOC immunoreactivity differs significantly from immunohistochemical localization of the OFQ/NOC receptor in that it appears much more restricted. Such discrepancies between ligand and receptor distribution have been described in other transmitter systems and may underline the differences between OFQ/NOC sites of synthesis and sites of action. The receptor is distributed throughout the rat CNS but is predominantly confined to fibers (Anton et al. 1996). Prominent receptor expression was observed in the neocortex, cingulate and piriform cortex, hippocampus, anterior olfactory nucleus, cortical amygdala, claustrum and endopiriform nucleus. Moderate contents of OFQ/NOC receptor was found in the central and medial amygdala, dentate gyrus, subiculum, entorhinal cortex, dorsal and ventral pallidum, triangular and medial septum, medial preoptic area, mammilary bodies and parafascicular and posterior thalamic nuclei but also at a lesser level in the olfactory system, lateral septum, basal forebrain, thalamus and hypothalamus.

In the brainstem, a dense network of OFQ/NOC immunoreactive fibers were observed in many areas (Neal et al. 1998), in particular in the substantia nigra, the locus coeruleus, the tegmental nuclei, several vestibular nuclei, the superior colliculus and the pons. Intense OFQ/NOC-positive staining has been detected in the superficial layers of the dorsal horn in the spinal cord whereas few immunoreactivity was found in the cerebellum. OFQ/NOC precursor mRNA expression was observed in most of the brainstem and spinal cord areas where OFQ/NOC-immunoreactivity had been detected. Together these data show that OFQ/NOC is widely expressed and can thus act at numerous sites within the CNS. In the periphery, the tissue localization of the OFQ/NOC system has been analyzed less systematically. OFQ/NOC receptor mRNA has been detected in the vas deferens, the intestine, the liver and the spleen (Wang et al. 1994). In addition, the OFQ/NOC precursor mRNA has been found expressed in spleen and fetal kidney (Nothacker et al. 1996) and at a low level in the ovaries (Mollerau et al. 1996).

5
Physiological Effects and Behavioral Responses of OFQ/NOC

The widespread distribution of the OFQ/NOC system in brain and periphery suggests that it will regulate many behavioral and other physiological responses. Many effects of OFQ/NOC on numerous behavioral responses have already been examined, sometimes with controversial conclusions (Meunier 1997). Here, we will review these studies and discuss some of the discrepancies, but we are aware of the fact, that the investigation of functional aspects of the OFQ/NOC system is still at a very early stage compared to decades of research accumulated for other transmitter systems. Nevertheless, in the three years since the first publication of OFQ/NOC an impressive amount of knowledge has been produced, illustrating the prospects and power of reverse physiology.

The first effect described was a inhibitory effect on voluntary locomotion when OFQ/NOC was injected intracerebroventricularly (i.c.v.). At high doses (> 1–3 nmol/mouse) it inhibits horizontal and vertical activities (Reinscheid et al. 1995; Devine et al. 1996a; Devine et al. 1996b). This inhibition is reversible and depends on the route of injection; intrathecally (i.t.) OFQ/NOC was reported to have no effect on locomotion (Tian et al. 1997a). In contrast, low doses of OFQ/NOC (< 0.05 nmol/mouse) produce a stimulatory effect on locomotion (Florin et al. 1996), although this effect was originally viewed as not significant (Reinscheid et al. 1995). This dual action of OFQ/NOC on locomotor activity has to be taken into account when behavioral assays relying on locomotion are employed to analyze other behavioral aspects.

Stimulated by the high degree of homology between OFQ/N and its receptor with the endogenous opioid peptides and their corresponding receptors, a lot of attention has been focused on a possible role of OFQ/N in nociception. The effects of OFQ/NOC on nociception have been intensely debated and have been reviewed (Henderson and McKnight 1997; Meunier 1997). Originally, OFQ/NOC was described to produce hyperalgesic activity when injected i.c.v. (Meunier et al. 1995; Reinscheid et al. 1995). However, even the original reports were conflicting, one group finding the activity in the hot plate assay at low doses (which may induce hyperactivity) (Meunier et al. 1995), the other described the hyperalgesic effect in the tail flick assay but not in the hot plate assay using a variable range of doses (Reinscheid et al. 1995). Interestingly, similar results were found in the snail (Kavaliers and Perrot-Sinal 1996). The significance of these results was clarified when further experiments demonstrated a reversal of stress-induced analgesia after central administration of OFQ/NOC (Mogil et al. 1996a). In light of these findings, the apparent pronociceptive effect of OFQ/NOC could be explained by the stress caused by the intracranial injections. It is well known, that stress-induced analgesia is mediated – at least in part – by activation of endogenous opioid systems. Consequently, OFQ/NOC was designated as an "anti-opioid" compound (Grisel et al. 1996, Xu et al. 1996; Dawson-Basoa and Gintzler 1997; Heinricher et al. 1997). This concept was further extended after demonstrating that OFQ/NOC is able to block supraspinal antinociception produced by the μ, δ, and κ opioid receptors (Mogil et al. 1996b) and that OFQ/NOC can act as a functional antagonist of morphine analgesia in various assays (Zhu et al. 1997; Tian et al. 1997a, b), in particular acupuncture. It is important to keep in mind, that opioid peptides or opiate compounds do not interact with the OFQ/NOC receptor and OFQ/NOC does not bind to opioid receptors.

On the other hand, in some reports OFQ/NOC was shown to produce analgesia after i.c.v and i.t. administration, in a manner similar to that of morphine and without motor impairment (Xu et al, 1996; Rossi et al. 1996; Tian et al. 1997b). However, these studies used high doses of OFQ/NOC and reported a mostly naloxone-reversible analgesia, which can therefore no longer be explained at the receptor level, since it contradicts one of the major pharmacological features of the OFQ/NOC system (see below). Another important functional aspect related to pain perception was discovered when OFQ/NOC was shown to induce allodynia in response to innocuous tactile stimuli (Hara et al. 1997; Minami et al. 1997). Although morphine can also induce a similar allodynic response, this requires high doses of morphine (5 orders of magnitude over that of OFQ/NOC) (Zhu et al. 1997). More recently, it was reported that mice genetically-engineered to lack the OFQ/NOC receptor show no significant differences in nociceptive threshold or locomotor behavior when compared to control mice (Nishi et al. 1997). However they have lost OFQ/NOC-induced behavioral responses such as pronociceptive effects and locomotor inhibition. Importantly, all these behavioral effects were unaffected by opiate antagonists, further underscoring the separation between the OFQ/NOC and opioid systems. In addition, development of tolerance to repeated morphine administration was found to be attenuated in mice devoid of the OFQ/NOC receptor as well as an elevated threshold for auditory brainstem responses (Nishi et al. 1997; Ueda et al. 1997).

Numerous other behavioral responses have been found to be modulated by central administration of OFQ/NOC. Because OFQ/NOC is sequentially related to the opioid peptides which analogs are addictive, it was tested for its ability to produce reinforcement behavior (Devine et al. 1996b). In contrast to morphine, OFQ/NOC, injected i.c.v., failed to produce conditioned place preference or aversion, even at high doses, indicating that it does not share morphine's motivational properties. These results are somewhat surprising since it has also been reported that OFQ/NOC is able to decrease dopamine release in the nucleus accumbens of rats (Murphy et al. 1996). Further studies investigating the effect of OFQ/NOC on the brain reward circuit will clearly be necessary to clarify this question. With respect to cognitive functions, OFQ/NOC has also been tested for its role in spatial learning. Using the Morris water maze assay, OFQ/NOC injected into the CA3 region of the hippocampus at high dose (10 nmol) produced an impairment of spatial learning (Sandin et al. 1997). Although this dose of OFQ/NOC induced a decrease in locomotion, it was reported to not affect swimming performance. This result has been supported by a similar experiment carried out on mice lacking the OFQ/NOC receptor which were shown to exhibit an enhancement of spatial attention (Mamiya et al. 1998). At the cellular level it was found that OFQ/NOC is able to impair LTP formation in hippocampal slices of rat brain, suggesting again an involvement in processes of neuronal plasticity (Yu et al. 1997). In addition, an increase of food consumption has been reported in satiated rats after injection of OFQ/NOC into the shell of the nucleus accumbens, the ventromedial hypothalamic nucleus or the right lateral ventricle (Pomonis et al. 1996; Stratford et al. 1997).

Several peripheral functions have been shown to be influenced by OFQ/NOC. OFQ/NOC produced marked changes in the renal excretion of water and sodium (Kapusta et al. 1997). Central administration of OFQ/NOC resulted in a concurrent diuresis and antinatriuresis, implicating an important role in the central control of water balance and possibly of blood pressure. Application onto arterial rings demonstrated vasorelaxant properties of OFQ/NOC (Gumusel et al. 1997). In the cardiovascular system, OFQ/NOC was shown to induce hypotension and increase cardiac output (Champion and Kadowitz 1997a,b). Finally, OFQ/NOC has been reported to modulate sexual behaviors by facilitating lordosis in rats (Sinchak et al. 1997) and penile erection in cats (Champion et al. 1997).

The observations, that OFQ/NOC is able to reverse stress-induced analgesia and its molecular components are expressed in CNS pathways known to regulate stress responses led to the investigation of a potential role of OFQ/NOC in other stress-related behaviors, e.g. anxiety. Using four different assays which measure anxiety-like behavior it was found that i.c.v. injections of OFQ/NOC induce a profound anxiolytic response in mice and rats (Jenck et al. 1997). This effect occurred at doses which did not affect locomotion and exhibited an inverted U shape-dose dependency, similar to anxiolytic effects produced by diazepam. In this respect, OFQ/NOC can be viewed as a natural modulator of anxiety states generated by acute stress. While the initial studies on physiological functions of OFQ/NOC mainly focused on the nociceptive system, obviously driven by the sequence homology to the opioid system, the discovery of anxiolytic properties may guide OFQ/NOC research to novel directions (Walker and Koob 1997). Assuming that in nature the sensation of fear and the likelihood of painful injury often coincide (e.g. "predator stress"), the opioids and OFQ/NOC may be viewed as synergistic neurotransmitter systems that may act to reduce stress vulnerability.

In summary, OFQ/NOC has been shown to affect a wide variety of behavioral responses after central administration. First, OFQ/NOC modulates locomotion differently at low vs. high doses in a reversible manner (< 0.05 nmol/mouse: activation; > 1–3 nmol/mouse: inhibition). The physiological basis of this effect remains unclear but it certainly has an impact on many other behavioral responses, especially tests employing movement readouts. Second, OFQ/NOC can block the analgesia produced by either stress, the stimulation of the opioid system, electroacupuncture or gestation at practically any dose. Third, OFQ/NOC induces allodynia to innocuous stimuli but does not induce analgesia unless injected i.t. at high doses. Fourth, in contrast to morphine, OFQ/NOC does not possess rewarding or addictive properties. Fifth, OFQ/NOC has also profound peripheral effects, in particular as a smooth muscle relaxant and a diuretic and antinatriuretic agent. Sixth, OFQ/NOC has a clear anxiolytic activity which is reminiscent of that induced by benzodiazepines. This last discovery may indeed help to explain many of the other OFQ/NOC behavioral activities and simultaneously emphasizes the fundamental difference to the opioid system. The broad spectrum of the OFQ/NOC physiological effects, which is bound to grow even more when OFQ/NOC effects will be analyzed under different experimental paradigms, is in fact not surprising in view of the broad distribution of the OFQ/NOC system. The central but still remaining question is, which of these effects is the most important physiological function regulated by the endogenous release of OFQ/NOC in the human body. It is immanent to a concept of reverse physiology that answering these questions will occur at the late stages of such a project (in contrast to the classical physiological approach which starts from a function) and will rely on the development and administration of synthetic agonists or antagonists.

6
Similarities and Differences Between the Opioid and the OFQ/NOC Systems

At least some of the behavioral effects found for OFQ/NOC indicate a close relation to the opioids. It is therefore necessary to investigate whether OFQ/NOC – under certain conditions – is maybe an opioid peptide in disguise or whether OFQ/NOC may compete with the opioid peptides in some responses. In the following, we will outline analogies, similarities and differences between the two systems.

A striking homology between the opioid and OFQ/NOC systems is found at the sequence level, because the OFQ/NOC receptor shares more than 60 % identity with the three opioid receptors (Bunzow et al. 1994; Mollerau et al. 1994). Second, OFQ/NOC itself displays a striking similarity to the opioid peptides in its aminoterminal tetrapeptide FGGF which is reminiscent of the canonical sequence YGGF of all opioid peptides (Fig. 2). In addition, two clusters of basic amino acids which are present in the C-terminal half of OFQ/NOC are reminiscent of the arrangement of multiple positively charged residues in dynorphin A or β-endorphin. In general, the OFQ/NOC precursor exhibits several analogous features as compared to the opioid precursors: the active peptides are located in the C-terminal part and seven Cys residues are found conserved at the N-terminus (Nothacker et al. 1996). Finally, the OFQ/NOC precursor gene has retained an organization similar to that of the opioid precursor genes (Mollerau et al. 1996). The coding sequence is divided over two exons, the smaller one containing the site for translational initiation (AUG), the other encoding the rest of the sequence. The OFQ/NOC precursor gene differs from this architecture in that it contains an additional exon for the 3'-untranslated region of the mRNA. These data suggest that the molecular components of the OFQ/NOC system are evolutionarily related to those of the opioid system.

The similarity between the opioid and the OFQ/NOC systems continues at the cellular level. Both systems convey an inhibitory signal on adenylyl cyclase resulting in a reduction of the common second messenger cAMP (Meunier et

Phe-**Gly-Gly-Phe**-Thr-Gly-Ala-Arg-**Lys-Ser**-Ala-Arg-**Lys-Leu**-Ala-**Asn-Gln**	OFQ/NOC
Tyr-**Gly-Gly-Phe**-Leu-Arg-Arg-Ile-Arg-Pro-Lys-Leu-**Lys**-Trp-Asp-**Asn-Gln**	Dynorphin A
Tyr-**Gly-Gly-Phe**-Met-Thr-Ser-Glu-**Lys-Ser**-Gln-Thr-Pro-**Leu**-Val-Thr-Leu	γ-Endorphin
Tyr-**Gly-Gly-Phe**-Leu-Arg-Lys-Tyr-Pro-Lys	α-Neoendorphin
Tyr-**Gly-Gly-Phe**-Leu	Leu-Enkephalin

Fig. 2. Alignment of OFQ/NOC and opioid peptides. Identical amino acid residues are printed in bold type

al. 1995; Reinscheid et al. 1995, 1996). In addition, the activities of both systems on ion channels regulating neuronal excitability are similar if not homologues. Both systems modulate nociceptive responses although the actions of OFQ/NOC are far less understood. While the opioids are natural analgesics, OFQ/NOC has been reported to be hyperalgesic, analgesic, allodynic and anti-opioid, depending on the assays, the site and the dose injected, and probably the species used to examine these effects (see above). In contrast to the evident structural relationships, the picture is less clear in functional terms. Future studies will hopefully help to increase the understanding of OFQ/NOC functions, especially in the regulation of nociceptive processing.

The analyses of tissue distribution revealed another similarity between the OFQ/NOC and opioid systems. A coexistence or colocalization of POMC peptides and OFQ/NOC and/or OFQ/NOC receptors was detected, for example, in the bed nucleus of the stria terminalis, the arcuate nucleus of the hypothalamus and the central and medial nuclei of the amygdala (Anton et al. 1996; Riedl et al. 1996; Schulz et al. 1996; Neal et al. 1998). Enkephalin-containing fibers in the nucleus accumbens, hippocampus and ventromedial nucleus of the amygdala are regions which contain high levels of OFQ/NOC receptor (Neal et al. 1998). Expression of both dynorphin and OFQ/NOC is found in the dentate gyrus, the central nucleus of the amygdala, the arcuate nucleus and the substantia nigra (Neal et al. 1998). In the spinal cord, OFQ/NOC-containing nerve fibers parallel the opioid-expressing pathways (Riedl et al. 1996; Schulz et al. 1996). Modulation of the opioid system by OFQ/NOC or vice versa is therefore likely to happen and has indeed been demonstrated in the myenteric plexus where OFQ/NOC modulates the release of enkephalin (Gintzler et al. 1997). Therefore, the possibility that OFQ/NOC could replace the opioid peptides in some functional responses clearly exists at the anatomical level.

A fundamental difference between the OFQ/NOC and opioid systems was described early on at the pharmacological level. Analyses of ligand-receptor interactions revealed that both systems are separated pharmacologically, i.e. no cross-activation of receptors by others than their cognate ligands can occur (Table 1). OFQ/NOC does not activate opioid receptors and vice versa. This has been studied in-depth using receptors expressed in cell lines (Reinscheid et al. 1995, 1996, 1998) and is reflecting the *in vivo* situation. Moreover, it has been shown recently that both at the peptide and the receptor level the two systems have evolved structural features which exclude their direct interaction. First, at the level of the receptors, it was found by *in vitro* mutagenesis that as few as four amino acid residues in the OFQ/NOC receptor insure that it will not bind dynorphin A (Meng et al. 1996), the opioid peptide most closely related to OFQ/NOC. At the level of the neuropeptides, it was found that OFQ/NOC and dynorphin A contain short stretches of amino acids which prevent their recognition by their non-cognate receptor (Reinscheid et al. 1998). Together, these two studies converge on the unified conclusion, that these two systems have evolved to be distinct and are not interchangeable at the level of the neuropep-

Table 1. Pharmacological selectivity in the OFQ/NOC and opioid system.

Biological activity of OFQ/NOC, dynorphin A (its closest counterpart among the opioid peptides), deltorphin I and dermorphin at the OFQ/NOC receptor and the κ-, δ- and μ-opioid receptors (KOR, DOR, MOR). All receptors had been stably expressed in CHO cells and the inhibition of forskolin-stimulated adenylyl cyclase was determined. Receptor activation is more closely reflecting the *in vivo* situation of ligand-receptor interaction than receptor binding alone. For example, dynorphin A is unable to activate the OFQ/NOC receptor (Inactive: >10,000 nM), although it can bind to the OFQ/NOC receptor at high (unphysiological) concentrations (700 nM), but all opioid peptides are promiscuously active at all three opioid receptors, albeit with different potencies. (adapted from: Reinscheid et al. 1998)

	Inhibition of forskolin-stimulated cAMP accumulation at [EC_{50} (nM)]			
	OFQ/NOC-R	KOR	DOR	MOR
OFQ/NOC	1.68	Inactive	Inactive	Inactive
Dynorphin A	Inactive	0.13	nd	nd
Deltorphin I	Inactive	nd	0.18	nd
Dermorphin	Inactive	nd	nd	0.41

EC_{50}: median effective concentration of peptide; inactive: no effect up to 10,000 nM; nd: not determined

tide-receptor interactions. Based on these pharmacological data, the OFQ/NOC system therefore has to be classified as a separate system and not as a fourth opioid category, a conclusion that is underscored by the observation that the prototypical opioid antagonist naloxone does not bind to the OFQ/NOC receptor.

7
Novel Peptides Derived from the OFQ/NOC Precursor

The primary structure of the OFQ/NOC precursor protein contains additional putative sites for precursor processing. In the human precursor, one other pair of basic amino acids is located 34 residues upstream of the OFQ/NOC sequence and a triplet of Arg residues is found 17 residues C-terminally to the OFQ/NOC sequence (Mollerau et al. 1996; Nothacker et al. 1996) (Fig. 1). The bovine precursor harbours an additional pair of basic amino acids 19 residues N-terminal to the OFQ/NOC sequence (Okuda-Ashitaka et al. 1998). The entire C-terminus of the precursor protein including the OFQ/NOC is 100% conserved among human, rat, mouse and bovine genes. This C-terminus could generate either a 28 residue long peptide or, after cleavage at the Arg triplet, a 17 residue long peptide, whose terminal amino acids are the same as OFQ/NOC. However, when these two peptides were synthesized and tested for their abilities to bind to the OFQ/NOC receptor, no binding or activation of intracellular second messengers could be found (Nothacker et al. 1996). If these peptides exist, receptors distinct from the OFQ/NOC receptor must be postulated. In this context it is worth mentioning, that the 17 residue long peptide was found almost insoluble. However very recently, it has been reported that the 17 amino acid peptide (termed NocII and OFQ II, respectively) exhibits some effect on locomotion and pain perception (Florin et al. 1997b; Rossi et al. 1998). Peptides generated from the part of the bovine precursor N-terminally preceding the OFQ/NOC sequence have been found to possess an anti-OFQ/NOC activity since they were able to block OFQ/NOC-induced allodynia and hyperalgesia (Okuda-Ashitaka et al. 1998). This peptide was isolated from bovine brain and termed nocistatin. It is present in the brain and acts via a receptor different from the OFQ/NOC receptor. Its mode of processing from the precursor is not completely understood because the presumed maturation in human and rat will require uncommon proteolytic cleavage.

8
Conclusions: The Strength and the Pitfalls of the Orphan Receptor Strategy

OFQ/NOC was the first example of a novel neurotransmitter identified by use of an orphan receptor. Its discovery paved the way for the subsequent analyses of many exciting functions and clearly demonstrates the power of the reverse physiological approach as a new strategy to identify bioactive molecules. OFQ/NOC was purified solely on the basis of the trust that its receptor was a G protein-coupled receptor and thus would have a ligand, produced and stored in the brain. Although in the case of the opioid and the cannabinoid receptors, the identification of their endogenous ligands was also following far behind characterization of the receptors, the search for their natural ligands could benefit from a rich pharmacology (Hughes et al. 1975; Devane et al. 1992). The large number of orphan GPCRs all have in common the absence of pharmacological tools for their study, so that the reverse physiological approach appears as the only possible way to identify their corresponding natural ligands. The strategy employed for the isolation of OFQ/NOC has already been widely adopted and successfully used to identify more novel neurotransmitters.

We can surely assume that reverse physiology will dramatically increase the number of novel neurotransmitters and neuropeptides in the next years, yielding fascinating new insights into biological processes. However, some critical points have to be kept in mind and should be carefully considered when describing a new ligand. The first question to be raised will be, whether a newly discovered molecule is the true ligand for the orphan receptor or not. This can – at least partially – be answered by analyses of the pharmacological constants. It can be expected that a ligand will bind its receptor with high affinity, in a saturable and reversible manner and will show the right pharmac-ological profile. Binding constants can be tested easily, provided the develop-ment of a radioligand. By definition of orphan receptors, the pharmacological profile is untestable. Therefore, most claims that a new ligand is specific to an orphan receptor will be based on binding and second messenger activation. From our knowledge of known transmitters, some guidelines can be deduced. We can expect higher affinity constants for peptides (in the low nanomolar range) than, for example, for biogenic amines (high nanomolar to micromo-lar). However, the examples of the proteinase-activated GPCRs already illus-trate the technical difficulties one can encounter in determining binding affin-ities (Dery et al. 1998). Another important issue to be considered is, whether the new ligand is a true ligand (i.e. synthesized, stored and released like a transmitter) or a product of degradation. Such false-positive ligands can be molecules derived from catabolic pathways or artificially produced by the harsh conditions during preparation of tissue extracts. With respect to novel peptides, this concern can be tested by cloning of the precursor proteins which should reveal typical structures suitable for maturation that surround the pep-tide. However, there are also many exceptions to the rules of precursor pro-cessing and the identification of specific cleaving enzymes may only give final proof. So in short, there is no general catalogue of rules to ensure that a newly isolated ligand is the only and true ligand.

Finally, the nomenclature of these new ligands will and has already been controversially debated (Henderson and McKnigth 1997). The first one already has two names, orphanin FQ and nociceptin. Ironically, the second natural ligand of an orphan receptor was also discovered and named independently by two research teams, one calling it orexin and the other hypocretin (Sakurai et al. 1998, de Lecea et al 1998). This coincidence probably exemplifies the high level of competition in the field of orphan receptors. Names were chosen on the basis of a physiological effect (nociceptin, orexin) or on a more physical basis (hypocretin, OFQ). Naming after function has the disadvantage that all ligands of orphan receptors will have many physiological roles, but the first chosen might not necessarily be the most important one(s). The second school is less poetic but closer to the few solid facts of the original data.

Facing the large number of orphan receptors that have been and are contin-uously cloned, it is justified to expect as many novel transmitters and peptides to be discovered via the orphan receptor approach in the near future. This will stimulate a surge of our knowledge on the diversity in intercellular communi-cation, which will particularly impact CNS research. Ultimately, the discovery of more natural ligands will greatly extend our knowledge about complex physiological functions and states of disease related to them. The newly dis-covered ligand/receptor systems will immediately offer novel targets for ther-apeutic intervention and provide the necessary tools for pharmaceutical development.

Acknowledgments: R.K.R. is supported by grants from the Deutsche Forschungsgemeinschaft and Fonds der Chemischen Industrie. O.C. is supported by a grant of Hoffmann-La Roche, Inc.

References

Abdulla FA, Smith PA (1997) Nociceptin inhibits T-type Ca^{2+} channel current in rat sensory neurons by a G protein-independent mechanism. J Neurosci 17:8721–8728

Anton B, Fein J, To T, Li X, Silberstein L, Evans CJ (1996) Immunohistochemical localization of ORL-1 in the central nervous system of the rat. J Comp Neurol 368:229–251

Ardati A, Henningsen RA, Higelin J, Reinscheid RK, Civelli O, Monsma FJ Jr (1997) Interaction of [^{3}H]orphanin FQ and ^{125}I-Tyr^{14}-orphanin FQ with the orphanin FQ receptor: kinetics and modulation by cations and guanine nucleotides. Mol Pharmacol 51:816–824

Berzetei-Gurske IP, Schwartz RW, Toll L (1996) Determination of activity for nociceptin in the mouse vas deferens. Eur J Pharmacol 302:R1–2

Bunzow JR, Van Tol HH, Grandy DK, Albert P, Salon J, Christie M, Machida CA, Neve KA, Civelli O (1988) Cloning and expression of a rat D2 dopamine receptor cDNA. Nature 336:783–787

Bunzow JR, Zhou QY, Civelli O (1992) Cloning of dopamine receptors: Homology approach. Meth Neurosci 9:441–453

Bunzow JR, Saez C, Mortrud M, Bouvier C, Williams JT, Low M, Grandy DK (1994) Molecular cloning and tissue distribution of a putative member of the rat opioid receptor gene family that is not a mu, delta or kappa opioid receptor type. FEBS Lett 347:284–288

Butour JL, Moisand C, Mazarguil H, Mollereau C, Meunier JC (1997) Recognition and activation of the opioid receptor-like ORL 1 receptor by nociceptin, nociceptin analogs and opioids. Eur J Pharmacol 321:97–103

Butour JL, Moisand C, Mollereau C, Meunier JC (1998) [Phe^{1}psi(CH$_2$-NH)Gly^2]nociceptin-(1-13)-NH$_2$ is an agonist of the nociceptin (ORL1) receptor. Eur J Pharmacol 349:R5–R6

Calo G, Rizzi A, Bogoni G, Neugebauer V, Salvadori S, Guerrini R, Bianchi C, Regoli D (1996) The mouse vas deferens: a pharmacological preparation sensitive to nociceptin. Eur J Pharmacol 311:R3–5

Champion HC, Kadowitz PJ (1997a) [Tyr^1]-nociceptin, a novel nociceptin analog, decreases systemic arterial pressure by a naloxone-insensitive mechanism in the rat. Biochem Biophys Res Commun 234:309–312

Champion HC, Kadowitz PJ (1997b) Nociceptin, an endogenous ligand for the ORL1 receptor, has novel hypotensive activity in the rat. Life Sci 60:PL 241–245

Champion HC, Wang R, Hellstrom WJ, Kadowitz PJ (1997) Nociceptin, a novel endogenous ligand for the ORL1 receptor, has potent erectile activity in the cat. Am J Physiol 273:E214–219

Chen Y, Fan Y, Liu J, Mestek A, Tian M, Kozak CA, Yu L (1994) Molecular cloning, tissue distribution and chromosomal localization of a novel member of the opioid receptor gene family. FEBS Lett 347:279–283

Connor M, Vaughan CW, Chieng B, Christie MJ (1996a) Nociceptin receptor coupling to a potassium conductance in rat locus coeruleus neurones in vitro. Br J Pharmacol 119:1614–1618

Connor M, Yeo A, Henderson G (1996b) The effect of nociceptin on Ca^{2+} channel current and intracellular Ca^{2+} in the SH-SY5Y human neuroblastoma cell line. Br J Pharmacol 118:205–207

Dawson-Basoa M, Gintzler AR (1997) Nociceptin (Orphanin FQ) abolishes gestational and ovarian sex steroid-induced antinociception and induces hyperalgesia. Brain Res 750:48–52

de Lecea L, Kilduff TS, Peyron C, Gao X, Foye PE, Danielson PE, Fukuhara C, Battenberg EL, Gautvik VT, Bartlett FS II, Frankel WN, van den Pol AN, Bloom FE, Gautvik KM, Sutcliffe JG (1998) The hypocretins: hypothalamus-specific peptides with neuroexcitatory activity. Proc Natl Acad Sci USA 95:322–327

Dery O, Corvera CU, Steinhoff M, Bunnett NW (1998) Proteinase-activated receptors: novel mechanisms of signaling by serine proteases. Am J Physiol 274:C1429–C1452

Devane WA, Hanus L, Breuer A, Pertwee RG, Stevenson LA, Griffin G, Gibson D, Mandelbaum A, Etinger A, Mechoulam R (1992) Isolation and structure of a brain constitutent that binds to the cannabinoid receptor. Science 258:1946–1949

Devine DP, Taylor L, Reinscheid RK, Monsma FJ Jr, Civelli O, Akil H (1996a) Rats rapidly develop tolerance to the locomotor-inhibiting effects of the novel neuropeptide orphanin FQ. Neurochem Res 21:1387–1396

Devine DP, Reinscheid RK, Monsma FJ Jr, Civelli O, Akil H (1996b) The novel neuropeptide orphanin FQ fails to produce conditioned place preference or aversion. Brain Res 727: 225–229

Dooley CT, Houghten RA (1996) Orphanin FQ: receptor binding and analog structure activity relationships in rat brain. Life Sci 59:PL23–29

Dooley CT, Spaeth CG, Berzetei-Gurske IP, Craymer K, Adapa ID, Brandt SR, Houghten RA, Toll L (1997) Binding and in vitro activities of peptides with high affinity for the nociceptin/orphanin FQ receptor, ORL1. J Pharmacol Exp Ther 283:735–741

Faber ES, Chambers JP, Evans RH, Henderson G (1996) Depression of glutamatergic transmission by nociceptin in the neonatal rat hemisected spinal cord preparation in vitro. Br J Pharmacol 119:189–190

Florin S, Suaudeau C, Meunier JC, Costentin J (1996) Nociceptin stimulates locomotion and exploratory behaviour in mice. Eur J Pharmacol 317:9–13

Florin S, Duauadeau C, Meunier JC, Costentin J (1997) Orphan neuropeptide NocII, a putative pronociceptin maturation product, stimulates locomotion in mice. Neuroreport 8:705–707

Florin S, Leroux-Nicollet I, Meunier JC, Costentin J (1997) Autoradiographic localization of [³H]nociceptin binding sites from telencephalic to mesencephalic regions of the mouse brain. Neurosci Lett 230:33–36

Fukuda K, Kato S, Mori K, Nishi M, Takeshima H, Iwabe N, Miyata T, Houtani T, Sugimoto T (1994) cDNA cloning and regional distribution of a novel member of the opioid receptor family. FEBS Lett 343:42–46

Fukuda K, Shoda T, Morikawa H, Kato S, Mori K (1997) Activation of mitogen-activated protein kinase by the nociceptin receptor expressed in Chinese hamster ovary cells. FEBS Lett 412: 290–294

Gintzler AR, Adapa ID, Toll L, Medina VM, Wang L (1997) Modulation of enkephalin release by nociceptin (orphanin FQ). Eur J Pharmacol 325:29–34

Giuliani S, Maggi CA (1996) Inhibition of tachykinin release from peripheral endings of sensory nerves by nociceptin, a novel opioid peptide. Br J Pharmacol 118:1567–1569

Grisel JE, Mogil JS, Belknap JK, Grandy DK (1996) Orphanin FQ acts as a supraspinal, but not a spinal, anti-opioid peptide. Neuroreport 7:2125–2129

Guerrini R, Calo G, Rizzi A, Bianchi C, Lazarus LH, Salvadori S, Temussi PA, Regoli D (1997) Address and message sequences for the nociceptin receptor: a structure-activity study of nociceptin-(1-13)-peptide amide. J Med Chem 40:1789–1793

Guerrini R, Calo G, Rizzi A, Bigoni R, Bianchi C, Salvadori S, Regoli D (1998) A new selective antagonist of the nociceptin receptor. Br J Pharmacol 123:163–165

Gumusel B, Hao Q, Hyman A, Chang JK, Kapusta DR, Lippton H (1997) Nociceptin: an endogenous agonist for central opioid like1 (ORL1) receptors possesses systemic vasorelaxant properties. Life Sci 60:PL141–145

Hao JX, Wiesenfeld-Hallin Z, Xu XJ (1997) Lack of cross-tolerance between the antinociceptive effect of intrathecal orphanin FQ and morphine in the rat. Neurosci Lett 223:49–52

Hara N, Minami T, Okuda-Ashitaka E, Sugimoto T, Sakai M, Onaka M, Mori H, Imanishi T, Shingu K, Ito S (1997) Characterization of nociceptin hyperalgesia and allodynia in conscious mice. Br J Pharmacol 121:401–408

Heinricher MM, McGaraughty S, Grandy DK (1997) Circuitry underlying antiopioid actions of orphanin FQ in the rostral ventromedial medulla. J Neurophysiol 78:3351–3358

Helyes Z, Nemeth J, Pinter E, Szolcsanyi J (1997) Inhibition by nociceptin of neurogenic inflammation and the release of SP and CGRP from sensory nerve terminals. Br J Pharmacol 121:
613–615

Henderson G, McKnight AT (1997) The orphan opioid receptor and its endogenous ligand–
nociceptin/orphanin FQ. Trends Pharmacol Sci 18:293–300

Henikoff S, Greene EA, Pietrokovski S, Bork P, Attwood TK, Hood L (1997) Gene families: the
taxonomy of protein paralogs and chimeras. Science 278:609–614

Horn F, Weare J, Beukers MW, Horsch S, Bairoch A, Chen W, Edvardsen O, Campagne F, Vriend
G (1998) GPCRDB: an information system for G protein-coupled receptors. Nucleic Acids
Research 26:275–279

Houtani T, Nishi M, Takeshima H, Nukada T, Sugimoto T (1996) Structure and regional distribution of nociceptin/orphanin FQ precursor. Biochem Biophys Res Commun 219:714–719

Hughes J, Smith TH, Kosterlitz JW, Fothergill LA, Morgan BA, Morris HR (1975) Identification
of two related pentapeptides from the rat brain with potent opiate agonist activity. Nature 258
:577–579

Ikeda K, Kobayashi K, Kobayashi T, Ichikawa T, Kumanishi T, Kishida H, Yano R, Manabe T
(1997) Functional coupling of the nociceptin/orphanin FQ receptor with the G-protein-acti-
vated K^+ (GIRK) channel. Mol Brain Res 45:117–126

Jenck F, Moreau JL, Martin JR, Kilpatrick GJ, Reinscheid RK, Monsma FJ Jr, Nothacker HP,
Civelli O (1997) Orphanin FQ acts as an anxiolytic to attenuate behavioral responses to stress.
Proc Natl Acad Sci USA 94:14854–14858

Kapusta DR, Sezen SF, Chang JK, Lippton H, Kenigs VA (1997) Diuretic and antinatriuretic
responses produced by the endogenous opioid-like peptide, nociceptin (orphanin FQ). Life
Sci 60:PL15–21

Kavaliers M, Perrot-Sinal TS (1996) Pronociceptive effects of the neuropeptide, nociceptin, in
the land snail, Cepaea nemoralis. Peptides 17:763–768

Knoflach F, Reinscheid RK, Civelli O, Kemp JA (1996) Modulation of voltage-gated calcium
channels by orphanin FQ in freshly dissociated hippocampal neurons. J Neurosci 16:
6657–6664

Lachowicz JE, Shen Y, Monsma FJ Jr, Sibley DR (1995) Molecular cloning of a novel G protein-
coupled receptor related to the opiate receptor family. J Neurochem 64:34–40

Lai C, Lemke G (1991) An extended family of protein-tyrosine kinase genes differentially
expressed in the vertebrate nervous system. Neuron 6:691–704

Lai CC, Wu SY, Dun SL, Dun NJ (1997) Nociceptin-like immunoreactivity in the rat dorsal horn
and inhibition of substantia gelatinosa neurons. Neuroscience 81:887–891

Laudet V (1997) Evolution of the nuclear receptor superfamily: early diversification from an
ancestral orphan receptor. J Mol Endocrinol 19:207–226

Libert F, Parmentier M, Lefort A, Dinsart C, Van Sande J, Maenhaut C, Simon MJ, Dumont JE,
Vassart G (1989) Selective amplification and cloning of four new members of the G protein-
coupled receptor family. Science 244:569–572

Liu QR, Mandiyan S, Lopez-Corcuera B, Nelson H, Nelson N (1993) A rat brain cDNA encoding
the neurotransmitter transporter with an unusual structure. FEBS Lett 315:114–118

Lou LG, Zhang Z, Ma L, Pei G (1998) Nociceptin/orphanin FQ activates mitogen-activated protein kinase in Chinese hamster ovary cells expressing opioid receptor-like receptor. J Neurochem 70:1316–1322

Makman MH, Lyman WD, Dvorkin B (1997) Presence and characterization of nociceptin
(orphanin FQ) receptor binding in adult rat and human fetal hypothalamus. Brain Res 762:
247–250

Mamiya T, Noda M, Takeshima H, Nabeshima T (1998) Enhancement of spatial attention in noc-
iceptin/orphanin FQ receptor-knockout mice. Brain Res 783:236–240

Marchese A, George SR, O'Dowd BF (1998) Cloning of G protein-coupled receptor genes. In:
Lynch KR (ed) Identification and expression of G protein-coupled receptors. Wiley-Liss, New
York, pp 1–26

Mathis JP, Ryan-Moro J, Chang A, Hom JS, Scheinberg DA, Pasternak GW (1997) Biochemical evidence for orphanin FQ/nociceptin receptor heterogeneity in mouse brain. Biochem Biophys Res Commun 230:462–465

Matthes H, Seward EP, Kieffer B, North RA (1996) Functional selectivity of orphanin FQ for its receptor coexpressed with potassium channel subunits in Xenopus laevis oocytes. Mol Pharmacol 50:447–450

Meng F, Taylor LP, Hoversten MT, Ueda Y, Ardati A, Reinscheid RK, Monsma FJ, Watson SJ, Civelli O, Akil H (1996) Moving from the orphanin FQ receptor to an opioid receptor using four point mutations. J Biol Chem 271:32016–32020

Meunier JC, Mollereau C, Toll L, Suaudeau C, Moisand C, Alvinerie P, Butour JL, Guillemot JC, Ferrara P, Monsarrat B, Mazarguil H, Vassart G, Parmentier M, Costentin J (1995) Isolation and structure of the endogenous agonist of opioid receptor-like ORL1 receptor [see comments]. Nature 377:532–535

Meunier JC (1997) Nociceptin/orphanin FQ and the opioid receptor-like ORL1 receptor. Eur J Pharmacol 340:1–15.

Minami T, Okuda-Ashitaka E, Nishizawa M, Mori H, Ito S (1997) Inhibition of nociceptin-induced allodynia in conscious mice by prostaglandin D2. Br J Pharmacol 122:605–610

Mogil JS, Grisel JE, Reinscheid RK, Civelli O, Belknap JK, Grandy DK (1996a) Orphanin FQ is a functional anti-opioid peptide. Neuroscience 75:333–337

Mogil JS, Grisel JE, Zhangs G, Belknap JK, Grandy DK (1996b) Functional antagonism of mu-, delta- and kappa-opioid antinociception by orphanin FQ. Neurosci Lett 214:131–134

Mollereau C, Parmentier M, Mailleux P, Butour JL, Moisand C, Chalon P, Caput D, Vassart G, Meunier JC (1994) ORL1, a novel member of the opioid receptor family. Cloning, functional expression and localization. FEBS Lett 341:33–38

Mollereau C, Simons MJ, Soularue P, Liners F, Vassart G, Meunier JC, Parmentier M (1996) Structure, tissue distribution, and chromosomal localization of the prepronociceptin gene. Proc Natl Acad Sci USA 93:8666–8670

Montiel JL, Cornille F, Roques BP, Noble F (1997) Nociceptin/orphanin FQ metabolism: role of aminopeptidase and endopeptidase 24.15. J Neurochem 68:354–361

Morgan MM, Grisel JE, Robbins CS, Grandy DK (1997) Antinociception mediated by the periaqueductal gray is attenuated by orphanin FQ. Neuroreport 8:3431–3434

Murphy NP, Ly HT, Maidment NT (1996) Intracerebroventricular orphanin FQ/nociceptin suppresses dopamine release in the nucleus accumbens of anaesthetized rats. Neuroscience 75:1–4

Neal CR, Mansour A, Nothacker HP, Reinscheid RK, Civelli O, Watson SJ (1998) Localization of orphanin FQ (nociceptin) peptide and messenger RNA in the forebrain of the rat. Submitted

Neal MJ, Cunningham JR, Paterson SJ, McKnight AT (1997) Inhibition by nociceptin of the light-evoked release of ACh from retinal cholinergic neurones. Br J Pharmacol 120:1399–1400

Nicholson JR, Paterson SJ, Menzies JR, Corbett AD, McKnight AT (1998) Pharmacological studies on the "orphan" opioid receptor in central and peripheral sites. Can J Physiol Pharmacol 76:304–313

Nicol B, Lambert DG, Rowbotham DJ, Smart D, McKnight AT (1996) Nociceptin induced inhibition of K$^+$ evoked glutamate release from rat cerebrocortical slices. Br J Pharmacol 119:1081–1083

Nishi M, Houtani T, Noda Y, Mamiya T, Sato K, Doi T, Kuno J, Takeshima H, Nukada T, Nabeshima T, Yamashita T, Noda T, Sugimoto T (1997) Unrestrained nociceptive response and disregulation of hearing ability in mice lacking the nociceptin/orphaninFQ receptor. EMBO J 16:1858–1864

Noble F, Roques BP (1997) Association of aminopeptidase N and endopeptidase 24.15 inhibitors potentiate behavioral effects mediated by nociceptin/orphanin FQ in mice. FEBS Lett 401:227–229

Nothacker HP, Reinscheid RK, Mansour A, Henningsen RA, Ardati A, Monsma FJ Jr, Watson SJ, Civelli O (1996) Primary structure and tissue distribution of the orphanin FQ recursor. Proc Natl Acad Sci USA 93:8677–8682

Okuda-Ashitaka E, Tachibana S, Houtani T, Minami T, Masu Y, Nishi M, Takeshima H, Sugimoto T, Ito S (1996) Identification and characterization of an endogenous ligand for opioid receptor homologue ROR-C: its involvement in allodynic response to innocuous stimulus. Mol Brain Res 43:96–104

Okuda-Ashitaka E, Minami T, Tachibana S, Yoshihara Y, Nishiuchi Y, Kimura T, Ito S (1998) Nocistatin, a peptide that blocks nociceptin action in pain transmission. Nature 392:286–289

Pan YX, Xu J, Pasternak GW (1996) Cloning and expression of a cDNA encoding a mouse brain orphanin FQ/nociceptin precursor. Biochem J 315:11–13

Patel HJ, Giembycz MA, Spicuzza L, Barnes PJ, Belvisi MG (1997) Naloxone-insensitive inhibition of acetylcholine release from parasympathetic nerves innervating guinea-pig trachea by the novel opioid, nociceptin. Br J Pharmacol 120:735–736

Pomonis JD, Billington CJ, Levine AS (1996) Orphanin FQ, agonist of orphan opioid receptor ORL1, stimulates feeding in rats. Neuroreport 8:369–371

Reinscheid RK, Nothacker HP, Bourson A, Ardati A, Henningsen RA, Bunzow JR, Grandy DK, Langen H, Monsma FJ Jr, Civelli O (1995) Orphanin FQ: a neuropeptide that activates an opioidlike G protein-coupled receptor. Science 270:792–794

Reinscheid RK, Ardati A, Monsma FJ Jr, Civelli O (1996) Structure-activity relationship studies on the novel neuropeptide orphanin FQ. J Biol Chem 271:14163–14168

Reinscheid RK, Higelin J, Henningsen RA, Monsma FJ Jr, Civelli O (1998) Structures that delineate orphanin FQ and dynorphin A pharmacological selectivities. J Biol Chem 273:1490–1495

Riedl M, Shuster S, Vulchanova L, Wang J, Loh HH, Elde R (1996) Orphanin FQ/nociceptin-immunoreactive nerve fibers parallel those containing endogenous opioids in rat spinal cord. Neuroreport 7:1369–1372

Ross EM (1992) G proteins and receptors in neuronal signaling. In: Hall ZW (ed) An introduction to molecular neurobiology. Sinauer Associates, Sunderland, MA, pp 181–206

Rossi GC, Leventhal L, Pasternak GW (1996) Naloxone sensitive orphanin FQ-induced analgesia in mice. Eur J Pharmacol 311:R7–8

Rossi GC, Mathis JP, Pasternak GW (1998) Analgesic activity of orphanin FQ2, murine prepro-orphanin FQ 141-157 in mice. Neuroreport 9:1165–1168

Rowen L, Mahairas G, Hood L (1997) Sequencing the human genome. Science 278:605–607

Saito Y, Maruyama K, Saido TC, Kawashima S (1995) N23K, a gene transiently up-regulated during neural differentiation, encodes a precursor protein for a newly identified neuropeptide nociceptin. Biochem Biophys Res Commun 217:539–545

Saito Y, Maruyama K, Kawano H, Hagino-Yamagishi K, Kawamura K, Saido TC, Kawashima S (1996) Molecular cloning and characterization of a novel form of neuropeptide gene as a developmentally regulated molecule. J Biol Chem 271:15615–15622

Sakurai T, Amemiya A, Ishii M, Matsuzaki I, Chemelli RM, Tanaka H, Williams SC, Richardson JA, Kozlowski GP, Wilson S, Arch JRS, Buckingham RE, Haynes AC, Carr SA, Annan RS, McNulty DE, Liu WS, Terrett JA, Elshourbagy NA, Bergsma DJ, Yanagisawa M (1998) Orexins and orexin receptors: A family of hypothalamic neuropeptides and G protein-coupled receptors that regulate feeding behavior. Cell 92:573–585

Sandin J, Georgieva J, Schott PA, Ogren SO, Terenius L (1997) Nociceptin/orphanin FQ microinjected into hippocampus impairs spatial learning in rats. Eur J Neurosci 9:194–197

Schuligoi R, Amann R, Angelberger P, Peskar BA (1997) Determination of nociceptin-like immunoreactivity in the rat dorsal spinal cord. Neurosci Lett 224:136–138

Schulz S, Schreff M, Nuss D, Gramsch C, Hollt V (1996) Nociceptin/orphanin FQ and opioid peptides show overlapping distribution but not co-localization in pain-modulatory brain regions. Neuroreport 7:3021–3025

Shimohigashi Y, Hatano R, Fujita T, Nakashima R, Nose T, Sujaku T, Saigo A, Shinjo K, Nagahisa A (1996) Sensitivity of opioid receptor-like receptor ORL1 for chemical modification on nociceptin, a naturally occurring nociceptive peptide. J Biol Chem 271:23642–23645

Sim LJ, Xiao R, Childers SR (1996) Identification of opioid receptor-like (ORL1) peptide-stimulated [^{35}S]GTP gamma S binding in rat brain. Neuroreport 7:729–733

Sinchak K, Hendricks DG, Baroudi R, Micevych PE (1997) Orphanin FQ/nociceptin in the ventromedial nucleus facilitates lordosis in female rats. Neuroreport 8:3857–3860

Standifer KM, Cheng J, Brooks AI, Honrado CP, Su W, Visconti LM, Biedler JL, Pasternak GW (1994) Biochemical and pharmacological characterization of mu, delta and kappa 3 opioid receptors expressed in BE(2)-C neuroblastoma cells. J Pharmacol Exp Ther 270:1246–55

Stratford TR, Holahan MR, Kelley AE (1997) Injections of nociceptin into nucleus accumbens shell or ventromedial hypothalamic nucleus increase food intake. Neuroreport 8:423–426

Tian JH, Xu W, Fang Y, Mogil JS, Grisel JE, Grandy DK, Han JS (1997a) Bidirectional modulatory effect of orphanin FQ on morphine-induced analgesia: antagonism in brain and potentiation in spinal cord of the rat. Br J Pharmacol 120:676–680

Tian JH, Xu W, Zhang W, Fang Y, Grisel JE, Mogil JS, Grandy DK, Han JS (1997b) Involvement of endogenous orphanin FQ in electroacupuncture-induced analgesia. Neuroreport 8:497–500

Tian JH, Zhang W, Fang Y, Xu W, Grandy DK, Han JS (1998) Endogenous orphanin FQ: evidence for a role in the modulation of electroacupuncture analgesia and the development of tolerance to analgesia produced by morphine and electroacupuncture. Br J Pharmacol 124:21–26

Ueda H, Yamaguchi T, Tokuyama S, Inoue M, Nishi M, Takeshima H (1997) Partial loss of tolerance liability to morphine analgesia in mice lacking the nociceptin receptor gene. Neurosci Lett 237:136–138

Vaughan CW, Christie MJ (1996) Increase by the ORL1 receptor (opioid receptor-like1) ligand, nociceptin, of inwardly rectifying K conductance in dorsal raphe nucleus neurones. Br J Pharmacol 117:1609–1611

Vaughan CW, Ingram SL, Christie MJ (1997) Actions of the ORL1 receptor ligand nociceptin on membrane properties of rat periaqueductal gray neurons in vitro. J Neuroscience 17:996–1003

Wagner EJ, Ronnekleiv OK, Grandy DK, Kelly MJ (1998) The peptide orphanin FQ inhibits beta-endorphin neurons and neurosecretory cells in the hypothalamic arcuate nucleus by activating an inwardly-rectifying K$^+$ conductance. Neuroendocrinology 67:73–82

Walker JR, Koob GF (1997) Orphan anxiety. Proc Natl Acad Sci USA 94:14217–14219

Wang JB, Johnson PS, Imai Y, Persico AM, Ozenberger BA, Eppler CM, Uhl GR (1994) cDNA cloning of an orphan opiate receptor gene family member and its splice variant. FEBS Lett 348:75–79

Wick MJ, Minnerath SR, Lin X, Elde R, Law PY, Loh HH (1994) Isolation of a novel cDNA encoding a putative membrane receptor with high homology to the cloned mu, delta, and kappa opioid receptors. Mol Brain Res 27:37–44

Xu IS, Wiesenfeld-Hallin Z, Xu XJ (1998) [Phe^{1}psi(CH$_2$-NH)Gly^2]-nociceptin-(1-13)NH$_2$, a proposed antagonist of the nociceptin receptor, is a potent and stable agonist in the rat spinal cord. Neurosci Lett 249:127–130

Xu XJ, Hao JX, Wiesenfeld-Hallin Z (1996) Nociceptin or antinociceptin: potent spinal antinociceptive effect of orphanin FQ/nociceptin in the rat. Neuroreport 7:2092–2094

Yu TP, Fein J, Phan T, Evans CJ, Xie CW (1997) Orphanin FQ inhibits synaptic transmission and long-term potentiation in rat hippocampus. Hippocampus 7:88–94

Zhang G, Murray TF, Grandy DK (1997) Orphanin FQ has an inhibitory effect on the guinea pig ileum and the mouse vas deferens. Brain Res 772:102–106

Zhu CB, Cao XD, Xu SF, Wu GC (1997) Orphanin FQ potentiates formalin-induced pain behavior and antagonizes morphine analgesia in rats. Neurosci Lett 235:37–40

Molecular Biology of the Receptors for Somatostatin and Cortistatin

Hans-Jürgen Kreienkamp

1
Introduction

The neuropeptide somatostatin (SST) has been known for more than 25 years, and investigations into its function have yielded a large variety of physiological effects that can be ascribed to the peptide. These effects go far beyond the role as an inhibitor of the release of pituitary growth hormone which was observed initially. Consistently, somatostatin (and its more recently identified relative, cortistatin) exerts its functions via a whole family of different receptor subtypes. Since the molecular identification of these subtypes in the early 1990's the assignment of specific functions to individual subtypes has been a major goal in somatostatin research. This review summarizes the recent advances that have been made with respect to the signal transduction by somatostatin receptors as well as the regulation of the receptors. In addition I will give an outlook on studies using mice deficient for individual receptor subtypes, which – in combination with newly deveolped subtype specific ligands – will hopefully provide a clearer picture of somatostatin receptor function in the near future.

2
The Peptides

Somatostatin was originally identified on the basis of its ability to inhibit growth hormone (GH) secretion from the pituitary. This *somatotropin release inhibiting factor* (SRIF) was isolated from bovine hypothalamus as a peptide containing 14 amino acid residues and was therefore named somatostatin 14 (SST14). It was obtained in a cyclic form due to a disulfide bond between cysteine residues (Brazeau et al., 1973, see Fig. 1). Later on, a larger peptide was discovered in the pancreas which carried the SST14 sequence plus an N-terminal extension of 14 additional residues. This peptide was accordingly named

Institut für Zellbiochemie und klinische Neurobiologie, Universitätskrankenhaus Eppendorf, Universität Hamburg, Martinistrasse 52, 20246 Hamburg, Germany

somatostatin 28 (SST28, Pradayrol et al., 1980; Schally et al., 1980). As SST28 carries a prohormone processing site that would lead to the release of SST14, it was assumed that both hormones can be derived from the same precursor molecule by alternative proteolytic processing. This was indeed confirmed by the cloning of a cDNA coding for the precursor molecule, preprosomatostatin (Goodman et al., 1980). This protein yields prosomatostatin after cleavage of the N-terminal signal peptide; prosomatostatin can then be processed by the appropriate prohormone processing enzymes to release either SST28 or SST14 (see also Epelbaum et al., 1994).

   While it was considered a fact for a long time that there are only two bio-active somatostatin peptides, Sutcliffe`s group recently isolated a cDNA clone from rat brain that codes for a somatostatin like precursor protein (De Lecea et al., 1996, see also the review by Sutcliffe and De Lecea, this volume). The C-terminal 14 amino acids of this protein show only two mismatches when com-pared to the SST14 sequence (Fig. 1). This molecule was called cortistatin because it is mainly expressed in the cortex and in the hippocampus of rats. The predicted active peptide derived from this precursor was synthesized and was shown to exhibit similar agonist potencies against the somatostatin recep-tors when compared to SST14. So far it is unclear if the predicted cortistatin sequence actually corresponds to the native cortistatin peptide present *in vivo*, because the peptide has not yet been isolated from brain extracts. The rat pre-

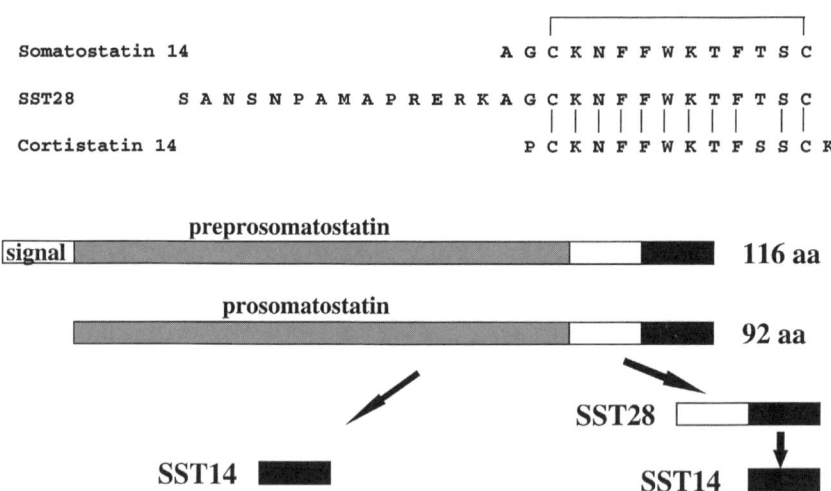

Fig. 1. Structure and biosynthesis of bioactive somatostatin peptides. Shown is the primary structure of the bioactive somatostatin peptides SST14, SST28 as well as cortistatin 14. The posi-tion of a disulfide loop is indicated in the sequence of SST14. Vertical lines show identity between the sequence of somatostatin and cortistatin. Prosomatostatin (92 amino acids) is derived from preprosomatostatin (116 amino acids) by cleavage of the signal peptide; further cleavage at typical prohormone cleavage sites yields either SST28 or SST14

cursor could give rise to another peptide of 29 amino acids, similar to SST28; however, the processing site for this peptide is not present in the mouse and human precursor proteins which have been cloned recently. The human peptide is actually predicted to carry an N-terminal extension by three amino acids, yielding a 17 residue sequence as the active peptide (Fukusumi et al., 1997; de Lecea et al., 1997b).

The physiological functions that have been ascribed to somatostatin are quite numerous. Somatostatin is present in pancreatic islets, the gastrointestinal tract and the thyroid gland, where it inhibits endocrine secretion of many hormones released from these tissues (Reichlin, 1983) in addition to the regulation of pituitary hormone release. Somatostatin also acts as an inhibitor of exocrine secretion; e.g. it reduces very efficiently the secretion of gastric acid.

In the central nervous system, the functions of SST are even more complex, due to the variety of brain regions in which SST and its binding sites have been detected. By using antibodies directed against either SST14 or SST28, it was shown that both ligands occur in many regions of the brain including the cortex, hippocampus, olfactory bulb, striatum and the hypothalamus (see Epelbaum et al., 1994, for review). In the hypothalamus the presence of somatostatin immunoreactivity in neurons projecting from the anterior periventricular nucleus to the median eminence is clearly related to its function in the regulation of growth hormone release from the pituitary (Hökfelt et al., 1975; Elde and Parsons, 1975). In the hippocampus and the cortex, SST-like immunoreactivity has been most frequently observed in interneurons which in many cases coexpress GABA or neuropeptide Y (Hendry et al., 1984; Somogyi et al., 1984; Schmechel et al., 1984). Cortistatin is also expressed in these interneurons (De Lecea et al., 1997a), and it appears possible that the immunostaining for somatostatin may in fact be due to staining for somatostatin and cortistatin. Behavioural studies revealed that SST may be involved in the control of body temperature, appetite, nociception, sleep, motricity and learning and memory. Surprisingly, the effect of central administration of cortistatin differs markedly from that of SST14 with respect to the modulation of sleep (De Lecea et al.,1996). Whereas somatostatin increases the time that rats spend in paradoxical REM (rapid eye movement) sleep, cortistatin decreases the time of REM sleep. In contrast to somatostatin, cortistatin appears to antagonize the effect of acetylcholine administration in the cortex and the hippocampus (De Lecea et al., 1996)

3
Clinical Aspects

The major clinical use of somatostatin and its peptide analogs is determined by the action on hormone release in different neuroendocrine tissues, as described above. SST may reduce hormone secretion from many hypersecretory neuroendocrine tumours similarly as it reduces secretion from the tissues where these tumors originate from (see Lamberts et al., 1995, for a review). Treatment of patients suffering from pituitary (acromegaly, TSH-secreting) or gastro-entero-pancreatic tumors with stable analogs of SST like the octapeptide octreotide (sandostatin, SMS 201-995) or the hexapeptide MK678 can provide substantial clinical improvement by reducing hormone secretion back to normal levels. In addition, a benign side effect of this treatment was the reduction of tumor growth that could be achieved by the octreotide therapy. This antiproliferative effect and the observation that SST-receptors are expressed not only on neuroendocrine tumors but also in many other tumor tissues including glioma and breast cancer spurred some optimism that treatment with SST-analogs might in fact be of much wider applicability. This lead to a search for somatostatin derivatives which preserve only the antiproliferative effects of SST-treatment but do no longer exhibit a reduction of hormone release, thus circumventing potential problems in endocrine function. One such analog, called TT-232, has in fact been shown to be very efficient in the treatment of a whole variety of tumors in mouse model systems (Kery et al., 1996). So far it is not clear if this is still a somatostatin receptor ligand or if this substance acts on a different type of receptor.

A further application of somatostatin analogs lies in the localization of neuroendocrine tumors by somatostatin receptor scintigraphy. This technique again relies on the presence of SST-receptors on many tumors; radioactive analogs are injected in high doses and are internalized by receptor expressing cells, allowing the visualization of the tumors.

A link to disorders of the CNS was established by the observation that the levels of SST in the human brain is consistently reduced in patients with Alzheimer's disease (see Epelbaum et al., 1994, for review). The loss of cortical SRIF correlates with an increase in the number of senile plaques and neurofibrillary tangles which are characteristic of the disease. However, a causal link between these two observations can not easily be established, and it remains unclear if the extensive neuronal loss is responsible for the loss of somatostatin, or if reduced SST-content contributes to the pathogenesis of Alzheimer's disease.

4
Somatostatin Receptors

Binding sites for somatostatin were originally classified in competition studies using the radioiodinated SST-derivative, ^{125}I-Tyr^{11}-SRIF (Schonbrunn and Tashjian, 1978). Whereas in the pituitary GH$_4$C1 cell line a single class of high affinity binding sites could be identified, the heterogeneity of somatostatin receptors became evident in studies on membranes derived from rat brain. The synthetic agonist SMS 201-995 competed with the binding of ^{125}I-Tyr^{11}-

SRIF in a biphasic manner, suggesting the existence of at least two different types of receptors (Reubi, 1984; Tran et al., 1985). A similar selectivity was displayed by the agonist MK678, and binding sites with high affinity for this substance were labeled SRIF1-receptors, while those with low affinity for MK678 were labeled SRIF2-receptors. These two different receptor populations did not discriminate between the two natural ligands, SST14 and SST28. Because the binding of somatostatin was in many (but not all) cases sensitive to treatment with guanine nucleotides or with pertussis toxin, it was inferred that somatostatin receptors couple to heterotrimeric G-proteins of the Ptx-sensitive G_i/G_o-type (e.g. Murray-Whelan and Schlegel, 1992; Lewis et al., 1986).

As was observed for many other types of G-protein coupled receptors (e.g. muscarinic acetylcholine receptors), the distinction between two classes of binding sites became too simple after the molecular cloning of somatostatin receptors, by which not only two but five different receptors were characterized. The cDNAs coding for the five somatostatin receptor subtypes (SSTR1-5, or alternatively sst_1-sst_5)which are currently known were identified either by expression cloning (Kluxen et al., 1992) or by a PCR based approach using degenerate oligonucleotides directed at the conserved transmembrane regions which are shared by all G-protein coupled receptors of the rhodopsin family (Meyerhof et al., 1991; Yamada et al., 1992a; Yasuda et al., 1992 Yamada et al., 1992b; Meyerhof et al., 1992; Li et al., 1992; Bruno et al., 1992; Rohrer et al., 1993; O'Carroll et al., 1992). In addition to these five subtypes, a splice variant of SSTR 2 was discovered in the mouse; alternative splicing occurs in the C-terminal part of the receptor. The two splice variants where named SSTR2A (which corresponds to the unspliced version of the mRNA) and SSTR2B (which carries the alternative exon coding for a slightly shorter C-terminus; Vanetti et al., 1992).

A comparative sequence analysis revealed that the five SSTRs exhibit the typical profile of G-protein coupled receptors with seven hydrophobic transmembrane domains, an extracellular N-terminus and an intracellular C-terminus. They constitute their own subfamily within the larger family of type I (Rhodopsin-like) G-protein coupled receptors, the closest relatives being the four known opioid receptors (μ, κ, δ and ORL, see Fig. 2). Functional expression of the SSTR cDNAs in HEK or COS cells yielded receptors which exhibit high affinity for the endogenous ligands SST14 and SST28. Only SSTR5 consistently showed some preference for SST28 over SST14 (O'Carroll et al., 1992), while in the case of SSTR1-4 the affinities were very similar for the native peptides. After the discovery of cortistatin, this peptide was also shown to be a high affinity agonist at all five receptor subtypes (Siehler et al., 1998).

By using the synthetic derivatives SMS 201 995 and MK678, the cloned SSTRs could be assigned to the SRIF1- and SRIF2-classes of binding sites which had been observed in brain membranes. Thus, SSTR2, 3 and 5 exhibit high to moderate affinity for these peptides and constitute the SRIF1 class of binding sites. SSTR1 and SSTR4 do not bind SMS and MK678 and therefore constitute the SRIF2 class of binding sites (Raynor et al., 1993a; Raynor et al., 1993b; Patel and Srikant, 1994; Hoyer et al., 1995). This classification is nicely supported by a sequence comparison, where the similarity between receptor subtypes is higher between SSTR1 and 4 on one hand and SSTR2, 3 and 5 on the other hand (see dendrogram in Fig. 2).

The regional and temporal expression patterns of the receptors have been studied extensively by *in-situ* hybridization, reverse transcriptase PCR and Northern blotting experiments (Wulfsen et al., 1993; Breder et al., 1993, Bruno et al., 1993; Senaris et al., 1994; Kong et al., 1994; Beaudet et al., 1995). It became obvious that the expression patterns of the subtypes are clearly distinct, but overlapping in many regions of the brain. Thus SSTR1-4 have been shown to be present in the hippocampus and the cortex of rats in various studies. In the cortex, SSTR2 is more restricted to the deeper layers, while SSTR1 mRNA is present in all cortical layers. SSTR3 was shown to be the only subtype present in the cerebellum of adult animals, while SSTR1 exhibits a transient expression in this brain region shortly before and after birth (Meyerhof et al., 1992). SSTR1 and SSTR2 are both expressed in various nuclei of the hypothalamus; here, SSTR1 mRNA is present in neurons which also contain somatostatin mRNA, whereas SSTR2 was found also on cells which presumably express growth hormone releasing factor (Beaudet et al., 1995).

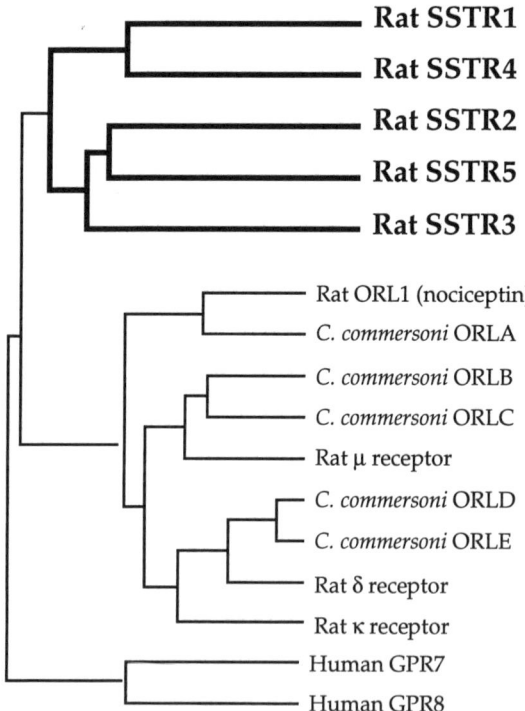

Fig.2. Sequence relationship between somatostatin and opioid receptors. The dendrogram was generated by aligning the cDNA sequences of the rat SSTRs (references see text) and the opioid receptors from rat (Nishi et al., 1993; Fukuda et al., 1993; Fukuda et al., 1994) and from the teleost fish Catostomus commersoni (Darlison et al., 1997). Sequences for the two orphan receptors GPR7 and GPR8 are taken from O'Dowd et al., 1995

SSTR4 appears to be the only receptor subtype that is relatively brain-specific; in contrast, the presence of SSTR5 in the brain of rats was disputed; its mRNA was not detected by *in situ* hybridization, but only by RNase protection and RT-PCR experiments. However, we have shown by using an affinity purified antibody against SSTR5 that this receptor is expressed in a few discrete neuronal populations of the hypothalamus and basal forebrain (Stroh T., Kreienkamp, H.-J., and Beaudet, A., submitted).

A major feature that becomes apparent from these studies is that many cell populations express not only one but several different somatostatin receptor subtypes. Thus it was shown by Perez and Hoyer (1995) using double-label in-situ hybridization that SSTR3 and SSTR4 are expressed in the same cells in the neurons of the CA1 and CA2 regions of the hippocampus of the rat. Similarly, SSTR4 and SSTR5 are coexpressed in somatotrophs in the pituitary (O'Carroll and Krempels, 1995).

5
The Regulation of SSTR Gene Expression

The 5'-flanking regions of the SSTR genes have been cloned for all subtypes in order to define the molecular basis for the distinct tissue-specific expression patterns that have been observed (see previous chapter). Two aspects became apparent from these studies: 1. the sequence homology between subtypes that can be observed throughout the coding regions is not existing in the 5'- (and 3') flanking regions of the SSTR genes; 2. all promoter regionslack TATA and CCAAT motifs 30 bp and 70 bp upstream of the transcriptional start site which are essential for transcription of a large number of genes (Hauser et al., 1994; Kraus et al., 1998; Glos et al., 1998;. Schwabe et al., 1996; Greenwood et al., 1994).

The rat SSTR3 is distinct from the other SSTRs because its mRNA contains a rather long 5'-untranslated region (UTR) of 1040 bp (Glos et al., 1998). In this 5'-UTR two introns were found when the sequence of the cDNA was compared to that of a corresponding genomic clone. This particular feature is shared only by the mouse SSTR2, for which it was shown recently that two previously unidentified exons exist that are separated from the coding region of the gene by exons which are larger than 25 kb (Kraus et al., 1998). Both additional exons are preceded by functional promoters, thus leading to a total of three promoters that are obviously all being used in a tissue-specific manner (Kraus et al., 1998).

It is unclear if three promoters also exist in the human SSTR2 gene, but promoter 3 of the mouse (the promoter most proximal to the coding region, driving transcription of a gene product containing none of the two 5' exons) is most similar to the human promoter which has been characterized functionally in some detail by Pscherer et al. (1996). These authors identified a novel initiator element (SSTR2inr), which contains a so-called E-box that is required for the binding of a novel basic helix-loop-helix transcription factor called SEF-2. SEF-2 interacts specifically with transcription factor IIB and may thus recruit the basal transcriptional machinery to the SSTR2 initiator element. As pointed out by Kraus et al. (1998), this may account for some but not all of the trancsriptional regulation of the SSTR2 gene. The alternative promoter 2 in the mouse gene, for example, contains an additional cyclic AMP responsive element, which may confer the enhancement of SSTR2 expression by cAMP that has been observed in AtT20 cells (Patel et al., 1993).

SSTRs are widely expressed in the pituitary, and the development of the pituitary is controlled by the transcription factor Pit1. This prompted an investigation into the importance of this factor for the regulation of the SSTR1 gene, which contains 3 putative Pit1 binding sites in its 5′ flanking region (Hauser et al., 1994). A functional analysis in the pituitary tumor cell line GH_3 which contains predominantly SSTR1 and SSTR2 mRNA (Hauser et al., 1994), showed that a 2 kb fragment encompassing all three sites is sufficient to drive transcriptional activity. The most distal Pit1 site (PitS1) actually attenuates promoter activity, as removal of this region from the promoter constructs leads to an increase in activity. Out of the remaining two sites, one (PitS3) is not functional with respect to Pit1-binding in a footprint assay. The other one (PitS2) is functional in footprint and electrophoretic mobility shift assays and is essential for SSTR1 promoter activity (Baumeister, H., Wegner, M., Richter, D. and Meyerhof, W., manuscript submitted). Dependence on Pit1 is relatively specific for SSTR1 as treatment of GH_3 cells with an antisense oligonucleotide directed at the Pit1 mRNA reduces the amount of ^{125}I-SST14 binding to GH_3 cells that is sensitive to competition with the SSTR1 specific compound CH275, but not the CH275 non-sensitive binding. In addition, the Pit1 site present in the 5′-flanking region of the rSSTR3 gene is not essential for transcriptional activity of this promoter in GH_3 cells (Glos et al., 1998).

6
Signal Transduction Through Somatostatin Receptor Subtypes

The signal transduction properties and the coupling to second messenger systems of the individual somatostatin receptor subtypes have been analyzed in heterologous expression systems like COS, CHO or HEK cells (Fig. 3). Though there was some initial controversy regarding SSTR1 and SSTR2 (Rens-Domiano et al., 1992; Yamada et al., 1992a), it was eventually shown that all somatostatin receptors can be coupled to the inhibition of adenylate cyclase via pertussis toxin sensitive G-proteins (Patel et al., 1994). The failure of SSTR1 and SSTR2 to do so in some studies could be ascribed to the lack of an appropriate G-protein α-subunit (Kagimoto et al., 1994). Similarly, SSTR1 (Roosterman et al., 1998), SSTR2 (Fujii et al., 1994; Tallent et al., 1996a), SSTR4

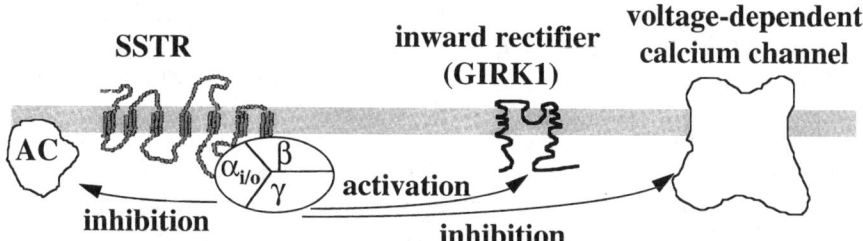

Fig. 3. Major signal transduction pathways used by somatostatin receptors. Shown are the three major pathways by which somatostatin decreases the excitability of neuronal and endocrine cells. AC, adenylate cyclase

(Zhang et al., 1996) and SSTR5 (Tallent et al., 1996a) were shown to inhibit the activity of voltage gated calcium channels of either the N- of the L-type. However, the work by Fujii et al. (1994) presents evidence that SSTR1 is not coupled to the inhibition of L-type channel in the experimental system used, suggesting again that the activity of SSTR1 is strongly dependent on the cellular environment and probably the presence of the appropriate G-protein subunits. The coupling of SSTR3 was not investigated in any of these experiments. The activation of G-protein gated inwardly rectifying potassium channels (GIRKs) constitutes a third major pathway by which somatostatin reduces the excitability of neuronal and neuroendocrine cells. As is true for the inhibition of N-type calcium channels, this effect is mediated by the βγ-subunits of the heterotrimeric G-proteins. By expression of SSTR subtypes and the GIRK1 subunit in Xenopus oocytes, it was shown that SSTR2-5 can activate this inward potassium current, with SSTR2 being most efficient (Kreienkamp et al. 1997). Again SSTR1 appears to be the receptor that is not in line with the other family members, as it does not couple to the GIRK channel in this system. Interestingly, Karschin (1995) has shown by single-cell RT-PCR analysis that the SSTR1 protein expressed in oligodendrocytes inhibits rather than activates an inwardly rectifying potassium conductance.

Coupling of all SSTRs to the activation of phospholipase C via non-Ptx sensitive G-proteins and the subsequent release of calcium from intracellular store has been observed in COS cells (Akbar et al., 1994). In contrast, Bito et al. (1994) did not observe this effect when expressing the rat SSTR4 in CHO cells. In our hands, we did not observe a mobilization of intracellular calcium and subsequent activation of calcium gated chloride channels when expressing SSTR1-5 in *Xenopus* oocytes. Activation of this pathway was possible only when the promiscuous G-protein subunit Gα16 was coexpressed (Schwärzler, A. and Kreienkamp, H.-J., unpublished data). It has been argued that this G-protein subunit may couple almost any receptor to the phospholipase C-pathway (Offermanns and Simon, 1995).

A special feature of SSTR signal transduction is the ability of somatostatin to induce antiproliferative effects in its target cells. This is an apparent paradox for a Gi/Go-coupled receptor, as it has been shown in numerous studies that activation of these G-proteins leads, again via the βγ-subunits (van Biesen et al., 1995), to the activation of the small GTP binding protein ras and to the eventual activation of mitogen activated protein kinases (MAP kinase). Activation of the MAP kinase was indeed induced by activation of SSTR4 expressed in CHO cells (Bito et al., 1994). Thus activation of SSTRs would be expected to stimulate proliferation rather than act in an antiproliferative fashion. A first clue towards the understanding of this mechanism came by the observation of Pan et al. (1992) that treatment of A431 tumor cells with SST14 results in activation of a tyrosine phosphatase activity which may dephosphorylate an active, tyrosine-phosphorylated epidermal growth factor receptor. This observation lead to extensive studies with respect to the ability of individual SSTR subtypes to activate this phosphatase activity. Buscail et al. (1994) presented evidence that SSTR1 and SSTR2 can activate such a tyrosine phosphatase activity after expression in COS cells. A similar study in NIH3T3 fibroblasts implied SSTR2, 3 and 4 in this signal transduction pathway (Reardon et al., 1997). These findings raised the question by which mechanism a G-protein coupled receptor could activate a tyrosine phosphatase. Sturgill and coworkers presented evidence that a tyrosine phosphatase may be an effector of an activated G-protein α-subunit, leading to the inactivation of tyrosine-phosphorylated raf and thus counteracting the effcets of G-protein βγ-subunits on the activation of ras (and subsequently on the activation of raf and the MAP kinase pathway; Reardon et al., 1997; Dent et al., 1996). Three independent studies provided evidence that the phosphatase involved here is an SH2-domain containing tyrosine phospatase called SHP (both isoforms of this enzyme called SHP-1 and SHP-2 were suggested to be involved in either one of these studies; Srikant and Shen, 1996; Lopez et al., 1997; Reardon et al., 1997). Treatment of MCF-7 human breast cancer cells with SST induces a translocation of this enzyme to the cell membrane, and Lopez et al. (1997) presented data that SHP-1 is physically associated with SSTR2 after activation with agonists. Taken together, these experiments might explain why specifically SSTRs (and not any other Gi/Go coupled receptor) can activate a tyrosine phosphatase. Whereas activation through the Gα-subunits is a common pathway shared by all receptors, the specific association between SSTRs and the SHP phosphatase is probably a prerequisite for tyrosine phosphatase signalling that is not fulfilled by other receptors. However, it is unclear at present if the interaction between SSTR and SHP is direct or indirect via an adaptor protein. As SSTR2 is not significantly phosphorylated at tyrosine residues (Schwartkop, C.-P., Kreienkamp H.-J. and Richter, D., unpublished observations; Hipkin et al., 1997), an interaction between the SH2 domain and a phosphotyrosine residue on SSTR2 seems unlikely to account for this association.

7
Do we Need Even More Receptor Subtypes?

At first glance, one would say no, because the known receptors correspond quite well to the binding profiles for somatostatin-like compounds that have been observed in brain. In addition, the major signal transduction pathways that were reported to be activated by somatostatin (i.e. activation of GIRK channels and tyrosine phosphatases; inhibition of voltage-gated potassium channels and adenylate cyclase) were also shown to be activated by several, if not all cloned SSTRs (see above). However, in the pituitary cell line AtT-20 an inwardly rectifying potassium channel can be activated by somatostatin; a pharmacological analysis shows that this activation can be induced also by hexapeptide analogs, but not by octapeptide analogs of somatostatin. As this pharmacological profile is not typical of any of the cloned SSTRs, the authors suggested that an additional SSTR subtype exists which mediates this effect (Tallent et al., 1996b). Also the differential effect of cortistatin on the modulation of sleep when compared to somatostatin (see above) could be explained by the presence of a specific cortistatin receptor. In addition to these findings, it is unclear which receptor is activated by the SST-analog TT-232 (Kery et al., 1996). It does not activate SSTR2-5 in the Xenopus oocyte expression system (Kreienkamp et al., 1997), and does not compete with octreotide binding to rat brain membranes (Kery et al., 1996). So far no new SST-receptor has been identified from AtT-20 cells or any other tissues by molecular cloning, but it is possible that further subtypes remain to be characterized. Potential candidates are among the several orphan G-protein coupled receptors that have been cloned by various forms of homology screening based on known receptor sequences and by analysis of the expressed sequence tag (EST) databases. Three genes coding for a somatostatin-like receptor (SLC1, Kolakowski et al., 1996) and GPR7 and GPR8 (O'Dowd et al., 1995; fig. 2) are relatively similar to the opioid and somatostatin receptors and may thus encode receptors with an overlapping pharmacology, which might also be activated by somatostatin-like compounds. In our hands neither SLC1 nor GPR7 could be activated by somatostatin, octreotide or TT-232 (Schwärzler, A. and Kreienkamp, H.-J., unpublished data).

8
Subtype Specific Differences in Receptor Regulation and Intracellular Distribution

From the previous two chapters it becomes apparent that individual somatostatin receptor subtypes are frequently expressed in the same cells and also share many signal transduction pathways. Thus, the question arises why so many different subtypes exist in mammals. Why would a single cell use two or more different receptors in order to respond to the same extracellular signal with the same intracellular response? One potential reason may be a differential regulation of receptor signalling by the mechanisms of receptor desensitization and internalization. The agonist-dependent regulation of signalling by G-protein coupled receptors has been studied in great detail in the β_2-adrenergic receptor by Lefkowitz and coworkers (Lefkowitz et al., 1992). A general scheme has emerged, in which a receptor is phosphorylated after activation by agonists through G$\beta\gamma$-dependent activation of a G-protein receptor kinase (Pitcher et al., 1992). This phosphorylation leads to a functional desensitization of the receptor and allows binding of a protein of the arrestin family to the receptor (Lohse et al., 1990; Lohse et al., 1992). Arrestins function as adaptor molecules to clathrin (Goodman et al., 1996; Ferguson et al., 1996), and once this contact is established the receptor (and in the case of peptide receptors also the ligand) is internalized via clathrin coated vesicles into endosomes. In the acidic environment of endosomes, the ligand dissociates from the receptor and the receptor is dephosphorylated by a specific phosphatase (Pitcher et al., 1995), thus allowing the receptor to recycle to the cell surface in a functional, resensitized state (Pippig et al. 1995).

The internalization and functional desensitization of the rat somatostatin receptor subtypes has been studied in detail by expression of epitope-tagged receptors in a human embryonic kidney (HEK) cell line (Roth et al., 1997a,b). Treatment with SST14 or SST28 lead to a rapid internalization of SSTR1, 2 and 3. SSTR4 was not internalized in these experiments whereas SSTR5 was internalized only in response to treatment with SST28 but not with SST14 (Roth et al., 1997a). These subtype- and agonist-specific differences could be confirmed also in a neuroendocrine cell line (Roosterman et al., 1997), but different results were obtained with the human SSTRs expressed in CHO cells. Here SSTR1 was the subtype that was not internalized in response to agonists, while SSTR4 was readily internalized (Hukovic et al., 1996). A closer look at the cellular pathway of internalization revealed that the rat SSTR3 is internalized via an endosomal compartment. Phosphorylation of this receptor at four serine/threonine residues in the intracellular carboxyterminal tail leads to a functional desensitization, and this phosphorylation is also required for internalization (Fig. 4; Roth et al., 1997b). Thus SSTR3 most closely resembles the prototypic β_2-adrenergic receptor in its regulatory behaviour.

Very much in contrast, the rat SSTR4 which is not internalized also is not phosphorylated and does not exhibit any functional desensitization after prolonged treatment with agonists (Kreienkamp et al., 1998). Internalization is prevented by a sequence element in the receptor C-terminus, as mutation of a single residue in the C-terminus (threonine 331) allows for a rapid agonist-dependent internalization of this receptor (Fig. 5; Kreienkamp et al., 1998). Interestingly, the only divergence between the sequence of the human and the rat SSTR4 in the C-terminus occurs around this position, providing an expla-

receptor/ mutant	agonist-dependent internalization	agonist-dependent phosphorylation
wt SSTR3	yes	2.31 mol/mol
S341A	no	n.d.
S346A	no	n.d.
S351A	no	n.d.
T357A	no	1.07 mol/mol
S341A/S346A/S351A	no	0.34 mol/mol

Fig. 4. Determinants of agonist-dependent internalization of the rat SSTR3. Shown is the sequence of the intracellular, C-terminal tail of the rat SSTR3. The amino acids Ser 341, S346, S351 and T 357 which are phosphorylated in response to agonist treatment and which are essential for internalization are indicated by black circles.Internalization was analyzed by whole-cell binding assays and by confocal microscopy of wild type and mutant receptors expressed in HEK cells. Phosphorylation was determined by immunoprecipitation of the receptors from cell extracts after loading with ^{32}P-phosphate and treatment with agonists (Roth et al., 1997b)

nation for the species specific differences mentioned above. The lack of functional desensitization of SSTR4 is accompanied by an extremely long-lasting signalling exhibited by this receptor. When coexpressed with the GIRK channel in Xenopus oocytes, the SSTR4 mediated signal decays with a halflife of approx. 7 minutes after washout of the agonist, whereas the signal mediated by SSTR3 decays about 10 times as quickly. The slow decay of the SSTR4 signal is not due to slow dissociation of the agonist, as ^{125}I-SST14 dissociates with similar kinetics from SSTR4 and from SSTR3 when expressed on the cell surface of HEK cells (Kreienkamp et al., 1997; 1998). In conclusion, the expression of SSTR4 in addition to e.g. SSTR3 would give a cell the ability to turn a rather short-lived signal by somatostatin into a rather long-lived signal.

The SSTR2A receptor appears to be similar to SSTR3 in its regulatory behaviour, i.e. it is rapidly phosphorylated, desensitized and internalized in various expression systems (Vanetti et al., 1993; Roth et al., 1997a; Hipkin et al., 1997). However, the analysis of C-terminal deletion constructs of SSTR2 suggests that phosphorylation is not required for internalization (Schwartkop et al., 1999).

The subcellular distribution of a receptor is of particular functional interest in neuronal cells, where a receptor in the somatodendritic compartment may contribute to postsynaptic signalling, whereas a receptor in the axon or nerve

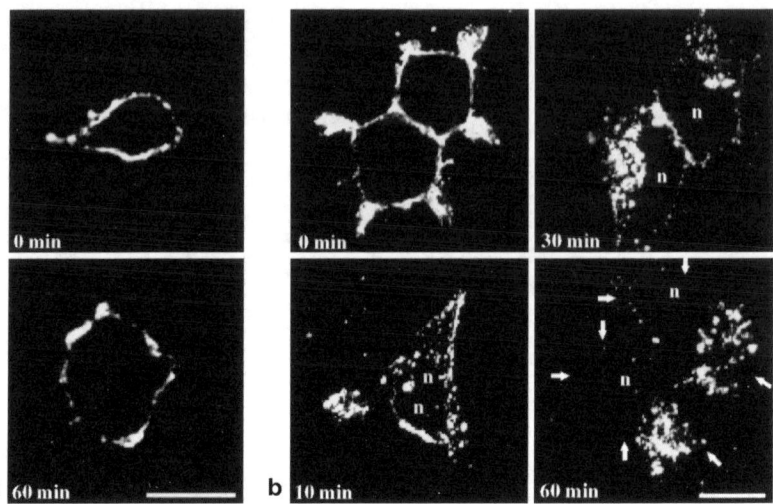

Fig. 5. Internalization of SSTR4 can be induced by a mutation in the intracellular carboxy-terminal tail. **a:** Intracellular distribution of the rat SSTR4 expressed in HEK cells before and after treatment with 1 μM SST14 for 60 minutes. **b:** intracellular distribution of the T331A mutant of the rat SSTR4 before and after treatment with 1 μM SST14 for various times.Bar: 10 μM. Arrows indicate the contours of the cells; n indicates the position of the nucleus

terminal may influence the amount of neurotransmitter that is released by an action potential. Data are finally beginning to emerge with respect to the localization of SSTRs in neuronal cells. By using a polyclonal antibody directed against a C-terminal segment of SSTR2A, Beaudet and coworkers have shown that this subtype is localized in the cell body and dendrites of many neuronal populations in the rat brain. In addition, axonal staining was observed in some tissues, suggesting that SSTR2 is not restricted to one neuronal compartment. (Dournaud et al., 1996). A correlation with SST14-containing nerve terminals by using an antibody against the ligand showed that the somatodendritic receptors are in close proximity to nerve terminals and my thus fulfil a role in normal synaptic transmission (Dournaud et al., 1998). This proximity was not observed for the axonal SSTR2A, leaving open the question if these axonal receptors fulfil a functional role in SST-signalling.

Using a polyclonal antibody directed against the entire C-terminal tail of the rat SSTR5, we showed in collaboration with Beaudets group that SSTR5 is similarly present in the somatodendritic compartment of nerve cells, with some staining again in axons/nerve terminals (Stroh, T., Kreienkamp, H.-J. and Beaudet, A., manuscript submitted). In contrast to SSTR2A which is very widespread in the brain and can account for a large proportion of the SRIF1-type binding in the rodent brain, SSTR5 immunoreactivity is present only in very few discrete neuronal populations in the hypothalamus and the basal forebrain. The presence of SSTR5 had actually been disputed, and SSTR5 mRNA was detected in brain only by RT-PCR and RNase protection analysis, but not by *in situ* hybridization or Northern blotting. However, the immunohisto-chemical data coincide with the RT-PCR analysis with respect to the distribution of this subtype, confirming that indeed some areas of the brain do express SSTR5.

SSTR1 exhibits a striking divergence in its subcellular distribution when compared to the other receptors studied so far. This receptor appears to be localized rather exclusively in axons and nerve terminal of neurons. In the hypothalamus SSTR1 is on those terminals which release somatostatin in the median eminence, suggesting that SSTR1 may function as an autoreceptor for SST14 (Helboe et al., 1998).

9
Towards an Understanding of the Functional Role of Individual Somatostatin Receptor Subtypes

Several recent developments have raised hope that the function of the individual somatostatin receptor subtypes will be elucidated more clearly in the near future. The first publication of a SSTR2 deficient mouse which was created by gene targeting techniques appeared recently. A second major break-through was the development of subtype-specific agonists (SSTR1, 2, 4) and antagonists (SSTR2); the SSTR4-specific compound is also a non-peptide compound, which should be very helpful in studying the function of this subtype.

The best characterized receptor with respect to its function is currently SSTR2; a careful correlation of the efficiency in biological assays and the affinity for expressed receptor subtypes has led to the suggestion that SSTR2 mediates the inhibtion of growth hormone release from the pituitary (Raynor et al., 1993a), pancreatic glucagon secretion (Rossowski and Coy, 1994) and gastric acid secretion from stomach (Rossowski et al., 1994). The presence of SSTR5 in a large proportion of growth hormone secreting cells in the pituitary makes it likely that SSTR5 is also involved in the regulation of this hormone (Kumar et al., 1997).

The pharmacological data are now complemented by the phenotype of the SSTR2 knock-out mice, which is manifested in two ways: first, a feedback control of growth hormone release by the pituitary is abolished due to lack of SSTR2 receptors on neurons in the arcuate nucleus in the hypothalamus. GH that has been secreted acts on somatostatinergic neurons in the hypothalamus which themselves act on the arcuate neurons which produce growth hormone releasing hormones. Activity of these neurons, as measured by the appearance of immunoreactivity for the immediated early gene fos, is reduced by somatostatin and growth hormone in wild type mice, but not in SSTR2 deficient mice (Zheng et al., 1997), suggesting that the pathway from growth hormone via somatostatin to the growth hormone secretagogue producing cells has been disrupted at the point of the somatostatin receptor.

A second clear phenotype was observed in the gastrointestinal tract, where somatostatin inhibits the release of gastric acid. Mice deficient in SSTR2 exhibit a gastric acid secretion which is elevated by more than tenfold when compared to control mice, suggesting that SSTR2 is the receptor subtype that mediates the effect of somatostatin on acid secretion (Martinez et al., 1998). These data are very nicely supported by studies using the newly-developed subtype-specific SSTR2 antagonist. In these experiments, application of the antagonist inhibited the effect of somatostatin on gastric acid secretion (Rossowski and Coy, 1998). Thus, at this stage studies using subtype deficient mice and subtype specific drugs complement each other very convincingly to assign a clear function to SSTR2.

The situation is much les clear for the other subtypes; based on the localization of the SSTR1 protein on axons of somatostatinergic neurons in the median eminence, it would be tempting to speculate that SST14 might regulate its own release via an ultra-short feedback loop using this subtype (Helboe et al., 1998). Indeed, such a feedback inhibition has been observed in hypothalamic neurons in culture (Peterfeund and Vale, 1984; Richardson and Twente, 1986). However, a study by Viollet et al. (1997) using the SSTR1-specific analog CH-275 showed that SSTR1 in hypotalamic neurons leads to an enhancement of glutamate-induced depolarizations rather than to an inhibition of neuronal activity. These data were complemented by RT-PCR analysis, which showed that an excitatory action of the non-selective SST14 correlated with the presence of SSTR1, whereas an inhibitory action of somatostatin or the SSTR2/3/5-selective agonist octreotide correlates with the presence of SSTR2. A lesson that can be learned from these studies is that the signal transduction properties of a given receptor *in vivo* or, more precisely, in its native environment, may differ substantially from that determined in heterologous expression experiments. Thus, whereas the inhibitory action of SSTR2 observed by Viollet et al. was predicted from various expression experiments, the excitatory action of SSTR1 is unexpected in the light of the inhibitory effect this receptor has shown towards adenylyl cyclase and voltage-gated calcium channels (see above). It appears possible that signal transduction *in vivo* may also be determined by other proteins than the set of G-protein subunits that is available in a given cell. Interestingly, it was shown for the $\beta_2$adrenergic receptor that the activation of a Na^+/H^+ exchanger by this receptor was entirely dependent on the presence of a regulatory protein that interacts directly with the receptor C-terminus after receptor activation. (Hall et al., 1998). Similar receptor-interacting proteins may dictate the outcome of SSTR signal transduction.

10
Conclusion

In conclusion, much progress has been made in understanding the regulation and the signal transduction properties of the SSTR subtypes. Voltage gated calcium channels, adenylate cyclases and G-protein gated inwardly rectifying potassium channels have been identified as the major effectors of all subtypes, but still many open questions remain. Thus it is still unclear how activation of tyrosine phosphateses can be achieved mechanistically.

An assignment of a physiological role has been possible in the case of SSTR2 by using gene targeting techniques as well as subtype specific drugs. Knockout mice deficient for the other subtypes have been generated by us (SSTR1) and others (Ute Hochgeschwender, personal communication), but have not revealed any phenotype so far. These mice will hopefully provide useful tools for studying the function of these subtypes, and might also help to explain the differential physiological effects that were observed after central application of cortistatin and somatostatin.

Acknowledgements: Work in the author's laboratory was supported by funds from the Deutsche Forschungsgemeinschaft (Ha 2445/1-1 and SFB 545/B7). I thank Dr. D. Bächner for critically reading the manuscript.

References

Akbar M, Okajima F, Tomura H, Abdul Majid M, Yamada Y, Seino S, Kondo Y (1994) Phospholipase C activation and Ca^{2+} mobilization by cloned human somatostatin receptor subtypes 1–5, in transfected COS-7 cells. FEBS Lett 348:192–196

Beaudet A, Greenspun D, Raelson J, Tannenbaum GS (1995) Patterns of expression of SSTR1 and SSTR2 somatostatin receptor subtypes in the hypothalamus of the adult rat: relationship to neuroendocrine function. Neuroscience 65:551–561

Bito H, Mori M, Sakanaka C, Takano T, Honda Z-i, Gotoh Y, Nishida E, Shimizu T (1994) Functional coupling of SSTR4, a major hippocampal somatostatin receptor, to adenylate cyclase inhibition, arachidonate release, and activation of the mitogen-activated protein kinase cascade. J Biol Chem 269:12722–12730

Brazeau P, Vale W, Burgus R, Ling N, Butcher M, Rivier J, Guillemin R (1973) Hypothalamic polypeptide that inhibits the secretion of immunoreactive pituitary growth hormone. Science 179:77–79

Breder CD, Yamada Y, Yasuda K, Seino S, Saper C, Bell GI (1993) Differential expression of somatostatin receptor subtypes in brain. J Neurosci 12:3920–3934

Bruno JF, Xu Y, Song J, Berelowitz M (1992) Molecular cloning and functional expression of a brain-specific somatostatin receptor. Proc Natl Acad Sci USA 89:11151–11155

Bruno JF, Xu Y, Song J, Berelowitz M (1993) Tissue distribution of somatostatin receptor messenger ribonucleic acid in the rat. Endocrinology 133:2561–2567

Buscail L, Delesque N, Esteve JP, Saint-Laurent N, Prats H, Clerc P, Robberecht P, Bell GI, Liebow C, Schally AV, Vaysse N, Susini C (1994) Stimulation of tyrosine phosphatase and inhibition of cell proliferation by somatostatin analogues: mediation by human somatostatin receptor subtypes SSTR1 and SSTR2. Proc Natl Acad Sci USA 91:2315–2319

Darlison MG, Greten FR, Harvey RJ, Kreienkamp HJ, Stühmer T, Zwiers H, Lederis K, Richter D (1997). Opioid receptors from a lower vertebrate (*Catostomus commersoni*): Sequence, pharmacology, coupling to G-protein-gated inward-rectifying potassium channel (GIRK1), and evolution. Proc Natl Acad Sci USA 94:8214–8219

Dent P, Reardon DB, Wood SL, Lindorfer MA, Graber SG, Garrison JC, Brautigan DL, Sturgill TW (1996) Inactivation of raf-1 by a protein-tyrosine phosphatase stimulated by GTP and reconstituted by Gαi/o subunits. J Biol Chem 271:3119–3123

De Lecea L, Criado JR, Prospero O, Gautvik KM, Schweitzer P, Danielson P, Dunlop CLM, Siggins GR, Henriksen SJ, Sutcliffe JG (1996) A cortical neuropeptide with neuronal depressant and sleep-modulating properties. Nature 381:242–245

De Lecea L, del Rio JA, Criado JR, Alcantara S, Morales M, Danielson PE, Henriksen SJ, Soriano E, Sutcliffe JG (1997a) Cortistatin is expressed in a distinct subset of cortical interneurons. J Neurosci 17:5868–5880

De Lecea L, Ruiz-Lozano P, Danielson PE, Peelle-Kirley J, Foye PE, Frankel WN, Sutcliffe JG (1997b) Cloning, mRNA expression and chromosomal maooing of human and mouse cortistatin. Genomics 42:499–506

Dournaud YZ, Schonbrunn A, Mazella J, Tannenbaum GS, Beaudet A (1996) Localization of the somatostatin receptor SST$_{2A}$ in rat brain using a specific anti-peptide antibody. J Neurosci 16:4468–4478

Dournaud P, Boudin H, Schonbrunn A, Tannenbaum G, Beaudet A (1998) Interrelationships between somatostatin sst2A receptors and somatostatin containing axons in rat brain: evidence for regulation of cell surface receptors by endogenous somatostatin. J Neurosci 18:1056–1071

Elde RP, Parsons JA (1975) Immunocytochemical localization of somatostatin on cell bodies of the rat hypothalamus. Am J Anat 144:541–548

Epelbaum J, Dournaud P, Fodor M, Viollet C (1994) The neurobiology of somatostatin. Crit Rev Neurobiol 8:25–44

Ferguson SSG, Downey WE III, Colapietro A-M, Barak LS, Menard L, Caron MG (1996) Role of β-arrestin in mediating agonist-promoted G protein-coupled receptor internalization Science 271:363–366

Fujii Y, Gono T, Yamad Y, Chihar K, Inagaki N, Seino S (1994) Somatostatin receptor subtype SSTR2 mediates the inhibition of high-voltage-activated calcium channels by somatostatin and its analogue SMS 201-995. FEBS Lett 355:117–120

Fukuda K, Kato S, Mori K, Nishi M, Takashima H (1993) Primary structures and expression from cDNAs of rat opioid receptor δ- and μ-subtypes. FEBS Lett 327:311–314

Fukuda K, Kato S, Mori K, Nishi M, Takeshima H, Iwabe N, Miyata M, Houtani T, Sugimoto T (1994) cDNA cloning and regional distribution of a novel member of the opioid receptor family. FEBS Lett 343:42–46

Fukusumi S, Kitada C, Takekawa S, Kizawa H, Sakamoto J, Miyamoto M, Hinuma S, Kitano K, Fujino M (1997) Identification and characterization of a novel human cortistatin-like peptide. Biochem Biophys Res Commun 232:157–163

Glos M, Kreienkamp H-J, Hausmann H, Richter D (1998) Characterization of the 5′-flanking promoter region of the rat somatostatin receptor subtype 3 gene. FEBS Lett 440:33–37

Goodman RH, Jacobs JW, Chin WW, Lun PK, Dee PC, Habener JF (1980) Nucleotide sequence of a cloned structural gene coding for a precursor of pancreatic somatostatin. Proc Natl Acad Sci USA 77:5869–5873

Goodman OB, Krupnick JG, Santini F, Gurevich VV, Penn RB, Gagnon AW, Kenn JH, Benovic JL (1996) β-arrestin acts as a clathrin adaptor in endocytosis of the β$_2$-adrenergic receptor Nature 383:447–451

Greenwood MT, Panetta R, Robertson LA, Liu, J-L, Patel YC (1994) Sequence analysisof the 5'-flanking promoter region of the human somatostatin receptor 5. Biochem Biophys Res Commun 205:1883–1890

Hall RA, Premont RT, Chow C-W, Blitzer JT, Pitcher JA, Claing A, Stoffel RH, Barak LS, Shenolikar S, Weinman EJ, Grinstein S, Lefkowitz RJ (1998) The β$_2$-adrenergic receptor interacts with the Na$^+$/H$^+$-exchanger regulatory factor to ontrol Na$^+$/H$^+$-exchange. Nature 392:626–630

Hauser F, Meyerhof W, Wulfsen I, Schönrock C, Richter D (1994) Sequence analysis of the promoter region of the rat somatostatin receptor subtype 1 gene. FEBS Lett 345:225–228

Hendry SHC, Jones EG, Emson PC (1984) Morphology, distribution and synaptic relations of somatostatin and neuropeptide Y immunoreactive neurons in rat and monkey neocortex. J Neurosci 4:2497–2517

Helboe L, Stidsen CE, Moeller M (1998) Immunohistochemical and cytochemical localization of the somatostatin receptor subtype sst$_1$ in the somatostatinergic parvocellular neuronal system of the rat hypothalamus. J Neurosci 18:4938–4945

Hipkin RW, Friedman J, Clark RB, Eppler CM, Schonbrunn A (1997) Agonist-induced desensitization, internalization, and phosphorylation of the sst2A somatostatin receptor. J Biol Chem 272:13869–13876

Hökfelt T, Efendic S, Hellerström C, Johansson O, Luft R, Arimura T (1975) Cellular localization of somatostatin in endocrine-like cells and neurons of the rat with special references to the A$_1$-cells of the pancreatic islets and the hypothalamus. Acta Endocrinol 200:5–41

Hoyer D, Bell GI, Berelowitz M, Epelbaum J, Feniuk W, Humphrey PP, O'Carroll AM, Patel YC, Schonbrunn A, Taylor JE (1995) Classification and nomenclature of somatostatin receptors. Trends Pharmacol Sci 16:86–88

Hukovic N, Panetta R, Kumar U, Patel YC (1996) Agonist-dependent regulation of cloned human somatostatin receptor types 1–5 (hSSTR1-5): subtype selective internalization or upregulation. Endocrinology 137:4046–4049

Kagimoto S, Yamada Y, Kubota A, Someya Y, Ihara Y, Yasuda K, Kozasa T, Imura H, Seino S, Seino Y (1994) Human somatostatin receptor,SSTR2, is coupled to adenylyl cyclase in the presence of Giα1 protein. Biochem Biophys Res Comm 202:1188–1195

Karschin A (1995) Molecular single-cell analysis identifies somatostatin type 1 (sst$_1$) receptors to block inwardly rectifying K$^+$ channels in rat brain oligodendrocytes. Neuroreport 7:121–124

Kery G, Erchegyi J, Horvath A, Mezo I, Idei M, Vantus T, Balogh A, Vadasz Zs, Bökönyi Gy, Seprodi J, Teplan I, Csuka O, Tejeda M, Gaal D, Szegedi Zs, Szende B, Roze C, Kalthoff H, Ullrich A (1996) A tumor-selective somatostatin analog (TT-232) with strong in vitro and in vivo antitumor activity Proc Natl Acad Sci USA 93:12513–12518

Kluxen F-W, Bruns C, Lübbert H (1992) Expression cloning of a rat brain somatostatin receptor cDNA. Proc Natl Acad Sci USA 89:4618–4622

Kong H, DePaoli AM, Breder CD, Yasuda K, Bell GI, Reisine T (1994) Differential expression of messenger RNAs for somatostatin receptor subtypes SSTR1, SSTR2, and SSTR3 in adult rat brain: analysis by RNA blotting and in situ hybridization histochemistry. Neuroscience 59:175–184

Kolakowski LF Jr, Jung BP, Nguyen T, Johnson MP, Lynch KR, Cheng R, Heng HH, George SR, O'Dowd BF (1996) Characterization of a human gene related to genes encoding somatostatin receptors. FEBS Lett 398:253–258

Kraus J, Wölte M, Schönwetter N, Höllt V (1998) Alternative promoter usage and tissue specific expression of the mouse somatostatin receptor 2 gene. FEBS Lett 428:165–170

Kreienkamp H-J, Hönck H-H, Richter D (1997) Coupling of rat somatostatin receptor subtypes to a G-protein gated inwardly rectifying potassium channel (GIRK1). FEBS Lett 419:92–94

Kreienkamp H-J, Roth A, Richter D (1998) Rat somatostatin receptor subtype 4 can be made sensitive to agonist-induced internalization by mutation of a single threonine (residue 331). DNA Cell Biol 17:869–878

Kumar U, Laird D, Srikant CB, Escher E, Patel YC (1997) Expression of the five somatostatin receptor (SSTR1-5) subtypes in rat pituitary somatotrophes: quantitative analysis by double-layer immunofluorescence confocal microscopy. Endocrinology 138:4473–4476

Lamberts SW, de Herder WW, van Koetsveld PM, Koper JW, van der Lely AJ, Visser-Wisselaar HA, Hofland LJ (1995) Somatostatin receptors: clinical implications for endocrinology and oncology. Ciba Found Symp 190:222–236

Lefkowitz RJ, Inglese J, Koch, WJ, Pitcher J, Attramadal H and Caron M (1992) G-protein-coupled receptors: regulatory role of receptor kinases and arrestin proteins. Cold Spring Harbor Symp Quant Biol LVII, pp 127–133

Lewis DL, Weight FF, Luini A (1986) A guanine nucleotide-binding protein mediates the inhibition of voltage-dependent calcium current by somatostatin in a pituitary cell line. Proc Natl Acad Sci USA 1986:9035–9039

Li X-J, Forte M, North RA, Ross CA, Snyder SH (1992) Cloning and expression of a rat somatostatin receptor enriched in brain. J Biol Chem 267:21307–21312

Lohse M, Andexinger S, Pitcher J, Trukawinski S, Codina J, Faure J-P, Caron MG, Lefkowitz RJ (1992) Receptor-specific desensitization with purified proteins. J Biol Chem 267:8558–8564

Lohse M, Benovic JL, Codina J, Caron MG, Lefkowitz RJ (1990) β-arrestin: a protein that regulates β-adrenerghic receptor function. Science 248:1547–1550

Lopez F, Esteve JP, Buscail L, Delesque N, Saint-Laurent N, Theveniau M, Nahmias C, Vaysse N, Susini C (1997) The tyrosine phosphatase SHP-1 associates with the sst2 somatostatin receptor and is an essential component of sst2-mediated inhibitory growth signaling. J Biol Chem 272:24448–24454

Martinez V, Curi AP, Torkian B, Schaeffer JM, Wilkinson HA, Walsh JH, Tache Y (1998) High basal gastric acid secretion in somatostatin receptor subtype 2 knockout mice. Gastroenterology 114:1125–1132

Meyerhof W, Paust H-J, Schönrock C, Richter D (1991) Cloning of a cDNA encoding a novel putative G-protein coupled receptor expressed in specific brain regions. DNA Cell Biol 10:689–694

Meyerhof W, Wulfsen I, Schönrock C, Fehr S, Richter D (1992) Molecular cloning of a somatostatin-28 receptor and comparison of its expression pattern with that of a somatostatin-14 receptor in rat brain. Proc Natl Acad Sci USA 89:10267–10271

Murray-Whelan R, Schlegel W (1992) Brain somatostatin receptor-G protein interaction. G alpha C-terminal antibodies demonstrate coupling of the soluble receptor with Gi(1–3) but not with Go. J Biol Chem 267:2960–2965

Nishi M, Takeshima H, Fukuda K, Kato S, Mori K (1993) cDNA cloning and pharmacological characterization of an opioid receptor with high affinities for κ-subtype-selective ligands. FEBS Lett 330:77–80

O'Carroll A-M, Lolait SJ, Köni M, Mahan LC (1992) Molecular cloning and expression of a pituitary somatostatin receptor with preferential affinity for somatostatin-28. Mol Pharmacol 42:939–946

O'Carroll A-M, Krempels K (1995) Widespread distribution of somatostatin receptor messenger ribonucleic acids in rat pituitary. Endocrinology 136:5224–5235

O'Dowd BF, Scheideler MA, Nguyen T, Cheng R, Rasmussen JS, Marchese A, Zastawny R, Heng HH, Tsui LC, Shi X (1995) The cloning and chromosomal mapping of two novel human opioid-somatostatin-like receptor genes, GPR7 and GPR8, expressed in discrete areas of the brain. Genomics 28:84–91

Offermanns S, Simon MI (1995) Gα15 and Gα16 couple a wide variety of receptors to phospholipase C. J Biol Chem 270:15175–15180

Pan GP, Florio T, Stork PJ (1992) G protein activation of a hormone stimulated phosphatase in human tumor cells. Science 256:1215–1217

Patel YC, Greenwood M, Kent G, Panetta R, Srikant C (1993) Multiple gene transcripts of the somatostatin receptor SSTR2: tissue selective distribution and cAMP regulation. Biochem Biophys Res Comm 192:288–294

Patel YC, Greenwood MT, Warszynska A, Panetta R, Srikant CB (1994) All five cloned human somatostatin receptor subtypes (hSSTR1-5) are functionally coupled to adenylyl cyclase. Biochem Biophys Res Comm 198:605–612

Patel YC, Srikant CB (1994) Subtype selectivity of peptide analogs for all five cloned human somatostatin receptors (hsstr1–5). Endocrinology 135:2814–2817

Perez J, Hoyer D (1995) Coexpression of somatostatin SSTR3 and SSTR4 messenger RNAs in the rat brain. Neuroscience 64:241–253

Peterfreund RA, Vale W (1984) Somatostatin analogs inhibit somatostatin secretion from cultured hypothalamus cells. Neuroendocrinology 39:397–402

Pippig S, Alexander S, Lohse MJ (1995) Sequestration and recycling of β_2-adrenergic receptors permit receptor resensitization. Mol Pharmacol 47:666–676

Pitcher JA, Inglese J, Higgins JB, Arriza JL, Casey PJ, Kim C, Benovic JL, Kwatra MM, Caron MG, Lefkowitz RJ (1992) Role of $\beta\gamma$ subunits of G proteins in targeting the β-adrenergic receptor kinase to membrane bound receptors. Science 257:1264–1267

Pitcher JA, Payne ES, Csortos C, DePaoli-Roach AA, Lefkowitz RJ (1995) The G-protein-coupled receptor phosphatase: a protein phosphatase type 2A with a distinct subcellular distribution and substrate specificity. Proc Natl Acad Sci USA 92:8343–8347

Pradayrol L, Jornvall H, Mutt V, Ribet A (1980) N-terminally extended somatostatin: the primary structure of somatostatin-28. FEBS Lett 109:55–58

Pscherer A, Dörflinger U, Kirfel J, Gawlas K, Rüschoff J, Buettner R, Schüle R (1996) The helix-loop-helix transcription factor SEF-2 regulates the activity of a novel initiator element in the promoter of the human somatostatin receptor II gene. EMBO J 15:6680–6690

Raynor K, Murphy WA, Coy DH, Taylor JE, Moreau J-P, Yasuda K, Bell GI, Reisine T (1993a) Cloned somatostatin receptors: identification of subtype-selective peptides and demonstration of high-affinity binding of linear peptides. Mol. Pharmacol. 43:838–844

Raynor K, O'Carroll A-M, Kong HY, Yasuda K, Mahan LC, Bell GI, Reisine T (1993b) Characterization of cloned somatostatin receptors SSTR4 and SSTR5. Mol Pharmacol 44:385–392

Reardon DB, Dent P, Wood SL, Kong T, Sturgill TW (1997) Activation in vitro of somatostatin receptor subtypes 2, 3, or 4 stimulates protein tyrosine phosphatase activity in membranes from transfected Ras-transformed NIH 3T3 cells: coexpression with catalytically inactive SHP-2 blocks responsiveness. Mol Endocrinol 11:1062–1069

Reichlin S (1983) Somatostatin. N Engl J Med 309:1495–1501

Rens-Domiano S, Law SF, Yamada Y, Seino S, Bell GI, Reisine T (1992) Pharmacological properties of two cloned somatostatin receptors. Mol Pharmacol 42:28–34

Reubi J-C (1984) Evidence of two somatostatin receptor types in rat brain cortex. Neurosci. Lett 4:259–263

Richardson SB, Twente S (1986) Inhibition of rat hypothalamic somatostatin release by somatostatin: evidence for somatostatin ultrashort feedback loop. Endocrinology 118:2076–2082

Rohrer L, Raulf F, Bruns C, Buettner R, Hofstaedter F, Schüle R (1993) Cloning and characterization of a fourth human somatostatin receptor. Proc Natl Acad Sci USA 90:4196–4200

Roosterman D, Glassmeier G, Baumeister H, Scherübl H, Meyerhof W (1998) A somatostatin receptor 1 selective ligand inhibits Ca^{2+} currents in rat insulinoma 1046-38 cells. FEBS Lett 425:137–140

Roostermann D, Roth A, Kreienkamp H-J, Richter D, Meyerhof W (1997) Distinct agonist-mediated endocytosis of cloned rat somatostatin receptor subtypes expressed in insulinoma cells. J Neuroendocrinology 9:741–751

Rossowski WJ, Coy DH (1994) Specific inhibition of rat pancreatic insulin or glucagon release by receptor-selective somatostatin analogs. Biochem Biophys Res Commun 205:341–346

Rossowski WJ, Coy DH (1998) New potent somatostatin recetor 2 antagonist inhibits SRIF, and SRIF-mediated effects on gastric acid and pancreatic secretion in rats. 25th Meeting of the Federation of European Biochemical Societies (FEBS), Abstracts, p 133

Rossowski WJ, Gu ZF, Akarca US, Jensen RT, Coy DH (1994) Characterization of somatostatin receptor subtypes controlling rat gastric acid and pancreatic amylase release. Peptides 15:1421–1424

Roth A, Kreienkamp H-J, Nehring R, Roostermann D, Meyerhof W, Richter D (1997a) Endocytosis of the rat somatostatin receptors: subtype discrimination, ligand specificity, and delineation of carboxy terminal positive and negative sequence motifs. DNA Cell Biol 16:111–119

Roth A, Kreienkamp H-J, Meyerhof W, Richter D (1997b) Phosphorylation of four amino acids in the carboxyl terminus of SSTR3 is crucial for its desensitization and internalization. J Biol Chem 272:23769–23774

Schally AV, Huang WY, Chang RC, Arimura A, Redding TW, Millar RP, Hunkapiller MW, Hood LE (1980) Isolation and structure of pro-somatostatin: a putative somatostatin precursor from pig hypothalamus. Proc Natl Acad Sci USA 1980:4489–4493

Schmechel DE, Vickrey BG, Fitzpatrick D, Elde RP (1984) GABAergic neurons of mammalian cerebral cortex: widespread subclass defined by somatostatin content. Neurosci Lett 47:227–232

Schonbrunn AH, Tasjian AJR (1978) Characterization of functional receptors for somatostatin in rat pituitary cells in culture. J Biol Chem 253:6473–6483

Schwabe W, Brennan MB, Hochgeschwender U (1996) Isolation and characterization of the mouse (Mus musculus) somatostatin receptor type-4-encoding gene (mSSTR4) Gene 168:233–235

Schwartkop C-P, Kreienkamp H-J, Richter D (1999) Agonist-independent internalization and activity of a C-terminally truncated somatostatin receptor subtype 2 (Δ349). J Neurochem, in press

Senaris RM, Humphrey PPA, Emson PC (1994) Distribution of somatostatin receptors 1, 2 and 3 mRNA in rat brain and pituitary. Eur J Neurosci 6:1883–1896

Siehler S, Seuwen K, Hoyer D (1998) [^{125}I]Tyr^{10}-cortistatin14 labels all five somatostatin receptors.Naunyn Schmiedebergs Arch Pharmacol 357:483–489

Somogyi P, Hodgson AJ, Smith AD, Nunzi MG, Gorio A, Wu J-Y (1984) Different populations of GABAergic neurons in the visual cortex and hippocampus of cat contain somatostatin- or cholecystokinin.immunoreactive material. J Neurosci 10:2590–2603

Srikant C, Shen S-H (1996) Octapeptide somatostatin analog SMS 201-995 induces translocation of intracellular PTP1C to membranes in MCF-7 human breast adenocarcinoma cells. Endocrinology 137:3461–3468

Tallent M, Dichter MA, Reisine T (1996a) Somatostatin receptor subtypes SSTR2 and SSTR5 couple to an L-type Ca^{++} current in the pituitary cell line AtT-20. Neuroscience 71: 1073–1081

Tallent M, Dichter MA, Riesine T (1996b) Evidence that a novel somatostatin receptor couples to an inward rectifier potassium current in AtT-20 cells. Neuroscience 73:855–864

Tran VT, Beal MF, Martin JB (1985) Two types of somatostatin receptors differentiated by cyclic somatostatin analogs. Science 228:492–494

van Biesen T, Hawes BE, Luttrell DK, Krueger KM, Touhara K, Porfiri E, Sakaue M, Luttrell LM, Lefkowitz RJ (1995) Receptor-tyrosine-kinase- and Gβγ-mediated MAP kinase activation by a common signalling pathway. Nature 376:781–784

Vanetti M, Kouba M, Wang X, Vogt G, Höllt V (1992) Cloning and functional expression of a novel mouse somatostatin receptor (SSTR2B) FEBS Lett 311: 290

Vanetti M, Vogt G, Höllt V (1993) The two isoforms of the mouse somatostatin (mSSTR2A and SSTR2B) differ in coupling efficiency to adenylate cyclase and in agonist-induced receptor desensitization. FEBS Lett 331:260–266

Viollet C, Lanneau C, Faivre-Baumann A, Zhang J, Djordjijevic D, Loudes C, Gardette R, Kordon C, Epelbaum J (1997) Distinct patterns of expression and physiological effects of sst1 and sst2 receptor subtypes in mouse hypothalamic neurons and astrocytes in culture. J Neurochem 68:2273–2280

Wulfsen I, Meyerhof W, Fehr S, Richter D (1993) Expression patterns of rat somatostatin receptor genes in pre- and postnatal brain and pituitary. J Neurochem 61:1549–1552
Yamada Y, Post SR, Wang K, Tager HS, Bell GI, Seino S (1992a) Cloning and functional characterization of a family of human and mouse somatostatin receptors expressed in brain, gastrointestinal tract and kidney. Proc Natl Acad Sci USA 89:251–255
Yamada Y, Reisine T, Law SF, Ihara Y, Kubota A, Kagimoto S, Seino M, Seino Y, Bell GI, Seino S
(1992b) Somatostatin receptors, an expanding gene family: cloning and functional characterization of human SSTR3, a protein coupled to adenylyl cyclase. Mol Endocrinol 6: 2136–2142
Yasuda K, Rens-Domiano S, Breder CD, Law SF, Saper CB, Reisine T, Bell GI (1992) Cloning of
a novel somatostatin receptor, SSTR3, coupled to adenylylcyclase. J Biol Chem 267:
20422–20428
Zhang J-F, Ellinor PT, Aldrich RW, Tsien RW (1996) Multiple structural elements in voltage-
dependent calcium channels support their inhibition by G-proteins. Neuron 17:991–1003
Zheng H, Bailey A, Jiang MH, Honda K, Chen HY, Trumbauer ME, Van der Ploeg LH, Schaeffer
JM, Leng G, Smith RG (1997) Somatostatin receptor subtype 2 knockout mice are refractory
to growth hormone-negative feedback on arcuate neurons. Mol Endocrinol 11:1709–1717

Novel Neurotransmitters for Sleep and Energy Homeostasis

J. Gregor Sutcliffe and Luis de Lecea

1
Summary

We have developed methodologies for identifying mRNAs with highly restricted expression within the brain. One postnatal-onset mRNA, restricted to sparse GABAergic interneurons of the cerebral cortex and hippocampus, encodes preprocortistatin, the precursor of a 14-residue peptide that shares 11 amino acids with somatostatin. Cortistatin binds to all five cloned somatostatin receptors when they are expressed in transfected cells and depresses neuronal activity, but, unlike somatostatin, it reduces locomotor activity and induces slow-wave sleep. Cortistatin, whose mRNA accumulates during sleep deprivation, apparently acts by antagonizing the effects of acetylcholine on cortical excitability, thereby causing synchronization brain slow waves. A single amino acid difference with somatostatin accounts for the dramatic differences in the effects of the two peptides on physiology and behavior.

A second postnatal-onset mRNA, restricted to 1100 large neuronal cell bodies of the dorsal-lateral hypothalamus, encodes preprohypocretin, the precursor of two peptides that share homology with each other and with members of the secretin peptide family. The peptides are detected immunohistochemically in secretory vesicles at synapses of fibers that project to posterior hypothalamus and diverse targets in other brain regions. The peptides are excitatory when applied to cultured hypothalamic neurons. Recent studies by Sakurai and colleagues (1998) have identified the hypocretin peptides (called the orexins by those workers) as ligands for two orphan receptors at which they stimulate food-intake behavior. Sakurai and collaborators showed that the mRNA for these peptides accumulates during food deprivation. The hypocretin projections suggest additional homeostatic roles for the peptides. These studies suggest the common mechanism of regulation for necessary, but voluntary, behaviors (sleep and feeding) by transcription-based accumulation of peptide transmitters that create a pressure for the voluntary activities.

Department of Molecular Biology, The Scripps Research Institute, 10550 N. Torrey Pines Rd., La Jolla, California 92037, USA

2
Introduction

Our group was among the earliest advocates of using what are now referred to as genomics approaches for identifying novel neural proteins (Milner and Sutcliffe 1983; Sutcliffe et al. 1983). Of the estimated 80000 mammalian genes, at least 30000 are expressed in the brain (Sutcliffe 1988). Some of the most interesting of these are those whose expression is restricted to discrete groups of neurons. The protein products of such genes are likely to have understandable functions highly important to the neurons by which they are expressed. Furthermore, their restricted expression patterns may make such proteins amenable targets for pharmaceutical compounds with minimal unwanted side effects.

Towards recognizing genes whose expression patterns are highly regional within the brain, we have refined the methodology of subtractive hybridization, a methodology that compares two populations of mRNAs, the target and the driver, by hybridization and enriches for RNA species present in the target but not in the driver. One implementation of this methodology, called directional tag PCR subtractive hybridization (Usui et al. 1994), was used in the two studies reviewed here to prepare cDNA libraries enriched for clones of mRNAs specific to either the hippocampus or the hypothalamus.

3
Cortistatin

3.1
A Novel Somatostatin-Like Neuropeptide

From a hippocampal subtraction, we isolated a rat cDNA clone whose sequence predicted it to encode a 112-residue protein with an N-terminal signal sequence for secretion or membrane insertion (de Lecea et al. 1996). Although the new sequence was novel, its C-terminal 14 residues, which were preceded by a pair of lysine residues that could act potentially as a site for proteolytic maturation, shared 11/14 identities with somatostatin-14, including two cysteine residues that are likely to render the peptide cyclic and the FWKT amino acid residues that are critical for somatostatin binding to its receptors (Veber et al. 1979). The new species, which we named preprocortistatin because of its predominant site of expression in the cerebral cortex and its neuronal depressant properties (see below), was the product of a gene distinct from that encoding somatostatin because (1) the alignment of the cortistatin and somatostatin 14mers was permuted by one amino acid (Fig. 1), (2) their cDNA sequences were completely distinct except for the regions encoding the 14mers, and (3) we mapped the mouse gene (gene symbol *Cort*) for cortistatin

K K P C K N F F W K T F S S C K   rodent cst

R R D R M P C R N F F W K T F S S C K   human cst

R K A G C K N F F W K T F T S C   sst

Fig. 1. Alignment of deduced amino acid sequences of the C-terminal portions of precursors of rodent (rat and mouse) cortistatin (*cst*), human cortistatin and somatostatin (*sst*). The tandem basic amino acids that are the presumptive sites of proteolytic cleavage and release from the precursor are included in the alignment. Notice that the human sequence contains three additional N-terminal amino acids and that the somatostatin sequence contains one additional N-terminal residue and is missing the C-terminal residue of cortistatin. Terminal sequences of endogenous peptides have not yet been determined

to chromosome 4, whereas somatostatin is encoded on chromosome 16 (de Lecea et al. 1997b).

We determined the sequence of the mouse homologue of the cortistatin mRNA (de Lecea et al. 1997b). After introduction of two gaps, the mouse and rat nucleotide sequences are 86% identical. The putative mouse translation product contains 108 amino acids. Again, after introduction of two gaps, the putative rat and mouse proteins share 82% identity. The mouse nucleotide sequence corresponding to the cortistatin 14mer and the adjacent lysine doublet that putatively serves as its site of proteolytic release from its precursor are identical to the same region in the rat sequence, thus supporting a functional conservation of the mature peptide. The cDNA sequence upstream from the processing site of the cortistatin 14mer showed several points of divergence, including some resulting in non-conservative amino acid substitutions. Two of these differences obliterate pairs of basic residues in the rat sequence, suggesting that the 14mer is the only active peptide processed from the preprocortistatin.

The human preprocortistatin is 114 amino acids in length. Its putative C-terminal cleavage product is a 17mer (Fig. 1), with 13 identities (including the single proline residue) and a conservative substitution with the rodent 14mer (de Lecea et al. 1997b; Fukusumi et al. 1997). Three additional residues are on the N-terminal end of the peptide. The 17mer peptide may be further processed to a 15mer by proteolytic cleavage at its arginine in position 2. The human gene was assigned by synteny to chromosome 1p36.

3.2
Cortistatin in Cortical Interneurons

By Northern blotting, we detected the 600 nucleotide cortistatin mRNA in brain, but not in any of several peripheral rat or mouse tissues (de Lecea et al. 1996, 1997b). In in situ hybridization studies in the rat (Fig. 2), we detected cortistatin mRNA in sparse neurons throughout the cerebral cortex and hippo-

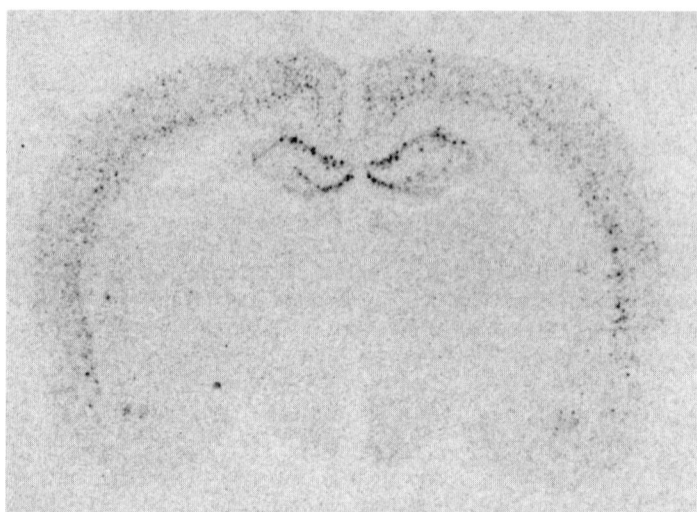

Fig. 2. Detection of cortistatin mRNA in sparse GABAergic interneurons of cerebral cortex and hippocampus by in situ hybridization to a coronal section from rat brain

campus (de Lecea et al. 1996). In the cortex they were especially abundant in layers II-III and VI (de Lecea et al. 1997a). The visual cortex displayed about twice as many cortistatin-positive neurons as other cortical regions, especially in layers II and III.

In the hippocampus, we found cortistatin mRNA expression in a small percentage of non-pyramidal cells in the subiculum and in the stratum oriens of the CA1 and CA3 fields, suggesting that cortistatin mRNA might be present in GABAergic interneurons. Less striking signals were observed in probable GABAergic striatal and olfactory granular neurons. We performed double in situ hybridization for preprocortistatin and GAD65/67, the enzymes that synthesize GABA. Cortistatin co-localized in every instance with either GAD65 or GAD67, demonstrating the GABAergic nature of cortistatin-expressing cells (de Lecea et al. 1997a).

3.3
Partial Coexpression with Somatostatin, Parvalbumin and Calbindin

Cortistatin and somatostatin have distinct, but partially overlapping, distributions in the rat brain, as evidenced by combined in situ hybridization for preprocortistatin mRNA and immunohistochemistry for somatostatin (de Lecea et al. 1997a). Cortistatin was present in a subset of somatostatin-containing neurons in layer VI of the neocortex (approximately 40% colocalization) and stratum oriens in the CA1 field in the hippocampus (approximately 60% colocalization). However, preprocortistatin mRNA and somatostatin rarely co-localized in layer II–III of the cerebral cortex. Approximately 75% of the layer II/III and hippocampal cortistatin interneurons were also positive for parvalbumin, whereas about half of the layer VI cortistatin interneurons were calbindin positive. We detected very few cortistatin-positive neurons in the hilar region of adult rats where somatostatin is highly expressed. These results suggest that preprocortistatin defines a distinct neurochemical subset of cortical interneurons.

3.4
Late Onset and Transient Expression

In Northern blots with mRNA samples prepared from rat brains at different developmental stages, preprocortistatin mRNA appeared very faintly at postnatal day (P) 5 and increased in concentration dramatically between P10 and P15 before declining slightly into adulthood (de Lecea et al. 1997a). By in situ hybridization, preprocortistatin mRNA was detected at P0 mainly in the subiculum and stratum oriens of the hippocampus, with very weak signals in the cortical plate. At P5, cortistatin mRNA was present in some neurons of the CA1 region and layer VI throughout the neocortex. By P10, the intensity of labeling in layer VI and hippocampal CA1 was still weak. At P12, cortistatin labeling extended to layers II–III of all cortical areas, but the signals were particularly strong in the visual/parietal area. The maximal levels of expression were seen at P16-21. At this age, very heavy labeling was observed in all layers of the cortex in all cortical areas. The distribution at this time was different and number of cortistatin mRNA-expressing cells was far greater than in the adult. There was transient expression of cortistatin mRNA in the hilar region of the hippocampus between P10 and P21, which paralleled electrophysiological changes in the response to cortistatin in this region. These findings may indicate that transient expressors turn off transcription of cortistatin.

3.5
Binding to Somatostatin Receptors

The physiological effects of somatostatin are mediated through the action of five different known receptor subtypes, named sst1–5, each belonging to the seven transmembrane G protein-coupled superfamily, and each having a distinct but overlapping distribution in the CNS (Patel et al. 1995). Somatostatin has been shown to depress neuronal activity in the hippocampus, in part by enhancing the M-current, a non-inactivating voltage-dependent potassium current (Moore et al. 1988; Schweitzer et al. 1990). Other pharmacological studies have shown that exogenous application of somatostatin has significant effects on behavior and state of arousal: intracerebroventricular administration of micromolar concentrations of somatostatin 14 causes hypermotility in rats (Vecsei et al. 1984; Bakhit and Swerdlow 1986); electroencephalographic (EEG) analysis of somatostatin-treated rats reveals that the animals spend significantly more time under paradoxical (REM) sleep than control rats (Danguir 1986). We conducted studies to compare synthetic cortistatin 14mer with these known properties of the somatostatin 14mer.

We examined the binding of synthetic cortistatin to the somatostatin receptors present in GH4 pituitary cells, a model system to study somatostatin binding to whole cells. Cortistatin displaced ^{125}I-somatostatin binding to these cells with an estimated K_d indistinguishable from that of somatostatin (de Lecea et al. 1996). We measured the cAMP levels of GH4 cells after stimulation with VIP or TRH and incubation with either somatostatin or cortistatin. The two 14mers acted indistinguishably in inhibiting the accumulation of VIP- and TRH- induced cAMP, indicating that cortistatin acts as an agonist to the somatostatin receptors present in GH4 cells. Although it is assumed that this pituitary cell line expresses the sst1 receptor, it is not clear whether other somatostatin receptor subtypes are present. Fukusumi and collaborators (1997) demonstrated that the human 17mer and the 15mer missing the amino-terminal pair of residues of the human peptide acted as agonists for sst2, sst3, sst4 and sst5 in transfected CHO cells.

The displacement of ^{125}I-somatostatin binding to cell lines transfected with each of the five cloned somatostatin receptors was used to measure the affinity of the human cortistatin 17mer and rodent cortistatin 14mer for these receptors (Fukusumi et al. 1997; Criado et al. 1999). The cortistatins bound to all five receptors with nanomolar affinities similar to those of somatostatin.

As has been previously found for somatostatin (Melacini et al. 1997), cortistatin exhibited no stable secondary structure in solution as measured by circular dichroism or two-dimensional nuclear magnetic resonance (Criado et al. 1999). Therefore, a series of synthetic cyclic peptide analogues with structures based on the well-defined, rigid conformation of the cyclic octapeptide sandostatin were prepared and their affinities for ssts determined. The conserved proline at the N terminus of cortistatin was found to confer specificity for sst3 and sst5 (Criado et al. 1999).

3.6
Neuronal Depressant Activity

We analyzed the effects of the administration of cortistatin on hippocampal physiology and state of arousal (de Lecea et al. 1996). We determined whether cortistatin had somatostatin-like effects on hippocampal neurons by means of current- and voltage-clamp recordings in the hippocampal slice preparation. Superfusion of the cortistatin 14mer, like somatostatin-14, hyperpolarized these neurons in association with inhibition of action potential firing, followed by recovery to control levels upon washout of the peptide. We assessed the effect of cortistatin on the M current (I_M), a non-inactivating voltage-dependent potassium current seen in hippocampal neurons. As had been described for somatostatin, cortistatin superfusion increased the amplitude of the I_M relaxation concomitantly with an outward steady-state current, with recovery to control levels upon washout. These results suggest that cortistatin may act on hippocampal neurons through mechanisms very similar to those of somatostatin, although in these studies cortistatin concentrations were probably saturating.

The inhibitory effects of cortistatin on the excitability of CA1 pyramidal neurons was paralleled in vivo in anesthetized rats (de Lecea et al. 1996). Field potentials were elicited in the CA1 region by stimulation of the commissural pathway. Stimulation of the monosynaptic afferent input to the CA1 region evoked a characteristic population spike (PS) that represents the synchronous firing of pyramidal cells, superimposed on a field synaptic potential waveform. Microiontophoretic application of cortistatin, like somatostatin, significantly decreased PS amplitudes both at half-maximal and maximal levels of stimulation.

3.7
Cortistatin Has Sleep-Promoting Properties

As cortistatin is expressed in interneurons located in the cerebral cortex and hippocampus, we determined its effects on cortical measures of neuronal excitability (de Lecea et al. 1996). We infused synthetic 14mer into the brain ventricles of rats and recorded the electroencephalogram (EEG) for 4 h after peptide injection (Fig. 3). Cortistatin-treated animals demonstrated a clear hypoactive behavior compared to saline- injected rats. In these animals, the EEG showed a dramatic increase in cortical slow waves (1–4 Hz). Polygraphic monitoring of arousal states subsequent to the administration of cortistatin also indicated that rats spent up to 75% of the 4-h recording time in slow-wave sleep compared to 40% in saline-treated control animals. Fukusumi and colleagues (1997) report analogous results with the human 17mer administered to rats. These findings are in clear contrast to the reported enhancement by somatostatin of high frequency EEG referred to as REM sleep. These results support the concept that cortistatin is a sleep-inducing molecule.

3.8
Cortistatin Antagonizes Acetylcholine

To investigate the mechanisms by which cortistatin might facilitate cortical slow waves and promote sleep, we explored the possibility that cortistatin produced these effects, in part, by modulating acetylcholine (ACh) activity. ACh has been shown to be at low concentrations in the cortex during slow-wave sleep, and higher concentrations are associated with wakefulness and REM sleep (Shiromani et al. 1987). We examined the interactions between cortista-

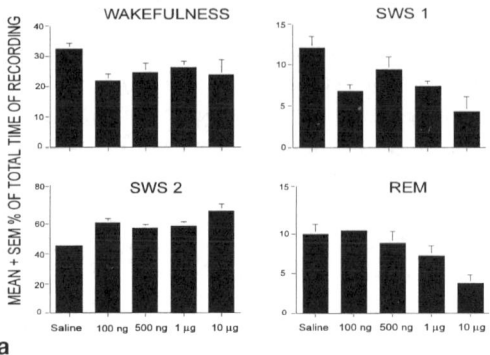

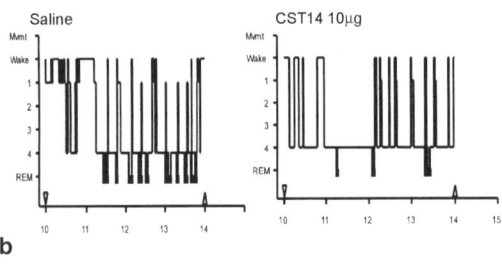

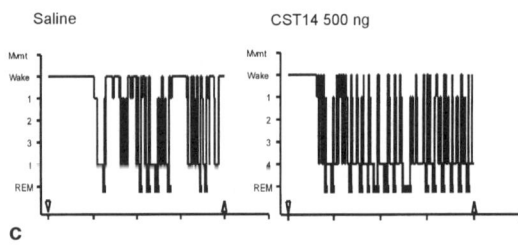

Fig. 3. Effect of introcerebroventricular administration of cortistatin on the sleep-wake cycle of the rat. **a** Synthetic cortistatin 14mer selectively increases the period of slow-wave sleep 2 (*SWS*). **b** Representative hypnographs from EEG recordings of a rat treated with control saline and with 10 μg peptide (*CST14*). At this dose, the peptide enhanced slow-wave activity and decreased REM (paradoxical) sleep. **c** Hypnographs from EEG recordings of a rat under reverse light/dark cycle treated with saline and with 500 ng CST14. The peptide induced both slow-wave and REM sleep in rats that had already experienced their normal physiological amount of sleep. These results support the concept that cortistatin antagonizes the effects of acetylcholine on cortical excitability

tin and ACh on hippocampal CA1 neurons using an evoked paired-pulse (PP) stimulation paradigm to test feed-forward and feed-back inhibitory processes mediated in part by hippocampal interneurons (de Lecea et al. 1996). Stimulation at half-maximal stimulus levels revealed a characteristic biphasic PP response curve. Microiontophoretic application of ACh significantly antagonized the typical inhibitory phase of PP responses seen at interstimulus intervals from 60 to 80 msec. Cortistatin itself had no significant effect on normalized PP responses in CA1, but completely antagonized the effects of ACh. The effect of somatostatin alone on PP responses was similar to or greater than that of ACh, and thus the opposite of the cortistatin effect.

To investigate further whether cortistatin also interacts with cholinergic systems that regulate cortical function, we measured the effects of cortistatin on ACh-induced desynchronization of EEG in the cerebral cortex of anesthetized rats. Slow waves (0.5–4 Hz) dominated the EEG baseline of these rats, whereas the iontophoretic administration of ACh desynchronized the local EEG by increasing the potency of the theta (4.5–8 Hz) and beta (13–20 Hz) bands. This effect was prevented by the simultaneous electrophoretic application of cortistatin (de Lecea et al. 1996). Cortistatin by itself had no effect. These results suggest that somatostatin and cortistatin have different functional properties in the cortex.

The proline-containing rigid peptide analogues mimicked the effects elicited by the cortistatin 14mer itself in these electrophysiological and behavioral assays (Criado et al. 1999). Because sst5 is not expressed in the cortical regions where cortistatin is found and where our recordings were made, these findings implicate sst3 as a potential target receptor for cortistatin's actions, although it is also possible that cortistatin acts through an as yet undiscovered receptor. The analogues provide a starting point for the generation of small molecules that have cortistatin-like activities.

3.9
Cortistatin: A Pressure for Sleep

In rats on a normal light/dark cycle, cortistatin mRNA shows a circadian fluctuation, with highest concentration at circadian time 21 (3 a.m.), late in the active period of the rat (de Lecea et al. 1998a). Rats deprived of sleep for 24 hours by gentle handling showed four-fold increases in cortistatin mRNA concentration. After an 8-h recovery period, mRNA concentrations returned to control levels. Rats deprived only of REM sleep by placement on a small platform over water did not have mRNA concentrations that were elevated over their control (matched rats on a larger platform). Good antibodies are not yet available for measuring cortistatin peptide concentrations, but, assuming that mRNA accumulation and peptide accumulation are coupled, cortistatin appears to serve as an endogenous regulator for sleep whose concentration is increased by transcription- mediated events when the physiological demand for sleep increases (de Lecea et al. 1998a).

3.10
Cortistatin: Not an Alternative Somatostatin

Cortistatin shares several properties with somatostatin: related amino acid sequence, partial coexpression by cortical inhibitory interneurons, functional binding to somatostatin receptors, and hyperpolarization of hippocampal neurons by enhancing the M current. However, cortistatin displays a series of distinct features: while cortistatin is selectively expressed in cells of the cerebral cortex and hippocampus, somatostatin-containing neurons have a wider, and only partially overlapping, distribution; there are subtle differences in the affinities of the two peptides for the five known somatostatin receptors in vitro; and its electrophysiological properties in vivo are different from those of somatostatin. Also, the effects of cortistatin on cortical activity, locomotor activity and sleep are distinct from, in fact contrasting to, those of somatostatin. Moreover, cortistatin antagonizes the effects of ACh on cortical measures of excitability, whereas somatostatin enhances ACh release and potentiates ACh responses. These observations suggest that cortistatin has distinct functional activities and thus is not merely an alternative somatostatin. Cortistatin regulates the onset of sleep by inhibition of the desynchronizing properties of endogenously released ACh. It accumulates as the physiological requirement for sleep increases. This new peptide could play a role in additional processes that depend upon inhibition of cholinergic input to the cortex or hippocampus. In this regard, Flood and colleagues (1997) have demonstrated that cortistatin has an amnestic activity when given to mice 3 min after avoidance training, an effect they speculate to result from inhibition of ACh activity in the hippocampus.

4
The Hypocretins (Orexins)

4.1
Two Hypothalamus-Specific Peptides

We used the directional tag PCR subtraction method to identify 38 rat mRNAs selectively expressed within the hypothalamus (Gautvik et al. 1996). In situ hybridization studies revealed that one of these was expressed exclusively by a bilaterally symmetric structure within the posterior hypothalamus (Fig. 4). Its 569 nucleotide sequence (Sutcliffe et al. 1997; de Lecea et al. 1998b) encoded a 130-residue putative secretory protein (preprohypocretin) with an apparent signal sequence and three additional sites for potential proteolytic maturation. Two of the four putative products of proteolysis had 14 amino acid identities across 20 residues (Fig. 5). This region of one of the peptides contains a 7/7 match with secretin, suggesting that the prepropeptide gave rise to two peptide products that are structurally related both to each other and to secretin. Thus, these peptides were named hypocretin (hcrt) 1 and 2 to reflect their hypothalamic origin and the similarity to secretin. A clone of rat preprohypocretin was subsequently isolated in an independent study by Sakurai and colleagues (1998) and called preproorexin (see below).

The mouse hcrt nucleotide sequence (Sutcliffe et al. 1997; de Lecea et al. 1998b; Sakurai et al. 1998) differed in 39 positions relative to the rat sequence. Of these differences, 19 were within the putative protein-coding region, only 7 of which affected the encoded protein sequence. One difference obliterated a potential proteolytic cleavage site, eliminating two of the four possible rat maturation products. However, hcrt1 and hcrt2 were absolutely preserved between

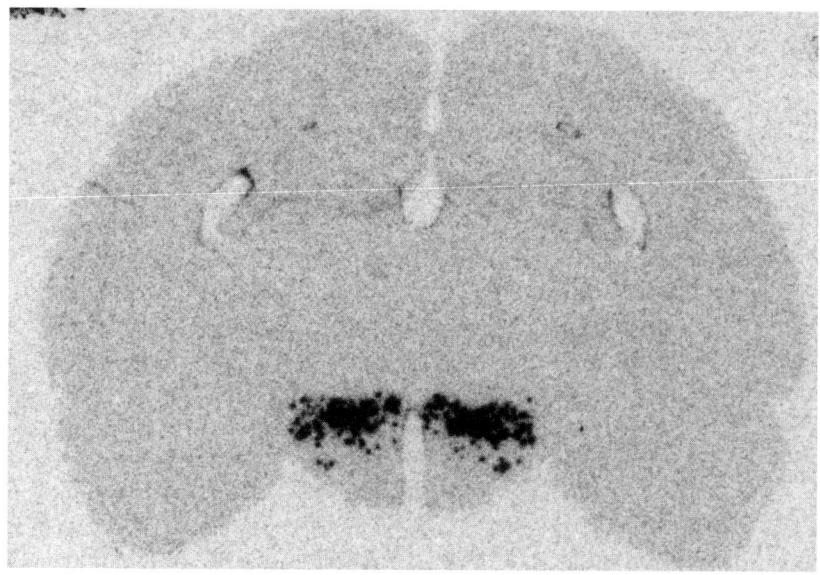

Fig. 4. Detection of hypocretin (orexin) mRNA in large neurons in the dorsal-lateral hypothalamus by in situ hybridization to a coronal section from rat brain

rat and mouse. Both hcrts terminated with glycine residues, which typically are substrates for peptidylglycine alpha-amidating monooxygenase, leaving a C-terminal amide in the mature peptide. We mapped the mouse *Hcrt* gene to chromosome 11, a region that shows conserved synteny with human chromosome 17q21–q24. Indeed, Sakurai et al. assigned the human gene by radiation hybrid mapping to 17q21. These workers isolated the human homologue:

```
consensus:              RL  LL   GNHAAGILT
HCRT1:      QPLPDCCRQKTCSCRLYELLHGAGNHAAGILTL-NH₂
HCRT2:      RPGPPGLQGRLQRLLQANGNHAAGILTM-NH₂
                       *  *******   **  **  *
SECRETIN:   HSDGTFTSELSRLQDSARLQRLLQGLV HSDGTFTSE
PACAP27:                                HSDGIFTSD
GLUCAGON:                               HSQGTFTSD
GLP-1:                                  HAEGTFTSD
```

Fig. 5. Alignment of the two hypocretin peptides with members of the secretin prptide family. Identical residues between hypocretin (*hcrt*) 1 and 2 are indicated above the alignment. Identities between at least one of the hypocretins and at least one of the secretin family members (secretin, PACAP27, glucagon, GLP-1) are indicated by *asterisk*. N-terminal regions of the secretin family members have beem aligned with the C-terminal hypocretin residues to extend the region of identity

hcrt1 is identical among rats, mice and humans; human hcrt2 differs from rodent hcrt2 at two residues.

In Northern blot studies with RNA from brain and different peripheral tissues, we detected the 700-nucleotide hcrt mRNA only in brain samples and within the brain only in hypothalamus. In samples of RNA from brains of developing rats, hcrt mRNA was detected at low concentrations as early as embryonic day 18, but increased in concentration dramatically after the third postnatal week (Sutcliffe et al. 1997; de Lecea et al. 1998b).

4.2
Hypocretin in Cell Bodies, Fibers and Synaptic Vesicles

We raised polyclonal antisera against chemically synthesized peptides corresponding to regions within the rat preprohcrt sequence and to bacterially expressed, histidine-tagged preprohcrt (Peyron et al. 1997, 1998; Sutcliffe et al. 1997; de Lecea et al. 1998b). In Western blots using these antisera and, as target extracts, bacteria expressing the fusion protein, we observed a single prominent immunoreactive band with a migration of approximately 19 kDa with the hyperimmune sera, but not with the preimmune sera. Control extracts contained no immunoreactive targets, confirming the specificity of the antisera for hcrt. Immunohistochemical studies detected prominently granular immunoreactivity within widely spaced, large polymorphic neurons exclusively in the perifornical nucleus, the dorsal, lateral and posterior hypothalamic areas, and the subincertal nucleus at the hypothalamic – thalamic border, coincident with the in situ hybridization-positive cells (Peyron et al. 1997, 1998; de Lecea et al. 1998b). This coincident staining and its elimination when the sera were preincubated with their immunogens, together with the Western blot studies, provided strong evidence for the specificity of the antiserum for Hcrt. Similar results have been obtained by other investigators using these and independently prepared antisera (Risold et al. 1999; Sakurai et al. 1998). The immunoreactive cell bodies have been shown to be the selfsame as those once recognized as cross reacting with an antiserum to ovine prolactin (Risold et al. 1999), and the cross reactivity of the anti-ovine prolactin was blocked with a synthetic peptide from the hypocretin sequence, thus resolving an earlier mystery in the literature.

Approximately 1100 reactive cell bodies were observed between the fornix and the mammillothalamic tracts, 1 mm lateral to the midline, at the level of the median eminence (Peyron et al. 1997, 1998; de Lecea et al. 1998b). The Hcrt neurons span the perifornical nucleus and the magnocellular nucleus of the lateral hypothalamus from the medial hypothalamus across the supra-fornical region at mid-to-posterior hypothalamic levels into the myelinated axons of the retrochiasmic optic radiation. In addition to the hypothalamic neurons, the antisera detected a prominent network of axons located within the posterior hypothalamus and beyond. Fiber projections were observed in apparent terminal fields within septal nuclei in the basal forebrain, the preoptic area, the paraventricular nucleus of the thalamus, the central gray, and the locus coeruleus. Less prominent fiber projections were observed in apparent terminal fields within the colliculi, the laterodorsal tegmental nucleus, and the nucleus of the solitary tract. A complete mapping of these extensive projections from a relatively small number of neurons is given by Peyron and colleagues (1998).

In electron microscopy studies (Sutcliffe et al. 1997; de Lecea et al. 1998b; Peyron et al. 1998), immunoreactivity was observed within cell soma in the lateral hypothalamus on the Golgi apparatus and dense-core granules. Similar vesicles were observed in dendrites, within myelinated axons and at presynaptic terminals opposite non-immunoreactive dendritic shafts.

4.3
Hcrt2 Is Neuroexcitatory

The putative structures of the hypocretins, their expression within the dorsolateral hypothalamus and accumulation within vesicles at axon terminals suggested that they may have intercellular signaling activity. To test this hypothesis, we applied a synthetic peptide corresponding to the amidated form of hcrt2 to rat hypothalamic neurons that had been cultured for 10 days, and recorded postsynaptic currents under voltage clamp (de Lecea et al. 1998b). At 1 µM the peptide evoked a substantial, but reversible, increase in the frequency of post-synaptic currents in 75% of the neurons tested, indicative of an excitatory effect. The other 25% of the cells showed no response to hcrt2. There was little response by hypothalamic neurons that had been in culture for only 3–5 days, suggesting that a certain degree of synaptic maturity was required for the effect. Hcrt2 elicited no response in cultures of synaptically coupled hippocampal dentate granule neurons (cells that do not express hcrt in vivo and where immunoreactive axons are rare), which demonstrated target selectivity and suggested that specific receptors for hcrt2 may exist.

4.4
Endogenous Peptides and Their Receptors

Sakurai and collaborators (1998) prepared transfected cell lines stably expressing each of 50 orphan G protein-coupled receptors. Calcium fluxes were measured in response to fractions from tissue extracts. A peptide from rat brain extracts called orexin A was found to act at one of the receptors, and its sequence matched that of the C-terminally amidated form of hcrt1 with the N-terminal glutamine derivatized as pyroglutamate. There were two intrachain disulfide bonds. A peptide from bovine brain had the same structure. A less active peptide in the rat brain extract, orexin B, matched hcrt2.

The initial orphan receptor, OX_1R, bound hcrt1 (orexin A) with high affinity, but hcrt2 (orexin B) with 100- to 1000-fold lower affinity. However, a related receptor, OX_2R, identified by searching database entries with the OX_1R sequence, had high affinity for both hcrt2 and hcrt1 (Sakurai et al. 1998). The mRNAs encoding the two receptors are both restricted to the brain and enriched in the hypothalamus (unpubl. observ.), but fine anatomical localizations have yet to be determined.

4.5
A Role in Feeding

Stereotactic ablation and physiological studies have previously implicated the dorsal-lateral hypothalamus in several homeostatic processes, including feeding behavior, blood pressure, thermoregulation and arousal. Sakurai and colleagues (1998) demonstrated that intracerebroventricular administration of either peptide increased food consumption in rats. Furthermore, rats fasted for 48 h increased the concentration of hypocretin mRNA by 2.4-fold. Based upon these two sets of observations the clever name, orexin, after the Greek word *orex* for appetite, was proposed for the hypocretins.

4.6
Hypocretins in Many Physiological Systems

The presumptive function of the hypocretins based on the cumulative evidence is as peptide neurotransmitters or neuromodulators. The 1100 hypocretin neurons project widely within the brain, suggesting that the hcrt peptides might affect several systems (de Lecea et al. 1998b; Peyron et al. 1998). Indeed, the finding that hcrt2 is excitatory for 75% of cultured hypothalamic neurons is consistent with a widespread dispatch arising from a limited set of neurons. The location of the neuronal soma and their major projections to the posterior hypothalamus are consistent with one of the roles involving regulation of food intake (Sutcliffe et al. 1997; de Lecea et al. 1998b; Peyron et al. 1998; Sakurai et al. 1998), and the feeding/fasting studies certainly support this notion.

The location of Hcrt-producing cells in the perifornical and lateral hypothalamus and their projections (considered in detail by Peyron et al. 1998) indicate that the peptides may have roles in other homeostatic processes. These cells receive inputs from brainstem areas associated with cardiovascular function, and output to the ventrolateral medulla, the locus coeruleus, the lateral paragigantocellular nucleus, the nucleus of the solitary tract and other areas that are consistent with a role in the regulation of blood pressure and heart rate (Dampney 1994). The projections to the arcuate nucleus suggest a role in the regulation of hormone release. The projections to the raphe magnus and subcoeruleus suggest a role in the regulation of body temperature. The dense projections to the tuberomammillary nucleus suggest involvement in regulation of the sleep-wake cycle.

Thus, at this stage we know that these new peptides function as intercellular messengers, and that they are involved in energy homeostasis. Further studies will be required to evaluate the involvement of these peptides in the several other physiological systems to which they appear to speak. Despite the appeal of the name orexin, which links the peptides with a single aspect of animal behavior, we suggest caution about this unintentionally pejorative term and recommend maintaining the original term, hypocretin, or using hypocretin/orexin until more complete information has been generated.

5
Peptides as Pressures for Voluntary but Necessary Behaviors

The two sets of studies reviewed here began as unrelated lines of investigation into mRNAs with highly restricted patterns of expression within the brain. Surprisingly, common themes have emerged. Both cortistatin and the hypocretins accumulate as the physiological requirement for a particular behavior increases: for cortistatin, sleep; for the hypocretins, feeding. Both of these behaviors are necessary, but they are voluntary in that an animal has considerable flexibility as to when these needs must be satisfied. Each of these peptides is expressed by cells with highly restricted locations. Each appears to be involved in more than a single system, and neither is the only signal for the behavior to which it has been most convincingly linked.

We have learned that cortistatin is involved in sleep, but that it also may function in short-term memory. Other substances have been implicated in sleep regulation: oleamide, delta sleep-inducing peptide, cholecystokinin vasointestinal peptide and some cytokines. Similarly, the hypocretins are involved in feeding but also, as discussed above, probably in several other processes. Furthermore, several additional neuropeptides have been implicated as promoters or inhibitors of food consumption: neuropeptide Y, galanin, CART and melanin-concentrating hormone.

One message that emerges from these two stories is that, to maintain flexibility in acceding to the multiplicity of demands imposed by internal physiology and the external natural and social environments, animals have evolved complex, overlapping neurohormonal signaling systems. One imagines that such overlapping systems allow both attention to individual demands and integration of several demands, some of which may have conflicting solutions. A second message is that we can expect that many additional signaling molecules remain to be found.

Acknowledgments. This work was supported in part by grants from the National Institutes of Health (GM32355, NS33396, MH58543) and Digital Gene Technologies. We thank our many excellent collaborators who contributed to the studies reviewed here.

References

Bakhit C, Swerdlow N (1986) Behavioral changes following central injection of cysteamine in rats. Brain Res 365:159–163

Criado JR, Li H, Jiang X, Spina M, Huitrón-Resendiz S, Liapakis G, Siehler S, Garcia C, Prospero-Garcia O, Henriksen SJ, Koob GF, Reisine T, Igarashi J, Hoyer D, Sutcliffe JG, Goodman M, de Lecea L (1999) Structural and compositional determinants of cortistatin activity. J Neurosci Res (in press)

Dampney RAL (1994) Functional organization of central pathways regulating the cardiovascular system. Physiol Rev 74:323–364

Danguir J (1986) Intracerebroventricular infusion of somatostatin selectively increases paradoxical sleep in rats. Brain Res 367:26–30

de Lecea L, Criado JR, Prospero O, Gautvik KM, Schweitzer P, Danielson P, Dunlop DLM, Siggins GR, Henriksen SJ, Sutcliffe JG (1996) A cortical neuropeptide with neuronal depressant and sleep-modulating properties. Nature 381:242–245

de Lecea L, del Rio JA, Criado JR, Alcantara S, Morales M, Henriksen SJ, Soriano E, Sutcliffe JG (1997a) Cortistatin is expressed in a distinct subset of cortical interneurons. J Neurosci 17:5868–5880

de Lecea L, Ruiz-Lozano P, Danielson PE, Peelle-Kirley J, Foye PE, Frankel, WN, Sutcliffe JG (1997b) Cloning, mRNA expression and chromosomal mapping of human and mouse cortistatin. Genomics 42:499–506

de Lecea L, Criado JR, Calbet M, Danielson PE, Henriksen SJ, Sutcliffe JG, Prospero-Garcia O (1998a) Sleep regulates accumulation of the cortical neuropeptide cortistatin. Soc Neurosci Abstr 24:1429

de Lecea L, Kilduff TS, Peyron C, Gao X-B, Foye PE, Danielson PE, Fukuhara C, Battenberg ELF, Gautvik VT, Bartlett FS, Frankel WN, van den Pol AN, Bloom FE, Gautvik KM, Sutcliffe JG (1998b) The hypocretins: hypothalamus-specific peptides with neuroexcitatory activity. Proc Natl Acad Sci USA 74:5463–5467

Flood JF, Uezu K, Morley JE (1997) The cortical neuropeptide, cortistatin-14, impairs post-training memory processing. Brain Res 775:250–252

Fukusumi S, Kitada C, Takekawa S, Kozawa H, Sakamoto J, Miyamoto M, Hinuma S, Kitano K, Fujino M (1997) Identification and characterization of a novel human cortistatin-like peptide. Biochem Biophys Res Commun 232:157–163

Gautvik KM, de Lecea L, Gautvik VT, Danielson PE, Tranque P, Dopazo A, Bloom FE, Sutcliffe JG (1996) Overview of the most prevalent hypothalamus-specific mRNAs, as identified by directional tag PCR subtraction. Proc Natl Acad Sci USA 93:8733–8738

Melacini G, Zhu Q, Goodman M (1997) Multiconformational NMR analysis of sandostatin (octreotide): equilibrium between beta- sheet and partially helical structures. Biochemistry 36:1233–1241

Milner RJ, Sutcliffe JG (1983) Gene expression in rat brain. Nucleic Acids Res 11:5497–5520

Moore SD, Madamba SG, Joëls M, Siggins GR (1988) Somatostatin augments the M-current in hippocampal neurons. Science 239:278–280

Patel YC, Greenwood MT, Panetta R, Demchyshyn L, Niznik H, Srikant CB (1995) The somatostatin receptor family. Life Sci 57:1249–1265

Peyron C, Tighe DK, Lee BS, de Lecea L, Heller HC, Sutcliffe JG, Kilduff TS (1997) Distribution of immunoreactive neurons and fibers for a hypothalamic neuropeptide precursor related to secretin. Soc Neurosci Abstr 23:2032

Peyron C, Tighe DK, van den Pol, de Lecea L, Heller HC, Sutcliffe JG, Kilduff TS (1998) Neurons containing hypocretin (orexin) project to multiple neuronal systems. J Neurosci 18: 9996–10015

Risold PY, Griffond B, Kilduff TS, Sutcliffe JG, Fellmann D (1998) Preprohypocretin (orexin) and Prolactin-like immunoreactivity are coexpressed by neurons of the rat lateral hypothalamic area. Neurosci Lett 259:153–156

Sakurai T, Amemiya A, Ishii M, Matsuzaki I, Chemelli RM, Tanaka H, Williams SC, Rickson JA, Kozlowski GP, Wilson S, Arch JRS, Buckingham RE, Haynes AC, Carr SA, Annan RS, McNulty DE, Liu W-S, Terrett JA, Elshourbagy NA, Bergsma DJ, Tanagisawa M (1998) Orexins and orexin receptors: A family of hypothalamic neuropeptides and G protein-coupled receptors that regulate feeding behavior. Cell 92:573–585

Schweitzer P, Madamba S, Siggins GR (1990) Arachidonic acid metabolites as mediators of somatostatin-induced increase of neuronal M-current. Nature 346:464–466

Shiromani P, Gillin JC, Henriksen SJ (1987) Acetylcholine and the regulation of REM sleep: Basic mechanisms and clinical implications for affective illness and narcolepsy. Annu Rev Pharmacol Toxicol 27:137–156

Sutcliffe JG (1988) mRNA in the mammalian central nervous system. Annu Rev Neurosci 11:157–198

Sutcliffe JG, Milner RJ, Shinnick TS, Bloom FE (1983) Identifying the protein products of brain-specific genes with antibodies to chemically synthesized peptides. Cell 33:671–682

Sutcliffe JG, Gautvik KM, Kilduff TS, Horn T, Foye PE, Danielson PE, Frankel WN, Bloom FE, de Lecea L (1997) Two novel hypothalamic peptides related to secretin derived from a single neuropeptide precursor. Soc Neurosci Abstr 23:2032

Usui H, Falk J, Dopazo A, de Lecea L, Erlander MG, Sutcliffe JG (1994) Isolation of clones of rat striatum-specific mRNAs by directional tag PCR subtraction. J Neurosci 14:4915–4926

Veber DF, Holly FE, Nutt RF, Bergstrand SJ, Brady SF, Hirschmann R, Glitzer MS, Saperstain R (1979) Highly active cyclic and bicyclic somatostatin analogues of reduced ring size. Nature 280:512–514

Vecsei L, Kiraly C, Bollok I, Nagy A, Varga J, Penke B, Telegdy G (1984) Comparative studies with somatostatin and cysteamine in different behavioral tests on rats. Pharmacol Biochem Behav 21:833–837

Galanin and Galanin Receptors

Tiina P. Iismaa and John Shine

1
Introduction

The neuropeptide galanin was isolated in 1983 and subsequently shown to have a widespread distribution in the central and peripheral nervous system of vertebrate and invertebrate species. Further studies have demonstrated that galanin has a broad range of neuroendocrine and physiological actions, some of which are species-specific. These include modulation of pituitary and glucoregulatory hormone secretion, inhibition of the release of neurotransmitters that may play a role in memory acquisition or contribute to anoxic damage in the brain, modulation of appetite and sexual behaviour, as well as effects on pain, gastrointestinal motility, heart rate and blood pressure. Galanin is also a potent growth factor for developing pituitary cells. It is expressed in a variety of tumour types and cancer cell lines, and can act either as a mitogenic agent in certain types of cancer or as an inhibitory factor in carcinogenesis.

Despite demonstration of significant and potent biological activity, the physiological role of galanin in developmental, neuromodulatory and neuroendocrine processes is only now beginning to be addressed, with the molecular cloning of genes encoding galanin and a number of galanin receptor subtypes. The molecular tools now available provide the means for assessment of sites and regulation of expression, design and screening for receptor subtype-specific agonists and antagonists, and the application of gene knockout approaches to elucidate the functional significance of galaninergic neurotransmission and receptor activity in normal and pathological situations.

Neurobiology Program, The Garvan Institute of Medical Research, 384 Victoria St., Sydney New South Wales 2010, Australia

2
Galanin

2.1
Peptide Sequence

Galanin was originally isolated from porcine intestine as a 29 amino acid C-terminally amidated peptide, and was named for its N-terminal Gly and C-terminal Ala residues (Tatemoto et al. 1983). The peptide has subsequently been shown to occur in a number of mammalian species, including humans, as well as in birds, reptiles, amphibia and fish (Fathi et al. 1998b), and galanin-like immunoreactivity has been observed in invertebrates including the mollusc, sea cucumber and blowfly (Roberts et al. 1989; Lundquist et al. 1991; Díaz-Miranda et al. 1996).

The sequence of galanin has been reported for 15 vertebrate species to date (Fig. 1a). In all species for which the mature peptide sequence is known, galanin is a 29 amino acid C-terminally amidated peptide, except for the human peptide which is 30 amino acids in length and, unlike all other galanins and unusual for bioactive peptides, is not C-terminally amidated. Amino acids 1–14 are conserved across species, except for the single substitution in tuna galanin of Ala for Ser[6]. Galanin has no significant sequence homology with any other known mammalian peptides or families of peptides, but the sequence of the N-terminal segment of galanin does show some conservation with myoinhibitory nonapeptides isolated from neuronal cell extracts of the locust (*Locusta migratoria*; Schoofs et al. 1991) and the tobacco hornworm (*Manduca sexta*; Blackburn et al. 1995; Fig. 1b).

The mature galanin peptide is proteolytically processed from a precursor peptide, designated preprogalanin, which comprises a hydrophobic signal peptide (von Heijne 1983) of 23 amino acids, a propeptide of nine amino acids, followed by galanin itself, and a 59–60 amino acid galanin message-associated peptide (GMAP; Fig. 2). Within preprogalanin, the galanin sequence is flanked by dibasic Lys-Arg residues, consistent with cleavage of galanin from the precursor by trypsin-like endoprotease activity. In all species except human, the last amino acid of galanin in the precursor peptide is Gly, which serves as substrate for the C-terminal amidating enzyme peptidyl-glycine α-amidating monooxygenase (Eipper et al. 1992). This enzyme catalyses post-translational modification of the galanin peptide, with transfer of the peptide backbone amino group of Gly to the penultimate residue and generation of a 29 amino acid C-terminally α-amidated mature peptide.

The sequence of GMAP exhibits overall conservation of the acidic nature of the peptide between species and high homology in the C-terminal segment of the peptide, extending from Arg[88] of human preprogalanin (Fig. 2).

(a)

```
Human          GWTLNSAGYLLGPHAVGNHRSFSDKNGLTS
Rat, Mouse     ---------------ID--------H---(amide)
Pig            ---------------ID-----H--Y--A(amide)
Cow            --------------LDS----Q--H--A(amide)
Sheep          ---------------ID-----H--H--A(amide)
Dog            ---------------ID-----HE-P---(amide)
Chicken, Quail ----------------D-----N--H-F-(amide)
Alligator      ---------------ID-----NE-H-IA(amide)
Frog           ---------------ID-----N--H--A(amide)
Bowfin         ----------------D----LN--H--A(amide)
Dogfish        ----------------D---
Trout          -------------GIDG--TL---H--A(amide)
Tuna           -----A--------GIDG--TLG--P--A(amide)
```

(b)

```
Human galanin  GWT LNSAGYLLGPHAVGNHRSFSDKNGLTS
Tuna galanin   --- --A--------GIDG--TLG--P--A(amide)
LOM-MIP        A-QD--A-W(amide)
Mas-MIP I      A-QD----W(amide)
Mas-MIP II     --QD----W(amide)
```

Fig. 1. (a) Amino acid sequences of endogenously occurring galanin peptides. Dashes represent identity with human galanin sequence. The sequence of dogfish galanin is known only for amino acid residues 1–20. (b) Comparison of amino acid sequence of human and tuna galanins with myoinhibitory peptides LOM-MIP, isolated from the locust, *Locusta migratoria*, and Mas-MIP I and Mas-MIP II, isolated from the tobacco hornworm, *Manduca sexta*. Dashes represent identity with human galanin sequence. A gap has been introduced into human and tuna galanin sequences to maximise homology in the alignment

Differential regulation of expression of galanin and GMAP peptides has been reported in vertebrate and invertebrate species (Bretherton-Watt et al. 1990; Hökfelt et al. 1992; Lundquist et al. 1992), raising the possibility of discrete biological function(s) attributable to GMAP. GMAP and fragments of GMAP have recently been shown to have biological activity in blocking spinal cord hyperexcitability and inhibiting adenylyl cyclase activity in rat insulinoma cells (X.-J. Xu et al. 1996; Andell-Jonsson et al. 1997; Andell-Jonsson and Bartfai 1998), but the nature of the receptor(s) mediating the effects of fragments of GMAP remains to be defined.

```
       1                                                          32
Human  MARGSALLLASLLLAAALSASAGLWSPA - KEKR
Rat    MARGSVILLAWLLLVATLSATLGLGMPT - KEKR
Mouse  MARGSVILLGWLLLVVTLSATLGLGMPT - KEKR
Pig    MPRGCALLLASLLLASALSATLGLGSPVVKEKR
Cow    MPRGSVLLLASLLLAAALSATLGLGSPV - KEKR
Dog                                      KEKR
```

MATURE GALANIN

```
       33                                                         62
Human  GWTLNSAGYLLGPHAVGNHRSFSDKNGLTS
Rat    GWTLNSAGYLLGPHAIDNHRSFSDKHGLT -
Mouse  GWTLNSAGYLLGPHAIDNHRSFSDKHGLT -
Pig    GWTLNSAGYLLGPHAIDNHRSFHDKYGLA -
Cow    GWTLNSAGYLLGPHALDSHRSFQDKHGLA -
Dog    GWTLNSAGYLLGPHAIDNHRSFHEKPGLT -
```

GALANIN MESSAGE-ASSOCIATED PEPTIDE (GMAP)

```
       63                                                         92
Human  - KRELRPE - DDMKPGSFDRSI - - PENNIMRTIIE
Rat    GKRELPLEVEEGRLGSVAVPL - - PESNIVRTIME
Mouse  GKRELQLEVEERRPGSVDVPL - - PESNIVRTIME
Pig    GKRELEPE - DEARPGGFDRLQ - - SEDKAIRTIME
Cow    GKRELEPE - DEARPGSFDRPL - - AENNVVRTIIE
Dog    GKRELPPE - DEGRSGGFAGPLSLSENAAVRTIME
```

```
       93                                                         123
Human  FLSFLHLKEAGALDRLLDLPAAASSEDIERS
Rat    FLSFLHLKEAGALDSLPGIPLATSSEDLEQS
Mouse  FLSFLHLKGYRALDSLPGIPLATSSEDLEKS
Pig    FLAFLHLKEAGALGRLPGLPSAASSEDAGQS
Cow    FLYFLHLKDAGALERLPSLPTAESACDAERS
Dog    FL
```

Fig. 2. Comparison of amino acid sequence of galanin and deduced amino acid sequence of preprogalanin in human, rat, mouse, pig, cow and dog (Evans and Shine 1991; Vrontakis et al. 1987; Kofler et al. 1996; Rökaeus and Brownstein 1986; Rökaeus and Carlquist 1988; Boyle et al. 1994). Only partial sequence of dog preprogalanin has been described to date. Dashes indicate gaps introduced into the alignment to maximise homology. Residues conserved in at least three of the aligned sequences are identified by shading. Numbering of residues is given for human preprogalanin

2.2
Preprogalanin Gene Structure and Chromosomal Localisation

The gene encoding preprogalanin, which has been isolated from human (*GALN*; Evans et al. 1993a; Kofler et al. 1995), mouse (*Galn*; Kofler et al. 1996) and rat (Kaplan et al. 1991; Corness et al. 1997), and as a partial clone from the cow genome (Rökaeus and Waschek 1994), exhibits conservation of exon:-

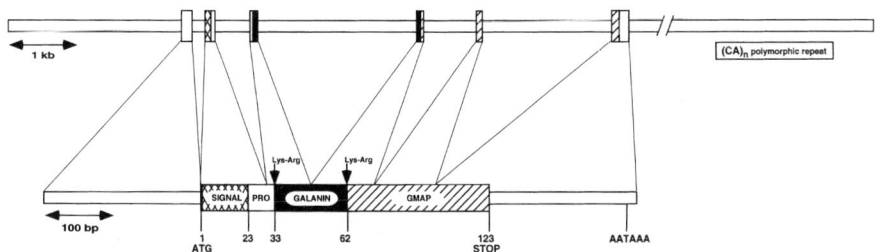

Fig. 3. Structural organisation of the human preprogalanin gene. The six exons comprising the gene are shown relative to segments of the mature cDNA transcript they encode: signal peptide, propeptide (PRO), mature galanin and galanin message-associated peptide (GMAP). Locations of the transcription start site (185 nucleotides upstream of the translation initiation codon; Kofler et al. 1995) and the polyadenylation signal (AATAAA, 172 nucleotides downstream of the translation termination codon; Evans and Shine 1991) in the cDNA transcript are indicated, as is the dibasic cleavage sites (Lys-Arg) flanking the mature galanin peptide sequence. Numbers refer to amino acids in the preprogalanin peptide. Existence of a $(CA)_n$ polymorphic repeat region approximately 10–15 kb downstream of the translation initiation codon (Kofler et al. 1998) is also indicated

intron organisation between species (Fig. 3). The gene spans approximately 6.5 kb in human, 4.5 kb in mouse and 5 kb in rat, and comprises six exons. Exon 1 encodes the 5′-untranslated region of the transcript, which ranges from approximately 150 to 190 nucleotides in length in different species. The second exon starts with the translation initiation codon of the signal sequence and ends before the dibasic cleavage site (Lys-Arg) preceding galanin. The dibasic cleavage site and the first 13 amino acids of galanin, which is the segment of the peptide conserved across most species, are encoded on exon 3. The remaining amino acids of galanin and most of GMAP are encoded on exons 4 and 5, while exon 6 encodes the C-terminal portion of GMAP and the 3′-untranslated region of the mRNA transcript.

The preprogalanin gene occurs as a single copy in the genome of normal diploid cells and has been localised to human chromosome 11q13.3-13.5 (Evans et al. 1993a) and mouse proximal chromosome 19 (Guida et al. 1998). This is in the vicinity of the breakpoint for translocation of 11q13 to the immunoglobulin heavy chain locus at 14q23 in chronic lymphocytic lymphoma, diffuse B-cell lymphomas and multiple myeloma (Croce 1986; Raffeld and Jaffe 1991), and to the gene encoding parathyroid hormone at 11p15 in benign parathyroid adenoma (Arnold et al. 1989; Rosenberg et al. 1991). In addition, the 11q13 region of the human genome contains a number of genes, such as those encoding cyclin D1 (*CCND1*) and fibroblast growth factors 3 (*FGF3*) and 4 (*FGF4*), whose activation or overexpression is tumourigenic. This region of the genome has been observed to be amplified in a proportion of bladder cancers, breast cancers, squamous cell carcinomas and in small cell lung cancer (Ormandy et al. 1998b).

2.3
Sites and Regulation of Galanin Expression

As determined by Northern blot analysis, *in situ* hybridisation, immunohisto-chemistry, and radioimmunoassay, galanin has been shown to be widely distributed in the central and peripheral nervous system. Within the central nervous system (CNS), preprogalanin mRNA is expressed at levels detectable by Northern hybridisation in the cerebral cortex, hippocampus, cerebellum, medulla and spinal cord, as well as in the amygdala, thalamus, sub-thalamic nuclei and substantia nigra (T.P. Iismaa, unpubl.). High levels of expression are also observed in the hypothalamus, median eminence and pituitary (Merchenthaler et al. 1993; Evans et al. 1993b; Crawley 1995). In the periphery, galanin is expressed in the heart and respiratory system, gastrointestinal, uri-nary and genital tracts, liver, pancreas, thyroid and adrenal glands, eye, skin and muscle (Rökaeus 1994).

2.3.1
Developmental Expression and Sexual Dimorphism

Expression of galanin in the rat has been reported as early as embryonic day 14 (E14) in the brain, gastroinestinal tract, developing sensory ganglia and in the eye, and in sensory epithelia in the developing ear and nose by E15. Its expression in fetal brain, gastrointestinal tract, ear and eye tissue is maximal at E15–E17 and reduced by E20, just prior to birth. Galanin is expressed at E19 in developing bone tissue in cells that are thought to represent osteoclasts (Gabriel et al. 1989; Z.-Q. Xu et al. 1996a). Galanin expression in the mammal-ian CNS increases progressively during postnatal growth and development (Sizer et al. 1990; Elmquist et al. 1992; Del Fiacco and Quartu 1994; Giorgi et al. 1995), with a temporal and spatial profile of expression of galanin and its receptors suggesting involvement of galaninergic neurotransmission in post-natal neurogenesis and pubertal development (Elmquist et al. 1993; Planas et al. 1994, 1995). Gender-specific differences, consistent with a role for sex ster-oids in regulation of the expression of galanin (Selvais et al. 1995; Rossmanith et al. 1996; Shen et al. 1998), are observed in adult rat CNS, including the pitui-tary gland and in hypothalamic nuclei involved in regulation of secretion of growth hormone (GH) and reproductive hormones (Delemarre-van de Waal et al. 1994; Cheung et al. 1996).

2.3.2
Transcriptional Regulation

The rat preprogalanin transcript was originally identified in the rat pituitary gland as an estrogen-inducible mRNA (Vrontakis et al. 1987), and has subse-quently been shown to be expressed in somatotrophs and thyrotrophs and to be upregulated in lactotrophs in response to estrogen (Steel et al. 1989; Hsu et al. 1990; O'Halloran et al. 1990; Hyde et al. 1991; Kaplan et al. 1991). Both basal and estrogen-induced expression of galanin in the rat pituitary is dependent on thyroid hormone (O'Halloran et al. 1990; Hooi et al. 1997). Its secretion from the pituitary is under both neuronal and endocrine control, being stimulated by adrenocorticotropic hormone (ACTH; Calvo et al. 1990), thyrotropin-releasing hormone (TRH; Hyde and Keller 1991) and growth hormone-releasing hormone (GHRH; Hemmer and Hyde 1992), and inhibited by dopamine (Hyde et al. 1992), somatostatin (Hyde and Howard 1992) and prolactin (Hammond et al. 1997). Estrogen responsiveness of the human preprogalanin gene has also been reported recently (Howard et al. 1997b; Ormandy et al. 1998a), but interestingly, lactotrophs obtained from either estrogen-treated men or pregnant women do not exhibit galanin-like immunoreactivity, and expression of galanin in human pituitary is restricted to corticotrophs (Vrontakis et al. 1990; Hsu et al. 1991). The functional significance of differences in cell-specific expression of galanin within human and rat pituitary is currently unknown.

The expression of galanin in nerve cell bodies is upregulated in response to nerve injury, by a mechanism that involves a stimulatory effect of leukemia inhibitory factor (LIF; Rao et al. 1993; Corness et al. 1996; Sun and Zigmond 1996), and potential modulation by target-derived growth factors, such as nerve growth factor (NGF; Corness et al. 1998). NGF-responsiveness of the rat preprogalanin promoter has been demonstrated *in vitro* in PC12 cells (Kaplan et al. 1991), and *in vivo* in rat basal forebrain neurons (Planas et al. 1997), while an inhibitory effect of NGF on expression of preprogalanin mRNA has been observed in cultured rat dorsal root ganglion (DRG) neurons (Corness et al. 1998). A role for activation of protein kinases C and A in transcription of bovine and rat preprogalanin genes is suggested by phorbol ester and forskolin stimulation of galanin gene expression in *in vitro* analyses (Rökaeus et al. 1990; Anouar et al. 1994; Rökaeus and Waschek 1994; Corness et al. 1997). The existence of a negative modulator, or silencer element, which may be of importance in temporal or tissue-specific regulation of expression, has been reported in bovine, human and rat preprogalanin genes (Rökaeus and Waschek 1994; Kofler et al. 1995; Corness et al. 1997), while a strong basal enhancer has also been identified in the rat preprogalanin promoter (Hooi et al. 1997).

3
Galanin Receptors

The actions of galanin are mediated by interaction with specific cell surface receptors that are members of the G protein-coupled receptor superfamily. The use of truncated forms and analogs of galanin, together with chimeric peptides that behave as agonists or antagonists in different physiological assays (Bartfai et al. 1992, 1993a; Kask et al. 1995), initially provided pharmacological evidence for the existence of galanin receptor subtypes and to date, three galanin receptor subtypes have been cloned.

3.1
Cloning, Pharmacology, Signalling Properties and Sites of Expression

Since the initial molecular cloning of a galanin receptor (GALR1) by an expression cloning strategy in 1994, DNA sequences encoding two other galanin receptor subtypes (GALR2 and GALR3) have been identified. Although all three receptor subtypes bind galanin with similar affinity, they only share amino acid identities of between 36 and 54% (Fig. 4). Such diversity of primary sequence has greatly facilitated the synthesis of receptor subtype-specific probes to evaluate the localisation of each subtype throughout the nervous system and promises to form a basis for future development of subtype-specific agonists and antagonists.

3.1.1
GALR1

Human Bowes melanoma (HMCB) cells were the source of mRNA for the cloning of the first high affinity galanin receptor, GALR1, which was identified by expression of cloned cDNA in COS cells (Habert-Ortoli et al. 1994). This initial report was soon followed by the subsequent cloning of cDNA sequences encoding GALR1 from rat Rin14B insulinoma cells (Parker et al. 1995) and from rat (Burgevin et al. 1995) and mouse brain (Wang et al. 1997a). Human,

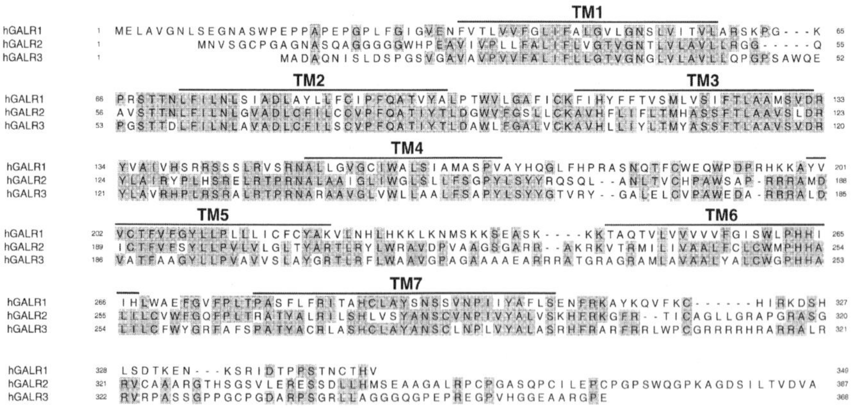

Fig. 4. Alignment of human GALR1, GALR2 and GALR3 galanin receptor amino acid sequences. Identical residues are shaded and putative transmembrane (TM) domains are indicated by overlining

rat and mouse GALR1 proteins comprise 349, 348 and 346 residues, respectively, are highly homologous ($\geq 92\%$ overall identity) and belong to the rhodopsin-like subfamily of G protein-coupled receptors (Iismaa et al. 1995). The genes encoding GALR1 in human (*GALNR1*) and mouse (*Galnr1*) comprise three exons, with a highly conserved exon:intron organisation (Jacoby et al. 1997; Wang et al. 1997a).

The availability of cloned GALR1 DNA sequences has facilitated examination of the pharmacology of galanin binding to this receptor subtype in both transiently and stably transfected mammalian cells (Habert-Ortoli et al. 1994; Burgevin et al. 1995; Parker et al. 1995; Sullivan et al. 1997; Wang et al. 1997a; Fathi et al. 1998a; Wang et al. 1998). Galanin binds to cloned GALR1 with high affinity (Kd = 0.02–0.8 nM), with the first two N-terminal residues of galanin required for high affinity binding to this receptor subtype. The rank order of potency for galanin peptides and analogs is [galanin(1–29) > galanin(1–16) >> galanin(2–29) >>> [D-Trp^2]galanin(1–29) $\cong$ galanin(3–29)]. The pharmacology of human galanin receptors endogenously expressed in Bowes melanoma (Heuillet et al. 1994) and CHP-12 neuroblastoma cells (Sullivan et al. 1997) is comparable to that of the cloned human and rat GALR1. Activation of heterologously expressed human and rat GALR1 by galanin causes a concentration-dependent decrease in forskolin-stimulated cyclic AMP (cAMP) accumulation. This response is sensitive to pertussis toxin and suggests the coupling of GALR1 to inhibitory G$_i$ proteins (Fig. 5). Signal transduction studies undertaken in native tissues as well as in cell lines endogenously expressing galanin receptors also demonstrate an inhibition of cAMP accumulation by galanin. These studies have been carried out in rat brain (Chen et al. 1992), hypothalamic and entorhinal cortical (Billecocq et al. 1994) and hippocampal (Valkna et al. 1995) membranes, as well as in rat pancreatic RINm5F (Amiranoff et al. 1988) and RIN14B (Amiranoff et al. 1991) cells and human Bowes melanoma cells (Heuillet et al. 1994). Heterologously expressed rat GALR1 has also recently been reported to mediate pertussis toxin-sensitive G$_{i\beta\gamma}$ signalling to activate mitogen-activated protein kinase (MAPK).

In situ hybridisation analysis of expression of GALR1 mRNA in rat CNS has demonstrated a widespread but highly specific distribution (Burgevin et al. 1995; Parker et al. 1995; Chan et al. 1996; Gustafson et al. 1996; Z.-Q. Xu et al. 1996b; Mitchell et al. 1997). Consistent with a role in many of the central effects of galanin, including cognition, feeding, neuroendocrine function and modulation of sensory processing, highest levels of GALR1 expression are seen throughout the basal forebrain, including the hypothalamus, especially in the medial preoptic area, paraventricular and supraoptic nuclei, ventral hippocampus and amygdala, especially in the bed nucleus of the accessory olfactory tract. Additionally, significant expression of GALR1 is also seen in the brain stem and dorsal horn of the spinal cord, and GALR1 is expressed in rat and human fetal brain and in sensory ganglia (Habert-Ortoli et al. 1994; Parker et al. 1995; Z.-Q. Xu et al. 1996a). Northern hybridisation of human and rodent

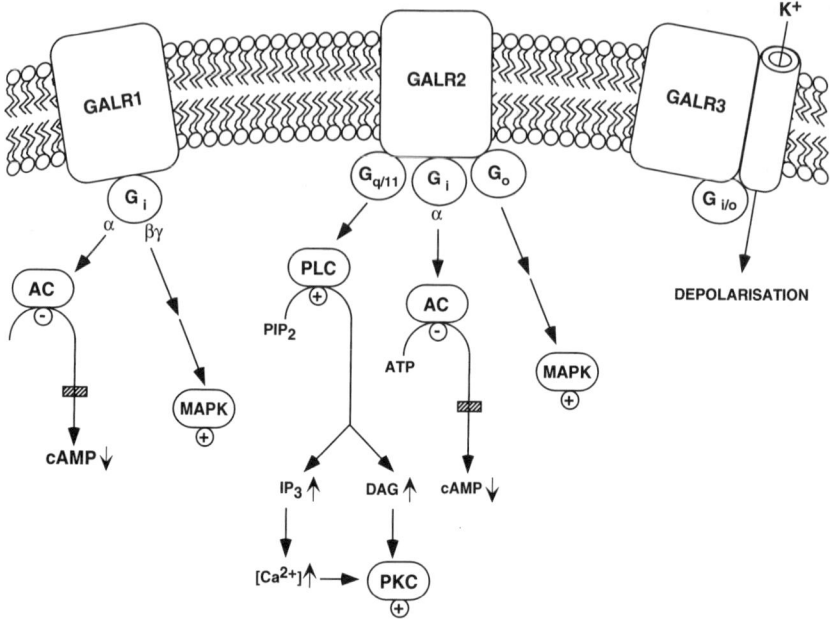

Fig. 5. G protein coupling and intracellular signalling of cloned galanin receptor subtypes. GALR1 couples through $G_{i\alpha}$ to inhibit adenylyl cyclase (AC) activity and reduce intracellular cyclic AMP (cAMP) levels, and through $G_{i\beta\gamma}$ and intermediates that have not yet been defined to activate mitogen-activated protein kinase (MAPK). Coupling of GALR2 is primarily through $G_{q/11}$ to activate phospholipase C (PLC), causing hydrolysis of phosphatidyl inositol 4,5-bisphosphate (PIP_2) to generate inositol 1,4,5-trisphosphate (IP_3) and diacylglycerol (DAG). Elevation of intracellular IP_3 levels results in mobilisation of Ca^{2+} from intracellular stores. Both DAG and Ca^{2+} may activate protein kinase C (PKC). GALR2 also couples through $G_{i\alpha}$ to inhibit AC activity and reduce cAMP levels, and through G_o and intermediates that have not yet been defined to activate MAPK. GALR3 couples to $G_{i/o}$ and in *Xenopus* oocytes activates co-transfected G protein-coupled inwardly rectifying K^+ channel subunits GIRK1 and GIRK4 to cause depolarisation of the cell

GALR1 (Parker et al. 1995; Sullivan et al. 1997; Wang et al. 1997a) has yielded contradictory data with respect to sites of expression of GALR1 in the periphery, with both widespread and restricted distribution having been reported. However, the use of reverse transcriptase-polymerase chain reaction (RT-PCR) analysis indicates expression of GALR1 mRNA in human gastrointestinal tract, heart, lung, kidney and testis (Lorimer and Benya 1996; Fathi et al. 1998a), rat pancreatic islets, bladder and uterus, and mouse placenta (T.P. Iismaa, unpubl.), and allows detection of GALR1 mRNA in human pituitary, but not rat pituitary (Fathi et al. 1997, 1998a). A summary of the results of localisation studies conducted to date is given in Table 1.

Table 1. Tissue distribution of galanin receptor subtypes. Detection of expression of human (h) or rat (r) GALR1, GALR2 or GALR3 by Northern hybridisation, ribonuclease protection, RT-PCR analysis or in situ hybridisation is denoted by plus signs, lack of signal is denoted by minus signs

	hGALR1	rGALR1	hGALR2	rGALR2	rGALR3
Brain	+	+	+	+	+
Hippocampus	+	+	+	+	−
Cerebellum	−	−	+	+	+
Hypothalamus	+	+		+	+
Pituitary	+	−	+	+	+
Spinal cord	+	+	−		+
Neuroblastoma	+				
Heart	+	+	+	+	+
Kidney	+	+	+	+	+
Liver	+	+	−	+	+
Lung	+	+	+	+	+
Small cell lung cancer (NCI-H69)	+	+			
Skeletal muscle	+	+	−	+	−
Stomach	+	+	+	+	+
Small intestine	+	+	+	+	
Large intestine		+		+	−
Colon	+		+		
Pancreas	+	+	−	+	+
Spleen	+	−	−	+	+
Thymus	+		−		
Leukocytes	−		−		
Adrenal		−			+
Prostate	+	+	−	+	+
Testis	+	+	+	+	+
Vas deferens		+		+	−
Ovary	+	+	−	+	+
Uterus	+	+	+	+	−
Placenta	+		−		

This table was collated from the following sources: Burgevin et al. (1995); Parker et al. (1995); Gustafson, et al. (1996); Lorimer and Benya, 1996; Fathi et al. (1997, 1998a); Howard et al. (1997a); Smith et al. (1997, 1998); Sullivan et al. (1997); Wang et al. (1997b,c); Bloomquist et al. (1998); Klaus et al. (1998); T.P. Iismaa (unpubl.)

3.1.2
GALR2

The second galanin receptor subtype, GALR2, has 40% overall amino acid identity with GALR1 (Fig. 4) and exhibits a distinct genomic structure, with the GALR2 coding sequence being interrupted by a single intron. Such a divergent primary sequence and gene structure suggest that GALR1 and GALR2 may have evolved independently, although the pharmacological properties of the two receptor subtypes, when expressed heterologously, are highly similar (Fathi et al. 1997, 1998a; Howard et al. 1997a; Smith et al. 1997; Wang et al. 1997b, 1998; Bloomquist et al. 1998). As with GALR1, a high degree of primary sequence identity (87%) exists between human and rat GALR2 sequences, which comprise 387 and 372 residues, respectively. Human GALR2 contains an additional 15 amino acids in the C-terminal tail of the receptor protein. GALR2 binds galanin with high affinity (Kd = 0.12–0.59 nM). The rank order of potency for galanin peptides and analogs is [galanin(1–29) ≅ galanin(2–29) > galanin(1–16) > [D-Trp^2]galanin(1–29) >>> galanin(3–29)], with GALR2 pharmacologically distinguishable from GALR1 based on significantly higher affinity for galanin(2–29) and [D-Trp^2]galanin(1–29). GALR2 stably expressed in transfected cell lines couples primarily through $G_{q/11}$ to pertussis toxin-insensitive activation of phospholipase C, causing elevation of intracellular inositol phosphate and Ca^{2+} levels (Fig. 5). This is consistent with demonstration of galanin-mediated phosphatidylinositol hydrolysis by endogenously expressed galanin receptors in human small cell lung carcinoma cells (Sethi and Rozengurt 1991). GALR2 also couples to inhibition of forskolin-stimulated cAMP accumulation through G_i and to mitogenic signalling pathways by G_o-mediated activation of MAPK.

Expression of GALR2 mRNA has been examined using a range of approaches including Northern hybridisation, ribonuclease protection and in situ hybridisation analysis (Fathi et al. 1997, 1998a; Howard et al. 1997a; Shi et al. 1997; Smith et al. 1997; Wang et al. 1997b; Bloomquist et al. 1998). In addition to expression in rat brain, spinal cord and anterior pituitary, GALR2 mRNA is widely expressed in peripheral tissues including heart, lung, kidney, liver, spleen, skeletal muscle, testis, uterus, stomach and small and large intestine (Table 1). Centrally, the regional distribution of GALR2 mRNA overlaps yet is distinctive from that of GALR1 mRNA, being detected in the cerebral cortex, hippocampus, hypothalamus, amygdala, thalamus, cerebellum and spinal cord. Such overlap in the pattern of expression of GALR1 and GALR2 mRNA, especially in the hippocampus, hypothalamus, amygdala and spinal cord, indicates that either or both receptor subtypes may be responsible for the central effects of galanin on cognition, feeding behaviour, neuroendocrine functions and sensory modulation.

3.1.3
GALR3

The third galanin receptor subtype, GALR3, shares 36 and 58% overall amino acid identity with human GALR1 and GALR2 sequences (Fig. 4), and exhibits a genomic organisation identical to that of GALR2 (Wang et al. 1997c; Iismaa et al. 1998; Smith et al. 1998). Human and rat GALR3 sequences comprise 368 and 370 residues, respectively, and share 92% identity. When expressed heterologously, they mediate saturable, high affinity binding of galanin (Kd =

0.55–2.23 nM), with preliminary binding experiments indicating that the rank order of potency for galanin peptides is [galanin(1-29) > galanin(2-29) > galanin(1-16) >>> galanin(3-29)]. Rat GALR3 may be distinguished from rat GALR1 by a higher affinity for galanin(2-29) and from rat GALR2 by a lower affinity for galanin(1-16). Galanin(3-29) is inactive at GALR3, as it is at the other two receptor subtypes. Functional coupling of human and rat GALR3 has been observed in *Xenopus* oocytes co-expressing G protein-coupled inwardly rectifying K^+ channel subunits GIRK1 and GIRK4, with pertussis toxin sensitivity of the response suggesting coupling through $G_{i/o}$ proteins (Fig. 5). Rat GALR3 mRNA was originally cloned from hypothalamic tissue and its expression in brain, spinal cord and pituitary is detectable using RT-PCR analysis. Results of Northern hybridisation, ribonuclease protection and RT-PCR studies indicate low levels of expression of rat GALR3 mRNA in peripheral tissues, with transcript detectable in heart, lung, liver, kidney, pancreas, spleen, testis, ovary, stomach, adrenal and prostate glands (Wang et al. 1997c; Smith et al. 1998; T.P. Iismaa, unpubl.; Table 1).

3.2
Other Galanin Binding Sites

Both radioligand binding and functional analyses have suggested that additional galanin receptor subtypes may exist, with discrete pharmacological profiles not accounted for by the three currently isolated receptors. For example, the use of [^{125}I]-galanin(1-15) as a radioligand in autoradiographic studies of rat brain has identified a specific high affinity binding site for this ligand in the dorsal hippocampal formation, neocortex and neostriatum where [^{125}I]Tyr^{26}-galanin(1-29) binding is minimal or absent (Hedlund et al. 1992). The functional significance of galanin binding sites that preferentially interact with the truncated peptide galanin(1-15) is not clear, and their relationship to galanin receptors that have been described in the dorsal hippocampus (Valkna et al. 1995) remains to be defined. In contrast to the rat, the human neocortex is rich in [^{125}I]Tyr^{26}-galanin(1-29) binding sites, with discrete layers of the visual cortex exhibiting a high density of galanin binding sites (Köhler and Chan-Palay 1990). This suggests that galanin may be important in human visual cortex function, but the identity of the receptor subtype(s) involved has not yet been determined. As is the case for a number of neurotransmitter and neuropeptide receptors, galanin receptor expression in the CNS is not restricted to neuronal cells. [^{125}I]Tyr^{26}-galanin(1-29) binding sites have recently been demonstrated on cultured astrocytes, with a subpopulation of cells that express galanin receptors also expressing muscarinic and nicotinic acetylcholine receptors (Hösli et al. 1997).

The existence of an additional galanin receptor subtype in smooth muscle cells of the gastrointestinal tract has been suggested by galanin induction of an elevation of intracellular cAMP levels in these cells (Gu et al. 1995).

Furthermore, galanin receptors that are responsive to the truncated peptide galanin(3–29) have also been described in rat hypothalamic and pituitary cells (Wynick et al. 1993b; Kinney et al. 1998). In contrast to the $[^{125}I]Tyr^{26}$-gala-nin(1–29) binding sites that have been demonstrated on human pituitary tumours (Hulting et al. 1993), and the galanin receptors occurring on guinea pig gastric smooth muscle cells that are activated by galanin(3-29) (Gu et al. 1995), the galanin receptors on dispersed rat pituitary cells do not bind $[^{125}I]Tyr^{26}$-galanin(1-29). They are detectable only with $[^{125}I$ Bolton Hunter]galanin(1-29), and are unresponsive to galanin(1-15).

The only endogenous ligand known to date for galanin receptors is galanin, but variant forms of the peptide, derived by differential proteolytic processing and cleavage of preprogalanin, have been reported. For example, purification of galanin-like immunoreactive peptides from human large intestine resulted in isolation of the full-length non-amidated human peptide galanin(1–30) and the non-amidated truncated peptide galanin(1–19) in approximately equimo-lar amounts (Bersani et al. 1991a). Given the potent biological activity of the truncated peptides galanin(1–15) and galanin(1–16) in a variety of physiolog-ical assays (Bartfai et al. 1993a), as well as in activation of heterologously expressed cloned galanin receptors, the occurrence of the truncated peptide galanin(1–19) in significant amounts in human tissues could be of biological relevance. An N-terminally truncated peptide, comprising galanin(5–29), has been isolated from porcine brain (Sillard et al. 1992). While galanin(5–29) could represent a degradation product of porcine galanin(1–29), N-terminally truncated galanin peptides have binding affinity in the rat hippocampus (Fisone et al. 1989) and biological activity in intestinal tissues (Rossowski et al. 1990; Fox et al. 1994), and the ability of galanin(5–29) to interact with cloned galanin receptor subtypes has not been specifically assessed to date. Similarly, the binding to cloned galanin receptors of N-terminally nine and seven resi-due elongated forms of galanin, which were originally isolated from porcine brain and adrenal medulla (Sillard et al. 1992; Bersani et al. 1991b), has not been investigated. These elongated galanin peptides have been shown to bind with high affinity and to have low agonist activity at galanin receptors express-sed in rat spinal cord (Bedecs et al. 1994).

Considerable use has been made of chimeric galanin peptides such as M15, M35 and M40 (Bartfai et al. 1992; Fig. 6) in efforts to elucidate the biological role of galanin. These peptides act as antagonists and inhibit galanin action in a range of functional assays, but they also exhibit agonist activity in a number of physiological assay systems (Bartfai et al. 1993b; Gu et al. 1993). Further-more, the chimeric peptides M15, M32, M35, M40 and C7 function as agonists for galanin receptors expressed endogenously on human Bowes melanoma cells (Heuillet et al. 1994) and for heterologously expressed cloned galanin receptors. The further characterisation of cloned receptors in heterologous expression systems, together with detailed localisation studies providing information on cell type-specific expression of discrete receptor subtypes, will

```
M15        GWTLNSAGYLLGPQQFFGLM(amide)
M32        ------------RHYINLITRQRY(amide)
M35        ------------PPGFSPFR(amide)
M40        ------------PPALALA(amide)
C7         ------------[D-R]PKPQQ[D-W]F[D-W]LL(amide)
```

Fig. 6. Amino acid sequences of chimeric galanin peptides. Galanin(1–13) is linked to the following C-terminal residues: M15, also known as galantide, substance P(5–11); M32, neuropeptide Y(25–36); M35, bradykinin(2–9); M40, PPALALA (synthetic sequence) and C7, spantide (substance P receptor antagonist). Sequence of galanin(1–13) is given only for M15; dashes represent identity with sequence of galanin(1–13) in other chimeric peptides

be necessary to achieve a greater understanding of the mechanisms of action of chimeric galanin peptides. Whether their biological activity reflects the existence of additional receptor subtypes with discrete pharmacological properties, or can be accounted for by modulation of the binding or signalling properties of existing galanin receptors, remains to be determined.

3.3
Galanin Receptor Gene Structure

The gene encoding GALR1 in human (*GALNR1*) and mouse (*Galnr1*) spans approximately 15–20 kb, and comprises three exons. Exon 1 encodes the N-terminal end and the first five transmembrane (TM) domains, exon 2 encodes the third intracellular loop and exon 3 encodes the remainder of the receptor, from TM6 to the C terminus (Jacoby et al. 1997; Wang et al. 1997a; Fig. 7). Encoding of the third intracellular loop on a separate exon raises the possibility that this segment of the receptor may undergo alternative splicing to generate functional diversity in receptor subtypes, as is the case for some other members of the G protein-coupled receptor superfamily. However, the characterisation of human and rat cDNA clones has not yet provided evidence for this (Iismaa et al. 1998). The human *GALNR1* and mouse *Galnr1* genes contain no consensus TATA or CCAAT boxes, or potential transcription factor recognition sites within approximately 1 kb upstream of the translation initiation codon.

The exon:intron organisation of the gene encoding human GALR2 (*GALNR2*) differs from that of *GALNR1*, with the coding sequence of *GALRN2* being contained on two exons that are separated by an intron of approximately 1.4 kb (Fathi et al. 1998a). Exon 1 encodes the N-terminal end and first three TM domains of GALR2, while exon 2 encodes the remainder of the receptor, from the second intracellular loop to the C terminus (Fig. 7). This genomic organisation is conserved in the rat, as indicated by the isolation and characterisation of partially processed hypothalamic cDNA transcripts encoding rat GALR2 (Howard et al. 1997a; Smith et al. 1997). This structural organisation is also conserved in the gene encoding human GALR3 (*GALNR3*) (Iismaa et al.

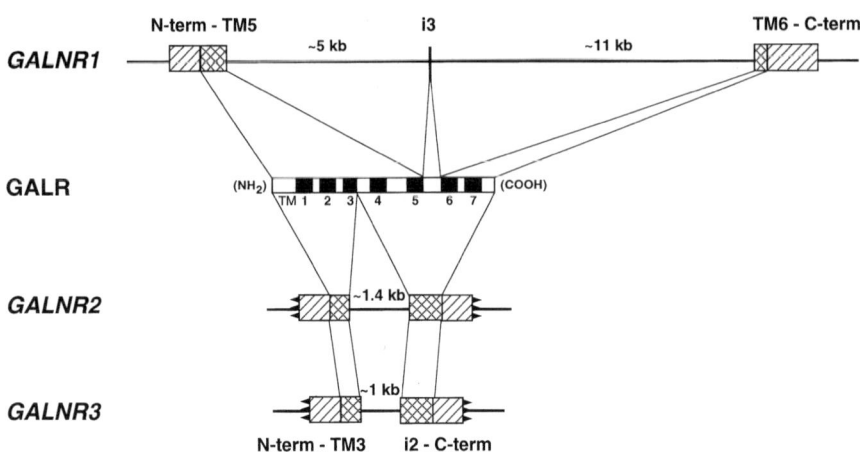

Fig. 7. Structural organisation of genes encoding human galanin receptors GALR1 (*GALNR1*), GALR2, (*GALNR2*) and GALR3 (*GALNR3*). Within schematic representations of the exon:intron organisation of the three genes, cross-hatched boxes represent coding sequence and lightly hatched boxes non-coding exon sequence. Precise limits of non-coding exon sequence are not known for *GALNR2* and *GALNR3*. Mature galanin receptor (GALR) protein is depicted schematically below representation of *GALNR1*. Putative transmembrane (TM) domains are shown as black boxes numbered 1–7, and other sparts of the receptor protein are shown as white boxes. Segments of mature GALR protein encoded by each of the exons are indicated

1998; Fig. 7). Conservation of genomic organisation, together with the high degree of sequence identity between GALR2 and GALR3 (58%), suggest a common evolutionary origin for *GALNR2* and *GALNR3*, which has presumably involved a gene duplication event. The different structural organisation of *GALNR1* suggests convergent evolution for this gene.

3.4
Chromosomal Localisation of Galanin Receptor Genes

The human *GALNR1* gene has been localised to 18q23, in the vicinity of genes encoding cytochrome b5 (*CYB5*) and peptidase A (*PEPA*), and telomeric of the gene encoding myelin basic protein (*MBP*) at 18q22.1 (Nicholl et al. 1995). Mouse *Galnr1* occurs on mouse chromosome 18E4 (Jacoby et al. 1997; Simoneaux et al. 1997), homoeologous with the human localisation. The human *GALNR2* gene occurs at 17q25.3 (Fathi et al. 1998a), in the vicinity of genes encoding two other G protein-coupled receptors, glucagon receptor (*GCGR*) at 17q25 and somatostatin receptor 2 (*SSTR2*) at 17q24. The human *GALNR3* gene has been localised to 22q13.1, in the vicinity of the gene encoding somatostatin receptor 3 (*SSTR3*), and a polymorphic *Pst*I site has been identified within the intron of this gene (Iismaa et al. 1998).

While no disease locus has yet been assigned to the location of *GALNR3*, the *GALNR1* locus at 18q23 co-localises with a locus for bipolar disorder (manic-depressive illness) identified by analysis of two Costa Rican pedigrees (Freimer et al. 1996) and a chromosomal deletion that gives rise to GH insufficiency syndrome (Cody et al. 1997). The *GALNR2* locus at 17q25 co-localises with the disease loci for hereditary neuralgic amyotrophy (HNA; Pellegrino et al. 1997) and Russell-Silver Syndrome [RSS; Genome Data Base – GBD™, Online Mendelian Inheritance in Man (NCBI)]. HNA is a rare autosomal dominant disorder of the peripheral nervous system. It is a recurrent focal neuropathy involving painful episodes of brachial palsy and is often associated with short stature, hypotelomerism and facial dysmorphic features. The main features of RSS are low-birthweight dwarfism and lateral asymmetry, where the majority of afflicted individuals apparently have normal levels of growth hormone (Tanner and Ham 1969; Galasso et al. 1995). The potential involvement of galanin and its receptors in these and other disorders may now be assessed by more precise mapping studies using polymorphic microsatellite markers.

4
Physiological Effects and Therapeutic Indications

4.1
Learning, Memory and Cognition

Galanin has been reported to modulate the activity of cholinergic neurons and may therefore play a role in memory processes. In the rat ventral hippocampus, which is an area of importance to spatial learning, galanin inhibits cholinergic neurotransmission, both presynaptically by inhibition of acetylcholine release (Fisone et al. 1987) and postsynaptically by inhibition of muscarinic acetylcholine receptor signalling (Palazzi et al. 1988). The administration of galanin into the lateral ventricles, hippocampus or medial septum of rats impairs performance in memory tasks, primarily impairing the acquisition but not retrieval of spatial memory (Mastropaolo et al. 1988; Sundstrom et al. 1988; Givens et al. 1992). It has recently been demonstrated that in normal aged human brain, galaninergic fibres occur in direct apposition to cholinergic basal forebrain neurons, suggestive of synaptic contact and providing a morphological basis for galanin modulation of the activity of these neurons (Bowser et al. 1997). In dementias such as Alzheimer's disease, Down's syndrome and in many patients with Parkinson's disease, where the cholinergic neurons projecting from the basal forebrain to the cerebral cortex degenerate, galaninergic fibres and terminals hypertrophy and hyperinnervate the surviving cholinergic neurons of the nucleus basalis of Meynert and diagonal band (Chan-Palay 1988; Bowser et al. 1997). This has led to the suggestion that the cognitive deficit of such neurodegenerative disorders may be exacerbated by inhibition of acetylcholine release by galanin. Consistent with this are observations of the preservation of galanin receptors (Ikeda et al. 1995; Rodríguez-Puertas et al. 1997) and elevated galanin levels in the cerebral cortex of afflicted individuals (Gabriel et al. 1994).

The potential contribution of galanin to the cognitive deficit of such disorders suggests therapeutic potential for galanin receptor antagonists, although the receptor subtype(s) involved in the effects of galanin on learning, memory and cholinergic neurotransmission remain to be defined. Limited pharmacological data indicate that the galanin receptor(s) mediating inhibition of acetylcholine release in ventral hippocampus may be activated by galanin(1–29) and galanin(1–16), and that the action of galanin at this receptor(s) is antagonised by the chimeric peptide M15 (Bartfai et al. 1991; Girotti et al. 1993), while the chimeric peptides M15, M35 and M40 have all been shown to function as antagonists at the galanin receptor(s) characterised in learning and memory paradigms (Ögren et al. 1992; McDonald and Crawley 1996; Arletti et al. 1997)

The cholinergic neurons that degenerate during Alzheimer's disease are dependent on NGF, which acts *in vivo* to stimulate the synthesis and secretion of acetylcholine (Rylett et al. 1993). As administration of NGF will reverse memory deficits associated with age and lesion-induced damage of cholinergic pathways in rats (Fischer et al. 1987; Montero and Hefti 1989), it has been identified as a potential therapeutic in the treatment of human neurodegenerative disorders that involve cholinergic dysfunction. However, NGF also induces galanin gene expression in cholinergic basal forebrain neurons (Planas et al. 1997), which could compromise the therapeutic efficacy of NGF. Thus, the use of NGF, acetylcholinesterase or cholinergic agonists for treatment of disorders arising from cholinergic deficit may be enhanced by concomitant administration of galanin receptor antagonists.

4.2
Feeding Behaviour and Obesity

Central administration of galanin into the paraventricular nucleus (Tempel et al. 1988; Crawley et al. 1990), amygdala (Kyrkouli et al. 1990; Corwin et al. 1993) or nucleus of the solitary tract (Koegler and Ritter 1998) of satiated rats has been reported to increase food intake, with a preferential increase in voluntary fat intake observed in a three macronutrient choice paradigm (Tempel et al. 1988). However, the stimulatory effect of galanin on fat ingestion is not observed in all experimental dietary choice paradigms (Smith et al. 1994; Corwin et al. 1995), and underlying macronutrient preferences may determine the feeding response to galanin (Smith et al. 1996). Repeated intracerebroventricular administration of galanin for 2 weeks has no effect on total daily food intake and does not result in any significant weight gain in rats (Smith et al. 1994), while reports of elevated levels of circulating galanin in obese human subjects are controversial (Invitti et al. 1995; Baranowska et al. 1997). The role of galanin in the regulation of feeding behaviour and body weight thus remains unclear, although high concentrations of galanin and galanin binding sites are present in hypothalamic nuclei associated with the control of feeding behaviour (Bonnefond et al. 1990; Evans et al. 1993b) and galanin mRNA levels in the hypothalamus appear to be regulated by leptin (Sahu, 1998). Both galanin(1–29) and galanin(1–16) stimulate feeding behaviour in rats (Crawley et al. 1990), consistent with the pharmacological properties and expression of GALR1 and GALR2 receptor subtypes in the hypothalamus. However, the chimeric peptides M40 and C7, which function as agonists at cloned galanin receptor subtypes, are able to block galanin-induced feeding (Corwin et al. 1993), indicating that additional galanin receptor subtype(s) that have not yet been cloned may also be involved. Delineation of the role of galanin in the regulation of food intake and body weight will require elucidation of the mechanism of action of the chimeric peptides in functional antagonism of the effects of galanin, together with the development of subtype-specific galanin receptor ligands.

4.3
Neuroendocrine Responses

Galanin has been demonstrated to have potent and species-specific effects on the regulation of pituitary and glucoregulatory hormone secretion (Bartfai et al. 1993a; Carey et al. 1993; Cheung et al. 1996), consistent with its expression in the hypothalamic-pituitary axis and in peripheral neuroendocrine tissues such as the pancreas. In humans, intravenous infusion of galanin stimulates the secretion of GH and causes elevation of prolactin levels (Bauer et al. 1986a; Carey et al. 1993). While galanin-like immunoreactivity has been observed in human pancreatic islets and exocrine parenchyma, and inhibition of insulin secretion by galanin has been demonstrated *in vitro* (Ahrén et al. 1991), galanin does not inhibit insulin secretion when infused intravenously into humans to achieve circulating levels of galanin sufficient to elicit a GH response (Carey et al. 1993). Similarly, galanin does not appear to play a significant role in modulation of secretion of follicle-stimulating hormone (FSH), luteinizing hormone (LH) or thyroid-stimulating hormone (TSH), but may play a minor role in regulating ACTH and cortisol levels in humans (Giustina et al. 1992b, 1994; Carey et al. 1993). This is in contrast to its actions in the rat, where its expression in hypothalamic neurons that also produce gonadotropin-releasing hormone (GnRH) is markedly upregulated by estrogen, and it is thought to play an important role in generation of the preovulatory LH surge (Cheung et al. 1996).

The potent effect of galanin in stimulating GH secretion in humans suggests potential for use of galanin agonists as GH secretagogues in both the diagnosis and treatment of GH insufficiency syndromes. Evidence from a range of immunological and pharmacological studies has suggested that galanin stimulation of GH secretion is mediated by modulation of the release of either GHRH (Murakami et al. 1987) or somatostatin (Davis et al. 1987; Tanoh et al. 1993). A tonic modulation by galanin of both GHRH release and somatostatinergic tone is indicated by the demonstration that intracerebroventricular administration of galanin antisera in rats decreases GH pulse amplitude and increases GH pulse frequency (Maiter et al. 1990). The ability of galanin to stimulate the release of GHRH and somatostatin has been demonstrated *in vitro* using rat brain median eminence fragments (Aguila et al. 1992) and the presence of GALR1 mRNA in a subset of somatostatin-synthesising neurons in the periventricular nucleus (Chan et al. 1996), together with demonstration of galaninergic innervation of somatostatin neurons (Liposits et al. 1993), provides morphological evidence for galaninergic modulation of somatostatin neuron activity. Galanin is co-expressed with GHRH in neurons whose cell bodies are located in the arcuate nucleus of the hypothalamus, where its levels are downregulated in response to reduction of circulating levels of GH (Chan et al. 1996). Although the presence of GALR1 mRNA has been reported in regions of the arcuate nucleus that also contain GHRH neurons (Mitchell et al. 1997), co-localisation studies have not yet been performed and the site of action of galanin synthesised in these neurons is currently unknown.

In addition to actions on hypothalamic neurons that may be mediated by GALR1, galanin released into the hypophysial portal circulation could act directly at the anterior pituitary. Although studies on galanin stimulation of pituitary hormone secretion by direct action on the pituitary gland are contradictory (Ottlecz et al. 1986; Gabriel et al. 1988; Hulting et al. 1991), the observed expression of GALR2 mRNA in a proportion of lactotrophs, somatotrophs, thyrotrophs and gonadotrophs in the rat anterior pituitary (Depczynski et al. 1998), together with expression of GALR3 mRNA in rat pituitary (Table 1), is consistent with autocrine or paracrine regulation of prolactin and GH secretion by galanin (Wynick et al. 1993a; Moore et al. 1997; Cai et al. 1998). The hypothesis that galanin acts as a paracrine regulator of prolactin gene expression is supported by the phenotype of mice homozygous for a loss of function mutation in the galanin gene. Galanin knockout mice apparently grow normally, although mutant females fail to lactate. Prolactin synthesis is markedly reduced by 30-40%, with an associated failure in mammary gland maturation (Wynick 1997; Wynick et al. 1998).

It is not clear at present how cell-specific differences in the expression of galanin within human and rat pituitary (Vrontakis et al. 1991) relate to pituitary function. While mRNA encoding both GALR1 and GALR2 is expressed in human pituitary (Table 1), the cell-specific distribution of galanin receptor subtypes in human hypothalamic or pituitary tissue has not yet been described. This will be of particular interest in the light of identification of the *GALNR1* gene as a candidate gene in GH insufficiency syndrome associated with haploinsufficiency at 18q23 in the human genome (Cody et al. 1997).

The available pharmacological data pertaining to galanin stimulation of pituitary hormone secretion do not allow for correlation with the pharmacological profiles of galanin receptor subtypes cloned to date. For example, GH secretion in the rat is inducible by intracerebroventricular administration of galanin(1–29), as well as galanin(1–19) (Murakami et al. 1989), while the chimeric peptide M15 acts as an antagonist to reduce GH pulse amplitude and frequency (Gabriel et al. 1993). The pharmacological profile of the galanin receptor reported to mediate prolactin release *in vitro* from dispersed rat pituitary cells does not correspond to that of GALR2, as it exhibits functional activation in response to galanin(3–29) as well as to galanin(1–29), and absence of functional response to galanin(1–15) and M15 (Wynick et al. 1993b).

In contrast to normal human subjects, galanin has been reported to inhibit GH secretion in some acromegalic patients (Giustina et al. 1992a; Cuerda et al. 1998), with restoration of a normal GH response following surgical removal of GH-secreting adenomas (Giustina et al. 1995). Galanin inhibition of GH secretion has also been demonstrated *in vitro* in a number of adenomas, with a significant proportion of such adenomas not exhibiting inhibition of GH secretion in response to the somatostatin analog, octreotride (Giustina et al. 1997). These studies suggest that galanin receptor agonists may be useful therapeutic agents for the treatment of acromegalic patients who are unresponsive to octreotride.

4.4
Analgesia and Neuropathic Pain

The presence of galanin binding sites on dorsal horn neurons of the spinal cord (Kar and Quirion 1994), together with the localisation of galanin-like immunoreactivity in primary afferent terminals and interneurons of the superficial dorsal horn (Wiesenfeld-Hallin et al. 1992a), suggests a role for galanin in the modulation of sensory processing and particularly in nociception. Although endogenously administered galanin has been reported to have antinociceptive, hyperalgesic and pro-nociceptive effects in tests of sensitivity to mechanical stimuli (Cridland and Henry, 1988; Kuraishi et al. 1991; Wiesenfeld-Hallin et al. 1993), galanin knockout mice exhibit hyperalgesia (D. Wynick, pers. commun.). This indicates that galanin acts as a tonic inhibitor of spinal cord transmission in intact animals. Furthermore, as both galanin(1–29) and galanin(1–15) potentiate the analgesic effects of morphine (Wiesenfeld-Hallin et al. 1990; Xu et al. 1990), there is growing interest in the possible use of spinal galanin receptor agonists in the treatment of chronic pain. All three cloned galanin receptor subtypes are expressed in neuronal populations involved in somatosensory neurotransmission (Parker et al. 1995; Shi et al. 1997; Table 1), and GALR2 mRNA levels in sensory ganglia are elevated in response to peripheral tissue inflammation (Shi et al. 1997). However, as the

chimeric peptides M15 and M35 exhibit functional antagonism of spinal galanin receptors (Reimann et al. 1994; Selve et al. 1996), identification of the
molecular target for the development of such therapeutic agents requires further analysis with receptor subtype-specific reagents.

That galanin may be an endogenous analgesic factor of particular importance after nerve injury (Wiesenfeld-Hallin et al. 1992b) is suggested by the
observation that the inhibitory effect of galanin on the nociceptive spinal flex-
or reflex is stronger after sciatic nerve axotomy than in intact animals (Wiesenfeld-Hallin et al. 1989). Furthermore, the addition of galanin antisense oligonucleotides to the proximal end of a transected sciatic nerve results in
increased self-mutilation (Verge et al. 1993; Ji et al. 1994). As this behaviour is
thought to reflect the development of neuropathic pain, a role for galanin in
suppressing this behaviour and in controlling the development of such pain
has been suggested. However, somewhat surprisingly, this response to nerve
injury is abolished in homozygous galanin knockout mice (Holmes et al. 1997;
Wynick 1997). Detailed characterisation of the galanin knockout mouse, particularly including investigation of potential developmental deficits, should
resolve this apparent discrepancy and allow definition of the exact role of galanin in neuropathic pain.

4.5
Nerve Injury and Regeneration

The synthesis of galanin is upregulated in response to axotomy of sensory,
motor and sympathetic neurons, and after trauma and mechanical or pharmacological insults to the brain (Crawley 1995). Interestingly, galanin synthesis is upregulated in neuronal and glial cells in response to certain pharmacological agents (Xu et al. 1992), but the significance of enhanced galanin expression is not clear. Galanin knockout mice exhibit significantly diminished
capacity for functional regeneration of sensory neurons after peripheral nerve
transection (Holmes et al. 1997; Wynick 1997), suggesting involvement of galanin in the regenerative process in injured sensory neurons, and indicating
therapeutic potential for galanin agonists to facilitate nerve regeneration.
However, both GALR1 and GALR2, which are normally expressed in rat DRG
neurons, are downregulated in response to sciatic nerve transection (Z.-Q. Xu
et al. 1996b), possibly representing complex adaptive mechanisms mediating
responses to nerve injury and associated neuropathic pain.

Blocking neuronal activity or an imbalance in excitatory or inhibitory neuronal inputs also results in elevated expression of galanin (Agoston et al. 1994;
Ohno et al. 1994). Galanin modulates the release of a range of neurotransmitters in the brain, including inhibition of the release of excitatory neurotransmitters such as glutamate and aspartate, whose elevated levels following
ischemic or epileptic episodes cause irreversible neuronal damage (Ben-Ari
and Ladzunski 1989; Zini et al. 1993). This suggests its involvement in *in vivo*

neuroprotective mechanisms that may be of relevance in strategies for treatment of acute brain trauma or hyperactivity. For example, galanin has a beneficial effect if administered prior to experimental traumatic brain injury, specifically reducing deficits in performance of sensory motor and memory tasks (Liu et al. 1994), and has anti-convulsant activity in decreasing seizure severity (Mazarati et al. 1992). The pharmacological characterisation of galanin receptors mediating inhibition of excitatory neurotransmitter release is limited (Ben-Ari and Ladzunski 1989), but they respond to galanin(1–29) and galanin(1–16), consistent with the pharmacological profile of all three cloned galanin receptors. Both detailed localisation studies and the development of subtype-specific ligands will contribute to delineation of the discrete functional properties of galanin receptors that would identify them as therapeutic targets.

4.6
Mitogenesis: Development and Tumourigenesis

Galanin does not promote survival in standard neurotrophic assay systems, such as primary culture of neonatal rat sympathetic neurons or neurite outgrowth of rat pheochromocytoma (PC12) cells (Klimaschewski et al. 1995). However, it is a potent growth factor for developing pituitary cells and lactotrophs in the galanin knockout mouse fail to proliferate in response to high doses of estrogen (Wynick et al. 1993a, 1998).

Galanin is expressed in pituitary adenoma, pheochromocytoma (Bauer et al. 1986b; Hulting et al. 1989; Sano et al. 1991), neuroblastoma (Klaus et al. 1998) and breast cancer cells (Ormandy et al. 1998a). A role for galanin in the development of neuroendocrine tumours has been suggested on the basis of correlation of increased galanin gene expression with pituitary hyperplasia and the development of estrogen-induced tumours in tumour-sensitive rats, and decreased galanin gene expression when tumour formation is inhibited (Hsu et al. 1990; Hyde and Howard 1992; Gregg et al. 1996). Both GALR1 and GALR2 have mitogenic signalling properties (Wang et al. 1998) and galanin stimulates the growth of rat 235-I lactotroph (Wynick et al. 1993a) and human small cell lung cancer cells (Sethi and Rozengurt 1991; Seufferlein and Rozengurt 1996). However, galanin has also been observed to inhibit experimentally induced gastric (Iishi et al. 1994, 1995) and pancreatic (Iishi et al. 1998) carcinogenesis, and is downregulated in the colon of patients with colon adenocarcinoma (El-Salhy et al. 1998). The mechanism for these effects has not been elucidated, but evaluation of galanin receptor subtype expression together with mitogenic signalling potential in different tumour types may identify novel therapeutic uses for receptor-specific agonist and antagonist compounds.

5
Summary and Conclusions

The development of a strain of galanin knockout mice has provided confirmation of a neuroendocrine role for galanin, as well as supporting results of previous physiological investigations indicating a role for galanin in analgesia and neuropathic pain, and potentially in neuronal growth and regeneration processes. Whether elevation of galanin expression in neurodegenerative disorders such as Alzheimer's disease represents a survival response or exacerbates functional deficit in afflicted individuals remains to be determined. More detailed analysis of the phenotype of the galanin knockout mouse should provide insights into the physiological role of galanin in memory and learning processes, as well as in hypothalamic function and other aspects of neuroendocrine regulation.

Biochemical and molecular cloning efforts have demonstrated that the multiplicity of actions of galanin is matched by complexity in the distribution and regulation of galanin and its receptors. A focus on characterisation of galanin receptors has resulted in the molecular cloning of three receptor subtypes to date. The distribution and functional properties of these receptors have not yet been fully elucidated, currently precluding assignment of discrete functions of galanin to any one receptor subtype. It is not currently possible to reconcile available pharmacological data using analogs of galanin and chimeric peptides in functional assay systems with the pharmacological properties of cloned receptor subtypes. This highlights the value of further knockout approaches targeting galanin receptor subtypes, but also raises the possibility of the existence of additional receptor subtypes that have yet to be cloned, or that receptor activity may be modulated by regulatory molecules that remain to be identified. The development of receptor subtype-specific compounds remains a high priority to advance work in this area. The ability to selectively modulate the many different actions of galanin, through a clearer understanding of receptor structure-function relationships and neuronal distribution, promises to provide important insights into the molecular and cellular basis of galanin action in normal physiology, and may provide lead compounds with therapeutic application in the prevention and treatment of a range of disorders.

References

Agoston DV, Komoly S, Palkovits M (1994) Selective up-regulation of neuropeptide synthesis by blocking the neuronal activity: galanin expression in septohippocampal neurons. Exp Neurol 126:247–255

Aguila MC, Marubayashi U, McCann SM (1992) The effect of galanin on growth hormone-releasing factor and somatostatin release from median eminence fragments in vitro. Neuroendocrinology 56:889–894

Ahrén B, Ar'Rajab A, Böttcher G, Sundler F, Dunning BE (1991) Presence of galanin in human pancreatic nerves and inhibition of insulin secretion from isolated human islets. Cell Tissue Res 264:263–267

Amiranoff B, Lorinet A-M, Lagny-Pourmir I, Laburthe M (1988) Mechanism of galanin-inhibited insulin release. Occurrence of a pertussis-toxin-sensitive inhibition of adenylate cyclase. Eur J Biochem 177:147–152

Amiranoff B, Lorinet A-M, Laburthe M (1991) A clonal rat pancreatic δ cell line (Rin14B) expresses a high number of galanin receptors negatively coupled to a pertussis-toxin-sensitive cAMP-production pathway. Eur J Biochem 195:459–463

Andell-Jonsson S, Bartfai T (1998) Identification of the spinal degradation products and inhibition of adenylate cyclase by recombinant rat galanin message-associated peptide. Neuropeptides 32:191–196

Andell-Jonsson S, Xu IS, Bartfai T, Xu XJ, Wiesenfeld-Hallin Z (1997) The effect of naturally occurring fragments of galanin message-associated peptide on spinal cord excitability in rats. Neurosci Lett 235:154–156

Anouar Y, MacArthur L, Cohen J, Iacangelo AL, Eiden LE (1994) Identification of a TPA-responsive element mediating preferential transactivation of the galanin gene promoter in chromaffin cells. J Biol Chem 269:6823–6831

Arletti R, Benelli A, Cavazzuti E, Bertolini A (1997) Galantide improves social memory in rats. Pharmacol Res 35:317–319

Arnold A, Goo Kim H, Gaz RD, Eddy RL, Fukushima Y, Byers MG, Shows TB, Kronenberg HM (1989) Molecular cloning and chromosomal mapping of DNA rearranged with the parathyroid hormone gene in a parathyroid adenoma. J Clin Invest 83:2034–2040

Baranowska B, Wasilewska-Dziubinska E, Radzikowska M, Plonowski A, Roguski K (1997) Neuropeptide Y, galanin, and leptin release in obese women and in women with anorexia nervosa. Metabolism 46:1384–1389

Bartfai T, Bedecs K, Land T, Langel Ü, Bertorelli R, Girotti P, Consolo S, Xu XJ, Wiesenfeld-Hallin Z, Pieribone V, Hökfelt T (1991) M-15: high-affinity chimeric peptide that blocks the neuronal actions of galanin in the hippocampus, locus coeruleus, and spinal cord. Proc Natl Acad Sci USA 88:10961–10965

Bartfai T, Fisone G, Langel Ü (1992) Galanin and galanin antagonists: molecular and biochemical perspectives. Trends Pharmacol Sci 13:312–317

Bartfai T, Hökfelt T, Langel Ü (1993a) Galanin – a neuroendocrine peptide. Crit Rev Neurobiol 7:229–274

Bartfai T, Langel Ü, Bedecs K, Andell S, Land T, Gregersen S, Ahrén B, Girotti P, Consolo S, Corwin R, Crawley J, Xu XJ, Wiesenfeld-Hallin Z, Hökfelt T (1993b) Galanin-receptor ligand M40 peptide distinguishes between putative galanin-receptor subtypes. Proc Natl Acad Sci USA 90:11287–11291

Bauer FE, Ginberg L, Venetikou M, McKay DJ, Burrin JM, Bloom SR (1986a) Growth hormone release in man induced by galanin, a new hypothalamic peptide. Lancet 2:192–195

Bauer FE, Hacker GW, Terenghi G, Adrian TE, Polak JM, Bloom SR (1986b) Localization and molecular forms of galanin in human adrenals: elevated levels in pheochromocytomas. J Clin Endocrinol Metab 63:1372–1378

Bedecs K, Langel Ü, Xu XJ, Wiesenfeld-Hallin Z, Bartfai T (1994) Biological activities of two endogenously occurring N-terminally extended forms of galanin in the rat spinal cord. Eur J Pharmacol 259:151–156

Ben-Ari Y, Ladzunski M (1989) Galanin protects hippocampal neurons from the functional effects of anoxia. Eur J Pharmacol 165:331–332

Bersani M, Johnsen AH, Højrup P, Dunning BE, Andreasen JJ, Holst JJ (1991a) Human galanin: primary structure and identification of two molecular forms. FEBS Lett 283:189–194

Bersani M, Thim L, Rasmussen TN, Holst JJ (1991b) Galanin and galanin extended at the N terminus with seven and nine amino acids are produced in and secreted from the porcine adrenal medulla in almost equal amounts. Endocrinology 129:2693–2698

Billecocq A, Hedlund PB, Bolaños-Jiménez F, Fillion G (1994) Characterization of galanin and 5-HT(1A) receptor coupling to adenylyl cyclase in discrete regions of the rat brain. Eur J Pharmacol 269:209–217

Blackburn MB, Wagner RM, Kochansky JP, Harrison DJ, Thomas-Laemont P, Raina AK (1995) The identification of two myoinhibitory peptides, with sequence similarities to the galanins, isolated from the ventral nerve cord of *Manduca sexta*. Regul Pept 57:213–219

Bloomquist BT, Beauchamp MR, Zhelnin L, Brown S-E, Gore-Willse AR, Gregor P, Cornfield LJ (1998) Cloning and expression of the human galanin receptor GalR2. Biochem Biophys Res Commun 243:474–479

Bonnefond C, Palacios JM, Probst A, Mengod G (1990) Distribution of galanin mRNA containing cells and galanin receptor binding sites in human and rat hypothalamus. Eur J Neurosci 2:629–637

Bowser R, Kordower JH, Mufson EJ (1997) A confocal microscopic analysis of galaninergic hyperinnervation of cholinergic basal forebrain neurons in Alzheimer's disease. Brain Pathol 7:723–730

Boyle MR, Verchere CB, McKnight G, Mathews S, Walker K, Taborsky GJ (1994) Canine galanin – sequence, expression and pancreatic effects. Regul Pept 50:1–11

Bretherton-Watt D, Kenny MJ, Ghatei MA, Bloom SR (1990) The distribution of galanin message-associated peptide-like immunoreactivity in the pig. Regul Pept 27:307–315

Burgevin MC, Loquet I, Quarteronet E, Habert-Ortoli E (1995) Cloning, pharmacological characterization, and anatomical distribution of a rat cDNA encoding for a galanin receptor. J Mol Neurosci 6:33–41

Cai A, Bowers RC, Moore JP, Hyde JF (1998) Function of galanin in the anterior pituitary of estrogen-treated Fischer 344 rats: autocrine and paracrine regulation of prolactin secretion. Endocrinology 139:2452–2458

Calvo JJ, De-Carvalho LF, Burnet PW, Ghatei MA, Bloom SR (1990) Effect of ACTH on VIP and galanin release from the pituitary. Endocrinology 126:1283–1287

Carey DG, Iismaa TP, Ho KY, Rajkovic IA, Kelly J, Kraegen EW, Ferguson J, Inglis AS, Shine J, Chisholm DJ (1993) Potent effects of human galanin in man: growth hormone secretion and vagal blockade. J Clin Endocrinol Metab 77:90–93

Chan YY, Grafstein-Dunn E, Delemarre-van de Waal HA, Burton KA, Clifton DK, Steiner RA (1996) The role of galanin and its receptor in the feedback regulation of growth hormone secretion. Endocrinology 137:5303–5310

Chan-Palay VL (1988) Galanin hyperinnervates surviving neurons of the human basal nucleus of Meynert in dementias of Alzheimer's and Parkinson's disease: a hypothesis for the role of galanin in accentuating cholinergic dysfunction in dementia. J Comp Neurol 273:543–557

Chen Y, Laburthe M, Amiranoff B (1992) Galanin inhibits adenylate cyclase of rat brain membranes. Peptides 13:339–341

Cheung CC, Clifton DK, Steiner RA (1996) Galanin - an unassuming neuropeptide moves to center stage in reproduction. Trends Endocrinol Metab 7:301–306

Cody JD, Hale DE, Brkanac Z, Kaye CI, Leach RJ (1997) Growth hormone insufficiency associated with haploinsufficiency at 18q23. Am J Med Genet 71:420–425

Corness J, Shi TJ, Xu Z-Q, Brulet P, Hökfelt T (1996) Influence of leukemia inhibitory factor on galanin/GMAP and neuropeptide Y expression in mouse primary sensory neurons after axotomy. Exp Brain Res 112:79–88

Corness JD, Burbach JPH, Hökfelt T (1997) The rat galanin-gene promoter – response to members of the nuclear hormone receptor family, phorbol ester and forskolin. Mol Brain Res 47:11–23

Corness J, Stevens B, Fields RD, Hökfelt T (1998) NGF and LIF both regulate galanin gene expression in primary DRG cultures. NeuroReport 9:1533–1536

Corwin RL, Robinson JK, Crawley JN (1993) Galanin antagonists block galanin-induced feeding in the hypothalamus and amygdala of the rat. Eur J Neurosci 5:1528–1533

Corwin RL, Rowe PM, Crawley JN (1995) Galanin and the galanin antagonist M40 do not change fat intake in a fat-chow choice paradigm in rats. Am J Physiol 38:R511–R518

Crawley JN (1995) Biological actions of galanin. Regul Pept 59:1–16

Crawley JN, Austin MC, Fiske SM, Martin B, Consolo S, Berthold M, Langel Ü, Fisone G, Bartfai T (1990) Activity of centrally administered galanin fragments on stimulation of feeding behavior and on galanin receptor binding in the rat hypothalamus. J Neurosci 10: 3695–3700

Cridland RA, Henry JL (1988) Effects of intrathecal administration of neuropeptides on a spinal nociceptive reflex in the rat: VIP, galanin, CGRP, TRH, somatostatin and angiotensin II. Neuropeptides 11:23–32

Croce CM (1986) Chromosome translocations and human cancer. Cancer Res 46:6019–6023

Cuerda C, Lucas T, Silvestre RA, Bretón I, García P, Marco J, Barceló B (1998) Growth hormone secretory response to intravenous galanin infusion in acromegalic patients. Exp Clin Endocrinol Diabetes 106:68–73

Davis, TM, Burrin JM, Bloom SR (1987) Growth hormone (GH) release in response to GH-releasing hormone in man is three-fold enhanced by galanin. J Clin Endocrinol Metab 65: 1248–1252

Delemarre-van de Waal HA, Burton KA, Kabigting EB, Steiner RA, Clifton DK (1994) Expression and sexual dimorphism of galanin messenger ribonucleic acid in growth hormone-releasing hormone neurons of the rat during development. Endocrinology 134:665–671

Del Fiacco M, Quartu M (1994) Somatostatin, galanin and peptide histidine isoleucine in the newborn and adult human trigeminal ganglion and spinal nucleus: immunohistochemistry, neuronal morphometry and colocalization with substance P. J Chem Neuroanat 7:171–184

Depczynski B, Nichol K, Fathi Z, Iismaa T, Shine J, Cunningham A (1998) Distribution and characterization of the cell types expressing GALR2 mRNA in brain and pituitary gland. Ann NY Acad Sci 863:120–128

Díaz-Miranda L, Pardo-Reoyo CF, Martínez R, García-Arrarás JE (1996) Galanin-like immunoreactivity in the sea cucumber Holothuria glaberrima. Cell Tissue Res 286:385–391

Eipper BA, Stoffers DA, Mains RE (1992) The biosynthesis of neuropeptides: peptide α-amidation. Annu Rev Neurosci 15:57–85

Elmquist JK, Fox CA, Ross LR, Jacobson CD (1992) Galanin-like immunoreactivity in the adult and developing Brazilian opossum brain. Dev Brain Res 67:161–179

Elmquist JK, Kao A, Kuehl-Kovarik C, Jacobson CD (1993) Developmental profile of galanin binding sites in the mammalian brain. Mol Cell Neurosci 4:354–365

El-Salhy M, Norrgård Ö, Boström A (1998) Low levels of colonic somatostatin and galanin in patients with colon carcinoma. GI Cancer 2:221–225

Evans HF, Shine J (1991) Human galanin: molecular cloning reveals a unique structure. Endocrinology 129:1682–1684

Evans H, Baumgartner M, Shine J, Herzog H (1993a) Genomic organization and localization of the gene encoding human preprogalanin. Genomics 18:473–477

Evans HF, Huntley GW, Morrison JH, Shine J (1993b) Localisation of mRNA encoding the protein precursor of galanin in the monkey hypothalamus and basal forebrain. J Comp Neurol 328:203–212

Fathi Z, Cunningham AM, Iben LG, Battaglino PM, Ward SA, Nichol KA, Pine KA, Wang JC, Goldstein ME, Iismaa TP, Antal Zimanyi I (1997) Cloning, pharmacological characterization and distribution of a novel galanin receptor. Mol Brain Res 51:49–59

Fathi Z, Battaglino PM, Iben LG, Li H, Baker E, Zhang D, McGovern R, Mahle CD, Sutherland GR, Iismaa TP, Dickinson KEJ, Antal Zimanyi I (1998a) Molecular characterization, pharmacological properties and chromosomal localisation of the human GALR2 galanin receptor. Mol Brain Res 58:156–169

Fathi Z, Church WB, Iismaa TP (1998b) Galanin receptors: recent advances and potential use as therapeutic targets. Annu Rep Med Chem 33:41–50

Fischer WK, Wictorin K, Bjorklund A, Williams LR, Varon S, Gage FH (1987) Amelioration of cholinergic neuron atrophy and spatial memory impairment in aged rats by nerve growth factor. Nature 329:65–68

Fisone G, Wu CF, Consolo S, Nordström O, Brynne N, Bartfai T, Melander T, Hökfelt T (1987) Galanin inhibits acetylcholine release in the ventral hippocampus of the rat: histochemical, autoradiographic, in vivo and in vitro studies. Proc Natl Acad Sci USA 84:7339–7343

Fisone G, Langel Ü, Carlquist M, Bergman T, Consolo T, Hökfelt T, Undén A, Andell S, Bartfai T (1989) Galanin receptor and its ligands in the rat hippocampus. Eur J Biochem 181:269–276

Fox MD, Hyde JF, Muse KN, Keeble SC, Howard G, London SN, Curry TE (1994) Galanin: a novel intraovarian regulatory peptide. Endocrinology 135:636–641

Freimer NB, Reus VI, Escamilla MA, McInnes LA, Spesny M, Leon P, Service SK, Smith LB, Silva S, Rojas E, Gallegos A, Meza L, Fournier E, Baharloo S, Blankenship K, Tyler DJ, Batki S, Vinogradov S, Weissenbach J, Barondes SH, Sandkuijl LA (1996) Genetic mapping using haplotype, association and linkage methods suggests a locus for severe bipolar disorder (BPI) at 18q22-q23. Nat Genet 12:436–441

Gabriel SM, Milbury CM, Nathanson JA, Martin JB (1988) Galanin stimulates rat pituitary growth hormone secretion in vitro. Life Sci 42:1981–1986

Gabriel SM, Kaplan LM, Martin JB, Koenig JI (1989) Tissue-specific sex differences in galanin-like immunoreactivity and galanin mRNA during development in the rat. Peptides 10:369–374

Gabriel SM, Rivkin A, Mercado J (1993) The galanin antagonist, M-15, inhibits growth hormone release in rats. Peptides 14:633–636

Gabriel SM, Bierer LM, Davidson M, Purohit DP, Perl DP, Harotunian V (1994) Galanin-like immunoreactivity is increased in the postmortem cerebral cortex from patients with Alzheimer's disease. J Neurochem 62:1516–1523

Galasso C, Scirè G, Boscherini B (1995) Growth hormone and dysmorphic syndromes. Horm Res 44:42–48

Giorgi S, Forloni G, Baldi G, Consolo S (1995) Gene expression and in vitro release of galanin in rat hypothalamus during development. Eur J Neurosci 7:944–950

Girotti P, Bertorelli R, Fisone G, Land T, Langel Ü, Consolo S, Bartfai T (1993) N-terminal galanin fragments inhibit the hippocampal release of acetylcholine in vivo. Brain Res 612: 258–262

Giustina A, Bodini C, Doga M, Schettino M, Pizzocolo G, Giustina G (1992a) Galanin decreases circulating growth hormone levels in acromegaly. J Clin Endocrinol Metab 74:1296–1300

Giustina A, Girelli A, Licini M, Schettino M, Negro-Vilar A (1992b) Thyrotropin and prolactin secretion are not affected by porcine and rat galanin in normal subjects. Horm Metab Res 24: 351–352

Giustina A, Licini M, Schettino M, Doga M, Pizzocolo G, Negro-Vilar A (1994) Physiological role of galanin in the regulation of anterior pituitary function in humans. Am J Physiol 266: E57–E61

Giustina A, Bresciani E, Bussi AR, Bollati A, Bonfanti C, Bugari G, Chiesa L, Giustina G (1995) Characterization of the paradoxical growth hormone inhibitory effect of galanin in acromegaly. J Clin Endocrinol Metab 80:1333–1340

Giustina A, Ragni G, Bollati A, Cozzi R, Licini M, Poiesi C, Turazzi S, Bonfanti C (1997) Inhibitory effects of galanin on growth hormone (GH) release in cultured GH-secreting adenoma cells – comparative study with octreotide, GH-releasing hormone, and thyrotropin-releasing hormone. Metabolism 46:425–430

Givens BS, Olton DS, Crawley JN (1992) Galanin in the medial septal area impairs working memory. Brain Res 582:71–77

Gregg DW, Galkin M, Gorski J (1996) Effect of estrogen on the expression of galanin mRNA in pituitary tumor-sensitive and tumor-resistant rat strains. Steroids 61:468–472

Gu Z-F, Rossowski WJ, Coy DH, Pradhan TK, Jensen RT (1993) Chimeric galanin analogs that function as antagonists in the CNS are full agonists in gastrointestinal smooth muscle. J Pharmacol Exp Therapeut 266:912–918

Gu Z-F, Pradhan TK, Coy DH, Jensen RT (1995) Interaction of galanin fragments with galanin receptors on isolated smooth muscle cells from guinea pig stomach: identification of a novel galanin receptor subtype. J Pharmacol Exp Therapeut 272:371–378

Guida LC, Charlton P, Gilbert DJ, Jenkins NA, Copeland NG, Nicholls RD (1998) Genetic mapping of the galanin-GMAP (Galn) gene to mouse chromosome 19. Mamm Genome 9: 240–242

Gustafson EL, Smith KE, Durkin MM, Gerald C, Branchek TA (1996) Distribution of a rat galanin receptor mRNA in rat brain. NeuroReport 7:953–957

Habert-Ortoli E, Amiranoff B, Loquet I, Laburthe M, Mayaux J-F (1994) Molecular cloning of a functional human galanin receptor. Proc Natl Acad Sci USA 91:9780–9783

Hammond PJ, Khandannia N, Withers DJ, Jones PM, Ghatei MA, Bloom SR (1997) Regulation of anterior pituitary galanin and vasoactive intestinal peptide by oestrogen and prolactin status. J Endocrinol 152:211–219

Hedlund PB, Yanaihara N, Fuxe K (1992) Evidence for specific N-terminal galanin fragment binding sites in the rat brain. Eur J Pharmacol 224:203–205

Hemmer A, Hyde JF (1992) Regulation of galanin secretion from pituitary cells in vitro by estradiol and GHRH. Peptides 13:1201–1206

Heuillet E, Bouaiche Z, Menager J, Dugay P, Munoz N, Dubois H, Amiranoff B, Crespo A, Lavayre J, Blanchard JC, Doble A (1994) The human galanin receptor: ligand-binding and functional characteristics in the Bowes melanoma cell line. Eur J Pharmacol 269:139–147

Hökfelt T, Åman K, Arvidsson U, Bedecs K, Ceccatelli S, Hulting A-L, Langel Ü, Meister B, Pieribone V, Bartfai T (1992) Galanin messsage-associated peptide (GMAP)- and galanin-like immunoreactivities: overlapping and differential distributions in the rat. Neurosci Lett 142: 139–142

Holmes FE, McMahon SB, Murphy D, Wynick D (1997) Targeted disruption of galanin reduces nerve regeneration and neuropathic pain. Soc Neurosci Abstr 23:1954

Hooi SC, Koenig JI, Abraczinskas DR, Kaplan LM (1997) Regulation of anterior pituitary galanin gene expression by thyroid hormone. Mol Brain Res 51:15–22

Hösli E, Ledergerber M, Kofler A, Hösli L (1997) Evidence for the existence of galanin receptors on cultured astrocytes of rat CNS: colocalization with cholinergic receptors. J Chem Neuroanat 13:95–103

Howard AD, Tan C, Shiao L-L, Palyha OC, McKee KK, Weinberg DH, Feighner SD, Cascieri MA, Smith RG, Van Der Ploeg LHT, Sullivan KA (1997a) Molecular cloning and characterization of a new receptor for galanin. FEBS Lett 405:285–290

Howard G, Peng LH, Jyde JF (1997b) An estrogen receptor binding site within the human galanin gene. Endocrinology 138:4649–4656

Hsu DW, El-Azouzi M, Black PM, Chin WW, Hedley-White ET, Kaplan LM (1990) Estrogen increases galanin-immunoreactivity in hyperplastic prolactin-secreting cells in Fischer 344 rats. Endocrinology 126:3159–3167

Hsu DW, Hooi SC, Hedley-White ET, Strauss RM, Kaplan LM (1991) Coexpression of galanin and adrenocorticotropic hormone in human pituitary and pituitary adenomas. Am J Pathol 138:897–909

Hulting A-L, Meister B, Grimelius L, Wersäll J, Änggård A, Hökfelt T (1989) Production of a galanin-like peptide by a human pituitary adenoma: immunohistochemical evidence. Acta Physiol Scand 137:561–562

Hulting A-L, Meister B, Carlsson L, Hilding A, Isaksson O (1991) On the role of the peptide galanin in regulation of growth hormone secretion. Acta Endocrinol 125:518–525

Hulting A-L, Land T, Berthold M, Langel Ü, Hökfelt T, Bartfai T (1993) Galanin receptors from human pituitary tumors assayed with human galanin as ligand. Brain Res 625:173–176

Hyde JF, Howard G (1992) Regulation of galanin gene expression in the rat anterior pituitary gland by the somatostatin analog SMS 201-995. Endocrinology 131:2097–2102

Hyde JF, Keller BK (1991) Galanin secretion from anterior pituitary cells in vitro is regulated by dopamine, somatostatin, and thyrotropin-releasing hormone. Endocrinology 128:917–922

Hyde JF, Engle MG, Maley BE (1991) Colocalization of galanin and prolactin within secretory granules of anterior pituitary cells in estrogen-treated Fischer 344 rats. Endocrinology 129: 270–276

Hyde JF, Keller BK, Howard G (1992) Dopaminergic regulation of galanin gene expression in the rat anterior pituitary gland. J Neuroendocrinol 4:449–454

Iishi H, Tatsuta M, Baba M, Uehara H, Nakaizumi A (1994) Protection by galanin against gastric carcinogenesis induced by N-methyl-N'-nitro-N-nitrosoguanidine in Wistar rats. Cancer Res 54:3167–3170

Iishi HM, Tatsuta M, Baba M, Uehara H, Yano H, Nakaizumi A (1995) Chemoprevention by galanin against colon carcinogenesis induced by azoxymethane in Wistar rats. Int J Cancer 61: 861–863

Iishi H, Tatsuta M, Baba M, Yano H, Iseko K, Uehara H, Nakaizumio A (1998) Inhibition by galanin of experimental carcinogenesis induced by azaserine in rat pancreas. Int J Cancer 75:396–399

Iismaa TP, Biden TJ, Shine J (1995) G protein-coupled receptors. RG Landes Company, Biomedical Publishers, Austin, USA

Iismaa TP, Fathi Z, Hort YJ, Iben LG, Dutton JL, Baker E, Sutherland GR, Shine J (1998) Structural organization and chromosomal localization of three human galanin receptor genes. Ann NY Acad Sci 863:56–63

Ikeda M, Dewar D, McCulloch J (1995) Galanin receptor binding sites in the temporal and occipital cortex are minimally affected in Alzheimer's disease. Neurosci Lett 192:37–40

Invitti C, Brunani A, Pasqualinotto L, Dubini A, Bendinelli P, Maroni P, Cavagnini F (1995) Plasma galanin concentrations in obese, normal weight and anorectic women. Int J Obesity 19:347–349

Jacoby AS, Webb GC, Liu ML, Kofler B, Hort YJ, Fathi Z, Bottema CDK, Shine J, Iismaa TP (1997) Structural organization of the mouse and human GALR1 galanin receptor genes (*Galnr* and *GALNR*) and chromosomal localization of the mouse gene. Genomics 45:496–508

Ji R-R, Zhang Q, Bedecs K, Arvidsson J, Zhang X, Xu X-J, Wiesenfeld-Hallin Z, Bartfai T, Hökfelt T (1994) Galanin antisense oligonucleotides reduce galanin levels in dorsal root ganglia and induce autotomy in rats after axotomy. Proc Natl Acad Sci USA 91:12540–12543

Kaplan LM, Hooi SC, Abraczinskas DR, Strauss RM, Davidson MB, Hsu DW, Koenig JI (1991) Neuroendocrine regulation of galanin gene expression. In: Hökfelt T, Bartfai T, Jacobowitz D, Ottoson D (eds) Galanin: a new multifunctional peptide in the neuro-endocrine system. Macmillan, Cambridge, pp 43–65

Kar S, Quirion R (1994) Galanin receptor binding sites in adult rat spinal cord respond differentially to neonatal capsaicin, dorsal rhizotomy and peripheral axotomy. Eur J Neurosci 6: 1917–1921

Kask K, Langel Ü, Bartfai T (1995) Galanin – a neuropeptide with inhibitory actions. Cell Mol Neurobiol 15:653–673

Kinney GA, Emmerson PJ, Miller RJ (1998) Galanin receptor-mediated inhibition of glutamate release in the arcuate nucleus of the hypothalamus. J Neurosci 18:3489–3500

Klaus C, Hametner R, Jones N, Jones R, Iismaa TP, Sperl W, Kofler B (1998) Galanin and galanin receptor expression in neuroblastoma. Ann NY Acad Sci USA 863:438–441

Klimaschewski L, Unsicker K, Heym C (1995) Vasoactive intestinal peptide but not galanin promotes survival of neonatal rat sympathetic neurons and neurite outgrowth of PC12 cells. Neurosci Lett 195:133–136

Koegler FH, Ritter S (1998) Galanin injection into the nucleus of the solitary tract stimulates feeding in rats with lesions of the paraventricular nucleus of the hypothalamus. Physiol Behav 63:521–527

Kofler B, Evans HF, Liu ML, Falls V, Iismaa TP, Shine J, Herzog H (1995) Characterization of the 5'-flanking region of the human preprogalanin gene. DNA Cell Biol 14:321–329

Kofler B, Liu ML, Jacoby AS, Shine J, Iismaa TP (1996) Molecular cloning and characterisation of the mouse preprogalanin gene. Gene 182:71–75

Kofler B, Lapsys N, Furler SM, Klaus C, Shine J, Iismaa TP (1998) A polymorphism in the 3' region of the human preprogalanin gene. Mol Cell Probes 12:431-432

Köhler C, Chan-Palay V (1990) Galanin receptors in the post-mortem human brain. Regional distribution of ^{125}I-galanin binding sites using the method of in vitro receptor autoradiography. Neurosci Lett 120:179-182

Kuraishi Y, Kawamura M, Yamaguchi T, Houtani T, Kawabata S, Futaki S, Fujii N, Satoh M (1991) Intrathecal injections of galanin and its antiserum affect nociceptive response of rat to mechanical, but not thermal, stimuli. Pain 44:321-324

Kyrkouli SE, Stanley BG, Seirafi RD, Leibowitz SF (1990) Stimulation of feeding by galanin: anatomical localization and behavioral specificity of this peptide's effects in the brain. Peptides 11:995-1001

Liposits Z, Merchenthaler I, Reid JJ, Negro-Vilar A (1993) Galanin-immunoreactive axons innervate somatostatin-synthesizing neurons in the anterior periventricular nucleus of the rat. Endocrinology 132:917-923

Liu S, Lyeth BG, Hamm RJ (1994) Protective effect of galanin on behavioural deficits in experimental traumatic brain injury. J Neurotrauma 11:73-82

Lorimer DD, Benya RV (1996) Cloning and quantification of galanin-1 receptor expression by mucosal cells lining the human gastrointestinal tract. Biochem Biophys Res Commun 222:379-385

Lundquist CT, Rökaeus Å, Nässel DR (1991) Galanin immunoreactivity in the blowfly nervous system: localization and chromatographic analysis. J Comp Neurol 312:77-96

Lundquist CT, Rökaeus Å, Nässel DR (1992) Galanin message-associated peptide-like immunoreactivity in the nervous system of the blowfly: distribution and chromatographic characterization. J Neuroendocrinol 4:605-616

Maiter DM, Hooi SC, Koenig JI, Martin JB (1990) Galanin is a physiological regulator of spontaneous pulsatile secretion of growth hormone in the male rat. Endocrinology 126:1216-1222

Mastropaolo JN, Nadi NS,Ostrowski NL, Crawley JN (1988) Galanin antagonizes acetylcholine on a memory task in basal forebrain-lesioned rats. Proc Natl Acad Sci USA 85:9841-9845

Mazarati AM, Halaszi E, Telegdy G (1992) Anticonvulsive effects of galanin administered into the central nervous system upon the picrotoxin-kindled seizure syndrome in rats. Brain Res 589:164-166

McDonald MP, Crawley JN (1996) Galanin receptor antagonist M40 blocks galanin-induced choice accuracy deficits on a delayed-nonmatching-to-position task. Behav Neurosci 110:1025-1032

Merchenthaler I, López FJ, Negro-Vilar A (1993) Anatomy and physiology of central galanin-containing pathways. Prog Neurobiol 40:711-769

Mitchell V, Habert-Ortoli E, Epelbaum J, Aubert J-P, Beauvillain J-C (1997) Semiquantitative distribution of galanin-receptor (GAL-R1) mRNA-containing cells in the male rat hypothalamus. Neuroendocrinol 66:160-172

Montero CN, Hefti F (1989) Intraventricular nerve growth factor administration prevents lesion-induced loss of septal cholinergic neurons in aging rats. Aging 10:739-743

Moore JP, Maley BE, Lennes J, Hyde JF (1997) Evaluation of the distribution, secretion, and function of the peptide galanin in the anterior pituitary glands of human growth hormone-releasing hormone transgenic and normal mice. Soc Neurosci Abstr 23:1250

Murakami Y, Kato Y, Koshiyama H, Inoue T, Yanaihara N, Imura H (1987) Galanin stimulates growth hormone (GH) secretion via GH-releasing factor (GRF) in conscious rats. Eur J Pharmacol 136:415-418

Murakami Y, Kato Y, Shimatsu A, Koshiyama H, Hattori N, Yanaihara N, Imura H (1989) Possible mechanisms involved in growth hormone secretion induced by galanin in the rat. Endocrinology 124:1224-1229

Nicholl J, Kofler B, Sutherland GR, Shine J, Iismaa TP (1995) Assignment of the gene encoding human galanin receptor (*GALNR*) to 18q23 by in situ hybridization. Genomics 30:629-630

Ögren SO, Hökfelt T, Kask K, Langel Ü, Bartfai T (1992) Evidence for a role of the neuropeptide galanin in spatial learning. Neuroscience 51:1–5

O'Halloran DJ, Jones PM, Steel JH, Gon G, Giaid A, Ghatei MA, Polak JM, Bloom SR (1990) Effect of endocrine manipulation on anterior pituitary galanin in the rat. Endocrinology 127:467–475

Ohno K, Takeda N, Kiyama H, Kubo T, Tohyama M (1994) Occurrence of galanin-like immuno-reactivity in vestibular and cochlear efferent neurons after labyrinthectomy in the rat. Brain Res 644:135–143

Ormandy CJ, Lee CSL, Ormandy HF, Fantl V, Shine J, Peters G, Sutherland RL (1998a) Amplification, expression, and steroid regulation of the preprogalanin gene in human breast cancer. Cancer Res 58:1353–1357

Ormandy CJ, Hui R, Sutherland RL (1998b) Galanin and 11q13 amplification in human breast cancer. Biotech Lab Int 3:14–16

Ottlecz A, Samson WK, McCann SM (1986) Galanin: evidence for a hypothalamic site of action to release growth hormone. Peptides 7:51–53

Palazzi E, Fisone G, Hökfelt T, Bartfai T, Consolo S (1988) Galanin inhibits the muscarinic stim-ulation of phophoinositide turnover in rat ventral hippocampus. Eur J Pharmacol 148:479–480

Parker EM, Izzarelli DG, Nowak HP, Mahle LG, Iben LG, Wang JC, Goldstein ME (1995) Cloning and characterization of the rat GALR1 galanin receptor from Rin14B insulinoma cells. Mol Brain Res 34:179–189

Pellegrino JE, George RAV, Biegel J, Farlow MR, Gardner K, Caress J, Brown MJ, Rebbeck TR, Bird TD, Chance PF (1997) Hereditary neuralgic amyotrophy: evidence for genetic heteroge-neity and mapping to chromosome 17q25. Hum Genet 101:277–283

Planas B, Kolb PE, Raskind MA, Miller MA (1994) Activation of galanin pathways across puber-ty in the male rat: assessment of regional densities of galanin binding sites. Neuroscience 63:859–867

Planas B, Kolb PE, Raskind MA, Miller MA (1995) Galanin-binding sites in the female rat brain are regulated across puberty yet similar to the male pattern in adulthood. Neuroendocrinol 61:646–654

Planas B, Kolb PE, Raskind MA, Miller MA (1997) Nerve growth factor induces galanin gene expression in the rat basal forebrain: implications for the treatment of cholinergic dysfunc-tion. J Comp Neurol 379:563–570

Raffeld M, Jaffe ES (1991) Bcl-1, t(11;14), and mantle cell-derived lymphomas. Blood 78:259–263

Rao MS, Sun Y, Escary JL, Perreau J, Tresser S, Patterson PH, Zigmond RE, Brulet P, Landis SC (1993) Leukemia inhibitory factor mediates an injury response but not a target-directed developmental transmitter switch in sympathetic neurons. Neuron 11:1175–1185

Reimann W, Englberger W, Friderichs E, Selve N, Wilffert B (1994) Spinal antinociception by morphine in rats is antagonised by galanin receptor antagonists. Naunyn-Schmiedebergs Arch Pharmacol 350:380–386

Roberts MH, Speh JC, Moore RY (1989) The central nervous system of *Bulla gouldiana*: peptide localization. Peptides 9:1323–1334

Rodríguez-Puertas R, Nilsson S, Pascual J, Pazos A, Hökfelt T (1997) ^{125}I-galanin binding sites in Alzheimer's disease – increases in hippocampal subfields and a decrease in the caudate nucle-us. J Neurochem 68:1106–1113

Rökaeus Å (1994) Galanin. In: Walsh JH, Dockray GJ (eds) Gut peptides: biochemistry and phys-iology. Raven Press, New York, pp 525–552

Rökaeus Å, Brownstein MJ (1986) Construction of a porcine adrenal medullary cDNA library and nucleotide sequence analysis of two clones encoding a galanin precursor. Proc Natl Acad Sci USA 83:1682–1684

Rökaeus Å, Carlquist M (1988) Nucleotide sequence analysis of cDNAs encoding a bovine gala-nin precursor protein in the adrenal medulla and chemical isolation of bovine gut galanin. FEBS Lett 234:400–406

Rökaeus Å,Waschek JA (1994) Primary sequence and functional analysis of the bovine galanin gene promoter in human neuroblastoma cells. DNA Cell Biol 13:845-855

Rökaeus Å, Pruss RM, Eiden LE (1990) Galanin gene expression in chromaffin cells is controlled by calcium and protein kinase signaling pathways. Endocrinology 127:3096-3102

Rosenberg CL, Goo Kim H, Shows TB, Kronenberg HM, Arnold A (1991) Rearrangement and overexpression of D11S287E, a candidate oncogene on chromosome 11q13 in benign parathyroid tumors. Oncogene 6:449-453

Rossmanith WG, Marks DL, Clifton DK, Steiner RA (1996) Induction of galanin mRNA in GnRH neurons by estradiol and its facilitation by progesterone. J Neuroendocrinol 8:185-191

Rossowski WJ, Rossowski TM, Zacharia S, Ertan A, Coy DH (1990) Galanin binding sites in rat gastric and jejunal smooth muscle membrane preparations. Peptides 11:333-338

Rylett RJ, Goddard S, Schmidt BM, Williams LR (1993) Acetylcholine synthesis and release following continuous intracerebral administration of NGF in adult and aged Fischer-344 rats. J Neurosci 13:3956-3963

Sahu A (1998) Evidence suggesting that galanin (GAL), melanin-concentrating hormone (MCH), neurotensin (NT), proopiomelanocortin (POMC) and neuropeptide Y (NPY) are targets of leptin signaling in the hypothalamus. Endocrinology 139:795-798

Sano T, Vrontakis ME, Kovacs K, Asa SL, Friesen HG (1991) Galanin immunoreactivity in neuroendocrine tumors. Arch Pathol Lab Med 115:926-929

Schoofs L, Holman GM, Hayes TK, Nachman RJ, De Loof A (1991) Isolation, identification and synthesis of locustamyoinhibiting peptide (LOM-MIP), a novel biologically active neuropeptide from *Locusta migratoria*. Regul Pept 36:111-119

Selvais PL, Denef J-F, Adam E, Maiter DM (1995) Sex-steroid control of galanin in the rat hypothalamic-pituitary axis. J Neuroendocrinol 7:401-407

Selve N, Englberger W, Friderichs E, Hennies H-H, Reimann W, Wilffert B (1996) Galanin receptor antagonists attenuate spinal antinociceptive effects of DAMGO, tramadol and non-opioid drugs in rats. Brain Res 735:177-187

Sethi T, Rozengurt E (1991) Galanin stimulates Ca^{2+} mobilization, inositol phosphate accumulation, and clonal growth in small cell lung cancer cells. Cancer Res 51:1674-1679

Seufferlein T, Rozengurt E (1996) Galanin, neurotensin, and phorbol esters rapidly stimulate activation of mitogen-activated protein kinase in small cell lung cancer cells. Cancer Res 56:5758-5764

Shen ES, Meade EH, Pérez MC, Deecher DC, Negro-Vilar A, López FJ (1998) Expression of functional estrogen receptors and galanin messenger ribonucleic acid in immortalized luteinizing hormone-releasing hormone neurons: estrogenic control of galanin gene expression. Endocrinology 139:939-948

Shi T-JS, Zhang X, Holmberg K, Xu Z-QD, Hökfelt T (1997) Expression and regulation of galanin-R2 receptors in rat primary sensory neurons - effect of axotomy and inflammation. Neurosci Lett 237:57-60

Sillard R, Rökaeus Å, Xu Y, Carlquist M, Bergman T, Jörnvall H, Mutt V (1992) Variant forms of galanin isolated from porcine brain. Peptides 13:1055-1060

Simoneaux DK, Leach RJ, O'Connell P (1997) Galanin receptor 1 gene (*Galnr1*) is tightly linked to the myelin basic protein gene on chromosome 18 in mouse. Mamm Genome 8:875

Sizer AR, Rökaeus Å, Foster GA (1990) Analysis of the ontogeny of galanin in the rat central nervous system by immunohistochemistry and radioimmunoassay. Int J Dev Neurosci 8:81-97

Smith BK, York DA, Bray GA (1994) Chronic cerebroventricular galanin does not induce sustained hyperphagia or obesity. Peptides 15:1267-1272

Smith BK, York DA, Bray GA (1996) Effects of dietary preference and galanin administration in the paraventricular or amygdaloid nucleus on diet self-selection. Brain Res Bull 39:149-154

Smith KE, Forray C, Walker MW, Jones KA, Tamm JA, Bard J, Branchek TA, Linemeyer DL, Gerald C (1997) Expression cloning of a rat hypothalamic galanin receptor coupled to phosphoinositide turnover. J Biol Chem 272:24612-24616

Smith KE, Walker MW, Artymyshyn R, Bard J, Borowsky B, Tamm JA, Yao, W-J, Vaysse PJ-J, Branchek TA, Gerald C, Jones KA (1998) Cloned human and rat galanin GALR3 receptors – pharmacology and activation of G-protein inwardly rectifying K$^+$ channels. J Biol Chem 273: 23321–23326

Steel JH, Gon G, O'Halloran DJ, Jones PM, Yanaihara N, Ishikawa H, Bloom SR, Polak JM (1989) Galanin and vasoactive intestinal polypeptide are colocalised with classical pituitary hormones and show plasticity of expression. Histochemistry 93:183–189

Sullivan KA, Shiao LL, Cascieri MA (1997) Pharmacological characterization and tissue distribution of the human and rat GALR1 receptors. Biochem Biophys Res Commun 233:823–828

Sun Y, Zigmond RE (1996) Leukaemia inhibitory factor induced in the sciatic nerve after axotomy is involved in the induction of galanin in sensory neurons. Eur J Neurosci 8:2213–2220

Sundstrom E, Archer T, Melander T, Hökfelt T (1988) Galanin impairs acquisition but not retrieval of spatial memory in rats in the Morris swim maze. Neurosci Lett 88:331–335

Tanner JM, Ham TJ (1969) Low birthweight dwarfism with asymmetry (Silver's syndrome): treatment with human growth hormone. Arch Dis Child 44:231–243

Tanoh T, Shimatsu A, Ishikawa Y, Ihara C, Yanaihara N, Imura H (1993) Galanin-induced growth hormone secretion in conscious rats – evidence for a possible involvement of somatostatin. J Neuroendocrinol 5:183–187

Tatemoto K, Rökaeus Å, Jörnvall H, McDonald TJ, Mutt V (1983) Galanin – a novel biologically active peptide from porcine intestine. FEBS Lett 164:124–128

Tempel DL, Leibowitz KJ, Leibowitz SF (1988) Effects of PVN galanin on macronutrient selection. Peptides 9:309–314

Valkna A, Juréus A, Karelson E, Zilmer M, Bartfai T, Langel Ü (1995) Differential regulation of adenylate cyclase activity in rat ventral and dorsal hippocampus by rat galanin. Neurosci Lett 187:75–78

Verge VMK, Xu X-J, Langel Ü, Hökfelt T, Wiesenfeld-Hallin Z, Bartfai T (1993) Evidence for endogenous inhibition of autotomy by galanin in the rat after sciatic nerve section - demonstrated by chronic intrathecal infusion of a high affinity galanin receptor antagonist. Neurosci Lett 149:193–197

von Heijne G (1983) Patterns of amino acids near signal-sequence cleavage sites. Eur J Biochem 133:17–21

Vrontakis ME, Peden LM, Duckworth ML, Friesen HG (1987) Isolation and characterization of a complementary cDNA (galanin) clone from estrogen-induced pituitary tumor messenger RNA. J Biol Chem 262:16755–16758

Vrontakis ME, Sano T, Kovacs K, Friesen HG (1990) Presence of galanin-like immunoreactivity in nontumorous corticotrophs and corticotroph adenomas of the human pituitary. J Clin Endocrinol Metab 70:747–751

Vrontakis ME, Torsello A, Friesen HG (1991) Galanin. J Endocrinol Invest 14:785–794

Wang SK, He CG, Maguire MT, Clemmons AL, Burrier RE, Guzzi MF, Strader CD, Parker EM, Bayne ML (1997a) Genomic organization and functional characterization of the mouse GALR1 galanin receptor. FEBS Lett 411:225–230

Wang S, Hashemi T, He CG, Strader C, Bayne M (1997b) Molecular cloning and pharmacological characterization of a new galanin receptor subtype. Mol Pharmacol 52:337–343

Wang SK, He CG, Hashemi T, Bayne M (1997c) Cloning and expressional characterization of a novel galanin receptor – identification of different pharmacophores within galanin for the three galanin receptor subtypes. J Biol Chem 272:31949–31952

Wang S, Hashemi T, Fried S, Clemmons AL, Hawes BE (1998) Differential intracellular signaling of the GalR1 and GalR2 galanin receptor subtypes. Biochemistry 37:6711–6717

Wiesenfeld-Hallin Z, Xu X-J, Villar MJ, Hökfelt T (1989) The effect of intrathecal galanin on the flexor reflex in rat: increased depression after sciatic nerve section. Neurosci Lett 105:149–154

Wiesenfeld-Hallin Z, Xu X-J, Villar MJ, Hökfelt T (1990) Intrathecal galanin potentiates the spinal analgesic effect of morphine: electrophysiological and behavioural studies. Neurosci Lett 109:217–22

Wiesenfeld-Hallin Z, Bartfai T, Hökfelt T (1992a) Galanin in sensory neurons in the spinal cord. Front Neuroendocrinol 13:319–343

Wiesenfeld-Hallin Z, Xu X-J, Langel Ü, Bedecs K, Hökfelt T, Bartfai T (1992b) Galanin-mediated control of pain: enhanced role after nerve injury. Proc Natl Acad Sci USA 89:3334–3337

Wiesenfeld-Hallin Z, Xu X-J, Hao J-X, Hökfelt T (1993) The behavioural effects of intrathecal galanin on tests of thermal and mechanical nociception in the rat. Acta Physiol Scand 147: 457–458

Wynick D (1997) Targeted disruption of galanin reduces nerve regeneration and neuropathic pain and abolishes lactation. Summer Neuropeptide Conference, Key West FL, USA 7

Wynick D, Hammond PJ, Askinanya KO, Bloom SR (1993a) Galanin regulates basal and oestrogen-stimulated lactotroph function. Nature 364:529–532

Wynick D, Smith DM, Ghatei M, Akinsanya K, Bhogal R, Purkiss P, Byfield P, Yanaihara N, Bloom SR (1993b) Characterization of a high-affinity galanin receptor in the rat anterior pituitary - absence of biological effect and reduced membrane binding of the antagonist M15 differentiate it from the brain/gut receptor. Proc Natl Acad Sci USA 90:4231–4235

Wynick D, Small CJ, Bacon A, Holmes FE, Norman M, Ormandy CJ, Kilic E, Kerr NCH, Ghatei M, Talamantes F, Bloom SR, Pachnis V (1998) Galanin regulates prolactin release and lactotroph proliferation. Proc Natl Acad Sci USA 90:12671–12676

Xu X-J, Wiesenfeld-Hallin Z, Fisone G, Bartfai, Hökfelt T (1990) The N-terminal 1–16, but not C-terminal 17–29, galanin fragment affects the flexor reflex in rats. Eur J Pharmacol 182: 137–141

Xu X-J, Andell S, Bartfai T, Wiesenfeld-Hallin Z (1996) Fragments of galanin message-associated peptide (GMAP) modulate the spinal flexor reflex in rat. Eur J Pharmacol 318:301–306

Xu Z, Cortés R, Villar M, Morino P, Castel M-N, Hökfelt T (1992) Evidence for upregulation of galanin synthesis in rat glial cells in vivo after colchicine treatment. Neurosci Lett 145: 185–188

Xu Z-Q, Shi T-J, Hökfelt T (1996a) Expression of galanin and a galanin receptor in several sensory systems and bone anlage of rat embryos. Proc Natl Acad Sci USA 93:14901–14905

Xu Z-Q, Shi T-J, Landry M, Hökfelt T (1996b) Evidence for galanin receptors in primary sensory neurones and effect of axotomy and inflammation. NeuroReport 8:237–242

Zini S, Roisin M-P, Langel Ü, Bartfai T, Ben-Ari Y (1993) Galanin reduces release of endogenous excitatory amino acids in the rat hippocampus. Eur J Pharmacol 245:1–7

The Cholecystokinin – Gastrin Family of Peptides and Their Receptors

Jens F. Rehfeld

1
Introduction

The two major gastrointestinal hormones, gastrin and cholecystokinin (CCK), were discovered at the beginning of this century. In the immediate wake of Bayliss and Starling's breakthrough discovery of secretin – the first hormone to be known (1902) – Edkins from the same institute in London suggested that the antral mucosa of the stomach released a specific gastric acid stimulating factor to blood. He named the factor gastrin, an abbreviation of "gastric secretin" (Edkins 1905). Hence, gastrin was the second peptide hormone to be discovered in the history of biology. Twenty three years after Edkins' discovery, Ivy and Oldberg in Chicago found the third gastrointestinal hormone, CCK, as a new gallbladder emptying factor present in extracts of the small intestinal mucosa (1928). The painstaking efforts to purify and identify the structure of secretin and CCK began in the middle of this century in Stockholm. During this work Jorpes and Mutt (1966) realized that CCK is identical to pancreozymin, the fourth gastrointestinal hormone, which was described by Harper and Raper from Newcastle-upon-Tyne (1943). Pancreozymin regulates the exocrine secretion of enzymes from the pancreas.

While Mutt and coworkers purified secretin (1965) and CCK (1968), gastrin was concurrently purified by R. Gregory and Tracy in Liverpool (1964). The structure of gastrin-17 was subsequently identified by H. Gregory et al. (1964). Comparison of the primary structures of gastrin-17 and CCK-33 disclosed that the C-terminal active sites of gastrin and CCK are identical (H. Gregory et al. 1964; Mutt and Jorpes 1968). Thus, biologically active gastrin and CCK peptides have the same C-terminal pentapeptide amide sequence, and this structure has been exceedingly well preserved over the last 500 million years of evolution (Anastasi et al. 1967; Larsson and Rehfeld 1977; Johnsen and Rehfeld 1990). The complete homology of the active site, as well as other structural similarities, define gastrin and CCK as members of the same hormone family (for review, see Johnsen 1998).

Department of Clinical Biochemistry, Rigshospitalet, University of Copenhagen, Denmark

During the last two decades, the concept of gastrin and CCK as simple gut hormones changed considerably (for reviews, see Dockray and Gregory 1989; Rehfeld 1989, 1998b; Liddle 1994; Walsh 1994). Now, both gastrin and CCK are known to occur in multiple hormonal forms in plasma and gastrointestinal tissue. Moreover, they are both expressed in a tissue-specific manner in a wide variety of cells outside the gastrointestinal tract. For instance, the CCK peptides constitute the most abundant peptide system in the brain, and, vice versa, the brain is the main production site of CCK (Rehfeld 1978c; Crawley 1985). Accordingly, CCK and gastrin are not only hormones. They are also potent transmitters in the central and peripheral nervous systems. Moreover, gastrin and CCK have been shown to be growth factors in the gastrointestinal tract, and perhaps elsewhere (for reviews, see Johnson 1989; Rehfeld and van Solinge 1994). Today, the view is even favoured that growth stimulation is the most important function of gastrointestinal CCK and gastrin (Johnson 1976, 1989). The growth effects of gastrin and CCK may well be exerted by paracrine and/or autocrine secretory mechanisms (Rehfeld and van Solinge 1994).

In synchrony with the increased focusing on neurotransmitter activities and growth effects, the clinical interest in gastrin and CCK has gradually expanded from their pathophysiological roles in duodenal ulcers, gallbladder, and pancreatic diseases to their significance in neuropsychiatric disorders and tumour development. So far, neuropsychiatric interest has been shown particularly in eating disorders, panic-anxiety disorders and schizophrenia, whereas the oncological interest has focused mainly on the role of gastrin and CCK in the development of carcinoid tumours and carcinomas in the stomach, in the pancreas and in the colorectal mucosa. However, gastrin and CCK may, in accordance with their widespread expression, also be involved in neoplasias elsewhere (for review, see Rehfeld and van Solinge 1994).

2
Structure – Function Relationships

Most biologically active peptides occur in families whose members display significant structural homology. The occurrence of peptide families is supposed to reflect evolution from a single ancestral peptide gene by gene duplication and subsequent mutations (for reviews, see Adelson 1971; Barrington 1982; Young 1992; Acher 1993). Also, gastrin and CCK are members of a family (for review, see Rehfeld 1981). As mentioned, CCK and gastrin have an identical active site that is extremely well preserved during evolution (Larsson and Rehfeld 1977; Johnsen and Rehfeld 1990, 1992; Björnskov et al. 1992; Johnsen 1998). It is possible that the putative common ancestor resembles a dityrosyl-sulphated peptide, cionin, which has been isolated from protochordean neurons (Johnsen and Rehfeld 1990). The frogskin peptides, caerulein and phyllo-caerulein (Anastasi et al. 1967, 1969) are also members of the gastrin - CCK family (Fig. 1). At our present state of knowledge, however, CCK and gastrin are the only members of the family in mammals.

All known biological effects of gastrin and CCK peptides reside in the conserved common C-terminal tetrapeptide amide (-Trp-Met-Asp-Phe-NH$_2$; Fig. 1). Modification of this sequence grossly reduces or abolishes its biological effects and receptor binding in terms of gastric acid stimulation and gallbladder contraction in mammals. While Trp, Asp and Phe-NH$_2$ cannot be modified at all, conservative substitutions of the methionyl residue with leucyl or Norleucyl residues are tolerated (Morley et al. 1965). However, even monooxidation of the methionyl residue reduces the bioactivity of gastrin and CCK peptides significantly, although some binding to the receptors apparently may still

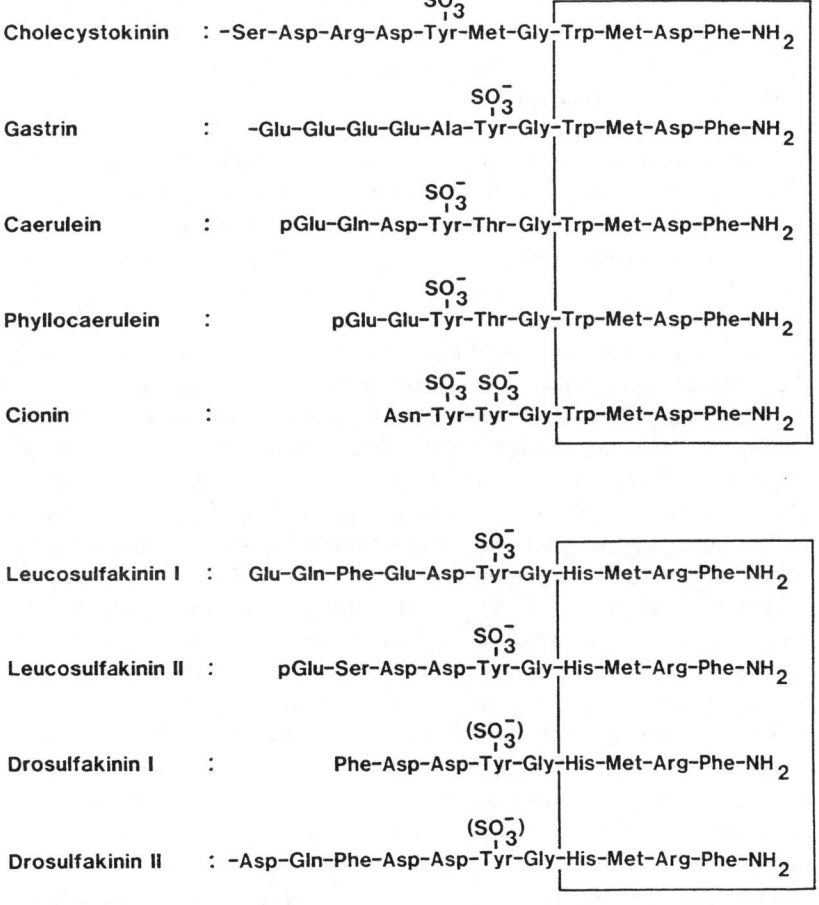

Fig. 1. C-terminal amino acid sequences of members of the gastrin-cholecystokinin family (*above*) and of the sulfakinins (*below*)

occur (Galleyrand et al. 1992). The different N-terminal extensions of the common pentapeptide amide increase the biological potency and the specificity for the CCK-A and gastrin/CCK-B receptor binding, respectively. Of particular importance in this respect is the tyrosyl residue in position six of mammalian gastrin and in position seven of CCK peptides, as counted from the C-terminal phenylalanyl amide (Fig. 1). Structurally, the gastrins are consequently defined as peptides, that stimulate gastric acid secretion and have the C-terminal sequence Tyr-X-Trp-Met-Asp-Phe-NH$_2$ (where X in most mammalian species is a glycyl residue). CCKs are defined as gallbladder emptying peptides having the C-terminal sequence Tyr-X$_1$-X$_2$-Trp-Met-Asp-Phe-NH$_2$ (where X$_1$ in most mammals is a methionyl residue and X$_2$ a glycyl residue). CCK peptides normally occur with the tyrosyl residue O-sulphated (Mutt and Jorpes 1968), which is important, since binding to the CCK-A receptor requires sulphation. In contrast, antral gastrin may occur with its tyrosyl residue either sulphated or unsulphated (Gregory and Tracy 1964; Gregory et al. 1964; Andersen 1985), and the gastrin/CCK-B receptor tolerates sulphated and unsulphated ligands equally well.

It has been suggested that sulphakinins, i.e. carboxyamidated and tyrosyl-sulphated peptides isolated from the cockroach, *Leucophaea maderae* (Nachman et al. 1986), or deduced from cDNA cloned from *Drosophila* (Nicols et al. 1988) might belong to the family. However, as shown in Fig. 1, the difference in the C-terminal active site comprises radical substitutions of two decisive residues (Trp→His and Asp→Arg). Such substitutions render a phylogenetic relationship less likely.

The active-site homology of gastrin with CCK has posed large problems in the study of these peptides, because the active site is strongly immunogenic (Rehfeld 1984, 1998a) and because the CCK-B/gastrin receptor binds peptides having the common sequence with almost equal affinity (Kopin et al. 1992; Wank et al. 1992). Consequently, evaluation of antibody and receptor binding of gastrin and CCK as well as studies of the physiological effects of gastrin and CCK require due consideration of functional interactions and immunochemical crossreactivity between the two systems of peptides in the family.

Human gastrin and CCK peptides display, not unexpectedly, extensive homology with corresponding peptides from other species. The extent of the species differences has been elucidated by cDNA deduction of the preprogastrin and preproCCK structures for various mammals (Yoo et al. 1982; Boel et al. 1983; Deschenes et al. 1984; Gubler et al. 1984; Takahashi et al. 1985; Fuller et al. 1987; Kang et al. 1989; Lund et al. 1989; Gantz et al. 1990). Comparison of the sequences is quite instructive. It shows how well the active site and structures around major processing sites are preserved (Figs. 2 and 3). The structure of some of the chemically identified bioactive peptides – gastrin-71, -34, -17 and CCK-83, -58, -33, -22 and -8 – can to some extent be deduced from the localization of mono- and dibasic cleavage sites in the prohormones (Figs. 2 and 3).

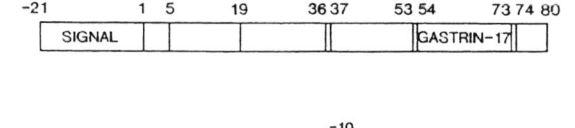

```
        -21           1  5      19        36 37     53 54      73 74 80
        ┌──────────────────────────────────────────────────────────────┐
        │  SIGNAL    │  │        │        ││         │GASTRIN-17││       │
        └──────────────────────────────────────────────────────────────┘
```

```
          -21                                                      -10
MAN  :  Met-Gln -Arg -Leu-Cys- Val - Tyr - Val -Leu- Ile - Phe- Ala -Leu-Ala -Leu- Ala - Ala -Phe-Ser-Glu -Ala
PIG  :                          - Ala -                       -His - Val -                        -Cys-
COW  :                          - Ala - His -                 -Leu- Val -                               -Cys-
DOG  :                                                        -Leu-                    - Thr -
RAT  :           -Pro-                       -Cys- Met-       - Val - Leu- Val -        - Thr -
MICE :           -Pro-                       - Met-           - Val - Leu- Val -        - Thr -
```

```
          1                                   10                                        20
MAN  :  Ser-Trp -Lys - Pro- Arg - Ser - Gln - Gln- Pro- Asp-Ala - Pro-Leu- Gly - Thr - Gly - Ala - Asn- Arg -Asp-Leu-
PIG  :                      - Gly -Phe-      -Leu- Gln -          - Ser- Ser -    - Pro-              Gly -Lys -
COW  :                              - His - Leu- Gln -            - Val - Ala - Pro- Arg-            Gly - Gln -
DOG  :                              - Arg- Leu- Gln -             - Ser-        - Pro-               Cys-
RAT  :                              - Leu- Gln -                  - Ser- Ser -  - Pro- Arg- Thr -    - Gly - Ala -
MICE :                              - Leu- Gln -                  - Ser- Ser -  - Pro-      - Thr -   - Glu -
```

```
                                             30                               40
MAN  :  Glu- Leu- Pro - Trp - Leu- Glu- Gln - Gln - Gly - Pro - Ala - Ser- His - His - Arg- Arg - Gln - Leu- Gly - Pro- Gln -
PIG  :  - Pro- His - Glu -      - Asp- Arg- Leu-                                               - Leu-
COW  :  - Pro- Leu- Arg- Met- Asn- Arg - Leu-                      - Asn- Pro -                 - Leu-
DOG  :  - Pro- His - Gly -      - Asp-    - Leu-                                                 - Leu-
RAT  :  - Gln - His - Gln -         - Lys - Leu-                                                - Leu-
MICE :  - Gln - Arg - Gln - Phe- Asn- Lys - Leu-    - Ser -                                     - Leu-
```

```
                                 50                          60
MAN  :  Gly - Pro - Pro - His - Leu- Val - Ala - Asp- Pro - Ser - Lys- Lys - Gln - Gly - Pro- Trp - Leu- Glu - Glu - Glu - Glu -
PIG  :                                              - Leu- Ala -                              - Met-
COW  :  Asp-                - Met-                   - Leu-                                    - Val -
Dog  :                      - Gln-                   - Leu-                                    - Met-
RAT  :              - Gln -    - Phe- Ile -          - Leu-                       - Arg -      - Pro - Met-
MICE :              - Gln -    - Phe- Ile -          - Leu-                       - Arg -      - Arg- Met-
```

```
                     70                               80
MAN  :  Glu- Ala - Tyr - Gly - Trp - Met- Asp- Phe- Gly - Arg - Arg - Ser- Ala - Glu -Asp- Glu -Asn
PIG  :                                                           - Glu - Gly - Asp- Gln - Arg- Pro
COW  :  Ala -                                                    - Glu - Gly - Asp- Gln - His - Pro
DOG  :  Ala -                                                    - Glu - Gly - Asp- Gln - Arg- Pro
RAT  :                                                           - Glu -      - Asp- Gln - Tyr- Asn
MICE :                                                           - Glu -Asp- Gln
```

Fig. 2. cDNA deduced amino acid sequences of preprogastrins from six mammalian species. *Boxes* indicate established proteolytic processing points, mainly at di- and monobasic sites. Note the preserved primary structures around the active site (residues 68–71) and around the processing sites

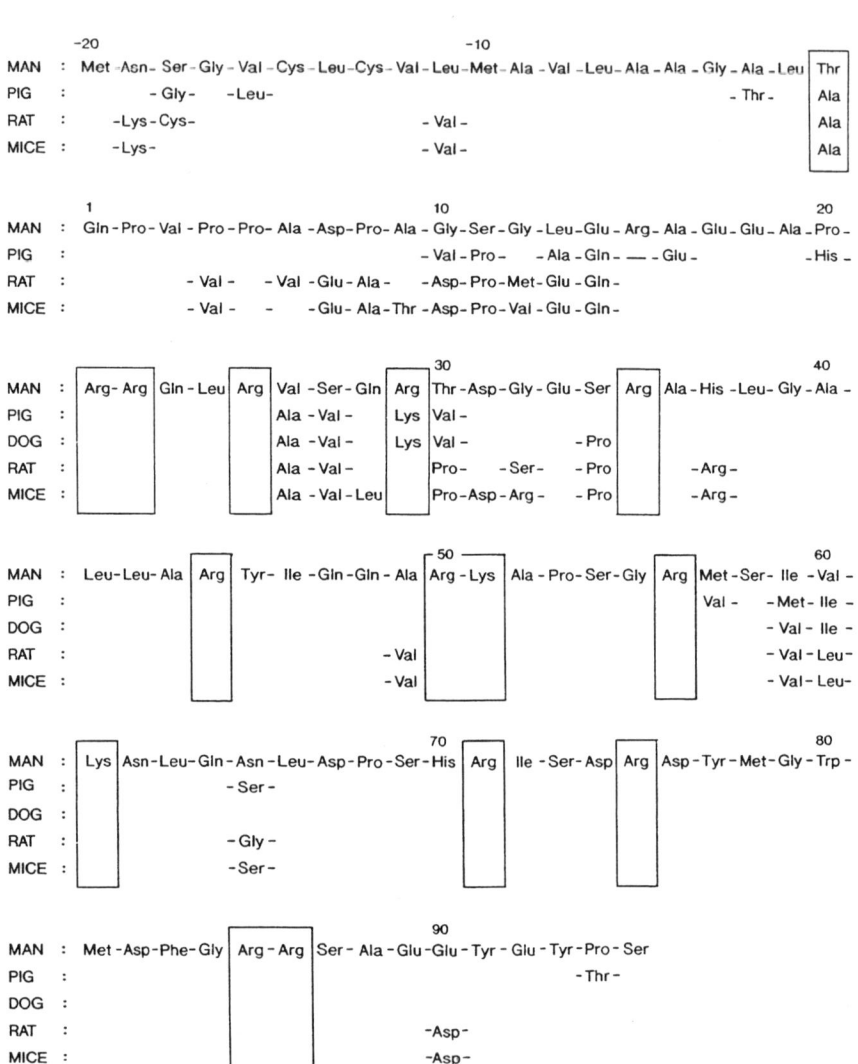

Fig. 3. cDNA deduced amino acid sequences of preprocholecystokinins from four mammalian species. In addition, the known canine sequence is also shown. Boxes indicate established proteolytic processing points, mainly at di- and monobasic sites. Note the preserved primary structures around the active site (residues 80–83) and around the processing sites

3
Gastrin and Cholecystokinin Genes and mRNAs

The single copy genes for human gastrin and CCK are located on chromosome 17q and chromosome 3 in the region 3q12-3pter, respectively (Lund et al. 1986; Takahashi et al. 1986). The cloning of the human gastrin gene (Kato et al. 1983; Ito et al. 1984; Wiborg et al. 1984) and CCK gene (Takahashi et al. 1985, 1986) showed that the genes are structurally similar, both in the overall exon intron organisation and in certain peptide coding sequences. The gastrin gene spans 4.1 kb chromosomal DNA and contains two introns of 3041 and 130 bp, respectively (Fig. 4). Antral G-cells generate a single mRNA of 0.7 kb which encodes the 101 amino acid preprogastrin in man and mice, whereas other known mammalian preprogastrins have 104 amino acids due to a prolonged C-terminal flanking peptide (Fig. 2; Boel et al. 1983; Wiborg et al. 1984). The first exon encodes the 5'-untranslated region (5'-UTR) (Fig. 4). The CCK gene spans 7 kb and contains, like the gastrin gene, two introns, although of ~1.1 and ~5 kb size. The I-cells in the small intestine express a single mRNA of 0.8 kb, coding for the 115 (114 in pigs) amino acid preproCCK (Takahashi et al. 1986). Also for CCK, the first exon encodes the 5'-UTR of the transcript (Fig. 5).

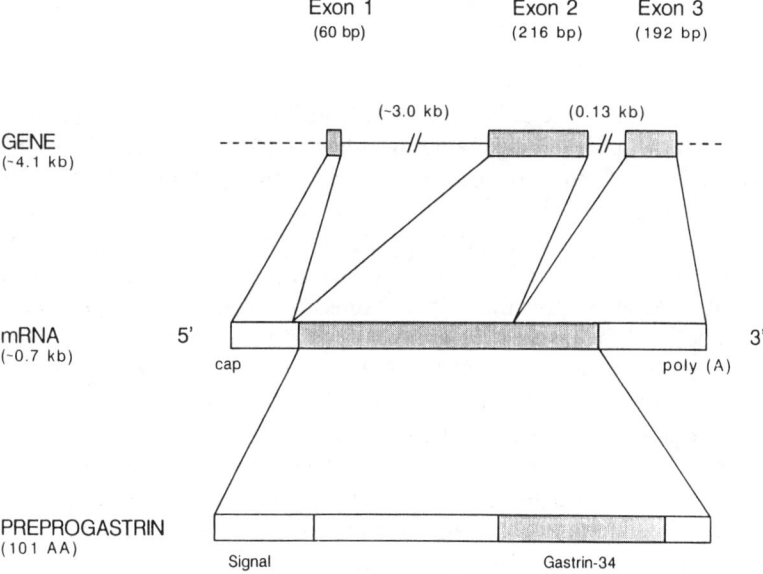

Fig. 4. Structure of the human gastrin gene, mRNA and prepropeptide. Exons are shown as *boxes*, introns as *straight lines*. The *grey area* of mRNA shows the coding region. *Numbers* indicate number of base pairs (bp) or kilobase pairs (kb) in each section of the gene. In preprogastrin position of gastrin-34 is shown

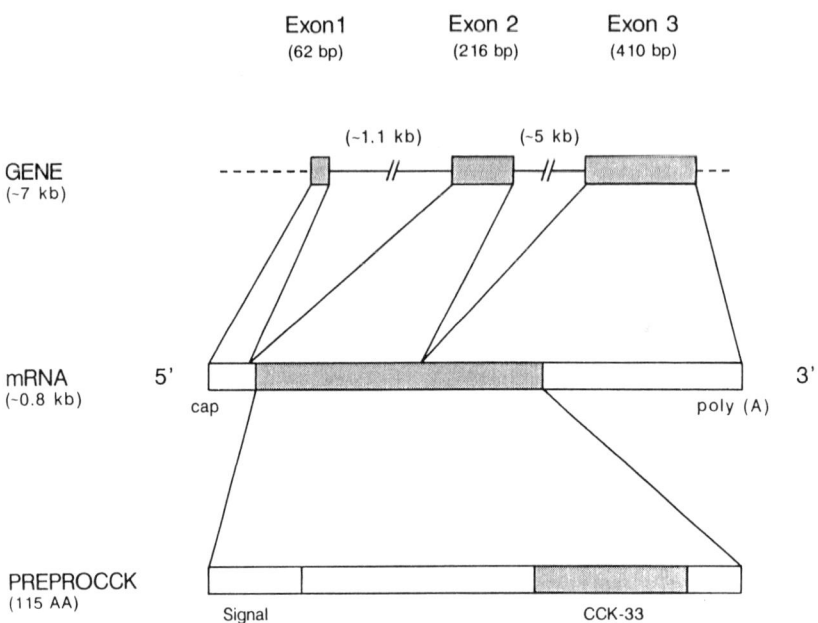

Fig. 5. Structure of the human cholecystokinin gene, mRNA and prepropeptide. Exons are shown as *boxes*, introns as *straight lines*. The *grey area* of mRNA shows the coding region. *Numbers* indicate number of base pairs (bp) or kilobase pairs (kb) in each section of the gene. In preprocholecystokinin (preproCCK) the position of *CCK-33* is shown

Gastrin and CCK cDNAs derived from tumor tissues are identical to those isolated from normal tissues (Kuwano et al. 1984; Kariya et al. 1986; Brand and Wang 1988). Studies on the promoters of the gastrin and CCK genes are of importance, since the differences in expression between normal and diseased tissues are likely to be caused entirely or partly by differences in transcriptional activation of the genes. Haun and Dixon (1990) found several *cis*-acting elements that are important for the expression of the rat CCK gene *in vitro*. They also defined the interacting *trans*-acting factors. Recent studies on the human CCK promoter (Nielsen et al. 1996) have revealed *cis*-acting elements in the human gene similar to that of the rat CCK gene (Haun and Dixon 1990).

Several studies have identified important regulatory domains in the human gastrin gene promoter using rodent-cell systems. Hence, a cell-specific regulatory element has been located in the cap-exon I region of the human gastrin gene (Theill et al. 1987). Wang and Brand (1990) have subsequently identified a pancreatic islet cell specific regulatory domain in the gastrin promoter, containing adjacent positive and negative DNA elements. The authors suggested that the regulatory domain is a switch controlling the transient transcription of the gastrin gene during fetal islet development (Larsson et al. 1976; Brand et al. 1984a). Godley and Brand (1989) reported that gastrin gene transcription was stimulated by epidermal growth factor (EGF) and inhibited by somatostatin. Later the EGF-responsive element was identified (Merchant et al. 1991). This element is of interest for the understanding of the growth promoting and oncological significance of gastrin, since EGF/TGF-α is expressed in a variety of carcinoma cell lines derived from the colorectal mucosa (Coffey et al. 1986; Huang et al. 1991; Karnes et al. 1992), ovary (Kurachi et al. 1991; Morishige et al. 1991; Stromberg et al. 1992; Kohler et al. 1992), pancreas (Smith et al. 1987) and stomach (Yasui et al. 1988). Therefore, neoplastic expression of the gastrin gene may be due to transcriptional activation of the gastrin gene by EGF and TGF-α. The transcriptional inhibition by somatostatin is probably mediated by a specific *cis* regulatory element located close to the EGF-responsive element (Brand et al. 1991). In addition to transcriptional inhibition by somatostatin, Karnik et al. (1989) have suggested that somatostatin may affect also the post-transcriptional processing of gastrin RNA.

So far, no tissue-specific splicing or use of alternative promoters have been described. Therefore, the cell-specific molecular patterns of gastrin and CCK peptides in different tissues are probably due to differences in the post-translational processing, rather than differential processing or transcription of the gene (Boel et al. 1983; Friedman et al. 1985; Brand and Fuller 1988;). However, in addition to the 0.7-kb gastrin mRNA, other transcripts have been found. Northern analysis of colorectal and testicular tissue extracts showed a transcript of ~3.3 kb hybridizing with a gastrin probe (Hoosein et al. 1990; Schalling et al. 1990; van Solinge et al. 1993a). It probably represents unprocessed gastrin RNA including sequences corresponding to the first intron of the gastrin gene. A minor band of 0.5 kb has also been observed (Schalling et al. 1990; van Solinge et al. 1993a). The nature of the latter mRNA is uncertain. It may encode a different but homologous peptide. The recently detected gastrin transcript containing the second 130 bp intron by reverse transcription polymerase chain reaction in colonic mucosal cell-lines (Baldwin et al. 1990) has now been found to be genomic DNA (Baldwin and Zhang 1992; van Solinge et al. 1993a).

Evidence of translational regulation of CCK gene expression has been reported by several authors. Thus, Gubler et al. (1987) found a major discrepancy between the levels of CCK mRNA and peptides in the cerebellum. It was later shown to be due partly to incomplete processing of proCCK (Rehfeld et al. 1992a). In other words, expression is in this particular case regulated both at the translational and post-translational level. Two additional cases of translational regulation have subsequently been reported: A human gastric carcinoma cell line (AGS) containing CCK mRNA but neither proCCK nor amidated CCK (van Solinge and Rehfeld 1992). The second observation was that the CCK mRNA concentrations in the rat colon increased after birth without concomitant peptide synthesis (Lüttichau et al. 1993).

In the developing rat colon, gastrin mRNA concentrations also increase from birth to adult apparently without a corresponding increase in peptide synthesis (Lüttichau et al. 1993). Therefore, the expression of the gastrin gene also appears to be regulated at the translational level. Comparison of rat, pig, and man cDNA sequences showed remarkably conserved base sequences in the 5'-and 3'-untranslated regions (Fuller et al. 1987). It is therefore possible that such sequence in the 5'-UTR may determine the translational efficiency in the expression cascade. The 3'-UTR may be involved in regulation of polyadenylation, termination of transcription or message stability. Further studies on gastrin and CCK mRNA are necessary to explain these observations. The expression of both the gastrin and the CCK genes is ontogenetically regulated (Friedman et al. 1985; Brand and Fuller 1988; Bardram et al. 1990; Rehfeld et al. 1992b; Lüttichau et al. 1993;). Consequently, expression of gastrin and CCK genes in tumours may involve factors that normally are expressed only in fetal life.

4
Progastrin Maturation

In normal mammals the antral G-cells are the main site of gastrin synthesis (for reviews, see Larsson 1980; Rehfeld 1981). G-cells are present also in the proximal duodenum. After antrectomy the duodenal G-cells "antralise" and increase their synthesis considerably (Nilsson and Brodin 1977). Gastrin biosynthesis studies have so far focused on antral tissue (Brand et al. 1984b; Sugano et al. 1985; Hilsted and Rehfeld 1987; S. Jensen et al. 1989). Combination of the results of these studies with general knowledge about peptide hormone synthesis (Kelly 1985; Schwartz 1990) provides a clear picture of the biosynthetic pathway of antral progastrin (Fig. 6).

After translation of gastrin mRNA in the RER (rough endoplasmic reticulum) and co-translational removal of the N-terminal signal peptide from preprogastrin, intact progastrin is transported to the Golgi apparatus. In the *trans*-Golgi network the first post-translational modifications occur. These are O-sulphation of the tyrosyl-66 residue neighbouring the active site, and the first of the prohormone convertase cleavages at two monobasic and three dibasic processing sites. From the *trans*-Golgi network, vesicles carry the processing intermediates towards the basal part of the G-cells, where the gastrin peptides are stored in characteristic secretory granules (Larsson and Rehfeld 1979a; Håkanson et al. 1982). Presumably, the endoproteolytic prohormone convertase and exoproteolytic carboxypeptidase E processings as well as the subsequent glutamyl cyclization, corresponding to the N termini of gastrin-34 and gastrin-17 (Figs. 2 and 6), continue during the transport from the Golgi to the early secretory granules. The last and decisive processing step in the synthesis of gastrin then occurs during storage and maturation in the secretory granules. The secretory granules contain the amidation enzyme (Murthy et al; 1986, Eipper et al. 1987), which removes glyoxylate from the immediate precursors, the glycine-extended gastrins, to complete the synthesis of bioactive

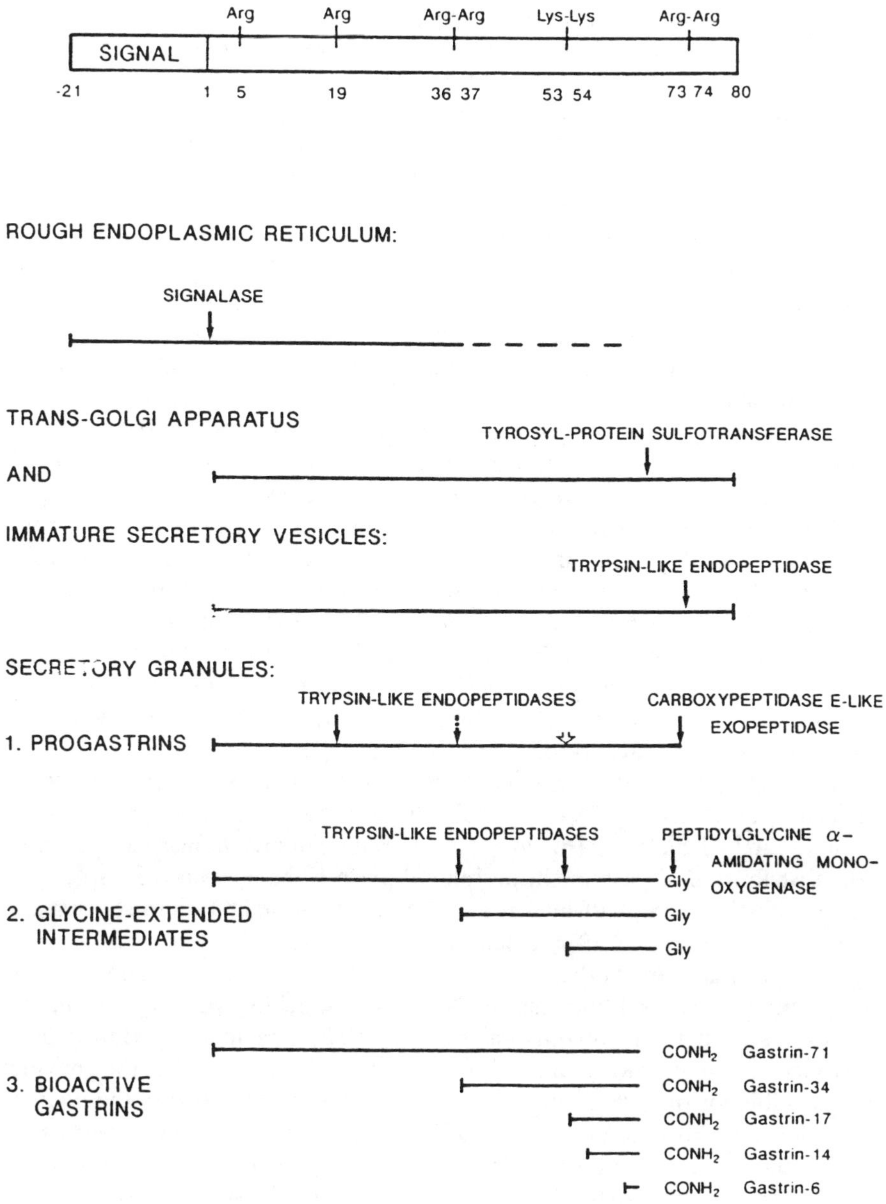

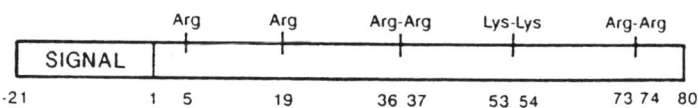

ROUGH ENDOPLASMIC RETICULUM:

SIGNALASE

TRANS-GOLGI APPARATUS

AND

TYROSYL-PROTEIN SULFOTRANSFERASE

IMMATURE SECRETORY VESICLES:

TRYPSIN-LIKE ENDOPEPTIDASE

SECRETORY GRANULES:

TRYPSIN-LIKE ENDOPEPTIDASES   CARBOXYPEPTIDASE E-LIKE
EXOPEPTIDASE

1. PROGASTRINS

TRYPSIN-LIKE ENDOPEPTIDASES   PEPTIDYLGLYCINE α –
AMIDATING MONO-
Gly   OXYGENASE

2. GLYCINE-EXTENDED
INTERMEDIATES   Gly

Gly

3. BIOACTIVE
GASTRINS

CONH$_2$	Gastrin-71
CONH$_2$	Gastrin-34
CONH$_2$	Gastrin-17
CONH$_2$	Gastrin-14
CONH$_2$	Gastrin-6

Fig. 6. Post-translational processing of progastrin in the antral G-cell

α-carboxyamidated peptides (Fig. 6). Amidation of gastrin is a crucial all-or-none activation process, which is carefully controlled (Hilsted et al. 1986, 1988; Hilsted and Rehfeld 1987; S. Jensen et al. 1989). Activation of the enzymatic amidation process requires copper, oxygen and ascorbic acid as cofactors and a pH around 5 (Eipper et al. 1987; for reviews see Hilsted 1991; Eipper et al. 1992). Carboxyamidation of peptides is known to require two sequentially acting enzymes: a copper and ascorbate-dependent peptidylglycine α-hydroxylating monooxygenase (PHM), derived from the N-terminal part of the amidation enzyme precursor, and a separate peptidyl-α-hydroxyglycine α-amidating lyase (PAL), derived from the remaining intragranular region of the same precursor (Katopodis et al. 1990; Perkins et al. 1990; Suzuki et al. 1990; Takahashi et al. 1990). Partial phosphorylation of serine in the C-terminal flanking fragment of progastrin also occurs (Dockray et al. 1987). The significance of this phosphorylation is not yet known and the kinase is so far unknown.

As a result of the elaborate biosynthetic pathway, the normal antral G-cells in mammals release a heterogenous mixture of progastrin products from the mature secretory granules. In man, a few per cent are non-amidated precursors, mainly glycine-extended gastrins, which in rodents constitute a considerably larger fraction of the progastrin product. However, in man more than 95% are α-amidated bioactive gastrins of which the longest molecular form is gastrin-71 (Rehfeld and Johnsen 1994). Of the amidated gastrins, 80–90% is gastrin-17, 5–10% gastrin-34 and the rest is a mixture of mainly gastrin-71 and gastrin-52, gastrin-14 and the short C-terminal sulphated hexapeptide amide (Malmström et al. 1976; Rehfeld and Uvnäs-Wallensten 1978; Rehfeld and Larsson 1979; Gregory et al. 1983; Hilsted and Hansen 1988; S. Jensen et al. 1989; Rehfeld and Johnsen 1994; Rehfeld et al. 1995). Approximately half the amidated gastrins are tyrosyl-sulphated (Gregory and Tracy 1964, 1972; Gregory et al. 1964; ; Andersen et al. 1983; Brand et al. 1984b; Andersen 1985; Hilsted and Rehfeld 1987), the completely sulphated gastrin-6 being an exception (Rehfeld et al. 1995). Due to gross differences in metabolic clearance rates, the distribution of gastrins in peripheral plasma changes, so that larger gastrins with their long half-lives predominate over gastrin-17 and shorter gastrins (Walsh et al. 1974,1976; S. Jensen et al. 1989; Rehfeld et al. 1995;).

Increased gastrin synthesis changes the molecular pattern further. Abnormally increased antral synthesis occurs in man by achlorhydria, as seen in pernicious anaemia. In antrum-sparing pernicious anaemia the translational activity of gastrin mRNA in the G-cells seems to be so high that the enzymes responsible for the processing of progastrin cannot keep up with the maturation, i.e. the carboxyamidation process (S. Jensen et al. 1989; Bundgaard et al. 1995, 1997). Consequently, G-cells release more unprocessed and incompletely processed non-amidated progastrin products when the synthesis is increased. Also, the carboxyamidated gastrins are less sulphated (Borch et al. 1986) and the N-terminus of progastrin cleaved to a lesser degree (S. Jensen et al. 1989). Precursors, processing intermediates and long chained carboxyamidated gastrins, such as gastrin-71, gastrin-52 and gastrin-34 are, as mentioned, cleared at a relatively slow rate from the circulation and therefore accumulate in plasma when synthesis and release are increased.

Gastrin peptides are synthesized in several cell types other than the antroduodenal G-cells. Quantitatively these other cells contribute only little to circulating gastrin in normal organisms partly because the secretion seems to serve local purposes rather than a general endocrine purpose, and partly because the biosynthetic processing is cell-specific, i.e. so different from that of antroduodenal G-cells that bioactive amidated gastrin may not even be synthesized. So far, expression of progastrin and its products has been encountered outside the antroduodenal mucosa in endocrine cells in ileum (Larsson and Rehfeld 1979a; Rehfeld et al. 1992b; Friis-Hansen and Rehfeld 1994); in unidentified cells in the colon (Lüttichau et al. 1993; van Solinge et al. 1993a); in endocrine cells in the fetal and neonatal pancreas (Larsson et al. 1976; Brand et al. 1984a; Bardram et al. 1990); in pituitary corticotrophs and melanotrophs (Rehfeld 1978a, 1986; Larsson and Rehfeld 1981); in oxytocinergic hypothalamo-pituitary neurons (Rehfeld 1978a; Rehfeld et al. 1984); in a few cerebellar (Rehfeld 1991) and vagal neurons (Uvnäs-Wallensten et al. 1977); in the adrenal medulla of some species (Rehfeld, unpublished results); in the bronchial mucosa (Rehfeld et al. 1989); in postmenopausal ovaria (van Solinge et al. 1993a); and in human spermatogenic cells (Schalling et al. 1990). As shown in Table 1, the

Table 1. Expression of the gastrin gene at peptide level in normal adult mammalian tissue.

Tissue	Total translation product (pmol/g tissue)	Precursor percentage
Gastrointestinal tract		
Antral mucosa	10,000	5
Duodenal mucosa	400	20
Jejunal mucosa	40	30
Ileal mucosa	20	85
Colonic mucosa	0.2	100
Neuroendocrine tissue		
Cerebellum	5	20
Vagal nerve	8	10
Adenohypophysis	200	98
Neurohypophysis	30	5
Adrenal medulla	2	100
Pancreas	2	95
Genital tract		
Ovaries	0.5	100
Testicles	6	100
Spermatozoa	2	55
Respiratory tract		
Bronchial mucosa	0.3	100

Orders of magnitude based on examination of different mammalian species.

concentrations and presumably also the synthesis in the extra-antral tissues are far below that of the antral "main factory".

The precise function of gastrin synthesized outside the antroduodenal mucosa is largely unknown. Several suggestions can be offered. First, an obvious possibility is paracrine or autocrine regulation of growth. Second, it is possible that the low concentration of peptides is without significant function in the adult, but is a relic of a more comprehensive fetal synthesis for local stimulation of growth (Bardram et al. 1990; Lüttichau et al. 1993). A third possibility is that the low cellular and, hence, low tissue concentration is due to constitutive rather than regulated secretion (Kelly 1985). With constitutive secretion there is no storage of peptides in the cells in spite of a considerable release per time unit.

Although it is possible that the extra-antral synthesis of gastrin is without function in the normal adult organism, recognition of the phenomenon nevertheless has considerable biomedical interest, namely in an oncofetal sense. Hence, tumours originating from cells that normally express the gastrin gene at low levels in adult organisms may produce gastrin in significant amounts. For instance, most gastrinomas originate in the pancreas, which has a substantial fetal expression of gastrin (Larsson et al. 1976; Brand et al. 1984a; Bardram et al. 1990).

5
Procholecystokinin Maturation

Although normal mammals synthesize most cholecystokinin in the central nervous system, an essential part is synthesized also in the I-cells of the small intestine (Rehfeld 1978c). Moreover, almost all CCK in plasma originates from the endocrine I-cells of the small intestine (Rehfeld 1998a). The I-cells are, however, so disseminated in the intestinal mucosa, and the intensity of the CCK synthesis per gram of small intestine therefore so modest that dynamic biosynthesis studies have so far been essentially impossible to perform. However, the dynamics of cerebral CCK synthesis have been studied in detail (Golterman et al. 1980; Stengaard-Pedersen et al. 1984). In addition, a variety of proCCK derived peptides of different chain length have been identified from extracts of both the small intestine and brain (Mutt and Jorpes 1968, 1971; Mutt 1976; Dockray et al. 1978; Eng et al. 1983; Reeve et al. 1986, 1991; Rehfeld and Hansen 1986; Shively et al. 1987 Eberlein et al. 1992; Blanke et al. 1993). Combination of these results with general knowledge about peptide hormone synthesis (for reviews, see Kelly 1985; Schwartz 1990) has provided a picture of the biosynthetic pathway of CCK (Fig. 7).

After translation of CCK mRNA in the RER and cotranslational removal of the N-terminal pre- or signal peptide from preproCCK, intact proCCK is

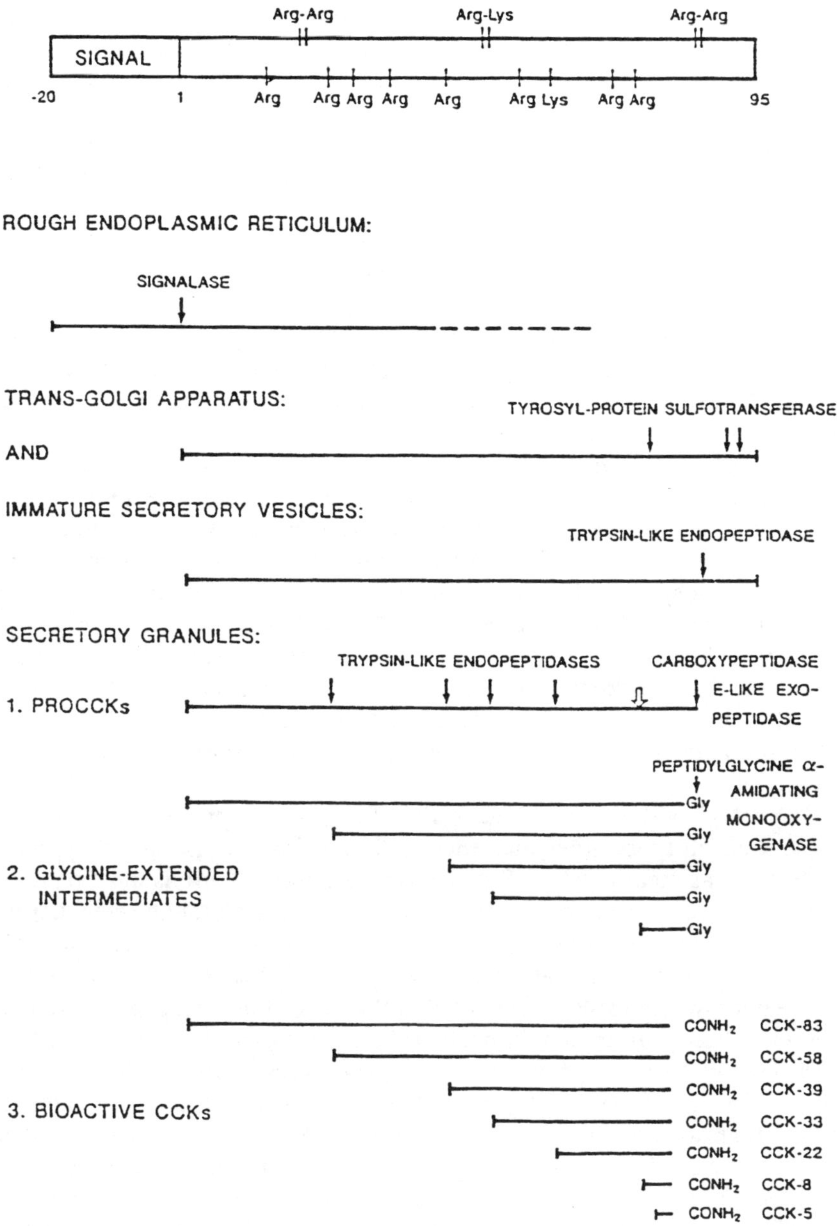

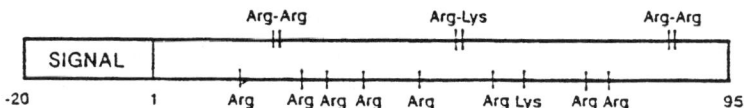

ROUGH ENDOPLASMIC RETICULUM:

SIGNALASE

TRANS-GOLGI APPARATUS:

AND

TYROSYL-PROTEIN SULFOTRANSFERASE

IMMATURE SECRETORY VESICLES:

TRYPSIN-LIKE ENDOPEPTIDASE

SECRETORY GRANULES:

TRYPSIN-LIKE ENDOPEPTIDASES   CARBOXYPEPTIDASE

1. PROCCKs

E-LIKE EXO-
PEPTIDASE

PEPTIDYLGLYCINE α-
AMIDATING
MONOOXY-
GENASE

2. GLYCINE-EXTENDED
    INTERMEDIATES

	-Gly	
	-Gly	
	-Gly	
	-Gly	
	-Gly	

3. BIOACTIVE CCKs

	CONH₂	CCK-83
	CONH₂	CCK-58
	CONH₂	CCK-39
	CONH₂	CCK-33
	CONH₂	CCK-22
	CONH₂	CCK-8
	CONH₂	CCK-5

Fig. 7. Post-translational processing of procholecystokinin in the I-cells of the small intestine

transported to the Golgi-apparatus. As for other peptide hormones, the first post-translational modifications occur in the Golgi apparatus, where proCCK is tyrosyl O-sulphated in three positions (Tyr-77, Tyr-92 and Tyr-95). The pro-hormone convertase cleavages at multiple monobasic and one dibasic process-ing sites also begin in the *trans*-Golgi apparatus, and continue in small vesicles carrying the processing intermediates towards the basal parts of the I-cells, where the processing continues in the secretory granules. The last and decisive processing-step in the synthesis of bioactive CCK peptides then, as for the gas-trins, occurs during storage and maturation in the secretory granules. The secretory granules contain the precursor for the two enzymes necessary for amidation (Murthy et al. 1986; Eipper et al. 1987; Katopodis et al. 1990; Perkins et al. 1990), which removes glyoxylate from the immediate precursors, the gly-cine-extended CCKs, to complete the synthesis of bioactive α-carboxyamidat-ed peptides (vide supra and Fig. 7).

As a result of the elaborate biosynthetic pathway, the normal CCK neurons and small intestinal I-cells release a heterogenous mixture of proCCK prod-ucts from the mature secretory granules. A few per cent are non-amidated pre-cursors. The amidated CCKs constitute a mixture of the longest possible bio-active product of proCCK, i.e. CCK-83 (Eberlein et al. 1992) in addition to the medium-sized CCK-58, CCK-39, CCK-33, and CCK22 (Mutt and Jorpes 1968; Mutt 1976; Rehfeld 1978c; Eng et al. 1984; Reeve et al. 1986) and the short CCK-8 and CCK-5 (Dockray et al. 1978; Rehfeld and Hansen 1986; Shively et al. 1987). The distribution and release patterns vary grossly between species (Larsson and Rehfeld 1979a; Eberlein et al. 1988; Cantor and Rehfeld 1989; Rehfeld 1994).

In analogy with the gastrin gene, the CCK gene is expressed at peptide level in several cell types other than the small intestinal I-cells. Entirely predomi-nating are, as mentioned, CCK neurons in the brain. Neurons in all regions of the central nervous system synthesize CCK peptides, although cerebellar neu-rons only in the fetal state (Rehfeld 1978c; Larsson and Rehfeld 1979b; Rehfeld et al. 1992a; Mogensen et al. 1990). The highest expression occurs in neocorti-cal regions, which explains why CCK is the most abundant peptide system in the mammalian brain and, in particular, in the human brain (Rehfeld 1978c; Crawley 1985). CCK peptides are also widely expressed in peripheral neurons, primarily in the intestinal tract, but also in the genitourinary tract and else-where (Larsson and Rehfeld 1979b). Low level expression has been found also in pituitary corticotrophs (Rehfeld 1986, 1987); in thyroid C-cells (Rehfeld et al. 1990); in the adrenal medulla (Bardram et al. 1989); in the bronchial muco-sa (Ghatei et al. 1982); and in spermatogenic cells of certain species (Persson et al. 1989). However, the processing pattern of proCCK in these secondary sites of expression generally deviates substantially from that of intestinal I-cells and cerebral CCK neurons (for reviews, see Rehfeld 1989, 1998b). Quantitative measures of the tissue-specific expression are presented in Table 2.

Table 2. Expression of the cholecystokinin gene at peptide level in normal adult mammalian tissue

Tissue	Total translation product (pmol/g tissue)	Precursor percentage
Gastrointestinal tract		
Duodenal mucosa	200	5
Jejunal mucosa	250	20
Ileal mucosa	20	50
Colonic mucosa	5	50
Neuroendocrine tissue		
Adenohypophysis	25	100
Neurohypophysis	20	10
Thyroid gland	2	20
Adrenal Medulla	1	50
Genital tract		
Testicles	5	80
Spermatozoa[a]	–	–
Central nervous system		
Cerebral cortex	400	2
Hippocampus	350	2
Hypothalamus	200	2
Cerebellum	2	80

Orders of magnitude based on examination of different mammalian species.

[a] Cholecystokinin peptides are present in spermatozoa of non-human mammals. The concentration, however, has not been quantitated.

6
Gastrin and Cholecystokinin Receptors

It was development of non-oxidative labelled CCK peptides that paved the way for the first characterization of the CCK receptors (Rehfeld 1978b; Sankaran et al. 1979; Jensen et al. 1980; Miller et al. 1981; for review, see Gardner and Jensen 1989). The biological and structural characterization was greatly advanced by subsequent design of receptor antagonists, of which the potent and selective benzodiazepine derivative L-364,718 and its further derivatives were first available (Chang and Lotti 1986; Evans et al. 1986). Using these tools, two subtypes of CCK receptors (CCK-A and CCK-B) have been classified (for reviews, see R. T. Jensen et al. 1989; Wank 1995).

The CCK-A receptors mediate the physiologic gallbladder contraction, pancreatic growth and enzyme secretion, delay of gastric emptying and relaxation of the sphincter of Oddi (R.T. Jensen et al. 1989). CCK-A receptors have been found also in the anterior pituitary, the myenteric plexus and important areas of the midbrain in which CCK-containing dopaminergic neurons have been implicated in the pathogenesis of schizophrenia and certain other neuropsychiatric diseases (Hökfelt et al. 1980; for review, see Crawley 1991). The CCK-A receptor binds with high affinity CCK peptides that are both carboxyamidated and tyrosyl-sulphated. In contrast, non-sulphated CCK peptides and gastrin peptides are bound with low affinity. The CCK-A receptor from rat pancreas has been purified and the corresponding cDNA cloned (Wank et al. 1992). The cDNA encodes a 444 amino acid protein containing seven transmembrane domains (Fig. 8). Thus, it belongs to the G-protein coupled receptor superfamily. The CCK-A receptor coupled G-protein activates phospholipase C, breaks down inositol phospholipids, mobilizes intracellular calcium and activates protein kinase C. Interestingly, the CCK-A receptor is overexpressed in some experimental pancreatic carcinomas (Bell et al. 1992).

The CCK-B receptor is the predominant CCK receptor of the central nervous system, the "brain receptor". It is expressed with particularly high density in the cerebral cortex (Saito et al. 1980). The CCK-B receptor is less selective than the CCK-A receptor, as it binds both tyrosyl-sulphated and non-sulphated CCK peptides as well as gastrins and short C-terminal fragments of CCK and gastrin, with almost similar affinity (Saito et al. 1980; Williams et al. 1986). The CCK-B receptor has not yet been purified as a protein, but cDNA encoding the human CCK-B receptor has been cloned and characterized (Lee et al. 1993, Fig. 9). Data on the cloned gastrin receptor from canine parietal cells (Kopin et al. 1992) show that the CCK-B receptor is identical with the gastrin receptor.

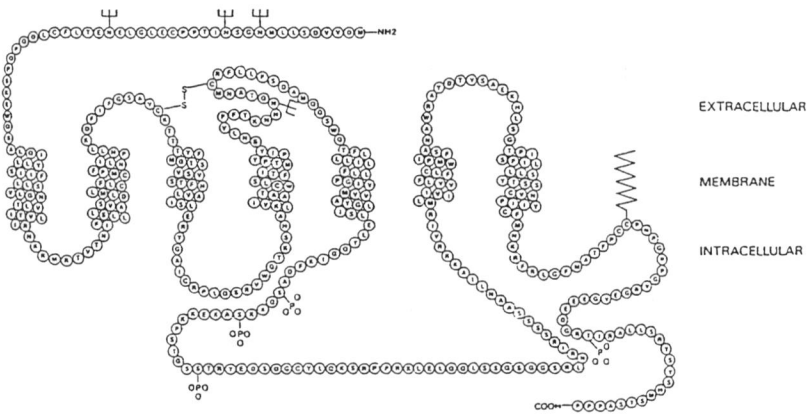

Fig. 8. Rat CCK-A receptor. The cDNA deduced amino acid sequence (*monoletter symbols*) shows putative transmembrane helices, consensus sites for N-linked glycosylation (tridents), serine and threonine phosphorylation (–PO₃), disulphide bridge (–S–S–) and cystein palmitoylation (*jagged line*). (Courtesy of Stephen A. Wank, NIH, Maryland, USA)

Procedures for mild labelling of gastrin peptides of high specific activity paved the way also for the initial characterization of the gastrin receptor (Stadil and Rehfeld 1972; Rehfeld 1980; Soll et al. 1984; for review, see Soll 1989). The gastrin receptor appears to be only moderately selective, as it binds carboxyamidated gastrin and CCK peptides with essentially similar affinities (Soll 1989). As mentioned, gastrin receptor cDNA from a canine parietal cDNA expression library has been cloned (Kopin et al. 1992) and shown to be identical with the CCK-B receptor cDNA. Thus, the gastrin/CCK-B receptor is abundantly expressed both in the brain, the stomach and also in the pancreas of man (Lee et al. 1993). The canine gastrin receptor is a 453 amino acid protein that also contains seven transmembrane domains, and belongs to the G-protein coupled receptor superfamily (Fig. 9). Like the CCK-A receptor it has a number of N-terminal glycosylation sites, several serine and threonine residues for phosphorylation, but no tyrosine phosphorylation site. The gastrin receptor coupled G-protein also activates phospholipase C, degrades inositolphospholipids, mobilizes intracellular calcium and activates protein kinase C (Kopin et al. 1992). Beinborn et al. (1993) have shown that valine-319 in the sixth transmembrane domain of the human gastrin/CCK-B receptor determines the specificity for binding of the nonpeptide receptor antagonists. This receptor site for antagonist binding differs from that of agonist binding, as shown also for other peptide receptors (Fong et al. 1993; Gether et al. 1993; Sachais et al. 1993). Hence, allosteric binding of non-peptide antagonists is a common feature of the G-protein coupled, seven transmembrane spanning family of peptide receptors. Song et al. (1993) demonstrated a few years ago

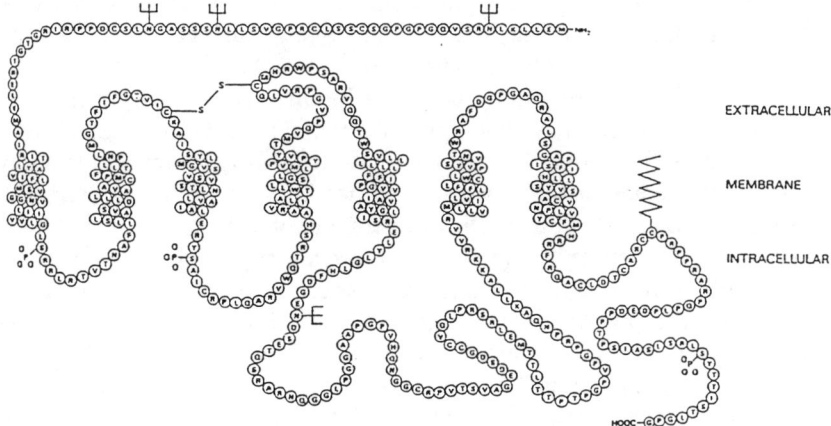

Fig. 9. Rat CCK$_B$ receptor. The cDNA deduced amino acid sequence (*monoletter symbols*) shows putative transmembrane helices, consensus sites for N-linked glycosylation (tridents), serine and threonine phosphorylation (–PO$_3$), disulphide bridge (–S–S–) and cysteine palmitoylation (*jagged line*). (Courtesy of Stephen A. Wank, NIH, Maryland, USA)

alternative splicing of CCK-B receptor mRNA, and suggested that the two receptor isoforms contribute to discriminate between CCK and gastrin signal transduction.

The identification of the CCK-A and gastrin/CCK-B receptors has opened new avenues for gastrin and CCK studies, and made it possible to adress a number of pertinent receptor questions. For instance, is the low affinity binding protein characterized by Baldwin et al. (1986) another gastrin receptor? Does a CCK-C receptor exist in certain endocrine tissues (Rehfeld 1992)? Can the CCK-A and gastrin/CCK-B receptors account for the binding of CCK and gastrin peptides as described in various cell-lines and carcinomas (Singh et al. 1985, 1986; Hoosein et al. 1988; Weinstock and Baldwin 1988; Sethi and Rozengurt 1991; Frucht et al. 1992; Ischizuka et al. 1992)? Are there specific receptors for glycine-extended processing intermediates, which mediate the growth effects of the gastrin and CCK peptide systems in normal and neoplastic tissue (Seva et al. 1994; Kaise et al. 1995; Singh et al. 1995)? Finally, are there oncoproteins homologous with the CCK and/or gastrin receptors?

7
Summary and Perspective

Gastrin and CCK are important regulatory peptide systems. Although originally discovered and named as gut hormones, they are now recognized as widespread intercellular messenger peptides in many compartments of the body. In addition to their classical blood-borne hormonal activity, they are also neurotransmitters and growth factors, while the significance of the acrosomal expression of CCK and gastrin peptides in mammals remains to be studied. In addition, their far-reaching phylogeny indicates vital roles in the evolution of multicellular life during more than 500 million years.

Gastrin and CCK are among the most intensively studied gut hormones. Probably more studies have been published about this family than about any other gut peptide system. Nevertheless, we are still ignorant about many of their functions. For instance, we still do not know the roles of gastrin CCK peptides and their receptors in the development of cancers in the gut and elsewhere. Also, the neurophysiological and psychiatric significances of cerebral CCK peptides are still largely unknown. Much research also remains to be done about the putative roles of gastrin and CCK in the fertility process. Finally, several pharmacological studies as well as recent knock-out studies (Koh et al. 1997; Langhans et al. 1997; Friis-Hansen et al. 1998) of gastrin, CCK and their receptors cannot be explained by today's knowledge about either peptides ligands or receptors. Hence, we may foresee an expansion of the family, with new peptides and/or new receptors recognized in coming decades.

Acknowledgements. The skilful secretarial assistance of Jette Jensen is gratefully acknowledged. The studies in the author's laboratory, upon which much of this review are based, have been supported by grants from the Danish Medical Research Council, the Danish Cancer Union, the Danish Biotechnology Program for Cellular Communication and the Vissing and the Novo Nordisk Foundations.

References

Acher R (1993) Neurophyseal peptide systems: processing machinery, adaptation and evolution. Regul Pept 45:1–13

Adelson JW (1971) Enterosecretory proteins. Nature 229:321–325

Anastasi A, Erspamer W, Endean R (1967) Isolation and structure of caerulein, an active decapeptide from the skin of *Hyla caerulea.* Experientia 23:699–700

Anastasi A, Bertaccini G, Cei JM, de Caro G, Erspamer V, Impicciatore M (1969) Structure and pharmacological actions of phyllocaerulein, a caerulein-like nonapeptide. Br J Pharmacol 37: 198–206

Andersen BN (1985) Species variation in the tyrosine sulfation of mammalian gastrins. Gen Comp Endocrinol 58:44–50

Andersen BN, Petersen B, Borch K, Rehfeld JF (1983) Variations in sulfation-ratios of circulating gastrins in gastrointestinal diseases. Scand J Gastroenterol 18:565–570

Baldwin GS, Zhang Q-X (1992) Measurement of gastrin and transforming growth factor α mRNA in colon carcinoma cell-lines by quantitative PCR. Cancer Res 52:2261–2267

Baldwin GS, Chandler R, Scanlon DB, Weinstock J (1986) Identification of a gastrin binding protein in porcine gastric mucosal membranes. J Biol Chem 261:12252–12257

Baldwin GS, Casey A, Mantamadiotis T, McBride K, Sizeland AM, Thumwood CM (1990) PCR cloning and sequence of gastrin mRNA from carcinoma cell-lines. Biochem Biophys Res Commun 170:691–697

Bardram L, Hilsted L, Rehfeld JF (1989) Cholecystokinin, gastrin and their precursors in phaechromocytomas. Acta Endocrinol 120:303–309

Bardram L, Hilsted L, Rehfeld JF (1990) Progastrin expression in the mammalian pancreas. Proc Natl Acad Sci USA 87:298–302

Barrington EJW (1982) Evolutionary and comparative aspects of gut and brain peptides. Br Med Bull 38:227–232

Bayliss WM, Starling EH (1902) The mechanism of pancreatic secretion. J Physiol 28:325–353

Beinborn M, Lee YM, McBride EW, Quinn SM, Kopin AS (1993) A single amino acid of the cholecystokinin-B/gastrin receptor determines specificity for non-peptide antagonists. Nature 362:348–350

Bell RH, Kuhlmann ET, Jensen RT, Longnecker DS (1992) Overexpression of cholecystokinin receptors in azaserine-induced neoplasms of the rat pancreas. Cancer Res 52:3295–3299

Björnskov I, Rehfeld JF, Johnsen AH (1992) Identification of four chicken gastrins obtained by processing at post-Phe bonds. Peptides 13:595–601

Blanke SE, Johnsen AH, Rehfeld JF (1993) N-terminal fragments of intestinal cholecystokinin: evidence for release of CCK-8 by cleavage on the carboxyl side of Arg-74 in proCCK. Regul Pept 46:575–583

Boel E, Vuust J, Norris K, Norris F, Wind A, Rehfeld JF, Marcker KA (1983) Molecular cloning of human gastrin cDNA: evidence for evolution of gastrin by gene duplication. Proc Natl Acad Sci USA 80:2866–2869

Borch K, Renvall H, Liedberg G, Andersen BN (1986) Relation between circulating gastrin and endocrine cell proliferation in the atrophic gastric mucosa. Scand J Gastroenterol 21:357–363

Brand SJ, Fuller PJ (1988) Differential gastrin gene expression in rat gastrointestinal tract and pancreas during development. J Biol Chem 263:5341–5347

Brand SJ, Wang TC (1988) Gastrin gene expression and regulation in rat islet cell lines. J Biol Chem 263:16597–16603

Brand SJ, Andersen BN, Rehfeld JF (1984a) Complete tyrosine O-sulphation of gastrin in neonatal rat pancreas. Nature 309:456–458

Brand SJ, Klarlund J, Schwartz TW, Rehfeld JF (1984b) Biosynthesis of tyrosine O-sulfated gastrins in rat antral mucosa. J Biol Chem 259:13246–13252

Brand SJ, Merchant J, Bachwich D (1991) Stimulation of gastrin gene transcription by epidermal growth factor/TGFα and inhibition by somatostatin. In: Håkanson R, Sundler F (eds) The stomach as an endocrine organ. Elsevier, Amsterdam, pp 233–249

Bundgaard JR, Vuust J, Rehfeld JF (1995) Tyrosine O-sulfation promotes proteolytic processing of progastrin. EMBO J 14:3073–3079

Bundgaard JR, Vuust J, Rehfeld JF (1997) New concensus features for tyrosine O-sulfation determined by mutational analysis. J Biol Chem 272:21700–21705

Cantor P, Rehfeld JF (1989) Cholecystokinin in plasma: Release of components devoid of a bioactive C-terminus. Am J Physiol 256:G53–G61

Chang RSL, Lotti VJ (1986) Biochemical and pharmacological characterization of an extremely potent and selective nonpeptide cholecystokinin antagonist. Proc Natl Acad Sci USA 83:4923–4926

Coffey RJ, Shipley D, Moses L (1986) Production of transforming growth factors by human colon cancer lines. Cancer Res 46:1164–1169

Crawley JE (1985) Comparative distribution of cholecystokinin and other neuropeptides: why is this peptide different from all other peptides? Ann NY Acad Sci 448:1–8

Crawley JE (1991) Cholecystokinin-dopamine interactions. Trends Pharmacol Sci 12:232–236

Deschenes RJ, Lorenz L, Haun R, Roos B, Collier K, Dixon JE (1984) Cloning and sequence analysis of a cDNA encoding rat preprocholecystokinin. Proc Natl Acad Sci USA 81:726–730

Dockray GJ, Gregory RA (1989) Gastrin. In: Makhlouf GM (ed) Handbook of physiology, sect 6, vol II. Amer Physiol Soc, Bethesda, Maryland, pp 311–336

Dockray GJ, Gregory RA, Hutchison JB, Harris JI, Runswick MJ (1978) Isolation, structure and biological activity of two cholecystokinin ocapeptides from sheep brains. Nature 274:711–713

Dockray GJ, Varro A, Desmond H, Young J, Gregory H, Gregory RA (1987) Posttranslational processing of the porcine gastrin precursor by phosphorylation of the C-terminal fragment. J Biol Chem 262:8643–8647

Eberlein GA, Eysselein VE, Goebell H (1988) Cholecystokinin-58 is the major molecular form in man, dog and cat, but not in pig, beef and rat intestine. Peptides 9:993–998

Eberlein GA, Eysselein VE, Davis MT, Lee TD, Shively JE, Grandt D, Niebel N, Williams R, Moessner J, Zeeh J, Meyer HE, Goebell H, Reeve JR (1992) Patterns of prohormone processing: order revealed by a new proCCK-derived peptide. J Biol Chem 267:1517–1521

Edkins JS (1905) On the chemical mechanism of gastric secretion. Proc R Soc London Sec B 76:376

Eipper BA, Park LP, Dickerson IM, Keutmann HT, Thiele EA, Rodriquez H, Schofield PR, Mains RE (1987) Structure of the precursor to an enzyme mediating C-terminal amidation in peptide biosynthesis. Mol Endocrinol 1:777–790

Eipper BA, Stoffers DA, Mains RE (1992) The biosynthesis of neuropeptides: peptide alpha-amidation. Annu Rev Neurosci 15:57–85

Eng J, Shiina Y, Pan YC, Blacher R, Chang M, Stein S, Yalow RS (1983) Pig brain contains cholecystokinin octapeptide and severel cholecystokinin desoctapeptides. Proc Natl Acad Sci USA 80:6381–6385

Eng J, Du BH, Pan YC, Chang M, Hulmes JD, Yalow RS (1984) Purification and sequencing of a rat intestinal 22 amino acid C-terminal CCK fragment. Peptides 5:1203–1206

Evans BE, Brock MG, Rittle KE, Di Pardo RM, Whittier WL, Veber DF, Anderson RS, Freidinger RM (1986) Design of orally nonpeptide antagonists for the peptide hormone cholecystokinin. Proc Natl Acad Sci USA 83:4918-4922

Fong TM, Cascieri MA, Yu H, Bansal A, Swain C, Strader CD (1993) Amino-aromatic interaction between histidine 197 of the neurokinin-1 receptor and CP 96345. Nature 362:350-353

Friedman J, Sneider BS, Powell D (1985) Differential expression of mouse cholecystokinin gene during brain and gut development. Proc Natl Acad Sci USA 82:5593-5597

Friis-Hansen L, Rehfeld JF (1994) Ileal expression of gastrin and cholecystokinin. FEBS Lett 343:115-119

Friis-Hansen L, Sundler F, Li Y, Gillespie PJ, Saunders TL, Greenson JK, Owyang C, Rehfeld JF, Samuelson LC (1998) Impaired gastric acid secretion in gastrin deficient mice. Am J Physiol 274:G561-G568

Frucht H, Gazda, AF, Park JA, Oie H, Jensen RT (1992) Characterization of receptors for gastrointestinal hormones on human colon cancer cells. Cancer Res 52:1114-1122

Fuller PJ, Stone D, Brand SJ (1987) Molecular cloning of rat preprogastrin cDNA. Mol Endocrinol 1:306-311

Galleyrand JC, Fulcrand P, Bali JP, Rodriguez M, Magous R, Laur J, Martinez J. (1992) Biological effects of human gastrin I and II chemically modified at the C-terminal tetrapeptide amide. Peptides 13:519-525

Gantz I, Takiuchi P, Yamada T (1990) Cloning of canine gastrin cDNA encoding variant amino acid sequences. Digestion 46 (Suppl 2):99-104

Gardner JF, Jensen RT (1989) Receptors for gut peptides and other secretagogous on pancreatic acinar cells. In: Makhlouf GM (ed) Handbook of physiology. The gastro-intestinal system: neural and endocrine biology, sect 6, vol II. Amer Physiol Soc, Bethesda, Maryland, pp 171-192

Gether U, Johansen TE, Snider RM, Lowe JA, Nakanishi S, Schwartz TW (1993) Different binding epitopes in the NK1 receptor for substance P and non-peptide antagonist. Nature 362:345-348

Ghatei MA, Sheppard MN, O'Shaughnessy DJ, Adrian TE, Gregor GM, Polak JM, Bloom SR (1982) Regulatory peptides in the mammalian respiratory tract. Endocrinology 111:1248-1254

Godley JM, Brand SJ (1989) Regulation of the gastrin promotor by epidermal growth factor and neuropeptides. Proc Natl Acad Sci USA 86:3036-3040

Golterman NR, Rehfeld JF, Petersen HR (1980) In vivo biosynthesis of cholecystokinin in rat cerebral cortex. J Biol Chem 255:6181-6185

Gregory H, Hardy PM, Jones DS, Kenner GW, Sheppard RC (1964) The antral hormone gastrin. Nature 204:931-933

Gregory RA, Tracy HJ (1964) The constitution and properties of two gastrins extracted from hog antral mucosa. Gut 5:103-117

Gregory RA, Tracy HJ (1972) Isolation of two big gastrins from Zollinger-Ellison tumour tissue. Lancet II:797-799

Gregory RA, Dockray GJ, Reeve JR, Shively JE, Miller C (1983) Isolation from porcine antral mucosa of a hexapeptide corresponding to the C-terminal sequence of gastrin. Peptides 4:319-323

Gubler U, Chua AO, Hoffman BJ, Collier KJ, Eng J (1984) Cloned cDNA to cholecystokinin mRNA predicts an identical preprocholecystokinin in pig brain and gut. Proc Natl Acad Sci USA 81:4307-4310

Gubler U, Chua AO, Young D, Fan ZW, Eng J (1987) Cholecystokinin mRNA in the porcine cerebellum. J Biol Chem 262:15242-15245

Håkanson R, Rehfeld J F, Ekelund M, Sundler F (1982) The life cycle of the gastrin cell. Cell Tissue Res 222:479-481

Harper AA, Raper HS (1943) Pancreozymin, a stimulant of secretion of pancreatic enzymes in extracts of the small intestine. J Physiol 102:115-125

Haun RS, Dixon JE (1990) A transscriptional enhancer essential for the expression of the rat cholecystokinin gene contains a sequence identical to the -296 element of the human C-fos gene. J Biol Chem 265:15455-15463

Hilsted L (1991) Glycine-extended gastrin precursors. Regul Pept 36:323-343

Hilsted L, Hansen CP (1988) Co-release of amidated and glycine-extended antral gastrins after a meal. Am J Physiol 255:G665-669

Hilsted L, Rehfeld JF (1987) Amidation of antral progastrin: relation to other posttranslational modifications. J Biol Chem 262:16953-16957

Hilsted L, Rehfeld Schwartz TW (1986) Impaired α-carboxyamidation of gastrin in vitamin C-deficient guinea pigs. FEBS Lett 196:151-154

Hilsted L, Bardram L, Rehfeld JF (1988) Progastrin maturation during ontogenesis: accumulation of glycine-extended gastrins in rat antrum at weaning. Biochem J 255:397-402

Hökfelt T, Rehfeld JF, Skirboll L, Ivemark B, Goldstein M, Markey K (1980) Evidence for coexistence of dopamine and CCK in mesolimbic neurons. Nature 285:476-478

Hoosein NM, Kiener PA, Curry RC, Rovati LC, McGilbra DK, Brattain MG (1988) Antiproliferative effects of gastrin receptor antagonists and antibodies to gastrin on human colon carcinoma cell lines. Cancer Res 48:7179-7183

Hoosein NM, Kiener PA, Curry RL, Brattain MG (1990) Evidence for autocrine growth stimulation of cultured colontumor cells by a gastrin/CCK-like peptide. Exp Cell Res 186:15-21

Huang S, Lin P, Fan D, Price JE, Truillo JM, Chakrabarty S (1991) Growth modulation by epidermal growth factor (EGF) in human colonic carcinoma cells: constitutive expression of the human EGF gene. J Cell Physiol 148:220-227

Ishizuka J, Martinez J, Townsend CM, Thompson JC (1992) The effect of gastrin on growth of human stomach cancer cells. Ann Surg 215:528-634

Ito R, Sato K, Helmer T, Jay G, Agarwal K (1984) Structural analysis of the gene encoding human gastrin: the large intron contains an Alu sequence. Proc Natl Acad Sci USA 81:4662-4666

Ivy AC, Oldberg E (1928) A hormone mechanism for gallbladder contraction and evacuation. Am J Physiol 86:599-613

Jensen RT, Lemp GF, Gardner JF (1980) Interaction of cholecystokinin with specific membrane receptors on pancreatic acinar cells. Proc Natl Acad Sci USA 77:2079-2083

Jensen RT, Wank SA, Rowley SH, Sato S, Gardner JD (1989) Interaction of CCK with pancreatic acinar cells. Trends Pharmacol Sci 10:418-423

Jensen S, Borch K, Hilsted L, Rehfeld JF (1989) Progastrin processing during antral G-cell hypersecretion in humans. Gastroenterology 96:1063-1070

Johnsen AH (1998) Phylogeny of the cholecystokinin/gastrin family. Front Neuroendocrinol 19:73-99

Johnsen AH, Rehfeld JF (1990) Cionin: a disulfotyrosyl hybrid of cholecystokinin and gastrin from the neural ganglion of the protochordate ciona intestinalis. J Biol Chem 265:3054-3058

Johnsen AH, Rehfeld JF (1992) Identification of cholecystokinin/gastrin peptides in frog and turtle: evidence that cholecystokinin is phylogenetically older than gastrin. Eur J Biochem 207:419-428

Johnson LR (1976) The trophic action of gastrointestinal hormones. Gastroenterology 70:278-288

Johnson LR (1989) Trophic effects of gut peptides. In: Makhlouf GM (ed) Handbook of physiology. The gastrointestinal system: neural and endocrine biology, sect 6, vol II. Amer Physiol Soc, Bethesda, Maryland, pp 291-310

Jorpes E, Mutt V (1966) Cholecystokinin and pancreozymin – a single hormone? Acta Physiol Scand 66:196-202

Kaise M, Muraoka A, Seva C, Takeda H, Dickinson CJ, Yamada T (1995) Glycine-extended gastrins induce H^2, K^+-ATPase alpha-subunit gene expression through a novel receptor. J Biol Chem 270:11155-11160

Kang SC, Agarwal KL, Yoo OJ (1989) Molecular cloning and sequence analysis of cDNA coding for canine gastrin. Biochem Int 18:613-636

Kariya Y, Kato K, Hayashizaki Y, Himeno S, Tarui S, Matsubara K (1986) Expression of human gastrin gene in normal and gastrinoma tissue. Gene 50:345-352

Karnes WE Jr, Walsh JH, Wu SW, Kim RS, Marinn MG, Wong HC, Mendelsohn J, Park J-G, Cuttitta F (1992) Autonomous proliferation of colon cancer cells that coexpress transforming growth factor alpha and its receptor. Gastroenterology 102:474-485

Karnik PS, Monahan SJ, Wolfe MM (1989) Inhibition of gastrin gene expression by somatostatin. J Clin Invest 83:367-372

Kato K, Hayashizaki Y, Takahashi Y, Himeno S, Matsubara K (1983) Molecular cloning of the human gastrin gene. Nucleic Acids Res 11:8197-8203

Katopodis AG, Ping D, May SW (1990) A novel enzyme from bovine neurointermediate pituitary catalyzes dealkylation of α-hydroxyglycine derivatives. Biochemistry 29:6115-6120

Kelly RB (1985) Pathways of protein secretion in eukaryotes. Science 230:25-32

Koh TJ, Goldenring JR, Ito S, Mashimo H, Kopin AS, Varro A, Dockray GJ, Wang TC (1997) Gastrin deficiency results in altered gastric differentiation and decreased colonic proliferation in mice. Gastroenterology 113:1015-1025

Kohler M, Bauknect T, Grimm M, Birmelin G, Kommoss F, Wagner E (1992) Epidermal growth factor receptor and transforming growth factor alpha expression in human ovarian carcinomas. Eur J Cancer 28A:1432-1437

Kopin AS, Lee Y-M, McBride EW, Miller LJ, Lu M, Lin HY, Kolakowski LF, Beinborn M (1992) Expression cloning and characterization of the canine parietal cell gastrin receptor. Proc Natl Acad Sci USA 89:3605-3609

Kurachi H, Morishige K, Amemiya K, Adachi H, Hirota K, Miyake A, Tanizawa O (1991) Importance of transforming growth factor alpha/epidermal growth factor receptor autocrine growth mechanism in an ovarian cancer cell line in vivo. Cancer Res 51:5956-5959

Kuwano R , Araki K, Usui H, Fukui T, Ohtsuka E, Ikehara M, Takahashi Y (1984) Molecular cloning and nucleotide sequence of cDNA coding for rat brain cholecystokinin precursor. J Biochem Tokyo 96:923-926

Langhans N, Rindi G, Chiu M, Rehfeld JF, Ardman B, Beinborn M, Kopin AS (1997) Abnormal gastric histology and decreased acid production in cholecystokinin-B/-gastrin receptor deficient mice. Gastroenterology 112:280-286

Larsson L-I (1980) Gastrointestinal cells producing endocrine, neurocrine and paracrine messengers. Clin Gastroenterol 9:485-516

Larsson L-I, Rehfeld JF (1977) Evidence for a common evolutionary origin of gastrin and cholecystokinin. Nature 269:335-338

Larsson L-I, Rehfeld JF (1979a) A peptide resembling the C-terminal tetrapeptide amide of gastrin from a new gastrointestinal endocrine cell-type. Nature 277:575-578

Larsson L-I, Rehfeld JF (1979b) Localization and molecular heterogeneity of cholecystokinin in the central and peripheral nervous system. Brain Res 165:201-218

Larsson L-I, Rehfeld JF (1981) Pituitary gastrins occur in corticotrophs and melanotrophs. Science 213:768-770

Larsson L-I, Rehfeld JF, Håkanson R, Sundler F (1976) Pancreatic gastrin in foetal and neonatal rats. Nature 262:607-611

Lee YM, Beinborn M, McBride EW, Lu M, Kolakowski LF, Kopin AS (1993) The human brain cholecystokinin-B/gastrin receptor: cloning and characterization. J Biol Chem 268: 8164-8169

Liddle RA (1994) Cholecystokinin. In: Walsh JH, Dockray GJ (eds) Gut peptides: biochemistry and physiology. Raven Press, New York, pp 175-216

Lund T, Geurts van Kessel AHM, Haun S, Dixon JE (1986) The genes for human gastrin and cholecystokinin are located on different chromosomes. Hum Genet 73:77-80

Lund T, Olsen J, Rehfeld JF (1989) Cloning and sequencing of bovine preprogastrin cDNA. Mol Endocrinol 3:1585-1589

Lüttichau HR, van Solinge WW, Nielsen FC, Rehfeld JF (1993) Developmental expression of the gastrin and cholecystokinin genes in rat colon. Gastroenterology 104:1092-1098

Malmström J, Stadil F, Rehfeld JF (1976) Gastrins in tissue: concentration and component pattern in gastric, duodenal and jejunal mucosa of normal human subjects and patients with duodenal ulcer. Gastroenterology 70:697–704

Merchant JL, Demediuk B, Brand SJ (1991) A GC-rich element confers epidermal growth factor responsiveness to transcription from the gastrin promotor. Mol Cell Biol 11:2686–2696

Miller LJ, Rosenzweig SA, Jamieson JD (1981) Preparation and characterization of a probe for the cholecystokinin octapeptide, N_2(125-I-desaminotyrosyl) CCK, and its interaction with pancreatic acini. J Biol Chem 256:12417–12423

Mogensen NW, Hilsted L, Bardram L, Rehfeld JF (1990) Procholecystokinin processing in rat cerebral cortex during development. Dev Brain Res 54:81–86

Morishige K, Kurachi H, Amemiya K, Adachi H, Inoue M, Miyake A, Tanizawa O, Sakoyama Y (1991) Involvement of transforming growth factor alpha/epidermal growth factor receptor autocrine growth mechanism in an ovarian cancer cell line in vivo. Cancer Res 51:5951–5955

Morley JS, Tracy HJ, Gregory RA (1965) Structure function relationships in the active C-terminal tetrapeptide amide sequence of gastrin. Nature 207:1356–1359

Murthy ASN, Mains RE, Eipper BA (1986) Purification and characterization of peptidyl glycine α-amidating monooxygenase from bovine neurointermediate pituitary. J Biol Chem 261:1815–1822

Mutt V (1976) Further investigations on intestinal hormonal polypeptides. Clin Endocrinol 5:175S–183S

Mutt V, Jorpes JE (1968) Structure of porcine cholcystokinin – pancreozymin. Eur J Biochem 6:156–162

Mutt V, Jorpes JE (1971) Hormonal polypeptides of the upper intestine. Biochem J 125:57P–58P

Mutt V, Magnusson S, Jorpes JE, Dahl E (1965) Structure of porcine secretin. Biochemistry 4:2358–2362

Nachman RJ, Holman GM, Haddon WF, Ling N (1986) Leucosulfakinin, a sulfated insect neuropeptide with homology to gastrin and cholecystokinin. Science 234:71–73

Nicols R, Schneuwly SA, Dixon JE (1988) Identification and characterization of a *Drosophila homologue* to the vertebrate neuropeptide cholecystokinin. J Biol Chem 263:12167–12170

Nielsen FC, Pedersen K, Hansen TvO, Rourke IJ, Rehfeld JF (1996) Transcriptional regulation of the human cholecystokinin gene: composite action of upstream stimulatory factor, Sp1, and members of the CREB/ATF-AP-1 family. DNA-Cell Biol 15:53–63

Nilsson G, Brodin K (1977) Increase of gastrin concentration in duodenal mucosa of dogs following resection of the gastric antrum. Acta Physiol Scand 99:510–512

Perkins SN, Husten EJ, Eipper BA (1990) The 108 kDa peptidylglycine α-amidating monooxygenase precursor contains two separate enzymatic activities involved in peptide amidation. Biochem Biophys Res Commun 171:926–932

Persson H, Rehfeld JF, Ericsson A, Schalling M, Pelto-Huikko M, Hökfelt T (1989) Transient expression of the cholecystokinin gene in male germ cells and accumulation of the peptide in the acrosomal granule. Proc Natl Acad Sci USA 86:6166–6170

Reeve JR, Eysselein V, Walsh JH, Ben-Avram CH, Shively JE (1986) New molecular forms of cholecystokinin. Microsequence analysis of forms previously characterized by chromatographic methods. J Biol Chem 261:16392-16397

Reeve JR, Eysselein VE, Eberlein GA, Chew P, Ho F-J, Huebner VD, Shively JE, Lee TD, Liddle RA (1991) Characterization of canine intestinal cholecystokinin-58 lacking its C-terminal nonapeptide. J Biol Chem 266:13770–13776

Rehfeld JF (1978a) Localization of gastrins to neuro- and adenohypophysis. Nature 272:771–773

Rehfeld JF (1978b) Immunochemical studies on cholecystokinin. I. Development of sequence-specific radioimmunoassays against porcine triacontatriapeptide cholecystokinin. J Biol Chem 253:4016–4021

Rehfeld JF (1978c) Immunochemical studies on cholecystokinin. II. Distribution and molecular heterogeneity in the central nervous system and small intestine of man and hog. J Biol Chem 253:4022–4030

Rehfeld JF (1980) NH$_2$-terminal monoiodination of hexadecapeptide gastrin for receptor studies. Clin Chim Acta 101:271–275

Rehfeld JF (1981) Four basic characteristics of the gastrin-cholecystokinin system. Am J Physiol 240:G255–266

Rehfeld JF (1984) How to measure cholecystokinin in plasma? Gastroenterology 87:434–438

Rehfeld JF (1986) Accumulation of nonamidated progastrin and procholecystokinin products in the porcine anterior pituitary. J Biol Chem 261:5841–5848

Rehfeld JF (1987) Procholecystokinin processing in the normal human anterior pituitary. Proc Natl Acad Sci USA 84:3019–3024

Rehfeld JF (1989) Cholecystokinin. In: Makhlouf GM (ed) Handbook of physiology. The gastrointestinal system: neural and endocrine biology, sect 6, vol II. Amer Physiol Soc, Bethesda, Maryland, pp 337–358

Rehfeld JF (1991) Progastrin and its products in the cerebellum. Neuropeptides 20:239–245

Rehfeld JF (1992) Introduction. In: Dourish CT, Cooper S, Iversen SD, Iversen LL (eds) Multiple cholecystokinin receptors in the CNS. Oxford University Press, pp 117–120

Rehfeld JF (1994) The molecular nature of cholecystokinin in plasma. An in vivo immunosorption study in rabbits. Scand J Gastroenterol 29:110–121

Rehfeld JF (1998a) Accurate measurement of cholecystokinin in plasma. Clin Chem 44:991–1001

Rehfeld JF (1998b) The new biology of gastrointestinal hormones. Physiol Rev 78:1087–1108

Rehfeld JF, Hansen HF (1986) Characterization of procholecystokinin products in the porcine cerebral cortex: evidence of different processing pathways. J Biol Chem 261:5832–5840

Rehfeld JF, Johnsen AH (1994) Identification of gastrin component-I as gastrin-71, the largest possible bioactive progastrin product. Eur J Biochem 223:765–773

Rehfeld JF, Larsson L-I (1979) The predominating form of gastrin and CCK in the gut is a small peptide corresponding to the C-terminal tetrapeptide amide. Acta Physiol Scand 105:117–119

Rehfeld JF, Uvnäs-Wallensten K (1978) Gastrins in cat and dog: evidence for a biosynthetic relationship between the large molecular forms of gastrin and heptadecapeptide gastrin. J Physiol 283:379–396

Rehfeld JF, van Solinge WW (1994) The tumor biology of cholecystokinin and gastrin. Adv Cancer Res 63:295–346

Rehfeld JF, Hansen HF, Larsson L-I, Stengaard-Pedersen K, Thorn NA (1984) Gastrin and cholecystokinin in pituitary neurons. Proc Natl Acad Sci USA 81:1902–1905

Rehfeld JF, Bardram L, Hilsted L (1989) Gastrin in bronchogenic carcinomas: constant expression but variable processing of progastrin. Cancer Res 49:2840–2843

Rehfeld JF, Johnsen AH, Ødum L, Bardram L, Schifter S, Scopsi L (1990) Nonsulfated cholecystokinin in human medullary thyroid carcinomas. J Endocrinol 124:501–506

Rehfeld JF, Mogensen NW, Bardram L, Hilsted L, Monstein HJ (1992a) Expression but failing maturation of procholecystokinin in cerebellum. Brain Res 76:111–119

Rehfeld JF, Bardram L, Hilsted L (1992b) Ontogeny of procholecystokinin maturation in rat duodenum, jejunum and ileum. Gastroenterology 103:424–430

Rehfeld JF, Hansen CP, Johnsen AH (1995) Post-poly(glu) cleavage and degradation modified by O-sulfated tyrosine: a novel post-translational processing mechanism. EMBO J 14:389–396

Sachais BS, Snider RM, Lowe JA, Krause JE (1993) Molecular basis for the species selectivity of the substance P antagonist CP-96,345. J Biol Chem 268:2319–2323

Saito A, Sankaran H, Goldfine ID, Williams JA (1980) Cholecystokinin receptors in the brain: characterization and distribution. Science 208:1155–1156

Sankaran H, Deveney CW, Goldfine ID, Williams JA (1979) Preparation of biologically active radioiodinated cholecystokinin. J Biol Chem 254:9349–9351

Schalling M, Persson H, Pelto-Huikko M, Ødum L, Ekman P, Gottlieb C, Hökfelt T, Rehfeld JF (1990) Expression and localization of gastrin mRNA and peptides in spermatogenic cells. J Clin Invest 86:660–669

Schwartz TW (1990) Biosynthesis of islet hormones. In: Okamoto H (ed) Molecular biology of islets of Langerhans. Cambridge University Press, Cambridge, pp 153–205

Sethi T, Rozengurt E (1991) Multiple neuropeptides stimulate clonal growth of small cell lung cancer: effects of bradykinin, vasopressin, cholecystokinin, galanin and neurotensin. Cancer Res 51:3621–3623

Seva C, Dickinson CJ, Yamada T (1994) Growth promoting effects of glycine-extended progastrin. Science 265:410–412

Shively J, Reeve JR, Eysselein VE, Ben-Avram C, Vigna SR, Walsh JH (1987) CCK-5: sequence analysis of a small cholecystokinin from canine brain and intestine. Am J Physiol 252: G272–G275

Singh P, Rae-Venter B, Townsend CM, Khalil T, Thompson JC (1985) Gastrin receptors in normal and malignant gastrointestinal mucosa: age-associated changes. Am J Physiol 249: G761–G769

Singh P, Walker JP, Townsend CM, Thompson JC (1986) Role of gastrin and gastrin receptors on the growth of a transplantable mouse colon carcinoma (MC-26) in BALB/c mice. Cancer Res 46:1612–1616

Singh P, Owlia A, Espeijo R, Dai B (1995) Novel gastrin receptors mediate mitogenic effects of gastrin and processing intermediates of gastrin on Swiss 3T3 fibroblasts. J Biol Chem 270: 8429–8438

Smith JJ, Derynck R, Korc M (1987) Production of transforming growth factor alpha in human pancreatic cancer cells. Proc Natl Acad Sci USA 84:7567–7570

Soll AH (1989) Gastrin mucosal receptors. In: Makhlouf GM (ed) Handbook of physiology. The gastrointestinal system: neural and endocrine biology, sect 6, vol II. Amer Physiol Soc, Bethesda, Maryland, pp 193–214

Soll AH, Amirian DA, Thomas L-P, Reedy TJ, Elashoff JF (1984) Gastrin receptors on isolated canine parietal cells. J Clin Invest 73:1434–1447

Song I, Brown DR, Wiltshire RN, Gantz I, Trent JM, Yamada T (1993) The human gastrin/CCK-B receptor gene: alternative splice donorsite in exon 4 generates two variant mRNAs. Proc Natl Acad Sci USA 90:9085–9089

Stadil F, Rehfeld JF (1972) Preparation of 125-I gastrin for radioimmunoanalysis. Scand J Clin Lab Invest 30:361–368

Stengaard-Pedersen K, Larsson L-I, Fredens K, Rehfeld JF (1984) Modulation of cholecystokinin concentrations in hippocampus by chelation of heavy metals. Proc Natl Acad Sci USA 81: 5867–5870

Stromberg K, Collins TJ, Gordon AW, Jackson CL, Johnson GR (1992) Transforming growth factor-alpha acts as an autocrine growth factor in ovarian carcinoma cell lines. Cancer Res 52:341–347

Sugano K, Aponte GW, Yamada T (1985) Identification and characterization of glycine-extended processing intermediates of progastrin in porcine stomach. J Biol Chem 260:11724–11729

Suzuki K, Shimoi H, Iwasaki Y, Kawahara T, Matsuura Y, Nishikawa Y (1990) Elucidation of amidating reaction mechanism by frog amidating enzyme expressed in insect cell culture. EMBO J 9:4259–4265

Takahashi K, Okamoto H, Seino H, Noguchi M (1990) Peptidylglycine α-amidating reaction: evidence for a two-step mechanism involving a stable intermediate at neutral pH. Biochem Biophys Res Commun 169:524–530

Takahashi Y, Kato K, Hayashizaki Y, Wakabayashi T, Ohtsuka E, Matsuki S, Ikehara M, Matsubara K (1985) Molecular cloning of the human cholecystokinin gene by use of a synthetic probe containing deoxyinosine. Proc Natl Acad Sci USA 82:1931–1935

Takahashi Y, Fukushige S, Korotsu T, Matsubara K (1986) Structure of human cholecystokinin gene and its chromocomal location. Gene 50:353–360

Theill LE, Wiborg O, Vuust J (1987) Cell-specific expression of the human gastrin gene: evidence for a control element located downstream of the TATA box. Mol Cell Biol 7:4329–4336

Uvnäs-Wallensten K, Rehfeld JF, Larsson L-I, Uvnäs B (1977) Heptadecapeptide gastrin in the vagal nerve. Proc Natl Acad Sci USA 74:5707–5710

van Solinge WW, Rehfeld JF (1992) Co-transcription of the gastrin and cholecystokinin genes with selective translation of gastrin mRNA in a human gastric carcinoma cell-line. FEBS Lett 309:47–50

van Solinge WW, Nielsen FC, Friis-Hansen L, Falkmer UG, Rehfeld JF (1993a) Expression but incomplete maturation of progastrin in colorectal carcinomas. Gastroenterology 104:1099–1107

van Solinge WW, Ødum L, Rehfeld JF (1993b) Ovarian cancers express and process progastrin. Cancer Res 53:1823–1830

Walsh JH (1994) Gastrin. In: Walsh JH, Dockray GJ (eds) Gut peptides: biochemistry and physiology. Raven Press, New York, pp 75–121

Walsh JH, Debas HT, Grossman MI (1974) Pure human big gastrin: immunochemical properties, disappearance half time and acid-stimulating action in dogs. J Clin Invest 54:477–485

Walsh JH, Isenberg JI, Ansfield J, Maxwell V (1976) Clearance and acid-stimulating action of human big and little gastrins in duodenal subjects. J Clin Invest 57:1125–1131

Wang TC, Brand SJ (1990) Islet-cell specific domain in the gastrin promotor contains adjacent positive and negative DNA elements. J Biol Chem 265:8908–8914

Wank SA, Harkins R, Jensen RT, Shapira H, de Weerth A, Slattery T (1992) Purification, molecular cloning and functional expression of the cholecystokinin receptor from rat pancreas. Proc Natl Acad Sci USA 89:3125–3129

Wank SA (1995) Cholecystokinin receptors. Am J Physiol 269:G628–G646

Weinstock J, Baldwin GS (1988) Binding of gastrin-17 to human gastric carcinoma cell lines. Cancer Res 48:932–937

Wiborg O, Berglund L, Boel E, Norris F, Rehfeld JF, Marcker KA, Vuust J (1984) Structure of a human gastrin gene. Proc Natl Acad Sci USA 81:1067–1069

Williams JA, Gryson KA, McChesney DJ (1986) Cholecystokinin induces the interaction of its receptor with a guamine nucleotide binding protein. Regul Pept 7:293–296

Yasui W, Hata J, Yokazaki H, Nakatani H, Ochiai A, Ito H, Tahara E (1988) Interaction between epidermal growth factor and its receptor in progression of human gastrin carcinoma. Int J Cancer 41:211–217

Yoo OJ, Powell CT, Agarwal KL (1982) Molecular cloning and nucleotide sequence of full-length cDNA coding for procine gastrin. Proc Natl Acad Sci USA 79:1049–1053

Young WS (1992) Expression of the oxytocin and vasopressin genes. J Neuroendocrinol 4:527–540

Function of the Neuropeptide Head Activator for Early Neural and Neuroendocrine Development

Wolfgang Hampe, Irm Hermans-Borgmeyer, and H. Chica Schaller

1
Introduction

Head activator (HA) is an undecapeptide (Fig. 1) identical in sequence from hydra to mammals (Bodenmüller and Schaller 1981; Schaller and Bodenmüller 1981). In hydra HA occurs as a gradient with maximal concentration in the head region. As a morphogen HA is required for head-specific growth and differentiation processes, hence its name (reviewed in Schaller 1996; Schaller et al. 1996). In mammals HA is a growth factor for early nerve and neuroendocrine development, and it has modulatory functions later in life (Kajiwara and Sato 1986; Schaller et al. 1989; Quach et al. 1992; Ulrich et al. 1996; Kayser et al. 1998).

2
Action of HA as Morphogen in Hydra

Stem cells in hydra can differentiate in a head- or foot-specific manner. Determination occurs, as in other organisms, in S phase and requires high concentrations of HA ($>10^{-11}$ M). In the presence of HA during the first 4–6 h of their S phase epithelials cells become determined to head-specific hypostomal or tentacle cells and interstitial stem cells enter the nerve-cell pathway. Low concentrations of HA ($>10^{-13}$ M) suffice to trigger final differentiation of these determined cells (Schaller 1976a; Holstein et al. 1986; Schaller et al. 1990).

Fig. 1. HA peptide as sequenced from coelenterates and mammals

Zentrum für Molekulare Neurobiologie, University of Hamburg, Martinistr. 52, 20246 Hamburg, Germany

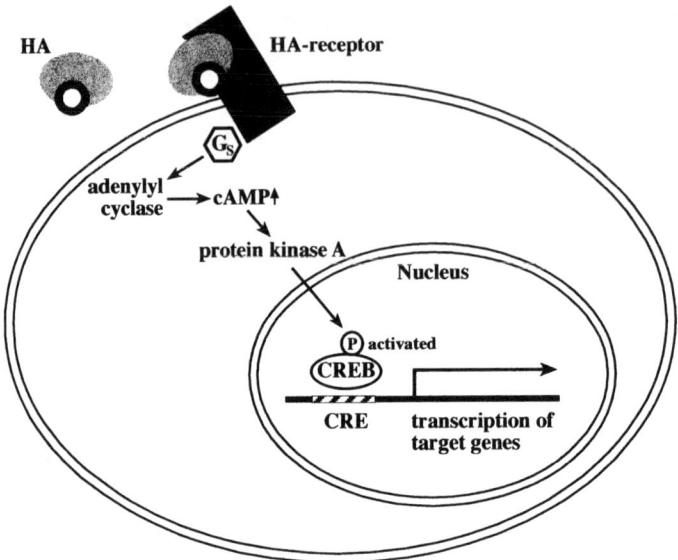

Fig. 2. Signal transduction to initiate determination and differentiation of nerve cells in hydra

Determination and differentiation to nerve cells by HA are mediated by cAMP as second messenger hinting at a G protein-coupled receptor (Fenger et al. 1994). Signaling involves the transcription factor cAMP response element binding protein (CREB) (Galliot et al. 1995). CREB activity is increased rapidly after HA treatment and early in head regeneration. Hydra is one of the oldest animals with a well-defined axial polarity. In other organisms anterior-posterior development is controlled by homeobox-containing transcription factors, of which plenty exist in hydra (Schummer et al. 1992; Galliot 1997; Gauchat et al. 1998). Since CREB activation occurs early, it might precede the transcription cascade of such homeobox-containing genes (Fig. 2).

3
HA Signal Transduction to Initiate Mitosis

In addition to its role in cell fate, HA also influences cell proliferation (Schaller 1976b). Treatment of hydra with low HA concentrations ($>10^{-13}$ M) results some hours later in an increase of cells in mitosis. This effect is not cell-type specific; all proliferation-competent cells respond to HA and enter M phase.

In contrast to the action of HA on determination and differentiation, its effect on mitosis stimulation was not mimicked by cAMP in hydra. This suggests that the mitogenic effect of HA is mediated by other second messengers.

To investigate HA signal transduction at the G2/mitosis transition, we used mammalian cell lines of neuroectodermal or neuroendocrine origin, for which we found that HA promotes cell proliferation by stimulating entry into mitosis. Some of these cell lines produced HA and used it as autocrine growth factor. These included the neuroblastoma-glioma cell lines NH15-CA2 and NG108-15, the pancreas carcinoid-derived cell line BON, and the neurogenic embryonal carcinoma cell lines NT2 and P19, the latter after induction with retinoic acid. Responsive to HA, but not autocrine HA producers, were the pheochromocytoma cell line PC12 and the pituitary-derived cell line AtT20. In these cell lines HA, in a dose-dependent manner, stimulated mitosis (Ulrich et al. 1996; Kayser et al. 1998). Mitosis stimulation, as analyzed by flow cytometry 2–4 h after HA application, was prevented by preincubation of cells with pertussis toxin, identifying the HA receptor as being coupled to inhibitory G proteins. As an immediate effect of HA, a calcium influx from the extracellular space into the cells was observed. The increase in intracellular calcium concentration led to an activation of calcium-dependent potassium channels resulting in hyperpolarization. Blocking calcium channels or potassium channels inhibited the effect of HA on cell proliferation. Mitosis was also prevented by treating cells with Forskolin, indicating that cAMP acts antagonistically to calcium as second messenger. In summary, stimulation of mitosis by HA is mediated by an inhibitory G protein and requires calcium influx, inhibition of the cAMP pathway, and hyperpolarization of cells (Fig. 3).

Cell cycle progression and proliferation of cells were most efficiently inhibited with specific inhibitors of the calcium-activated potassium channel, which

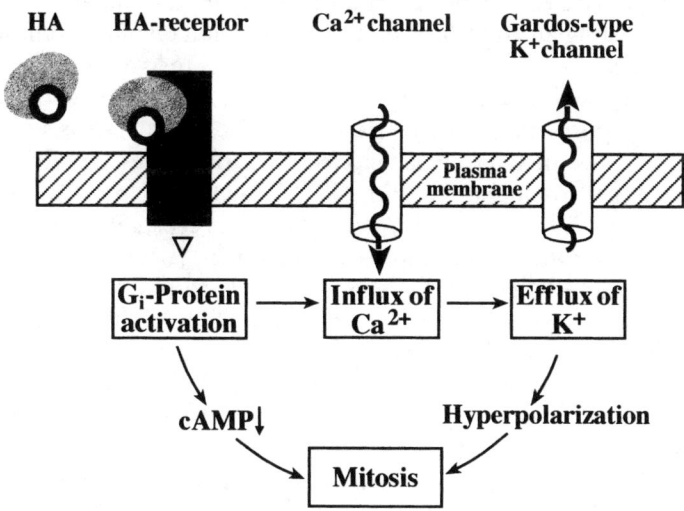

Fig. 3. Signal transduction to initiate mitosis in mammalian cell lines

by pharmacology and RNA analysis was identified as a Gardos-type potassium channel (Ishii et al. 1997; Kayser et al. 1998). Two imidazole derivatives, clotrimazole and SK&F 96365, which specifically target this channel, were found to be potent inhibitors of BON-cell proliferation. Because of its physiological compatibility with the human organism, clotrimazole is a suitable lead structure to design drugs for *in vivo* therapy of neural and neuroendocrine tumors.

4
Corelease of HA with a Carrier Complex

In hydra one counterpart of HA is the head inhibitor which is also found in a head to foot concentration gradient and which controls HA release (Schaller 1996). Both activator and inhibitor are small molecules very similar in molecular mass. Nevertheless, they differ drastically in their range of action, HA being a local morphogen with a very limited diffusion and head inhibitor controlling a larger area. As explanation of the different diffusion properties we found that head inhibitor is released in its naked, low molecular weight form, whereas HA is bound to a large molecular weight complex (Schaller et al. 1986). If HA is extracted from hydra tissue with aqueous solvents or if medium is collected from animals regenerating a head or from cell lines producing HA, all HA activity elutes from Sepharose S-300 columns with an apparent molecular mass of approximately 800 kDa. Binding of HA to the carrier complex is noncovalent, as demonstrated by quantitative extractability with organic solvents such as methanol and/or high salt. The HA binding component in hydra is a protein of approximately 200 kDa; in mammals it is in the 250-kDa range. The HA-carrier complex from hydra and mammals binds to heparin, suggesting an additional barrier to diffusion by interaction with extracellular matrix components. This carrier-bound HA in hydra is active at 10^{-13} M, indicating that binding to the carrier does not inhibit the biological effect. The free HA peptide, once released from the carrier or offered as synthetic HA, has little chance of long persistence, since it dimerizes rapidly or is degraded by enzymes. We found that dimeric HA has no biological action on either hydra (Bodenmüller et al. 1986) or mammalian cells, and that it is an inadequate ligand for the HA receptor (Neubauer et al. 1991). The rapid inactivation fulfils an additional prerequisite for a local morphogen, namely that a molecule once released cannot diffuse far. Binding of HA to the carrier, therefore, fulfils two functions: it inhibits diffusion and it prevents degradation, thus ensuring the longer half-life of activator over inhibitor. It also explains why substances of low molecular weight, by definition diffusible, can nevertheless be morphogens with local action.

We found that HA is present in the blood of all mammals including humans. When trying to analyze its action in the blood, we discovered that injected HA was degraded within minutes (Roberge et al. 1984). Most of this degradation was due to the angiotensin-converting enzyme. An astonishing finding was that endogenous HA was stable and unavailable for the protease. This was due to the fact that HA in the blood, like in hydra, is bound to a high molecular weight carrier from which it can be released by salt treatment or methanol extraction, indicating that HA is not covalently bound, but only attached to the carrier. Similarly, HA as released by HA-producing mammalian cell lines, is always coupled to a carrier complex.

5
HA Binding Protein

As the very first step of the signaling cascade, HA has to interact with a membrane-bound receptor. Two classes of HA binding sites were characterized in hydra membranes: a "low-affinity" site with a K_d in the nanomolar range, and a 50 times less abundant "high-affinity" site with a K_d in the picomolar range (Neubauer et al. 1991). The two types of binding sites were assumed to mediate different effects of HA at the cellular level. For further characterization of the binding sites we synthesized photoaffinity ligands which were cross-linked to a 200-kDa protein. Non-labeled HA derivatives, but no other neuropeptides, inhibited the cross-linking (Hampe et al. 1996).

For purification of the 200-kDa protein a multiheaded mutant of *Chlorohydra viridissima* was used, which not only contained more HA than normal hydra, but also overexpressed the low-affinity binding site (Neubauer et al. 1991). As a first step in the purification protocol the low-affinity HA-receptor was functionally solubilized using Triton X-100 or Chaps. For affinity chromatography HA was coupled to Sepharose. After rigorous washing, a 200-kDa protein was eluted from HA Sepharose (Fig. 4). Ligand binding with nanomolar affinity was preserved in the eluate from the HA sepharose, and the 200-kDa protein could be photoaffinity labeled with HA ligands. By Edman degradation of the purified protein, sequence information was obtained for the amino terminus and, after protease digestion, for several internal peptides (Franke et al. 1997).

Using these partial sequences for designing oligonucleotides, the cDNA of the HA binding protein (HAB) was cloned (Hampe et al., submitted). The open reading frame codes for a protein of 1661 amino acids, very well in agreement with the 200-kDa, as found by SDS-PAGE, for the endogenous glycosylated hydra HAB (Hampe et al. 1996; Franke et al. 1997). Hydrophobicity analysis revealed that HAB, in addition to the amino-terminal signal peptide, contains only a single transmembrane segment located near the carboxy terminus (Fig. 5). HAB is synthesized as a proprotein which is cleaved posttranslationally behind amino acid 84 to yield the mature form. Sequence analysis revealed that HAB is a novel type of mosaic protein consisting of structural domains, so far not found in this combination and alignment in other proteins. The

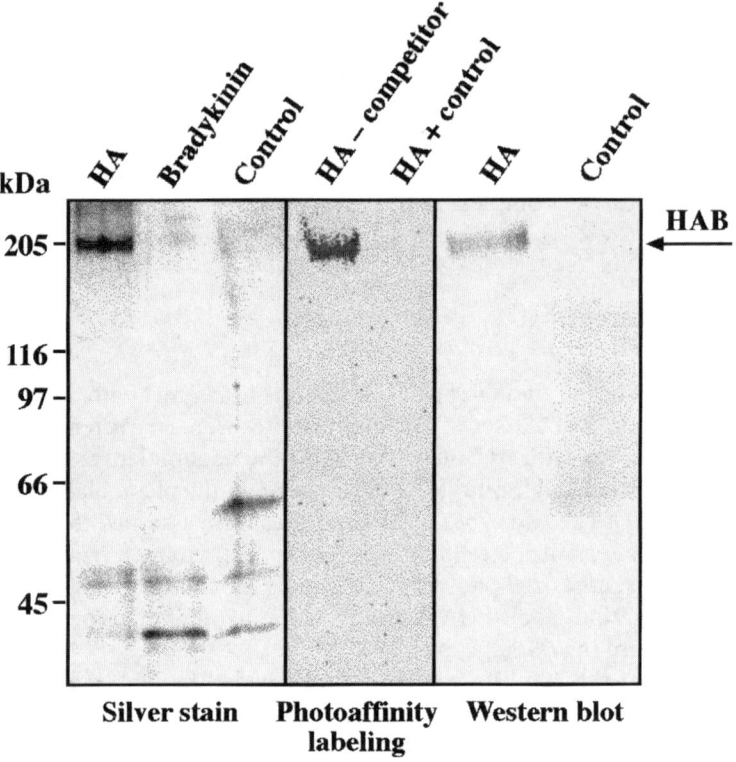

kDa

205 –

116 –
97 –

66 –

45 –

HA · Bradykinin · Control · HA – competitor · HA + control · HA · Control

HAB

Silver stain   **Photoaffinity**   **Western blot**
                   labeling

Fig. 4. Affinity purification of HA binding protein (*HAB*). Solubilized hydra membranes were incubated with Sepharoses coupled to HA or as control to bradykinin or to no peptide. After rigorous washing, bound proteins were eluted with 2 M acetic acid or 4 M $MgCl_2$. Analysis by reducing SDS-PAGE and silver staining showed that HAB bound to HA Sepharose, but not to control Sepharoses. Photoaffinity labeling with ^{125}I-Bpa-HA-HA bipeptide (Hampe et al. 1996) in the presence or absence of an excess of unlabeled HA ligand as competitor proved HA binding to HAB. A specific antiserum raised against recombinant hydra HAB recognized HAB in the eluate from HA sepharose

amino-terminal half of hydra HAB shares homology with the lumenal domain of human sortilin (Petersen et al. 1997), and with the yeast protein VPS10 (Marcusson et al. 1994). The latter is involved in trafficking proteins between different cellular compartments. Characteristic for the VPS10-like domain is the spacing of twelve cysteine residues which is conserved in hydra HAB. The VPS10-like domain is followed by repeats with homology to members of the low-density lipoprotein (LDL) receptor family. Members of this family are known to bind and internalize extracellular ligands via cysteine-rich class A repeats. Some reports hint at an important signaling function for members of the LDL-receptor family and their ligands during development (Schaefer et al. 1993; Shackelford et al. 1995; Pahl et al. 1996; Allen et al. 1998; Farese and Herz

Hydra HAB

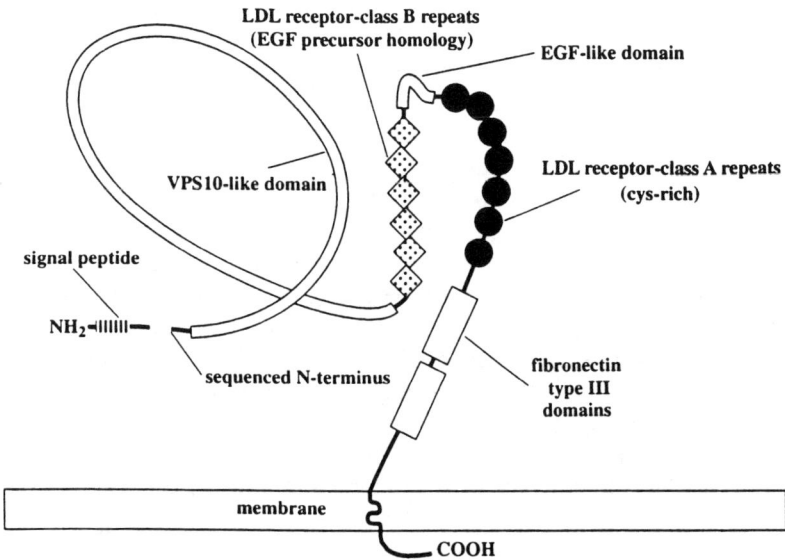

Vertebrate HAB

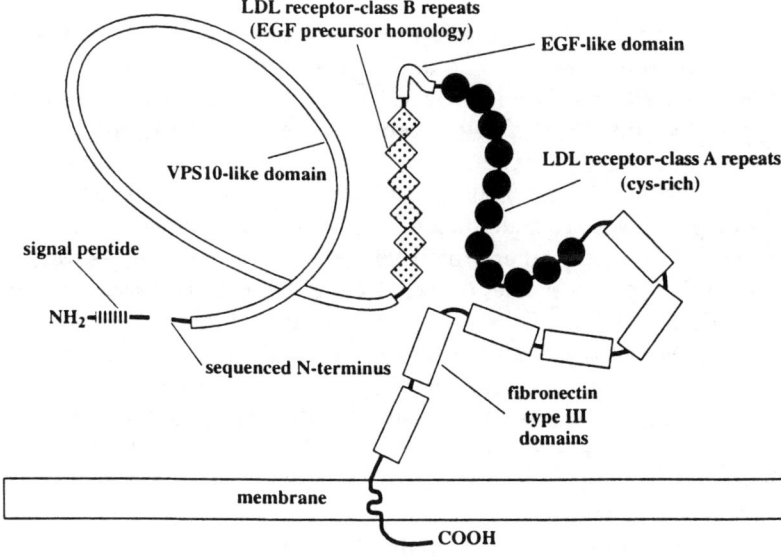

Fig. 5. Domain structure of hydra and vertebrate HAB. Mature protein is derived by posttranslational cleavage of an amino-terminal propeptide. Vertebrate protein has more LDL receptor-class A repeats and more fibronectin type III domains than hydra HAB

1998; Fong et al. 1998). The transmembrane domain with a short 55-residues-long intracellular carboxy terminus in hydra HAB is preceded by two consecutive fibronectin type III domains, as found in many transmembrane receptors, neural all adhesion molecules, and extracellular matrix proteins. They are highly flexible, thus allowing HAB to interact with cell surface molecules at different distances (Oberhauser et al. 1998). In addition, they may play a role in receptor dimerization. The intracellular domain is devoid of known catalytic functions, but contains motifs for G-protein coupling, acidic clusters, and putative casein kinase II phosphorylation sites.

To study function and expression we raised an antiserum against hydra HAB. On Western blots the resulting HAB antiserum recognized in solubilized hydra membranes a protein of 200 kDa (Fig. 4). The immunoreactive 200-kDa protein bound strongly to HA Sepharose, but not to control Sepharose, showing that HAB binds HA. Further evidence for HA binding to HAB comes from the finding that the HAB antiserum was able to remove the HA-binding activity from solubilized hydra membranes by immunoprecipitation. Full-length hydra HAB, expressed in CHO cells, bound HA ligands. We conclude that HA is the first morphogen that is able to interact with an LDL receptor-like molecule.

6
HAB Expression in Mouse Development

While sequencing of hydra HAB was completed, homologues were discovered in chicken, rabbit, and human and named SorLA and LR11 (Jacobsen et al. 1996; Yamazaki et al. 1996; Mörwald et al. 1997). These HAB homologues were isolated by homology to LDL receptors or by binding to the LDL receptor-associated protein RAP. Combination and alignment of domains are identical between hydra and mammals, but differ in the number of repeats (Fig. 5). The homology between each domain of hydra HAB and its vertebrate counterparts is higher than to any other protein. Since no other homologues were found by PCR analysis in any species or by searching the EST database, we assume that hydra HAB and the vertebrate homologues are orthologues. Preliminary data, which show that HA binds to cells expressing the human homologue, support this notion. Hydra or coelenterates in general and vertebrates diverged from a common ancestor more than 500 million years ago. The high conservation of 37% identical and 58% similar amino acids and the strict conservation and alignment of motifs hint at an identical function for these proteins.

To gain insight into the function of this new type of protein we isolated the mouse HAB homologue and used it to study the expression pattern in the adult organism and during embryonic development (Hermans-Borgmeyer et al. 1998). Northern blot analysis showed that HAB transcripts were present in a variety of adult organs, namely brain, lung, muscle, and kidney, with most prominent expression in the adult brain. In situ hybridization on frozen sections of adult mouse and rat brain revealed that HAB expression was restricted to specific neuronal cell populations of the brain. Highest transcript levels were detected over the pyramidal cell layer of the cerebral cortex, the pyramidal cell layer of the hippocampal formation, and the Purkinje cells of the cerebellum (Fig. 6). Thus, HAB transcripts were most abundant in large neurons with long processes.

The *in situ* analysis of embryonal sections showed that the adult pattern of HAB expression was established during development. A unique pattern of embryonic expression was observed in the telencephalon (Fig. 7), where HAB transcripts were detected starting at E11.5, when the first neurons of the cerebral cortex had become postmitotic. Up to E16.5 transcripts were restricted to the cortical area. At no developmental stage transcripts were detected in the striatum (Figs. 7, 8). Neuronal progenitor cells are able to move within both dorsal cortical and basal striatal ventricular zones, but they are unable to cross the boundary area between them (Fishell 1997). This hints at the existence of molecular and cellular signals that establish the borders between cortical and basal zones. The very restricted expression of HAB in the developing telencephalon points to an involvement in this process. Two areas with high levels of HAB expression became distinguishable at E16.5: the subventricular zone and the outer aspects of the cortical plate (Figs. 7, 8E-J). With ongoing development, the cortical plate showed stronger hybridization signals than the sub-

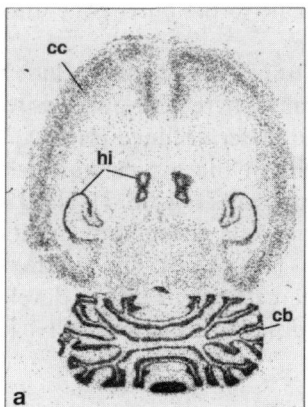
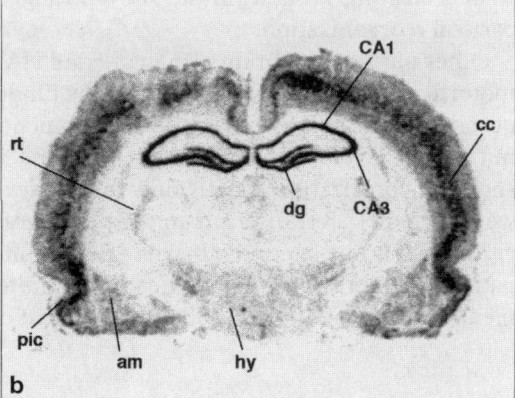

Fig. 6. Expression of HAB in distinct areas of the adult brain. Autoradiograms of a horizontal (**a**) and coronal section (**b**) through rat brain hybridized to a HAB riboprobe. Strong hybridization signals were detected in layers 3–5 of the cerebral cortex (*cc*), the piriform cortex (*pic*), the pyramidal (*CA*) cell layer of the hippocampal formation (*hi*), and the Purkinje cells of the cerebellum (*cb*). Less intense hybridization signals were detected over the other layers of the cerebral cortex (*cc*), the granular cells of the dentate gyrus (*dg*), the amygdala (*am*), the reticular nucleus of the thalamus (*rt*), and a variety of nuclei of the hypothalamus (*hy*)

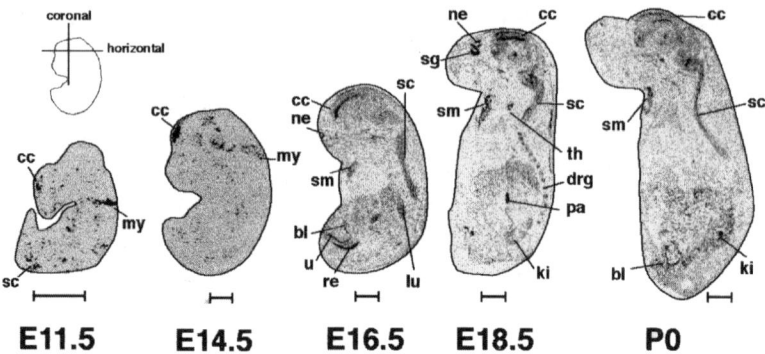

E11.5   E14.5   E16.5   E18.5   P0

Fig. 7. Expression of HAB during organogenesis and brain development. Autoradiograms of parasagittal sections through mouse embryos from E11.5 to E18.5 and from a newborn mouse. Ventral surface is to the *left*, dorsal to the *right*. At E11.5 strong hybridization signals were seen over the cerebral cortex (*cc*), the myelencephalon (*my*) and the spinal cord (*sc*). At E14.5 the signal in the hindbrain region was less intense and from E16.5 towards P0 a switch in hybridization intensity between the subventricular zone and outer aspects of the cortical plate of the cerebral cortex became apparent. In addition to signals observed in the central nervous system, the autoradiograms show hybridization over the thyroid (*th*) and submandibullary gland (*sm*), as well as over the serous gland (*sg*), nasal epithelium (*ne*), dorsal root ganglia (*drg*), pancreas (*pa*), lung (*lu*), kidney (*ki*), bladder (*bl*), ureter (*u*), and rectum (*re*). Scale bar is 2 mm

ventricular zone. The most likely explanation for this is that HAB is expressed in neurons migrating from the subventricular zone to the outer zone during cortical reorganization.

Other parts of the brain also expressed HAB transcripts. These include the motorneurons of the spinal cord and the hindbrain, where strong signals were observed during early phases of development and later declined (Fig. 7). The mitral cell layer of the olfactory bulb and cells in the geniculate nucleus showed hybridization signals only late in development (Fig. 8G–J). Similarly, we detected HAB transcripts in the peripheral nervous system only in differentiated neurons late in embryonic development (Fig. 9A,B). The differential pattern of expression either in early postmitotic neurons or late in development in differentiated neurons is suggestive of different functions of HAB in

Fig. 8. High power analysis of HAB expression in the developing brain. Brightfield photomicrographs (**A,C,E,G,I**) of the Giemsa stained sections and their corresponding dark-field pictures revealing the hybridization signals (**B,D,F,H,J**). A parasagittal section through a head region at E11.5 (**A,B**) demonstrates that transcripts of HAB were most abundant in the outer aspects of developing cerebral cortex (*cc*) and in motorneurons of the myelencephalon (*my*). In the rhombic lip of the metencephalon (*me*) a thin layer of cells also exhibited strong hybridization signals. A coronal section through the forebrain region at E14.5 (**C,D**) reveals that expression of HAB is restricted to the lateral aspects of the cerebral cortex. In a coronal section through a telenceph-

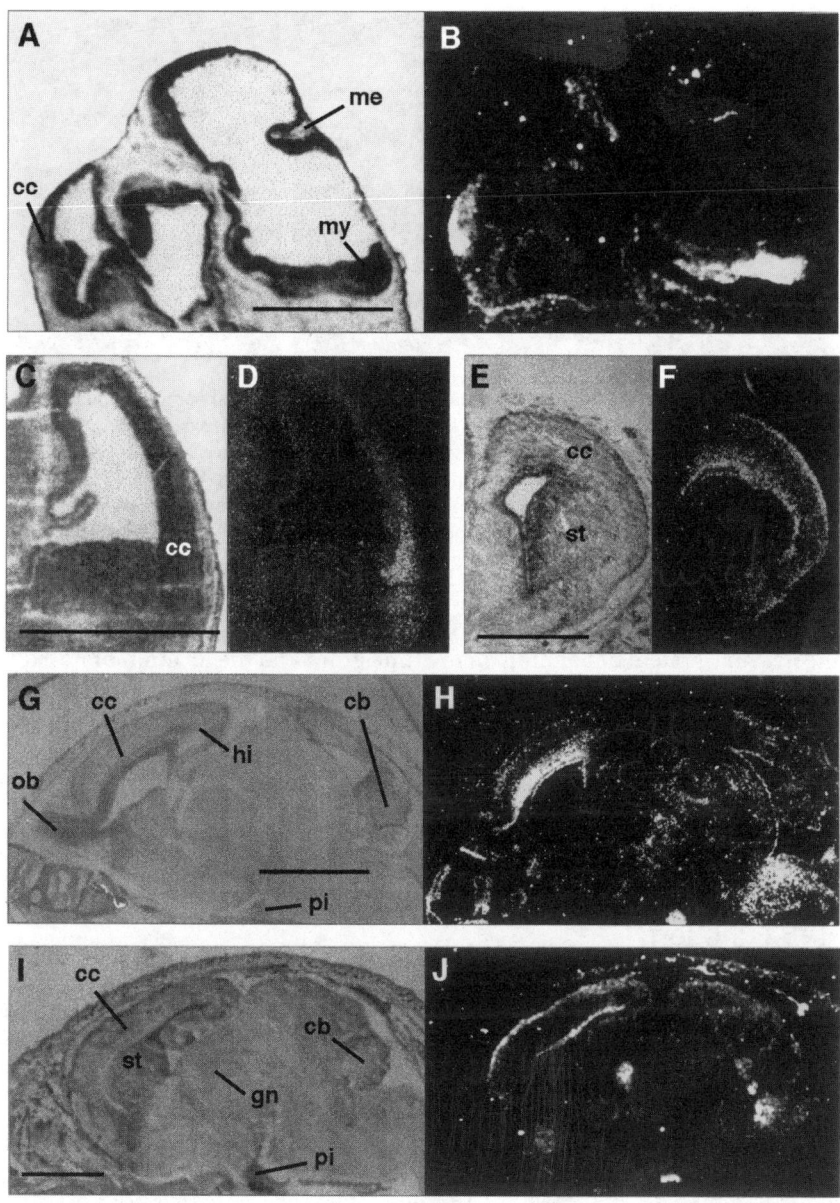

alon at E16.5 (**E,F**) also no signals were seen over the striatum (*st*). The signal has spread dorsally compared to E14.5. The subventricular zone and outer aspect of the cerebral cortex exhibited strong hybridization signals. At E18.5 (**G,H**) and at P0 (**I,J**), hybridization intensity in the subventricular zone decreased compared to that observed in the cortical plate. In addition to the pattern already described, transcripts were also detected over the mitral cell layer of the olfactory bulb (*ob*), Purkinje cell layer of the cerebellum (*cb*), geniculate nucleus of the thalamus (*gn*), and pituitary gland (*pi*). In all developmental stages analyzed, HAB expression was never detected over the neuroepithelium. Scale bar is 2 mm

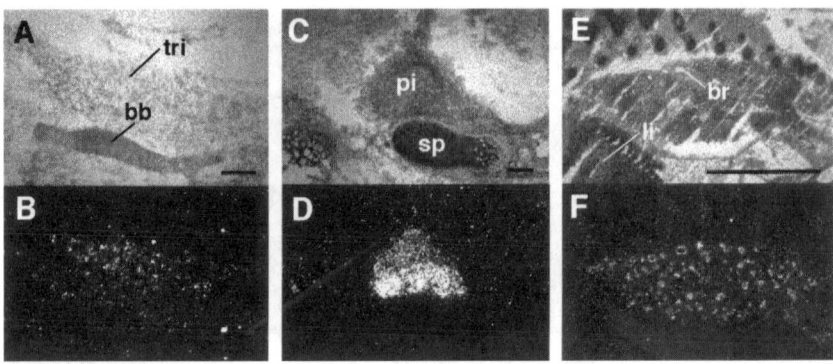

Fig. 9. HAB expression outside the central nervous system. A sagittal section through a trigeminal ganglion (*tri*) shows expression of HAB in the peripheral nervous system beginning at E16.5 (**A,B**). HAB transcripts were not only present at E12.5 in the pituitary, but also visible in a sagittal section (**C,D**) at P0 in the pituitary above the sphenoid cartilage (*sp*). HAB expression was detected in epithelia surrounding the bronchii (*br*) of the lung throughout embryonic development. A sagittal section taken at E14.5 is shown (**E,F**). The liver (*li*) did not exhibit hybridization signals. Scale bar is 0.2 mm

the nervous system depending on the time point and the location of its expression.

Outside of the nervous system HAB was detected in a variety of glands and organs, for example in the pituitary gland (Figs. 8I, J, 9C, D), the pancreas (Fig. 7), the lung (Fig. 9E, F) and the kidney which rely on epithelial-mesenchymal interactions for their proper development.

7
HAB Function

Since HA signaling includes activation of G proteins, we expected a seven-transmembrane receptor as immediate signal transducer. The most conserved part between hydra and vertebrate HAB is the short cytoplasmic tail. The abundance and spacing of several basic residues resemble a motif important for G-protein coupling, found in receptors with a single transmembrane domain (Nishimoto et al. 1987, 1993; Okamoto et al. 1990; Anand-Srivastava et al. 1996; Yamatsuje et al. 1996). HAB might, therefore, be a receptor for HA, able to transduce the HA signal directly to second messenger systems (Fig. 10). Another interpretation is that HAB is part of a receptor complex and modifies HA function positively or negatively. Both as signaling and accessory receptor, HAB should be present on target cells. Since HAB transcripts were not detected in premitotic cells, e.g. during active phases of proliferation in the neuro-epithelium, but rather in specific subpopulations of newborn postmitotic neu-

Fig. 10. Models for HAB function

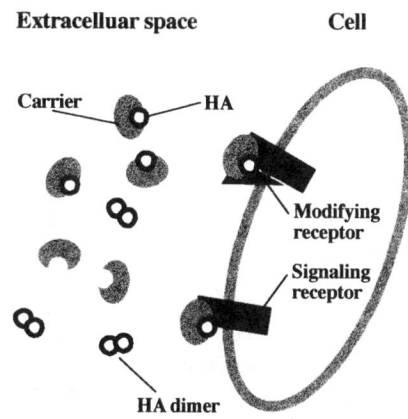

rons and in a variety of fully differentiated mature neurons, a different inter-pretation is more likely. Therefore, as the third and as most likely function we suggest that HAB could be the HA-binding part of the carrier complex (Schaller et al. 1986, 1996) which presents HA to its receptor (Fig. 10). Depending on HAB location either as membrane-bound or released molecule the range of action of HA could thus be regulated.

References

Allen S, Khan S, Futwan-Al-Mohanna, Batten P, Yacoub M (1998) Native low density lipopro-tein-induced calcium transients trigger VCAM-1 and E-selectin expression in cultured human vascular endothelial cells. J Clin Invest 101:1064–1075

Anand-Srivastava MB, Sehl PD, Lowe DG (1996) Cytoplasmic domain of natriuretic peptide receptor-C inhibits adenylyl cyclase. J Biol Chem 271:19324–19329

Bodenmüller H, Schaller HC (1981) Conserved amino acid sequence of a neuropeptide, the head activator, from coelenterates to humans. Nature 293:579–580

Bodenmüller H, Schilling E, Zachmann B, Schaller HC (1986) The neuropeptide head activator loses its biological acitivity by dimerization. EMBO J 5:1825–1829

Farese RV Jr, Herz J (1998) Cholesterol metabolism and embryogenesis. Trends Genet 14: 115–120

Fenger U, Hofmann M, Galliot B, Schaller HC (1994) The role of the cAMP pathway in mediat-ing the effect of head activator on nerve-cell determination and differentiation in hydra. Mech Dev 47:115–125

Fishell G (1997) Regionalization in the mammalian telencephalon. Curr Opin Neurobiol 7:62–69

Fong RL, Vadgama JV, Laguer MR, Ward HJ (1998) Low density lipoprotein receptors mediate intracellular calcium signals in microalbuminuric hypertensives. Am J Hypertens 11:113–116

Franke I, Buck F, Hampe W (1997) Purification of a head-activator receptor from hydra. Eur J Biochem 244:940–945

Galliot B (1997) Signaling molecules in regenerating hydra. BioEssays 19:37–46

Galliot B, Welschof M, Schuckert O, Hoffmeister SAH, Schaller HC (1995) The cAMP response element binding protein is involved in hydra regeneration. Development 121:1205–1216

Gauchat D, Kreger S, Holstein T, Galliot B (1998) prdl-a, a gene marker for hydra apical differentiation related to triploblastic paired-like head-specific genes. Development 125:1737–1645

Hampe W, Frank RW, Schulze C, Dehning I, Schaller HC (1996) Photoaffinity labeling of the head-activator receptor from hydra. Eur J Biochem 235:814–820

Hampe W, Franke I, Urny J, Petersen CM, Schaller HC (submitted) The neuropeptide head activator binds to a new member of the low density lipoprotein receptor family.

Hermans-Borgmeyer I, Hampe W, Schinke B, Methner A, Nykjaer A, Süsens U, Fenger U, Herbarth B, Schaller HC (1998) Unique expression pattern of a novel mosaic receptor in the developing cerebral cortex. Mech Dev 70:65–76.

Holstein T, Schaller HC, David CN (1986) Nerve cell differentiation in hydra requires two signals. Dev Biol 115:9–17

Ishii TM, Silvia C, Hirschberg B, Bond CT, Adelman JP, Maylie J (1997) A human intermediate conductance calcium-activated potassium channel. Proc Natl Acad Sci USA 94:11651–11656

Jacobsen L, Madsen P, Moestrup SK, Lund AH, Tommerup N, Nykjaer A, Sottrup-Jensen L, Gliemann J, Petersen CM (1996) Molecular characterization of a novel human hybrid-type receptor that binds the α_2-macroglobulin receptor-associated protein. J Biol Chem 271:31379–31383

Kajiwara S, Sato T (1986) Growth promoting effect of head activator in cultured chick embryo brain cells. Acta Endocrinol 113:604–608

Kayser ST, Ulrich H, Schaller HC (1998) Involvement of a Gardos-type potassium channel in head activator-induced mitosis of BON cells. Eur J Cell Biol 76:119–124

Marcusson EG, Horazdovsky BF, Cereghino JL, Gharakhanian E, Emr SD (1994) The sorting receptor for yeast vacuolar carboxypeptidase Y is encoded by the VPS10 gene. Cell 77:579–586

Mörwald S, Yamazaki H, Bujo H, Kusunoki J, Kanaki T, Seimiya K, Morisaki N, Nimpf J, Schneider WJ, Saito Y (1997) A novel mosaic protein containing LDL receptor elements is highly conserved in humans and chickens. Arterioscler Thromb Vasc Biol 17:996–1002

Neubauer KH, Christians S, Hoffmeister SAH, Kreger S, Schaller HC (1991) Characterization of two types of head activator receptor on hydra cells. Mech Dev 33:39–47

Nishimoto I, Hata Y, Ogata E, Kojima I (1987) Insulin-like growth factor II stimulates calcium influx in competent BALB/c 3T3 cells primed with epidermal growth factor. J Biol Chem 262:12120–12126

Nishimoto I, Okamoto T, Matsuura Y, Takahashi S, Okamoto T, Murayama Y, Ogata E (1993) Alzheimer amyloid protein precursor complexes with brain GTP-binding protein G$_o$. Nature 362:75–79

Oberhauser AF, Marszalek PE, Erickson HP, Fernandez JM (1998) The molecular elasticity of the extracellular matrix protein tenascin. Nature 393:181–185

Okamoto T, Katada T, Murayama Y, Ui M, Ogata E, Nishimoto I (1990) A simple structure encodes G protein-activating function of the IGF-II/mannose 6-phosphate receptor. Cell 62:709–717

Pahl MV, Wang J, Zhou XJ, Wang XQ, Tehranzadeh A, Vaziri ND (1996) Effects of low density lipoprotein on cytosolic [Ca++] in cultured rat mesangial cells. J Invest Med 44:556–560

Petersen CM, Nielsen MS, Nykjær A, Jacobsen L, Tommerup N, Rasmussen HH, Røigaard H, Gliemann J, Madsen P, Moestrup SK (1997) Molecular identification of a novel candidate sorting receptor purified from human brain by receptor-associated protein affinity chromatography. J Biol Chem 272:3599–3605

Quach TT, Duchemin AM, Oliver AP, Schrier BK, Wyatt RJ (1992) Hydra head activator peptide has trophic activity for eukaryotic neurons. Dev Brain Res 68:97–102

Roberge M, Escher E, Schaller HC, Bodenmüller H (1984) The hydra head activator in human blood circulation. Degradation of the synthetic peptide by plasma angiotensin-converting enzyme. FEBS Lett 173:307–313

Schaefer HIMB, Höld KM, Egas-Kenniphaas JM, van der Laarse A (1993) Intracellular calcium signalling after binding of low-density lipoprotein to confluent and nonconfluent cultures of an endothelial cell line, EA.hy 926. Cell Calcium 14:507-516

Schaller HC (1976a) Action of the head activator on the determination of interstitial cells in hydra. Cell Differ 5:13-25

Schaller HC (1976b) Action of the head activator as a growth hormone in hydra. Cell Differ 5:1-11

Schaller HC (1996) Evolution of neuronal and endocrine traits: a network of neuroendocrine cells close to the base of the phylogenetic tree. In: Unsicker K (ed) Autonomic-endocrine interactions, vol 10. Harwood Academic Publishers, Amsterdam, pp 1-12

Schaller HC, Bodenmüller H (1981) Isolation and amino acid sequence of a morphogenetic peptide from hydra. Proc Natl Acad Sci USA 78:7000-7004

Schaller HC, Roberge M, Zachmann B, Hoffmeister SAH, Schilling E, Bodenmüller H (1986) The head activator is released from regenerating *Hydra* bound to a carrier molecule. EMBO J 5:1821-1824

Schaller HC, Hoffmeister SAH, Dübel S (1989) Role of the neuropeptide head activator for growth and development in hydra and mammals. Development (Suppl) 107:99-107

Schaller HC, Hofmann M, Javois LC (1990) Effect of head activator on proliferation, head-specific determination and differentiation of epithelial cells in hydra. Differentiation 43:157-164

Schaller HC, Hermans-Borgmeyer I, Hoffmeister SAH (1996) Neuronal control of development in hydra. Int J Dev Biol 40:339-344

Schummer M, Scheurlen I, Schaller HC, Galliot B (1992) HOM/HOX homeobox genes are present in hydra (*Chlorohydra viridissima*) and are differentially expressed during regeneration. EMBO J 11:1815-1823

Shackelford RE, Misra UK, Florine-Casteel K, Thai S-F, Pizzo SV, Adams DO (1995) Oxidized low density lipoprotein suppresses activation of NFκB in macrophages via a pertussis toxin-sensitive signaling mechanism. J Biol Chem 270:3475-3478

Ulrich H, Tarnok A, Schaller HC (1996) Head activator induced mitosis of NH15-CA2 cells requires calcium influx and hyperpolarization. J Physiol (Paris) 90:85-94

Yamatsuje T, Matsui T, Okamoto T, Komatsuzaki K, Takeda S, Fukumoto H, Iwatsubo T, Suzuki N, Asami-Odaka A, Ireland S, Kinane B, Giambarella U, Nishimoto I (1996) G protein-mediated neuronal DNA fragmentation induced by familial Alzheimer's disease-associated mutants of APP. Science 272:1349-1352

Yamazaki H, Bujo H, Kusunoki J, Seimiya K, Kanaki T, Morisaki N, Schneider WJ, Saito Y (1996) Elements of neural adhesion molecules and a yeast vacuolar protein sorting receptor are present in a novel mammalian low density lipoprotein receptor family member. J Biol Chem 271:24761-24768

Invertebrate Neurohormones and Their Receptors

Cornelis J. P. Grimmelikhuijzen, Frank Hauser, Kathrine Krageskov Eriksen
and Michael Williamson

1
Introduction

During the "boom" of neuropeptide discoveries in the 1970s and 1980s, it was sometimes assumed that neuropeptides were mainly present in mammals, where they were needed for the "fine-tuning" (neuromodulation) of the highly complicated mammalian brain; or, in other words, that neuropeptides made the mammalian brain what it was: the crown on nervous system evolution. However, during the last 15 years, it has become increasingly clear that neuropeptides are not a characteristic of higher nervous systems, but, instead, have a long evolutionary history. It might even be true that neuropeptides evolved before the "classical" neurotransmitters (acetylcholine, monoamines and amino acids), which would mean that neuropeptides are the original (slow) transmitters and that the smaller "classical" transmitters evolved later as a response to the need for a faster neurotransmission (Grimmelikhuijzen 1986; Grimmelikhuijzen et al. 1996).

In the first part of this chapter, we focus on neurohormones and neurohormone receptors in cnidarians, because they are the lowest group in the animal kingdom having a nervous system and because they probably are closely related to the first animal group in evolution that evolved a nervous system (Mackie 1990). In the second part, we give some examples of neurohormone receptors in insects, because they comprise by far the largest group of invertebrate animals.

2
Cnidarians

Cnidarians are primitive invertebrate animals such as sea anemones, corals, *Hydra* and jellyfishes and comprise a phylum, the Cnidaria, consisting of four classes: the Hydrozoa (such as *Hydra*), Cubozoa (box jellyfishes), Scyphozoa (true jellyfishes) and Anthozoa (animals such as sea anemones and corals).

Department of Cell Biology, Zoological Institute, University of Copenhagen, Universitetsparken 15, DK-2100 Copenhagen Ø, Denmark; e-mail: cgrimmelikhuijzen@zi.ku.dk

Cnidarians are not only beautiful animals capable, for example, of building large (up to 1000-km-long) coral reefs that form the basis of impressive eco-systems, but they are also interesting model animals for several biological disciplines. This is for the following reasons: (1) cnidarians have an anatomically simple nervous system and a simple behaviour. They can, therefore, be used by electrophysiologists as a model to understand the cellular basis of behaviour, for example swimming (Anderson and Schwab 1982; Spencer and Arkett 1984; Anderson and Spencer 1989); (2) cnidarians have an amazingly high regeneration capacity. *Hydra*, for example, can be cut into 10–20 small slices, from which, after a few days, new animals originate. One can even disintegrate *Hydra* into its separate cells and subsequently reaggregate them by centrifugation, after which new animals develop from the cell clumps (Gierer et al. 1972; Gierer 1977). *Hydra* consists of less than ten different cell types. By several kinds of manipulation, however, it is possible to obtain a *Hydra* variant that only consists of one cell type, the epitheliomuscular cells (Campbell 1976; Minobe et al. 1995). These "epithelial *Hydra*" are not capable of feeding any more and must be force-fed and subsequently internally rinsed with the help of a small glass capillary. It is possible to introduce stem cells for neurones into these epithelial *Hydra* and follow the development of a completely new nervous system in these originally nerve-free animals (Minobe et al. 1995). Thus, it is easy to manipulate *Hydra* and other hydrozoan polyps and for these reasons they are widely used model systems for developmental biologists to study processes such as cell differentiation, regeneration, and pattern formation (Gierer 1977; Bode 1992; Müller 1996).

3
Neurohormones in Cnidarians

The basic organization of the cnidarian nervous system is that of a nerve net. In some regions of the animal, this nerve net can condense to form a nervous plexus or circular or longitudinal nerve tracts (Mackie 1973, 1989; Grimmelikhuijzen and Spencer 1984; Spencer and Arkett 1984; Grimmelikhuijzen 1985; Grimmelikhuijzen et al. 1986). Thus, although ganglia and brains are absent in cnidarians, some primitive forms of neuronal centralization already exist.

At the ultrastructural level, many of the cnidarian neurones appear to be multifunctional: they have sensory cilia (are "sensory neurones"), make synaptic contacts with two or more other neurones (are "interneurones"), make synaptic contacts with epitheliomuscular cells (are "motorneurones") and have secretory dense-cored vesicles at nonsynaptic release sites (are "neurosecretory neurones") (Westfall 1973a; Westfall and Kinnamon 1978). Westfall (1973a) proposes that this primitive type of cnidarian nerve cell is the ancestor from which the more specialized neurones evolved that we find in higher animals today.

The existence of synapses and nonsynaptic release sites in the nervous systems of cnidarians has been known for more than 35 years (Horridge and Mackay 1962), but for a long time uncertainty has remained concerning the nature of the neurotransmitters or neurohormones. We ourselves, for example, have been unable to demonstrate acetylcholine, the catecholamines or serotonin in the nervous system of *Hydra* and the same negative results have been obtained by our colleagues (O. Koizumi, pers. comm.). However, the morphology of the presynaptic or neurosecretory vesicles in cnidarian neurones (most of which are dense-cored with a diameter of 100–200 nm; Westfall 1973a,b; Westfall and Kinnamon 1978) suggests that neuropeptides could be good candidates for being the long searched-for transmitters, and this is exactly what we found. Using various radioimmunoassays against the C-terminal sequence Arg-X-NH$_2$, we have isolated 16 different peptides from a single sea anemone species, *Anthopleura elegantissima* (13 are given in Table 1), and another research group, using a certain bioassay (see below), has recently added an additional peptide to this list (Grimmelikhuijzen and Graff 1986; Graff and Grimmelikhuijzen 1988a,b; Grimmelikhuijzen et al. 1990, 1996; Nothacker et al. 1991a,b; Carstensen et al. 1992, 1993; Leitz et al. 1994). Some of the peptides in Table 1 are confined to sea anemones, but other peptides, such as Antho-RFamide and Metamorphosin A (also called MMA or Antho-LWamide II; Table 1), have their relatives in other anthozoans and other cnidarian classes (Grimmelikhuijzen and Groeger 1987; Grimmelikhuijzen et al. 1988, 1992a, 1996; Moosler et al. 1996, 1997; Takahashi et al. 1997). This is exemplified for the cnidarian Arg-Phe-NH$_2$ (RFamide) peptides in Table 2. Peptide families that occur in all the classes of cnidarians, such as the RFamide and the Leu-Trp-NH$_2$ (LWamide) families, are especially interesting, because one could expect that members of these peptide families were present in the first nervous systems, and thus could be among the first neurotransmitter-like substances that were used in evolution.

We have raised antibodies against the peptides of Tables 1 and 2 and found that these peptides were localized in neurones and, thus, were neuropeptides (Grimmelikhuijzen and Spencer 1984; Grimmelikhuijzen 1985; Grimmelikhuijzen et al. 1988, 1992a,b, 1996; Nothacker et al. 1991b; Carstensen et al. 1992). Furthermore, we found that peptides belonging to one group (for example the Antho-RFamides; Table 1) were produced by one specific set of sea anemone neurones, whereas peptides belonging to another group (for example the Antho-RWamides) were produced by another specific set of neurones (Grimmelikhuijzen et al. 1992b, 1996). Therefore, sea anemones have at least six populations of neurones that are neurochemically and probably also functionally different. Thus, the presence of multifuntionality in cnidarian neurones (as described earlier) does not mean that all neurones in cnidarians are the same.

Finally, we found, using immuno-electronmicroscopy, that the cnidarian neuropeptides (at least the Hydra-RFamides, Antho-RWamides and Antho-

Table 1. Neuropeptide families in anthozoans

Species	Structure	Name
Anthopleura elegantissima	L-3-phenyllactyl-Phe-Lys-Ala-NH₂	Antho-KAamide
Anthopleura elegantissima	L-3-phenyllactyl-Tyr-Arg-Ile-NH₂	Antho-RIamide I
Anthopleura elegantissima	Tyr-Arg-Ile-NH₂	Antho-RIamide II
Anthopleura elegantissima	L-3-phenyllactyl-Leu-Arg-Asn-NH₂	Antho-RNamide I
Anthopleura elegantissima	Leu-Arg-Asn-NH₂	Antho-RNamide II
Anthopleura elegantissima	<Glu-Ser-Leu-Arg-Trp-NH₂	Antho-RWamide I
Anthopleura elegantissima	<Glu-Gly-Leu-Arg-Trp-NH₂	Antho-RWamide II
Anthopleura elegantissima	<Glu-Asn-Phe-His-Leu-Arg-Pro-NH₂	Antho-RPamide II
Anthopleura elegantissima Leu-Pro-Pro-Gly-Pro-Leu-Pro-Arg-Pro-NH₂		Antho-RPamide I
Anthopleura elegantissima	Gly-Pro-Hyp-Ser-Leu-Phe-Arg-Pro-NH₂	Antho-RPamide IV
Anthopleura elegantissima	<Glu-Val-Lys-Leu-Tyr-Arg-Pro-NH₂	Antho-RPamide III
Anthopleura elegantissima	Tyr-Arg-Pro-NH₂	Antho-RPamide V
Anthopleura elegantissima	<Glu-Gly-Arg-Phe-NH₂	Antho-RFamide
Renilla köllikeri	<Glu-Gly-Arg-Phe-NH₂	Antho-RFamide
Anthopleura elegantissima	<Glu-Gln-Pro-Gly-Leu-Trp-NH₂	Metamorphosin A (MMA or Antho-LWamide II)

RFamide) were present in neuronal dense-cored vesicles (Koizumi et al. 1989; Westfall and Grimmelikhuijzen 1993; Westfall et al. 1995).

We have investigated the action of the sea anemone neuropeptides by inject-ing them into intact sea anemones or by adding them to several types of mus-cle preparations or isolated muscle cells. These experiments showed that the neuropeptides of Table 1 are biologically active (at 10^{-9} to 10^{-8} M) and either stimulate or inhibit contraction of the different muscle preparations (McFar-lane et al. 1987, 1991, 1992, 1993; McFarlane and Grimmelikhuijzen 1991; Nothacker et al. 1991b; Carstensen et al. 1992, 1993; Grimmelikhuijzen et al. 1996). All the above-mentioned findings together, then, suggest that the sea anemone neuropeptides are neurotransmitters or neuromodulators at neuro-neuronal or neuromuscular junctions.

The neuropeptide MMA has been isolated from sea anemones because it causes metamorphosis from a planula larva into a primary polyp in the hydro-zoan *Hydractinia echinata* (Leitz et al. 1994). This means that cnidarian neu-ropeptides, in addition to being neurotransmitters or neuromodulators, also can be neurohormones that control developmental processes such as meta-morphosis.

Table 2. The RFamide neuropeptide family in cnidarians

Species	Structure	Name
Anthopleura elegantissima	<Glu-Gly-Arg-Phe-NH$_2$	Antho-RFamide
Renilla köllikeri	<Glu-Gly-Arg-Phe-NH$_2$	Antho-RFamide
Cyanea lamarckii	Gly-Arg-Phe-NH$_2$	Cyanea-RFamide II
Cyanea lamarckii	<Glu-Pro-Leu-Trp-Ser-Gly-Arg-Phe-NH$_2$	Cyanea-RFamide II
Polyorchis penicillatus	<Glu-Leu-Leu-Gly-Gly-Arg-Phe-NH$_2$	Pol-RFamide I
Hydra magnipapillata	<Glu-Trp-Leu-Gly-Gly-Arg-Phe-NH$_2$	Hydra-RFamide I
Hydra magnipapillata	<Glu-Trp-Phe-Asn-Gly-Arg-Phe-NH$_2$	Hydra-RFamide II
Polyorchis penicillatus	<Glu-Trp-Leu-Lys-Gly-Arg-Phe-NH$_2$	Pol-RFamide II
Cyanea lamarckii	<Glu-Trp-Leu-Arg-Gly-Arg-Phe-NH$_2$	Cyanea-RFamide I
Hydra magnipapillata	Lys-Pro-His-Leu-Arg-Gly-Arg-Phe-NH$_2$	Hydra-RFamide III
Hydra magnipapillata	His-Leu-Arg-Gly-Arg-Phe-NH$_2$	Hydra-RFamide IV

4
Biosynthesis of Neurohormones in Cnidarians

In higher animals, neuropeptides are produced as large precursor proteins, the preprohormones, which consist of an N-terminal hydrophobic signal sequence [for translocation across the rough endoplasmic reticulum (RER) membrane; the "pre" part] and a prohormone part, containing one or several copies of the immature neuropeptide. After removal of the "pre" part during RER membrane translocation, the prohormone is transported via the ER into the Golgi system, sorted and packaged into neurosecretory, dense-cored vesicles together with several processing enzymes that convert the hormone precursor into its biologically active peptides. From the prohormones in higher animals it is known that the immature neuropeptide sequence, both at its N- and C-terminal sides, is flanked by basic residues, and that these basic residues function as primary cleavage signals (Fig. 1). After an endoproteolytic cleavage at these basic residues by a prohormone convertase (PC1/PC3 or PC2; Seidah et al. 1990, 1991; Smeekens and Steiner 1990; Smeekens et al. 1991), other processing enzymes convert the immature propeptide sequence into a biologically active hormone (Fig. 1). If the biologically active peptide is protected by a C-terminal amidation, a Gly residue is found in this position in the prohormone. This C-terminal Gly residue is converted into an amide by the concerted actions of two enzymes, peptidylglycine α-hydroxylating monooxygenase (or PHM) and peptidyl-α-hydroxyglycine α-amidating lyase (or PAL) (Fig. 2; Bradbury and Smyth 1991; Eipper et al. 1992). If the biologically active peptide is protected by an N-terminal <Glu group, a Gln residue is found in

this position in the prohormone, which is converted into a <Glu residue by the enzyme glutaminyl cyclase (Fischer and Spiess 1987; Pohl et al. 1991). Thus, in higher animals, about four to five processing enzymes are known to be involved in the generation of biologically active peptides from a prohormone (Fig. 1).

We have cloned the preprohormones for nearly all of the neuropeptides listed in Tables 1 and 2 (Darmer et al. 1991, 1998; Schmutzler et al. 1992, 1994; Reinscheid and Grimmelikhuijzen 1994; Leviev and Grimmelikhuijzen 1995; Grimmelikhuijzen et al. 1996; Leviev et al. 1997). These cnidarian preprohormones are characterized by the following properties: (1) in general, there is a very high copy number of the immature neuropeptide sequences in the precursor. While in mammals the copy number varies between one and seven, this number can be up to 38 in cnidarians (Figs. 3 and 4). This suggests that neuropeptide biosynthesis in cnidarians is very efficient; (2) the structures of the immature neuropeptide sequences in the cnidarian precursors show that the same types of processing enzymes that are found in mammals (Fig. 1) must also be present in cnidarians: there are basic residues at the C-terminal sides of the immature neuropeptide sequences (Figs. 3, 4), pointing to the presence of a PC1/PC3- or PC2-like enzyme; Gly or Gln residues are found at positions

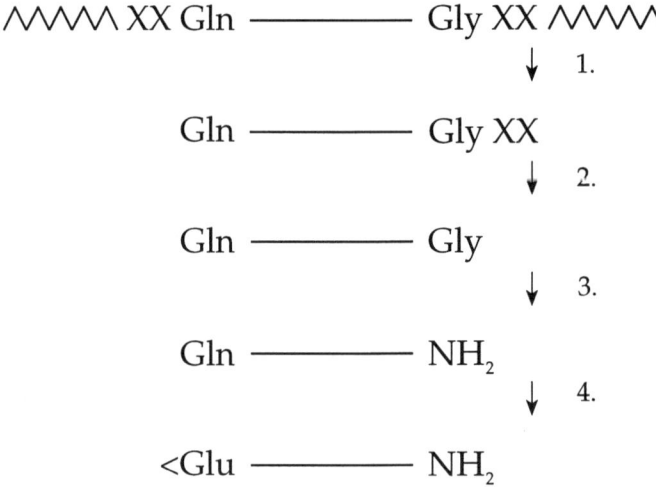

Fig. 1. Neuropeptide precursor processing in higher invertebrates and vertebrates. *Top* The precursor protein contains one or more copies of an immature neuropeptide sequence, which is flanked by basic residues (marked by *XX*). The cleavage or processing steps are catalyzed by the following enzymes: **1** An endoproteinase cleaving at the C-terminal sides of pairs or single basic residues; **2** a carboxypeptidase B-like enzyme; **3** peptidyl-glycine hydroxylase (also called peptidylglycine α-hydroxylating monooxygenase or PHM) and peptidyl-hydroxyglycine N-C lyase (also called peptidyl α-hydroxyglycine α-amidating lyase or PAL); **4** glytaminyl cyclase. (From Grimmelikhuijzen et al. 1996)

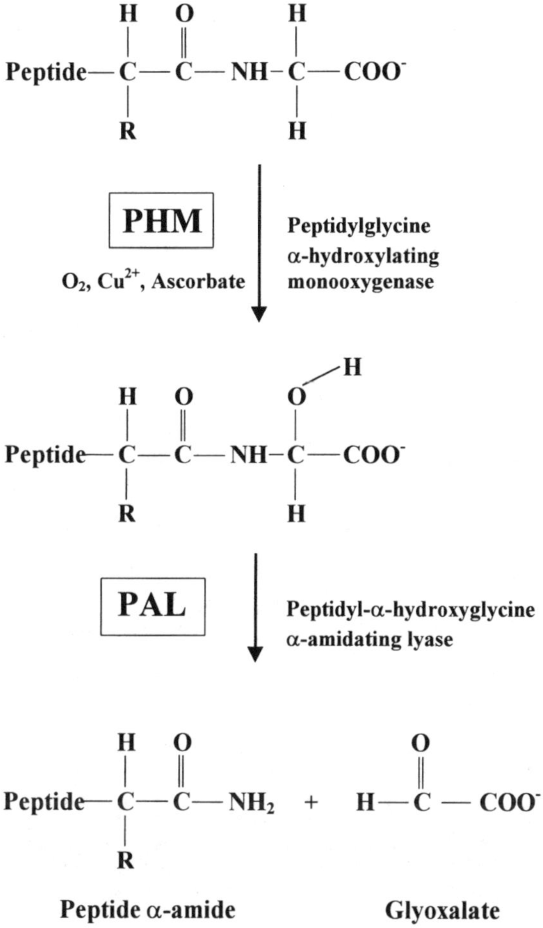

Fig. 2. Actions of the two enzymes that convert a peptidyl-glycine into a peptidyl-amide (step 3 of Fig. 1). PHM hydroxylates peptidylglycine (*top*), whereas PAL splits peptidylhydroxyglycine, yielding peptidyl-amide and glyoxylic acid (*bottom*)

$$\text{Peptide} - \underset{\underset{R}{|}}{\overset{\overset{H}{|}}{C}} - \overset{\overset{O}{\|}}{C} - NH - \underset{\underset{H}{|}}{\overset{\overset{H}{|}}{C}} - COO^{-}$$

PHM   Peptidylglycine
            α-hydroxylating
O₂, Cu²⁺, Ascorbate   monooxygenase

$$\text{Peptide} - \underset{\underset{R}{|}}{\overset{\overset{H}{|}}{C}} - \overset{\overset{O}{\|}}{C} - NH - \underset{\underset{H}{|}}{\overset{\overset{O-H}{|}}{C}} - COO^{-}$$

PAL   Peptidyl-α-hydroxyglycine
            α-amidating lyase

$$\text{Peptide} - \underset{\underset{R}{|}}{\overset{\overset{H}{|}}{C}} - \overset{\overset{O}{\|}}{C} - NH_2 \quad + \quad H - \overset{\overset{O}{\|}}{C} - COO^{-}$$

Peptide α-amide   **Glyoxalate**

where there are amide or <Glu groups in the biologically active peptides, showing that enzymes similar to mammalian PHM, PAL and glutaminyl cyclase must be present in neurones of cnidarians; (3) however, in addition to the four to five processing enzymes known from mammals, there must be many more processing enzymes present in cnidarians. This becomes evident from unusual processing sites located at the N-terminal sides of the immature neuropeptide sequences in the cnidarian preprohormones. These unusual cleavage sites are Asp/Glu, Ser/Thr, Asn and Tyr residues, and N-terminal X-Pro or X-Ala sequences. The enzymes that are responsible for cleavage at the C-terminal sites of Asp/Glu, Ser/Thr, Asn and Tyr residues can either be endoproteinases or aminopeptidases (Grimmelikhuijzen et al. 1996), whereas the enzyme that is responsible for the removal of N-terminally located X-Pro or X-Ala sequences is most probably a dipeptidyl aminopeptidase (Leviev and

```
                                        TTTAGTATTGCGTCGTCGTGTGTGTGTGTGTGTGTG     35
TTTGGCTCATTTTCTACGATCAACGTTGCTCTACACATTTGTATGGAGAAAACCTGTCAGTCTATAACGTATATACTTC    114
AAAATAATGGACTTAGTAATATAAGTATCAACTGCATTATTGTTGTCAATGTCCA ATG GTT AGT CTG GGT TTT    187
                                                        Met Val Ser Leu Gly Phe      6

TTC GTT CGT GAT GTT ACT CCA ACA TTC ATT GTA GAT CAT ATG TTC TAC ATG CTA TTC CCG    247
Phe Val Arg Asp Val Thr Pro Thr Phe Ile Val Asp His Met Phe Tyr Met Leu Phe Pro     26

ATG GAT TTT ACG TGT TAC GTT GCC GGG CTG TTG TTG ATA TTA AAT ACT TAC AGT TTG GCC    307
Met Asp Phe Thr Cys Tyr Val Ala Gly Leu Leu Leu Ile Leu Asn Thr Tyr Ser Leu Ala     46

GGG CCC TCA ACT AGC GAA GGA CTA AAC GAA CGG AAC TTG CTG GAT AAA ACA GAG TTG TCG    367
Gly Pro Ser Thr Ser Glu Gly Leu Asn Glu Arg Asn Leu Leu Asp Lys Thr Asp Leu Ser     66

ATA AAT GAC GAG ATA TTT AGC GAA GAT GAT GAT ATG CTG GCA AGG GAC GCT GAA GAC AAA    427
Ile Asn Asp Glu Ile Phe Ser Glu Asp Asp Asp Met Leu Ala Arg Asp Ala Glu Asp Lys     86

CAA GGA CGA TTT AAT CGT AAA TTA AAT AAC AAG TTG AAT GAA GCG GTA CAA GGA CGC TTC    487
Gln Gly Arg Phe Asn Arg Lys Leu Asn Asn Lys Leu Asn Glu Ala Val Gln Gly Arg Phe    106

GGG AGA AAT GAG AGA AAA GAA TCG GAA GAA GAA CAA GGA AGG TTC GGG GGA GAA AAT GAA    547
Gly Arg Asn Glu Arg Lys Glu Ser Glu Glu Glu Gln Gly Arg Phe Gly Arg Glu Asn Glu    126

AAA CAA GGA AGA TTT GGA AGG GAA AGC GAG GAG CAA GGG AGA TTT GGA CGA GAA AAC AAA    607
Lys Gln Gly Arg Phe Gly Arg Glu Ser Glu Glu Gln Gly Arg Phe Gly Arg Glu Asn Lys    146

GAA CAA GGA AGA TTT GGA AGG GAA AAC AAA GAA CAA GGA AGG TTT GGA CGA GAA AAC GAG    667
Glu Gln Gly Arg Phe Gly Arg Glu Asn Lys Glu Gln Gly Arg Phe Gly Arg Glu Asn Glu    166

GAA CAA GGA AGA TTT GGA AGG GAA AGC GAG GAG CAA GGG AGA TTT GGA AGA GAA AAC GAA    727
Glu Gln Gly Arg Phe Gly Arg Glu Ser Glu Glu Gln Gly Arg Phe Gly Arg Glu Asn Glu    186

GAA CAA GGA AGA TTT GGA CGA GAA AAC GAA GAA CAA GGA AGA TTT GGA CGA GAA AAC GAA    787
Glu Gln Gly Arg Phe Gly Arg Glu Asn Glu Glu Gln Gly Arg Phe Gly Arg Glu Asn Glu    206

GAA CAA GGA AGG TTT GGA CGA GAA AAC GAG GAG CAA GGG AGA TTT GGA AGG GAG AAC GAA    847
Glu Gln Gly Arg Phe Gly Arg Glu Asn Glu Glu Gln Gly Arg Phe Gly Arg Glu Asn Glu    226

GTT CAA GGA AGA TTT GGA CGA GAG AAC GAG GAG CAA GGA AGA TTT GGA CGA GAA AAC GAA    907
Val Gln Gly Arg Phe Gly Arg Glu Asn Glu Glu Gln Gly Arg Phe Gly Arg Glu Asn Glu    246

GAA CAA GGA AGG TTT GGA AGA GAA AAC GAA GAA CAA GGA CGA TTT GGA CGA GAA AAC GAG    967
Glu Gln Gly Arg Phe Gly Arg Glu Asn Glu Glu Gln Gly Arg Phe Gly Arg Glu Asn Glu    266

GAG CAA GGA AGA TTT GGA CGA GAA AAT GAA GAA CAA GGA AGA TTT GGA CGA GAA AAC GAG   1027
Glu Gln Gly Arg Phe Gly Arg Glu Asn Glu Glu Gln Gly Arg Phe Gly Arg Glu Asn Glu    286

GAG CAA GGA AGA TTT GGA CGA GAA AAC GAA GAA AAA CAA GGA CGA TTT GGA CGA GAA AAC GAA   1087
Glu Gln Gly Arg Phe Gly Arg Glu Asn Glu Lys Gln Gly Arg Phe Gly Arg Glu Asn Glu    306

GAA CAA GGA AGA TTT GGA CGA GGA AAC GAG GAG CAA GGG AGA TTT GGA AGG GAA AAC GAA   1147
Glu Gln Gly Arg Phe Gly Arg Gly Asn Glu Glu Gln Gly Arg Phe Gly Arg Glu Asn Glu    326

GAA CAA GGA AGA TTT GGA CGA GAA AAC GAA GAA CAA GGA AGA TTT GGA CGA GAA AAC GAA   1207
Glu Gln Gly Arg Phe Gly Arg Glu Asn Glu Glu Gln Gly Arg Phe Gly Arg Glu Asn Glu    346

GAA CAA GGA CGA TTT GGG CGA GAA AAC GAA GAA CAA GGA AGG TTT GGG CGA GAA AAC GAA   1267
Glu Gln Gly Arg Phe Gly Arg Glu Asn Glu Glu Gln Gly Arg Phe Gly Arg Glu Asn Glu    366

GAA CAA GGA AGG TTT GGA AGG GAG AAC GAG AAG CAA GGG AGA TTT GGA AGA GGG GAC GAA   1327
Glu Gln Gly Arg Phe Gly Arg Glu Asn Glu Lys Gln Gly Arg Phe Gly Arg Gly Asp Glu    386

GAA CAA GGA AGG TTT GGA AGG GAG AAC GAG GAG CAA GGG AGA TTT GGA AGA GGG GAC GAA   1387
Glu Gln Gly Arg Phe Gly Arg Glu Asn Glu Glu Gln Gly Arg Phe Gly Arg Gly Asp Glu    406

GAA CAA GGA AGA TTT GGA CGA GAA AAC GAA GAG CAA GGA AGA TTT GGA AGA GAA AAC AAA   1447
Glu Gln Gly Arg Phe Gly Arg Glu Asn Glu Glu Gln Gly Arg Phe Gly Arg Glu Asn Lys    426

GAA CAA GGA AGA TTT GGA AGA GAA AAC GAA GAA CAA GGA AGA TTT GGA AGG GGG AAC AAA   1507
Glu Gln Gly Arg Phe Gly Arg Glu Asn Glu Glu Gln Gly Arg Phe Gly Arg Gly Asn Lys    446

GAA CAA GGA AGA TTT GGA CGA GAA AAC GAA CAA GGA AGA TTT GGA AGA GAA AAC GAA GTT   1567
Glu Gln Gly Arg Phe Gly Arg Glu Asn Glu Gln Gly Arg Phe Gly Arg Glu Asn Glu Val    466

CAA GGA AGG TTT GGA AGA TTC AGT CGG GAG TTG GCG AAA GGT TTA AAG ATT GAC GAT GTT   1627
Gln Gly Arg Phe Gly Arg Phe Ser Arg Glu Leu Ala Lys Gly Leu Lys Ile Asp Asp Val    486

CTC TGA   CAATGAACTAATTACGTGAATTACTAGAAACAAACTTAGAGAAAGTTGTTTATGAATCTATCAATAGTAT   1703
Leu  *                                                                               487

TTAAAAAGCGTTTCCAAATATTTAGTGTGAAATGATATTTTAAAAAAATATACACCGAAAAAAAAAAAAAAGGCCGATCG   1782
GGCCG(A)32                                                                           1819
```

Fig. 3. cDNA and deduced amino acid sequence of the *Renilla* Antho-RFamide precursor. Nucleotide residues are numbered from 5′ to 3′ end, and amino acid residues are numbered starting with the first ATG codon in the open reading frame. Neuropeptide sequences are *underlined*. Polyadenylylation signals are marked by *broken lines*. (Modified from Reinscheid and Grimmelikhuijzen 1994)

```
CAATGAACTGAGTGGAACACAAGTAATACATATTCTTCACTTCGGTTGATA ATG GCC CTC AAG TGT CAT CTA GTT CTA CTG    81
                                                    Met Ala Leu Lys Cys His Leu Val Leu Leu     10

GCC ATT ACT TTA CTA TTA GCA CAG TGT TCA GGG TCA GTA GAC AAG AAG GAT AGT ACG ACG AAT CAC TTA   150
Ala Ile Thr Leu Leu Leu Ala Gln Cys Ser Gly Ser Val Asp Lys Lys Asp Ser Thr Thr Asn His Leu    33

GAT GAG AAG AAA ACA GAT TCC ACA GAA GCA CAT ATT GTA CAA GAA ACA GAC GCG TTA AAA GAA AAT TCT   219
Asp Glu Lys Lys Thr Asp Ser Thr Glu Ala His Ile Val Gln Glu Thr Asp Ala Leu Lys Glu Asn Ser    56

TAT CTT GGC GCC GAG GAG GAA TCT AAA GAA GAA GAC AAG AAG AGA TCC GCC GCT CCT CAG CAG CCT GGC   288
Tyr Leu Gly Ala Glu Glu Glu Ser Lys Glu Glu Asp Lys Lys Arg Ser Ala Ala Pro **Gln Gln Pro Gly**    79

CTC TGG GGG AAA CGC CAG AAA ATA GGA CTA TGG GGA AGA TCC GCT GAC GCA GGA CAG CCA GGC CTC TGG   357
**Leu Trp Gly** Lys Arg Gln Lys Ile Gly Leu Trp Gly Arg Ser Ala Asp Ala Gly Gln Pro Gly Leu Trp   102

GGC AAA CGA CAA AGT CCC GGA TTA TGG GGA ACA TCC GCT GAC GCA GGA CAG CCA GGC CTC TGG GGC AAA   426
Gly Lys Arg Gln Ser Pro Gly Leu Trp Gly Arg Ser Ala Asp Ala Gly Gln Pro Gly Leu Trp Gly Lys    125

CGT CAA AAT CCC GGA TTA TGG GGA AGA TCC GCT GAC GCA GGA CAG CCA GGC CTC TGG GGC AAA CGT CAA   495
Arg Gln Asn Pro Gly Leu Trp Gly Arg Ser Ala Asp Ala Gly Gln Pro Gly Leu Trp Gly Lys Arg Gln    148

AAT CCC GGA TTA TGG GGA AGA TCG GCT GAC GCA GGA CAG CCA GGC CTC TGG GGC AAA CGT CAA AAT CCC   564
Asn Pro Gly Leu Trp Gly Arg Ser Ala Asp Ala Gly Gln Pro Gly Leu Trp Gly Lys Arg Gln Asn Pro    171

GGA TTA TGG GGA AGG TCC GCT GAC GCA AGA CAA CCC GGA CTC TGG GGC AAA CGT GAA ATC TAC GCA TTA   633
Gly Leu Trp Gly Arg Ser Ala Asp Ala Arg Gln Pro Gly Leu Trp Gly Lys Arg Glu Ile Tyr Ala Leu   194

TGG GGA GGA AAA CGT CAA AAT CCC GGA CTT TGG GGA AGA TCC GCT GAC GCA GGA CAG CCC GGC CTC TGG   702
Trp Gly Gly Lys Arg Gln Asn Pro Gly Leu Trp Gly Arg Ser Ala Asp Pro Gly Gln Pro Gly Leu Trp   217

GGC AAA CGT GAA CTC GTC GGA TTA TGG GGG GGA AAA CGT CAA AAC CCC GGA TTG TGG GGA AGA TCG GCT   771
Gly Lys Arg Glu Leu Val Gly Leu Trp Gly Gly Lys Arg Gln Asn Pro Gly Leu Trp Gly Arg Ser Ala   240

GAA GCA GGA CAG CCA GGA CTT TGG GGA AAA CGC CAA AAA ATA GGA TTG TGG GGA CGT TCG GCT GAC CCA   840
Glu Ala Gly Gln Pro Gly Leu Trp Gly Lys Arg Gln Lys Ile Gly Leu Trp Gly Arg Ser Ala Asp Pro   263

CTT CAG CCT GGC CTC TGG GGC AAA CGT CAA AAT CCC GGA TTA TGG GGA AGA TCT GCT GAC CCG CAG CAG   909
Leu Gln Pro Gly Leu Trp Gly Lys Arg Gln Asn Pro Gly Leu Trp Gly Arg Ser Ala Asp Pro **Gln Gln**   286

CCT GGC CTC TGG GGC AAA CGT CAA AAT CCC GGA TTA TGG GGA AGA TCT GCT GAC CCG CAG CAG CCT GGC   978
**Pro Gly Leu Trp Gly** Lys Arg Gln Asn Pro Gly Leu Trp Gly Arg Ser Ala Asp Pro **Gln Gln Pro Gly**   309

CTC TGG GGC AAA CGT CAA AAT CCC GGA TTA TGG GGA AGA TCT GCT GAC CCG CAG CAG CCT GGC CTC TGG  1047
**Leu Trp Gly** Lys Arg Gln Asn Pro Gly Leu Trp Gly Arg Ser Ala Asp Pro **Gln Gln Pro Gly Leu Trp**   332

GGC AAA CGT CAA AAT CCC GGA TTA TGG GGA AGA TCT GCT GAC CCG CAG CAA CCT GGC CTC TGG GGC AAA  1116
**Gly** Lys Arg Gln Asn Pro Gly Leu Trp Gly Arg Ser Ala Asp Pro **Gln Gln Pro Gly Leu Trp Gly** Lys   355

AGC CCC GGT TTA TGG GGA CGA TCC GCT GAC GCA CAA CAG CCT GGA CTT TGG GGG AAA CGC CAA AAT CCC  1185
Ser Pro Gly Leu Trp Gly Arg Ser Ala Asp Pro **Gln Gln Pro Gly Leu Trp Gly** Lys Arg Gln Asn Pro   378

GGA TTT TGG GGA AGA TCT GCT GAC CCG CAG CAG CCT GGC CTC TGG GGC AAA CGT CAA AAT CCC GGA TTA  1254
Gly Phe Trp Gly Arg Ser Ala Asp Pro **Gln Gln Pro Gly Leu Trp Gly** Lys Arg Gln Asn Pro Gly Leu   401

TGG GGA AGA TCT GCT GAC CCG CAG CAA CCT GGC CTC TGG GGC AAA CGT CAA AAT CCC GGA TTA TGG GGA  1323
Trp Gly Arg Ser Ala Asp Pro **Gln Gln Pro Gly Leu Trp Gly** Lys Arg Gln Asn Pro Gly Leu Trp Gly   424

AGA TCT GCT GAC CCG CAG CAA CCT GGC CTC TGG GGC AAA CGT CAA AAC CCC GGT TTA TGG GGA CGA TCC  1392
Arg Ser Ala Asp Pro **Gln Gln Pro Gly Leu Trp Gly** Lys Arg Gln Asn Pro Gly Leu Trp Gly Arg Ser   447

GCT GAC CCA GAA CAA CAG CCT GGA CTT TGG GGG AAA CGC CAA AAT CCA GGA CTA TGG GGA AGA AGT GCT  1461
Ala Asp Pro **Gln Gln Pro Gly Leu Trp Gly** Lys Arg Gln Asn Pro Gly Leu Trp Gly Arg Ser Ala Gly   470

TCC GGT CAA CTC GGA CTT TGG GGT AAA AGG CAA TCA CGC ATT GGA TTA TGG GGA AGA TCT GCC GAG CCT  1530
Ser Gly Gln Leu Gly Leu Trp Gly Lys Arg Gln Ser Arg Ile Gly Leu Trp Gly Arg Ser Ala Glu Pro   493

CCA CAA TTT GAA GAT TTA GAA GAT TTA AAG AAA AAA TCA GCA ATT CCC CAA CCA AAA GGA CAA TGA TAA  1599
Pro Gln Phe Glu Asp Leu Glu Asp Leu Lys Lys Lys Ser Ala Ile Pro Gln Pro Lys Gly Gln Stop Stop   514

TATCCTAGGATCTTCAAAAGTTATCCCGATCATCAATCCCCGGACAAGAGATATTTTAATTTCTGCCGCACGATTGACAGTTCCATTCCAT       1690
TACGAAGAACAAAAAGCTACGTTTCTTTAAGATAATAAATCAAATTCAATATTGTTTGAAGCAATGCACTTCAGGTTTTCACACAAAACTA       1781
ATACAAAAGTTATAAACATAAATAATAAACAAAAAAGGGGTAAGAAACCTGGTTTTTCGTTTTAGACTTTTCAAATTGGTCCTGCATGTCA       1872
TTGTAAGAGGTCGGCAGTAGAAGCTCTTTGAAACACTGTAAGTAGTCATTGAATGCACATTTATGGCAATAAAGATCACGCAGTGGTCTGC       1963
GGCCGCGAA                                                                                       1972
```

Fig. 4. cDNA and deduced amino acid sequence of the precursor for the metamorphosis-inducing peptide MMA from *Anthopleura*. Copies of authentic MMA are *bold and underlined*, whereas the numerous related but putative neuropeptide sequences are *underlined only*. (From Leviev and Grimmelikhuijzen 1995)

Grimmelikhuijzen 1995). An example of a prohormone where processing must occur at X-Pro and X-Ala sequences is given in Fig. 4. Fig. 3 gives an example of a very regularly built- up preprohormone with 36 copies of the same immature Antho-RFamide (Gln-Gly-Arg-Phe-Gly) sequence, of which 31 are preceded by acidic (Glu) residues. Two copies of immature Antho-RFamide are preceded by a Lys residue and, of course, one could argue that only these sequences are cleaved from the precursor by a conventional PC-like processing enzyme, whereas the rest of the precursor is not used. However, cloning work on the precursors of the other cnidarian neuropeptides gives examples of preprohormones having only one copy of the immature neuropeptide in question, as is the case with Antho-RPamide I (Table 1), where the immature peptide sequence together with its flanking regions have the following structure: Gln-Tyr-Asp-Gly-Pro-Glu-Gly-Asp-Tyr-Glu-Asp-**Leu-Pro-Pro-Gly-Pro-Leu-Pro-Arg-Pro-Gly**-Arg-Lys-Arg (Carstensen and Grimmelikhuijzen, unpubl.). Here, it is clear that there is no other way to produce mature Antho-RPamide I (which we have isolated and sequenced) from its precursor than by cleaving at the C-terminal side of acidic residues. The novel processing enzyme that is responsible for this cleavage is, based on the structure of the N-terminal flanking sequence, most likely to be an endoproteinase.

Thus, although the primitive cnidarians have an anatomically simple nervous system, the processing of their preprohormones appears to be more complex than in mammals. This finding seems to be a paradox, but it could be explained in several ways. One explanation is that the first cnidarians, or their closely related ancestors that evolved a nervous system, simply used an ordinary collection of degradation enzymes that was also present in other parts of the body, as a set of processing enzymes for their prohormones. During subsequent evolution, the most efficient of these processing enzymes were retained, whereas the expression of the others was abolished in neurones. A second explanation is a variant of this first hypothesis: synapses are relatively scarce in cnidarians and, therefore, most neuropeptides in cnidarians are released in a paracrine way, only reaching their target cells after a relatively long route of diffusion. For this reason, it is important that the cnidarian neuropeptides are resistant against the collection of degradation enzymes that is present in the intercellular space. One way for the cnidarian neurones to make their neuropeptides resistant against these enzymes is to express these enzymes themselves, package them into neurosecretory vesicles, and use them as processing enzymes for their prohormones. This would yield the processing endproducts (the biologically active neuropeptides) fully resistant. This second hypothesis could be tested by cloning the cnidarian processing enzymes and comparing them with the enzymes present in the intercellular space. This work, however, has not been completed yet.

Since all cnidarian neuropeptides isolated so far are amidated, we have started to clone the cnidarian processing enzymes that are involved in peptide amidation (step 3 in Fig. 1). In mammals, it is known that PHM and PAL are contained within a bifunctional protein PAM (peptidyl α-amidating mono-oxygenase) that is coded for by a single gene (Kato et al. 1990; Eipper et al. 1992; Quafik et al. 1992). In the tetraploid claw frog *Xenopus laevis*, there are two, very closely related genes coding for PAM (Mizuno et al. 1987; Ohsuye et al. 1988). In the fruitfly *Drosophila melanogaster*, however, a single PHM occurs, and no bifunctional PAM. On the genomic level, one gene for PHM could be cloned, but PAL appears to be encoded by a separate and yet unidentified gene (Kolhekar et al. 1997).

The situation in cnidarians is the following: in sea anemones there are at least two separate genes coding for PHM. PAL appears to be encoded for by a third gene that we have not yet identified. The two PHM genes yield PHMs that show 40% amino acid residue identity between each other. There is a similar 40% identity when each of the two sea anemone PHMs are aligned with the PHM part of rat PAM (Hauser et al. 1997a). This example of the cnidarian α-amidating enzymes shows, again, that prohormone processing in cnidarians appears to be much more complicated than in mammals. Whereas in mammals one gene is sufficient to yield both the PHM and PAL activities residing in PAM, there are at least three different genes in cnidarians yielding two separate PHMs and one PAL.

5
Neurohormone Receptors in Cnidarians

So far, 17 neuropeptides have been isolated and sequenced from a single sea anemone species (*Anthopleura elegantissima*; 14 of them are shown in Table 1). When we cloned the preprohormone for one of these neuropeptides, we found between one and eight additional, different, putative neuropeptide copies located on the precursor (Figs. 3, 4; Table 3). This means that already now, we have the sequences of about 50–60 established and putative neuropeptides present in one sea anemone species and this number will certainly increase when we use additional methods to isolate and clone these substances. Because there are so many different neuropeptides, one could also expect an equally large number of neuropeptide receptors in sea anemones and other cnidarians. In order to clone these receptors from sea anemones, we applied PCR, using cDNA from *A. elegantissima* as a template and primers, coding for conserved sequences of mammalian G protein-coupled receptors. This approach gave us a surprising result, because we pulled out a receptor that clearly was not a neuropeptide receptor, but a receptor that was closely related to the glycoprotein hormone (thyroid-stimulating hormone, TSH; follicle-stimulating hormone, FSH; luteinizing hormone, LH; and choriogonadotropin, CG) receptors from mammals (Nothacker and Grimmelikhuijzen 1993). The mammalian glycoprotein hormone receptors are G protein-coupled receptors that are characterized by a very large, extracellular N terminus,

Table 3. Established and putative, mature neuropeptides that could be released from the MMA precursor protein (see Fig. 4)

Copies	Structure	Name
14	<Glu-Asn-Pro-Gly-Leu-Trp-NH$_2$	Antho-LWamide I
10	<Glu-Gln-Pro-Gly-Leu-Trp-NH$_2$	MMA (Antho-LWamide II)
6	Gly-Gln-Pro-Gly-Leu-Trp-NH$_2$	Antho-LWamide III
1	Leu-Gln-Pro-Gly-Leu-Trp-NH$_2$	Antho-LWamide V
1	Arg-Gln-Pro-Gly-Leu-Trp-NH$_2$	Antho-LWamide VI
1	<Glu-Ser-Pro-Gly-Leu-Trp-NH$_2$	Antho-LWamide VII
2	<Glu-Lys-Ile-Gly-Leu-Trp-NH$_2$	Antho-LWamide IV
1	<Glu-Ser-Arg-Ile-Gly-Leu-Trp-NH$_2$	Antho-LWamide VIII
1	Gly-Ser-Gly-Gln-Leu-Gly-Leu-Trp NH$_2$	Antho-LWamide IX
37		

Amino acid residues that are identical to those of MMA are highlighted by boxes. Residues that might possibly not be contained in the mature peptides are underlined with a dashed line.

which comprises about half of the receptor molecule. This receptor N terminus contains nine to ten "Leu-rich" repeats that are probably arranged in a horseshoe-type of 3D-structure with α-helices directed towards the outside (the convex side) and β-sheets covering the inside (the concave side) of the horseshoe (Jiang et al. 1995; Kajava et al. 1995). The β-sheets at the concave side are probably functioning as the glycoprotein hormone binding site (Fig. 5).

The seven transmembrane region of the sea anemone receptor shows about 50% amino acid residue identity with the TSH, FSH and LH/CG receptors from rat (Nothacker and Grimmelikhuijzen 1993). This is high if one takes into account that the sequence identity between the transmembrane regions of the rat TSH and LH/CG receptors is 70% and that between the rat TSH and FSH receptors is 67% (Salesse et al. 1991). The N terminus of the sea anemone receptor has only 18–25% sequence identity with those of the mammalian receptors. However, it has the same number and the same type of Leu-rich repeats. Moreover, we have also cloned a part of the genomic sequence of the sea anemone receptor and found the existence of two introns that occurred at exactly the same positions and had the same intron phasing as two introns in the mammalian receptor genes (Nothacker and Grimmelikhuijzen 1993). All these data suggest that the sea anemone receptor is both evolutionarily and functionally related to the glycoprotein hormone receptors from mammals.

The presence of receptors in cnidarians that are related to the glycoprotein hormone receptors from mammals is fascinating, because it suggests that glycoprotein hormone receptors and possibly all processes that are mediated by

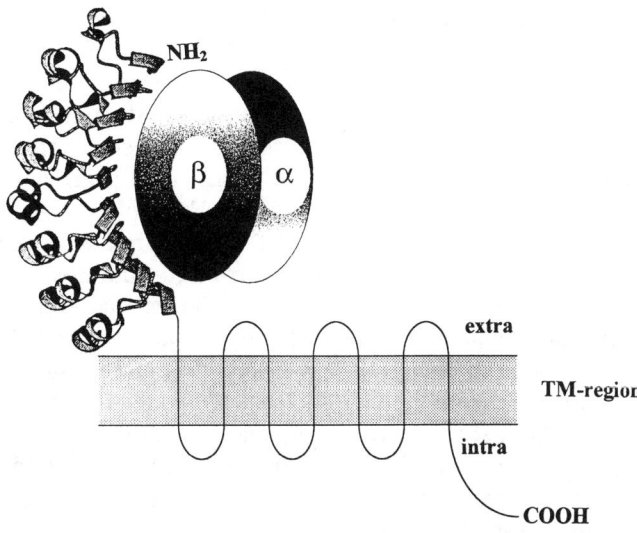

Fig. 5. Simplified model of a glycoprotein hormone receptor from mammals. This receptor is a G protein-coupled, seven transmembrane (*TM*) receptor, having a large extracellular N terminus. This N terminus contains about nine Leu-rich repeats that probably form a horseshoe-like structure. The innerside of the horseshoe is lined with β-sheets that are assumed to bind the β-subunit of the ligand. (Drawn after data from Jiang et al. 1995 and Kajava et al. 1995)

these types of receptors (such as certain steps in reproduction) are conserved throughout the animal kingdom, from cnidarians to mammals. Our results also mean that the endocrinology of sea anemones must already be quite complex, involving at least 50–60 neuropeptides and, in addition, a new class of presumed glycoprotein hormones. Since there are no endocrine cells in cnidarians, but only neurones, these presumed glycoprotein hormones are likely to be neurohormones.

6
Neurohormone Receptors in Insects

As explained above, we expected that glycoprotein hormones and glycoprotein hormone receptors are present throughout the animal kingdom. This hypothesis was confirmed shortly after our own discovery in sea anemones, by the cloning of another glycoprotein hormone receptor from the pond snail *Lymnaea stagnalis* (Tensen et al. 1994). We ourselves decided to focus our attention on insects, because (1) insects are by far the largest animal group on earth (75% of all animals or 80% of all invertebrates are insects), (2) insects are very successful and adaptive animals and are of extreme ecological and economical importance, since about 70% of all flowering plants depend on insects for their pollination and insects can be severe pest animals, destroying about 30% of our potential annual harvest, and (3) little or nothing is known about the molecular mechanisms that control reproduction in insects. This last point is in contrast to the situation in mammals, where reproduction is known to be controlled by (1) gonadal steroid hormones, (2) glycoprotein hormones produced by the anterior pituitary (FSH, LH) or placenta (CG), and (3) the hypothalamic-releasing hormone, gonadotropin-releasing hormone (GnRH) that controls the release of LH and FSH from the anterior pituitary.

Using PCR and oligonucleotide primers that were based on conserved amino acid sequences of several mammalian glycoprotein hormone receptors, we were able to clone the first insect glycoprotein hormone receptor from a cDNA library from the fruit fly *Drosophila melanogaster* (Hauser et al. 1997b). This receptor shows a striking similarity with the TSH, FSH and LH/CG receptors from mammals. This similarity includes a very large extracellular N terminus, containing the same type of Leu-rich repeats as found in the mammalian glycoprotein hormone receptors, and a transmembrane region that has 50–53% amino acid sequence identity with the rat TSH, FSH and LH/CG receptors (Figs. 6, 7, 8). We also cloned the gene of the receptor and found that

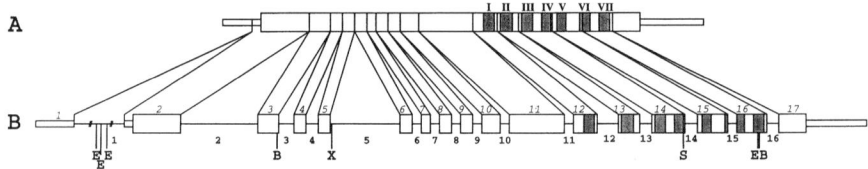

Fig. 6. Schematic representation of the cDNA and gene coding for the first glycoprotein hormone receptor from *Drosophila*: **A** Receptor cDNA. The *broad bar* represents the coding region, whereas the two *narrow bars* represent the 5'- and 3'-noncoding domains. *Boxes I–VII* highlighted in *grey* indicate the seven membrane-spanning domains. **B** Receptor gene. Introns are shown as *lines* and numbered 1–16, whereas exons are represented by *boxes* and numbered *1–17* (exon numbers are in *italic type*). E, *Eco*RI; B, *Bam*HI; X, *Xba*I; S, *Sal*I restriction sites. (Modified from Hauser et al. 1997b)

Fig. 7A, B. Amino acid sequence comparison of the first *Drosophila* receptor (GHRD), the rat TSH (TSHR), FSH (FSHR), and LH (LHR) receptors and putative glycoprotein hormone receptor from the sea anemone *Anthopleura elegantissima* (GHRA). Amino acid residues that are identical between the *Drosophila* receptor and at least one of the other receptors are *boxed*. Known intron-exon transitions in the genes coding for the four receptors are highlighted in *grey* at corresponding amino acid residues. Positions of the aliphatic residues characteristic of Leu-rich repeats in the *Drosophila* receptor (see Fig. 8) are marked by *asterisks* or *filled circles*. *Filled circles* also mark intron-exon transitions in the *Drosophila* receptor gene that occur at the same position and have the same intron phasing as in mammalian receptor genes. The seven membrane-spanning domains are indicated by *I–VII*. *Dashed lines* represent spaces introduced to optimize alignment. Amino acid positions are given on the *right*. (From Hauser et al. 1997)

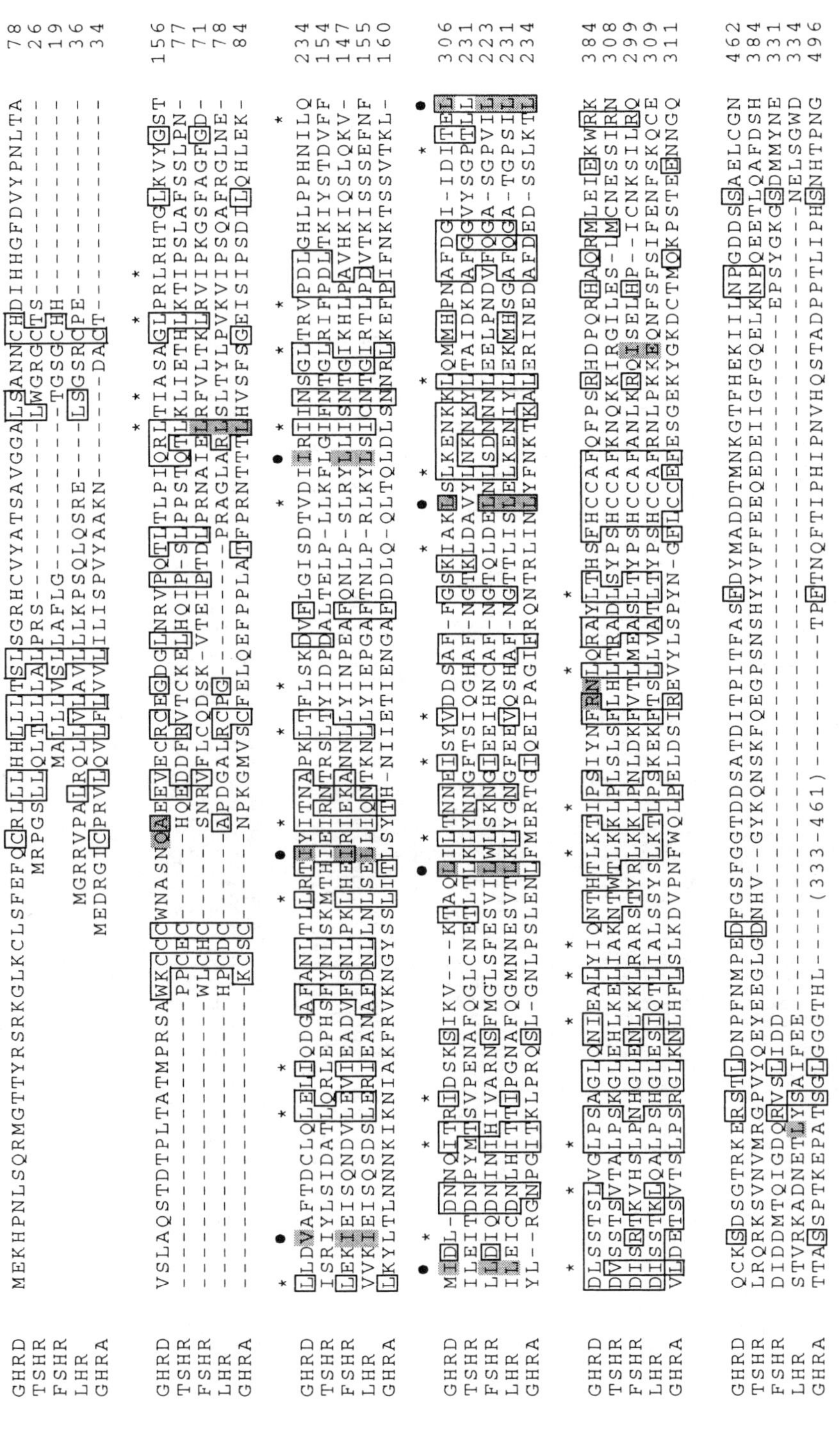

GHRD
TSHR
FSHR
LHR
GHRA

Fig. 7A

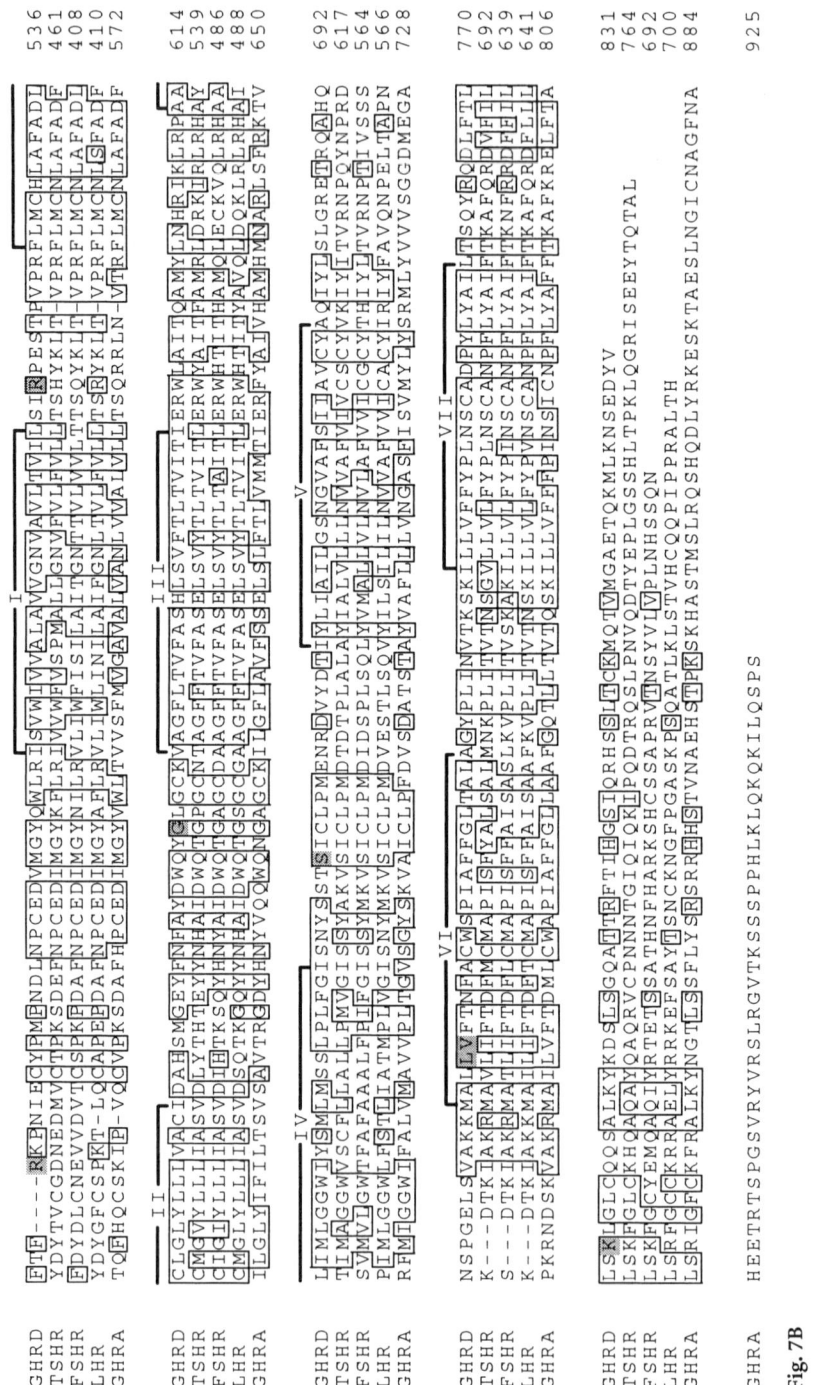

Fig. 7B

A

L1	L T I A S A G - L P R L R H T G L K V Y G S T L L D	159
L2	V A F T D C L Q L E L I Q D G A F A N L T - L L R T	184
L3	I Y I T N A P K L T F L S K D V F L G I S D T V D I	210
L4	I R I I N S G - L T R V P D L G H L P P H N I L Q M	235
L5	I D L D N N Q - I T R I D S K S I K V - - - K T A Q	257
L6	L I L T N N E - I S Y V D D S A F F G S - - K I A K	280
L7	L S L K E N K K L Q M M H P N A F D G I I - D I T E	305
L8	L D L S S T S - L V G L P S A G L Q - - - - N I E A	326
L9	L Y I Q N T H T L K T I P S I Y N F - - - R N L Q R	349
L10	A Y L T H S F H C C A F Q - - - - F P S R H D P Q R	371

B

L1	L R F V L T K - L R V I P K G S F A G F G - D L E K	74
L2	I E I S Q N D V L E V I E A D V F S N L P - K L H E	99
L3	I R I E K A N N L L Y I N P E A F Q N L P - S L R Y	124
L4	L L I S N T G - I K H L P A V H K I Q S L Q K V - L	148
L5	D I Q D N I N I H I V A R N S F M G L S F E S V I	174
L6	L W L S K N G - I E E I H N C A F N G T - - Q L D E	197
L7	L N L S D N N N L E E L P N D V F Q G A S - G P V I	222
L8	D I S R T K - V H S L P N H G L E - - - - N L K K	243
L9	L R A R S T Y R L K K L P N L D K F V - - - T L M E	266
L10	A S L T Y P S H C C A F A N L K R Q I - - S E L H P	290

Fig. 8. Leu (Ile/Val)-rich repeats in the N-terminal region of the *Drosophila* receptor and the rat FSH receptor. **A** Consecutive segments of the *Drosophila* receptor N terminus (*L1–L10*) were aligned, and small gaps (*dashed lines*) were introduced to show that many of the Leu (Ile-Val) residues in one segment occur at a similar position in the other segments. These residues are *boxed*. Aliphatic residues highlighted in *grey* correspond to intron-exon transitions in the receptor gene and are given at the start of seven repeating segments. Segments *L2–L7*, therefore, each correspond to a distinct exon. Amino acid positions are given at the *right*. **B** Similar alignment of Leu(Ile/Val)-rich repeats of the N terminus of the rat FSH receptor. Segments *L1–L7* each correspond to a distinct exon. (From Hauser et al. 1997)

seven intron positions of the *Drosophila* receptor (introns 2–8 of Fig. 6) occur at exactly the same positions and have the same intron phasing as seven introns in the mammalian genes (Hauser et al. 1997b). All these data strongly suggest that the *Drosophila* and the mammalian receptors are evolutionarily and functionally related.

Because two of the three mammalian glycoprotein hormone receptors are involved in reproduction, we expected (and hoped) that the *Drosophila* receptor would be involved in reproduction as well. However, it was a surprise when we found that the receptor was already expressed during early embryonic development (starting from 8 to 16 h after oviposition) and that the expression stayed high during all subsequent developmental stages. Adult male flies contained high levels of receptor mRNA, whereas female flies contained about six times less (Hauser et al. 1997b). These expression data in embryos, larvae and pupae suggest that the receptor must play an important role during development. The expression in adult male flies, on the other hand, does not exclude a role for the receptor in reproduction, which might be a role in spermatogenesis or other male-specific reproductive processes.

Mammals have three glycoprotein hormone (TSH, FSH, LH/CG) receptors. We continued, therefore, to look for a second and possibly third *Drosophila* glycoprotein hormone receptor. We were successful in finding a second receptor that we discovered in a very interesting way. Nowadays, molecular biologists are in the lucky situation of having access to the genomic sequences of several "genome projects". Also for *Drosophila*, there exists a "Berkeley *Drosophila* Genome Project", which, so far, has only a limited quantity of genomic information. Instead of screening a cDNA library, one of us (F.H.) screened the Berkeley database with several segments of the first *Drosophila* receptor. This electronic screening led to the alignment with a sequence from the database that showed 50% amino acid residue identity with the first *Drosophila* receptor. By using the information we had on the genomic organization of the first *Drosophila* receptor and the three mammalian receptors, we were able to make a putative allocation of the intron and exon positions within the novel receptor gene sequence. By using PCR, primers directed against the putative exons of the second *Drosophila* receptor, and *Drosophila* cDNA as a template, we could subsequently amplify and clone the cDNA of this receptor, showing that the second *Drosophila* receptor did indeed exist and that its gene was not a pseudo-gene. Figure 9C shows a schematic representation of the second *Drosophila* receptor. It came as a surprise that this receptor has 19–20 Leu-rich repeats instead of the 9–10 Leu-rich repeats found in the first *Drosophila* receptor and in the mammalian receptors (Figs. 8, 9). This large number of Leu-rich repeats might make the horseshoe-like shape of the receptor N terminus look more circular.

In the transmembrane region, the second *Drosophila* receptor shows 47–51% sequence identity with the TSH, FSH, LH/CG receptors from rat. Four introns in the *Drosophila* gene occur at exactly the same positions and have the same intron phasing as four introns in the mammalian genes and in the gene of the first *Drosophila* receptor (Fig. 9B). This strongly suggests that also the second *Drosophila* receptor is evolutionarily and functionally related to the three glycoprotein hormone receptors from mammals (K. Krageskov Eriksen et al. unpubl. results).

We have also investigated the developmental expression of the second *Drosophila* receptor in the hope that this would be confined to the adult stages, thus pointing to a possible role in reproduction. However, we found that the receptor was mainly expressed in the late embryonic and pupal stages and that it was absent in adults. This suggests that the role of the second *Drosophila* receptor is confined to development.

Using the same type of electronic screening method as described for the second *Drosophila* receptor, we discovered an additional, potential *Drosophila* receptor that shows a striking similarity with the GnRH receptors from vertebrates (Hauser et al. 1998). In the same way as mentioned for the second *Drosophila* receptor, also the expression of this putative GnRH receptor was verified with PCR, which resulted in the cloning of its full-length cDNA. Figure 10B gives a schematic representation of this novel receptor from *Drosophila*. Its transmembrane region shows 36% amino acid residue identity with the transmembrane region of the catfish and 31% amino acid residue identity

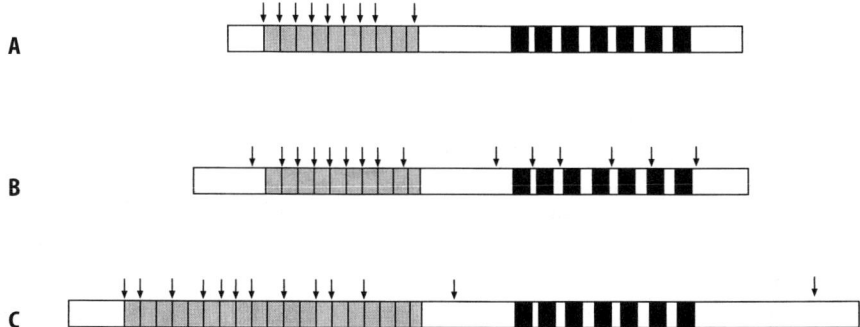

Fig. 9. Schematic representation of the rat FSH receptor (**A**), the first putative *Drosophila* glyco-protein hormone receptor (**B**) and the second putative *Drosophila* glycoprotein hormone receptor (**C**). The seven transmembrane regions are shown as *black boxes* and Leu-rich repeats as *grey boxes*. Intron positions in the corresponding receptor genes are indicated by *arrows*

with that of the rat GnRH receptor. The *Drosophila* receptor contains six introns, whereas the rat gene contains two; one intron in the *Drosophila* gene occurs at exactly the same position and has the same intron phasing as one intron in the rat gene (Fig. 10), suggesting that the *Drosophila* and mammalian GnRH receptor genes are evolutionarily related. Northern blot analyses showed that the *Drosophila* receptor gene is progressively expressed during larval development and that adult male flies contain about five times more mRNA than adult female flies (Hauser et al. 1998). The expression pattern of the GnRH receptor gene, therefore, follows roughly that of the first glycoprotein hormone receptor gene from *Drosophila*, which means that the GnRH receptor could possibly be involved in the release of a glycoprotein hormone. The expression of the GnRH receptor in adult male flies also points to a possible role of the GnRH receptor in spermatogenesis or other male-specific processes. Therefore, these findings open the possibility that reproduction in insects is controlled by the same type of mechanisms as those found in mammals.

Fig. 10. Schematic representation of the rat GnRH receptor (**A**) and the putative *Drosophila* GnRH receptor (**B**). *Black boxes* represent the transmembrane regions; *arrows* represent introns in the corresponding genes

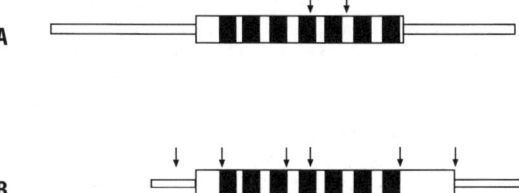

7
Conclusion

Primitive nervous systems, such as those of cnidarians, use a large variety of neuropeptides as neurotransmitters or neurohormones, suggesting that neuropeptides were among the earliest transmitter-like substances used in evolution. The biosynthesis of neuropeptides in cnidarians appears to be very efficient and requires many more processing enzymes than we know from higher animals, so far. The receptors for the numerous cnidarian neuropeptides are not known. On the other hand, a glycoprotein hormone receptor has been cloned from cnidarians, showing that also other hormonal systems, such as the glycoprotein hormone/receptor couple have a long evolutionary history. This conclusion was corroborated by the cloning of two glycoprotein hormone receptors from *Drosophila*. All of this shows that the endocrinology of invertebrates is much more complex and sophisticated than we might have previously anticipated.

Acknowledgements. We want to thank Astrid Juel Jensen for typing the manuscript and the Danish Natural Science Research Council, the Danish Biotechnological Research and Developmental Program of the Danish Research Councils and the Novo Nordisk Foundation for financial support.

References

Anderson PAV, Schwab WE (1982) Recent advances and model systems in coelenterate neurobiology. Prog Neurobiol 19:213–236

Anderson PAV, Spencer AN (1989) The importance of cnidarian synapses for neurobiology;. J Neurobiol 20:435–457

Bode HR (1992) Continuous conversion of neuron phenotype in *Hydra*. Trends Genet 8:279–284

Bradbury AF, Smyth DG (1991) Peptide amidation. Trends Biochem Sci 16:112–115

Campbell RD (1976) Elimination of *Hydra* interstitial and nerve cells by means of colchicine. J Cell Sci 21:1–14

Carstensen K, Rinehart KL, McFarlane ID, Grimmelikhuijzen CJP (1992) Isolation of Leu-Pro-Pro-Gly-Pro-Leu-Pro-Arg-Pro-NH₂ (Antho-RPamide), an N-terminally protected, biologically active neuropeptide from sea anemones. Peptides 13:851–857

Carstensen K, McFarlane ID, Rinehart KL, Hudman D, Sun F, Grimmelikhuijzen CJP (1993) Isolation of <Glu-Asn-Phe-His-Leu-Arg-Pro-NH₂ (Antho-RPamide II), a novel, biologically active neuropeptide from sea anemones. Peptides 14:131–135

Darmer D, Schmutzler C, Diekhoff D, Grimmelikhuijzen CJP (1991) Primary structure of the precursor for the sea anemone neuropeptide Antho-RFamide (<Glu-Gly-Arg-Phe-NH₂). Proc Natl Acad Sci USA 88:2555–2559

Darmer D, Hauser F, Nothacker H-P, Bosch TCG, Williamson M, Grimmelikhuijzen CJP (1998) Three different prohormones yield a variety of Hydra-RFamide (Arg-Phe-NH₂) neuropeptides in *Hydra magnipapillata*. Biochem J 332:403–412

Eipper BA, Stoffers DA, Mains RE (1992) The biosynthesis of neuropeptides: peptide α-amidation. Annu Rev Neurosci 15:57–85

Fischer WH, Spiess J (1987) Identification of a mammalian glutaminyl cyclase converting glutaminyl into pyroglutaminyl peptides. Proc Natl Acad Sci USA 84:3628–3632

Gierer A (1977) Biological features and physical concepts of pattern formation exemplified by *Hydra*. Curr Top Dev Biol 11:17–59

Gierer A, Berking S, Bode H, David CN, Flick K, Hansmann G, Schaller H, Trenckner E (1972) Regeneration of *Hydra* from reaggregated cells. Nat New Biol 239:98–101

Graff D, Grimmelikhuijzen CJP (1988a) Isolation of <Glu-Ser-Leu-Arg-Trp-NH₂, a novel neuropeptide from sea anemones. Brain Res 442:354–358

Graff D, Grimmelikhuijzen CJP (1988b) Isolation of <Glu-Gly-Leu-Arg-Trp-NH₂ (Antho-RWamide II), a novel neuropeptide from sea anemones. FEBS Lett 239:137–140

Grimmelikhuijzen CJP (1985) Antisera to the sequence Arg-Phe-amide visualize neuronal centralization in hydroid polyps. Cell Tissue Res 241:171–182

Grimmelikhuijzen CJP (1986) FMRFamide-like peptides in the primitive nervous systems of coelenterates and complex nervous systems of higher animals. In: Stephano G (ed) Handbook of comparative opioid and related neuropeptide mechanisms. CRC Press, Boca Raton, pp 103–115

Grimmelikhuijzen CJP, Graff D (1986) Isolation of <Glu-Gly-Arg-Phe-NH₂ (Antho-RFamide), a neuropeptide from sea anemones. Proc Natl Acad Sci USA 83:9817–9821

Grimmelikhuijzen CJP, Groeger A (1987) Isolation of the neuropeptide pGlu-Gly-Arg-Phe-amide from the pennatulid *Renilla köllikeri*. FEBS Lett 211:105–108

Grimmelikhuijzen CJP, Spencer AN (1984) FMRFamide immunoreactivity in the nervous system of the medusa *Polyorchis penicillatus*. J Comp Neurol 230:361–371

Grimmelikhuijzen CJP, Spencer AN, Carré D (1986) Organization of the nervous system of physonectid siphonophores. Cell Tissue Res 246:463–479

Grimmelikhuijzen CJP, Hahn M, Rinehart KL, Spencer AN (1988) Isolation of <Glu-Leu-Leu-Gly-Gly-Arg-Phe-NH₂ (Pol-RFamide), a novel neuropeptide from hydromedusae. Brain Res 475:198–203

Grimmelikhuijzen CJP, Rinehart KL, Jacob E, Graff D, Reinscheid RK, Nothacker H-P, Staley AL (1990) Isolation of L-3-phenyllactyl-Leu-Arg-Asn-NH₂ (Antho-RNamide), a sea anemone neuropeptide containing an unusual amino-terminal blocking group. Proc Natl Acad Sci USA 87:5410–5414

Grimmelikhuijzen CJP, Rinehart KL, Spencer AN (1992a) Isolation of the neuropeptide <Glu-Trp-Leu-Lys-Gly-Arg-Phe-NH₂ (Pol-RFamide II) from the hydromedusa *Polyorchis penicillatus*. Biochem Biophys Res Commun 183:375–382

Grimmelikhuijzen CJP, Carstensen K, Darmer D, Moosler A, Nothacker H-P, Reinscheid RK, Schmutzler C, Vollert H, McFarlane ID, Rinehart KL (1992b) Coelenterate neuropeptides: structure, action and biosynthesis. Am Zool 32:1–12

Grimmelikhuijzen CJP, Leviev I, Carstensen K (1996) Peptides in the nervous systems of cnidarians: structure, function and biosynthesis. Int Rev Cytol 167:37–89

Hauser F, Williamson M, Grimmelikhuijzen CJP (1997a) Molecular cloning of a peptidylglycine α-hydroxylating monooxygenase from sea anemones. Biochem Biophys Res Commun 241:509–512

Hauser F, Nothacker H-P and Grimmelikhuijzen CJP (1997b) Molecular cloning, genomic organization and developmental regulation of a novel receptor from *Drosophila melanogaster* structurally related to members of the thyroid-stimulating hormone, follicle-stimulating hormone, luteinizing hormone/choriogonadotropin receptor family from mammals. J Biol Chem 272:1002–1010

Hauser F, Søndergaard L, Grimmelikhuijzen CJP (1998) Molecular cloning, genomic organization and developmental regulation of a novel receptor from *Drosophila melanogaster* structurally related to gonadotropin-releasing hormone receptors from vertebrates. Biochem Biophys Res Commun 249:822–828

Horridge GA, Mackay B (1962) Naked axons and symmetrical synapses in coelenterates. Q J Microsc Sci 103:531–541

Jiang X, Dreano M, Buckler DR, Cheng S, Ythier A, Wu H, Hendrickson WA, El Tayar N (1995) Structural predictions for the ligand-binding region of glycoprotein hormone receptors and the nature of hormone-receptor interactions. Structure 3:1341–1353

Kajava AV, Vassart G, Wodak SJ (1995) Modeling of the three-dimensional structure of proteins with the typical leucine-rich repeats. Structure 3:867–877

Kato I, Yonekura H, Tajima M, Yanagi M, Yamamoto H, Okamoto H (1990) Two enzymes concerned in peptide hormone α-amidation are synthesized from a single mRNA. Biochem Biophys Res Commun 172:197–203

Koizumi O, Wilson JD, Grimmelikhuijzen CJP, Westfall JA (1989) Ultrastructural localization of RFamide-like peptides in neuronal dense-cored vesicles in the peduncle of Hydra. J Exp Zool 249:17–22

Kolhekar AS, Roberts MS, Jiang N, Johnson RC, Mains RE, Eipper BA, Taghert PH (1997) Neuropeptide amidation in Drosophila: separate genes encode the two enzymes catalyzing amidation. J Neurosci 17:1363–1376

Leitz T, Morand K, Mann M (1994) Metamorphosin A: a novel peptide controlling development of the lower metazoan Hydractinia echinata (Coelenterata, Hydrozoa). Dev Biol 163: 440–446

Leviev I, Grimmelikhuijzen CJP (1995) Molecular cloning of a preprohormone from sea anemones containing numerous copies of a metamorphosis inducing neuropeptide: a likely role for dipeptidyl aminopeptidase in neuropeptide precursor processing. Proc Natl Acad Sci USA 92:11647–11651

Leviev I, Williamson M, Grimmelikhuijzen CJP (1997) Molecular cloning of a preprohormone from Hydra magnipapillata containing multiple copies of Hydra-LWamide (Leu-Trp-NH$_2$) neuropeptides: evidence for processing at Ser and Asn residues. J Neurochem 68:1319–1325

Mackie GO (1973) Report on giant nerve fibres in Nanomia. Publ Seto Mar Biol Lab 20:745–756

Mackie GO (1989) Evolution of cnidarian giant axons. In: Anderson PAV (ed) Evolution of the first nervous systems. Plenum Press, New York, pp 395–407

Mackie GO (1990) The elementary nervous system revisited. Am Zool 30:907–920

McFarlane ID, Grimmelikhuijzen CJP (1991) Three anthozoan neuropeptides, Antho-RFamide and Antho-RWamides I and II, modulate spontaneous tentacle contractions in sea anemones. J Exp Biol 155:669–673

McFarlane ID, Graff D, Grimmelikhuijzen CJP (1987) Exitatory actions of Antho-RFamide, an anthozoan neuropeptide, on muscles and conducting systems in the sea anemone Calliactis parasitica. J Exp Biol 133:157–168

McFarlane ID, Anderson PAV, Grimmelikhuijzen CJP (1991) Effects of three anthozoan neuropeptides, Antho-RWamide I, Antho-RWamide II and Antho-RFamide, on slow muscles from sea anemones. J Exp Biol 156:419–431

McFarlane ID, Reinscheid RK, Grimmelikhuijzen CJP (1992) Opposite actions of the anthozoan neuropeptide Antho-RNamide on antagonistic muscle groups in sea anemones. J Exp Biol 164:295–299

McFarlane ID, Hudman D, Nothacker H-P, Grimmelikhuijzen CJP (1993) The expansion behaviour of sea anemones may be coordinated by two inhibitory neuropeptides, Antho-KAamide and Antho-RIamide. Proc R Soc B (Lond) 253:183–188

Minobe S, Koizumi O, Sugiyama T (1995) Nerve cell differentiation in epithelial Hydra from precursor cells introduced by grafting. I. Tentacles and hypostome. Dev Biol 172:170–181

Mizuno K, Ohsuye K, Wada Y, Fuchimura K, Tanaka S, Matsuo H (1987) Cloning and sequencing of a cDNA encoding a peptide C-terminal α-amidating enzyme from Xenopus laevis. Biochem Biophys Res Commun 148:546–552

Moosler A, Rinehart KL, Grimmelikhuijzen CJP (1996) Isolation of four novel neuropeptides, the Hydra-RFamides I–IV, from Hydra magnipapillata. Biochem Biophys Res Commun 229:596–602

Moosler A, Rinehardt KL, Grimmelikhuijzen CJP (1997) Isolation of three novel neuropeptides, the Cyanea-RFamides I–III, from scyphomedusae. Biochem Biophys Res Commun 236: 743–749

Müller WA (1996) Pattern formation in the immortal Hydra. Trends Genet 12:91–96

Nothacker H-P, Grimmelikhuijzen CJP (1993) Molecular cloning of a novel, putative G protein-coupled receptor from sea anemones structurally related to members of the FSH, TSH, LH/CG receptor family from mammals. Biochem Biophys Res Commun 197:1062–1069

Nothacker H-P, Rinehart KL, Grimmelikhuijzen CJP (1991a) Isolation of L-3-phenyllactyl-Phe-Lys-Ala-NH₂ (Antho-KAamide), a novel neuropeptide from sea anemones. Biochem Biophys Res Commun 179:1205–1211

Nothacker H-P, Rinehart KL, McFarlane ID, Grimmelikhuijzen CJP (1991b) Isolation of two novel neuropeptides from sea anemones: the unusual, biologically active L-3-phenyllactyl-Tyr-Arg-Ile-NH₂ and its des-phenyllactyl fragment Tyr-Arg-Ile-NH₂. Peptides 12: 1165–1173

Ohsye K, Kitano K, Wada Y, Fuchimura K, Tanaka S, Mizuno K, Matsuho H (1988) Cloning of cDNA encoding a new peptide C-terminal α-amidating enzyme having a putative membrane-spanning domain from Xenopus laevis skin. Biochem Biophys Res Commun 150:1275–1281

Pohl T, Zimmer M, Mugele K, Spiess J (1991) Primary structure and functional expression of glutaminyl cyclase. Proc Natl Acad Sci USA 88:10059–10063

Quafik L, Stoffers DA, Campbell TA, Johnson RC, Bloomquist BT, Mains RE, Eipper BA (1992) The multifunctional peptidylglycine α-amidating monooxygenase gene: exon/intron organization of catalytic, processing, and routing domains. Mol Endocrinol 6:1571–1584

Reinscheid RK, Grimmelikhuijzen CJP (1994) Primary structure of the precursor for the anthozoan neuropeptide Antho-RFamide from Renilla köllikeri: evidence for unusual processing enzymes. J Neurochem 62:1214–1222

Salesse R, Remy JJ, Levin JM, Jallal B, Garnier J (1991) Towards understanding the glycoprotein hormone receptors. Biochimie 73:109–120

Schmutzler C, Darmer D, Diekhoff D,Grimmelikhuijzen CJP (1992) Identification of a novel type of processing sites in the precursor for the sea anemone neuropeptide Antho-RFamide (<Glu-Gly-Arg-Phe-NH₂) from Anthopleura elegantissima. J Biol Chem 267:22534–22541

Schmutzler C, Diekhoff D, Grimmelikhuijzen CJP (1994) The primary structure of the Pol-RFamide neuropeptide precursor protein from the hydromedusa Polyorchis penicillatus indicates a novel processing proteinase activity. Biochem J 299:431–436

Seidah NG, Gaspar L, Mion P, Marcinkiewicz M, Mbikay M, Chrétien M (1990) cDNA sequence of two distinct pituitary proteins homologous to Kex2 and furin gene products: tissue-specific mRNAs encoding candidates for pro-hormone processing proteinases. DNA Cell Biol 9:415–424

Seidah NG, Marcinkiewicz M, Benjannet S, Gaspar L, Beaubien G, Mattei MG, Lazure C, Mbikay M, Chrétien M (1991) Cloning and primary sequence of a mouse candidate prohormone convertase PC1 homologous to PC2, furin and Kex2: distinct chromosomal localization and messenger RNA distribution in brain and pituitary compared to PC2. Mol Endocrinol 5:111–122

Smeekens SP, Steiner DF (1990) Identification of a human insulinoma cDNA encoding a novel mammalian protein structurally related to the yeast dibasic processing protease Kex2. J Biol Chem 265:2997–3000

Smeekens SP, Avruch AS, LaMendola J, Chan SJ, Steiner DF (1991) Identification of a cDNA encoding a second putative prohormone convertase related to PC2 in AtT20 cells and islets of Langerhans. Proc Natl Acad Sci USA 88:340–344

Spencer AN, Arkett SA (1984) Radial symmetry and the organization of central neurones in a hydrozoan jellyfish. J Exp Biol 110:69–90

Takahashi T, Muneoka Y, Lohmann J, Lopez de Haro MS, Solleder G, Bosch TCG, David CN, Bode HR, Koizumi O, Shimizu H, Hatta M, Fujisawa T, Sugiyama T (1997) Systematic isolation of peptide signal molecules regulating development in Hydra; LWamide and PW families. Proc Natl Acad Sci USA 94:1241–1246

Tensen CP, van Kesteren ER, Planta RJ, Cox KJA, Burke JF, van Heerikhuizen H, Vreugdenhil E (1994) A G protein-coupled receptor with low density lipoprotein-binding motifs suggests a role for lipoproteins in G-linked signal transduction. Proc Natl Acad Sci USA 91:4816–4820

Westfall JA (1973a) Ultrastructural evidence for a granule-containing sensory-motor interneuron in *Hydra littoralis*. J Ultrastruct Res 42:268–282

Westfall JA (1973b) Ultrastructural evidence for neuromuscular systems in coelenterates. Am Zool 13:237–246

Westfall JA, Grimmelikhuijzen CJP (1993) Antho-RFamide immunoreactivity in neuronal synaptic and nonsynaptic vesicles of sea anemones. Biol Bull 185:109–114

Westfall JA, Kinnamon JC (1978) A second sensory-motor-interneuron with neurosecretory granules in *Hydra*. J Neurocytol 7:365–379

Westfall JA, Sayyar KL, Elliott CF,Grimmelikhuijzen CJP (1995) Ultrastructural localization of Antho-RWamide I and II at neuromuscular synapses in the gastrodermis and oral sphincter muscle of the sea anemone *Calliactis parasitica*. Biol Bull 189:280–287

Index

Printing: Saladruck, Berlin
Binding: Buchbinderei Lüderitz & Bauer, Berlin